SONET

SONET

Second Edition
Walter Goralski

McGraw-Hill
New York · San Francisco · Washington, D.C. · Auckland
· Bogotá · Caracas · Lisbon · London · Madrid
· Mexico City · Milan · Montreal · New Delhi · San Juan
· Singapore · Sydney · Tokyo · Toronto

McGraw-Hill

*A Division of The **McGraw·Hill** Companies*

1 2 3 4 5 6 7 8 9 0   AGM/AGM   0 5 4 3 2 1 0

ISBN 0-07-212570-5

The executive editor for this book was Steven Elliot, the developmental editor was Jennifer Perillo, the associate developmental editor was Frances Kelly, and the production manager was Claire Stanley. It was set in Vendome by North Market Street Graphics.

Printed and bound by Quebecor/Martinsburg.

Throughout this book, trademarked names are used. Rather than put a trademark symbol after every occurrence of a trademarked name, we use names in an editorial fashion only, and to the benefit of the trademark owner, with no intention of infringement of the trademark. Where such designations appear in this book, they have been printed with initial caps.

 This book is printed on recycled, acid-free paper containing a minimum of 50% recycled, de-inked fiber.

CONTENTS

v

Contents

Contents

Part IV   SONET ADVANTAGES   363

Chapter 12   Customer and Carrier Advantages   367

Chapter 13   SONET OAM&P   389

Contents

PREFACE TO THE SECOND EDITION

The first edition of SONET was written in late 1996 and early 1997. In the three years or so since, optical networking in general and SONET in particular (I should include SDH here as well) have come a long way. Much of the information contained in the first edition has become dated, and I wanted to update the book in several areas. These areas are especially optical fibers, optical multiplexing (DWDM) and networking, SONET/SDH gear, and network management. For example, DWDM, relegated to the last chapter in the first edition on SONET/SDH "futures," now deserves consideration in a new chapter.

Some of the updates were easy. There is much more information about SONET network elements available online than there was four years ago. Some updates were more challenging. If TL1 is not to be the network management language of choice for SONET networks, should it be TMN or some more exotic, object-based, Web-driven application?

All chapters have been updated, of course, some quite extensively. Two new chapters have been added, the first on WDM and DWDM and the second on SONET payloads. These chapters are necessary because optical networking has evolved rapidly beyond simple SONET links and service providers have begun stuffing all manner of things, mostly IP packets, inside SONET/SDH frames. Also, this edition makes more of an effort to include SDH issues throughout the text itself, rather than relegating all SDH discussion to a concluding chapter.

Rather than rewrite the Acknowledgments section, let me just add here that I am thankful to all readers who support my work.

Finally, I have to admit that this was a very easy book to research. Besides slogging my way through various Telcordia (Bellcore), ITU-T, ANSI, and ETSI specifications, and talking to as many people involved with SONET/SDH as I could discover, I found that I filled in everything else I needed to know from just three books. I would be remiss not to mention them right away, since they have been reliable guides throughout the making of this edition.

Understanding Optical Communications by Harry J. R. Dutton is the best book on pure fiber optic issues. It is very long but very readable, and is one of the well-known IBM Redbook series. It is even downloadable as a *very* large file from *www.redbooks.ibm.com* (the link to the book itself changes, so it is best to search for it).

Understanding Fiber Optics, third edition by Jeff Hetch has all the facts on modern optical networks in one package. Written as a self-paced study guide, it has questions at the end of each chapter.

Understanding SONET/SDH by Ming-Chwan Chow is becoming dated, but is still very valuable for many basics. Written specifically for technicians working on or with SONET/SDH circa 1992, it also has questions at the end of each chapter.

More details on each book appear in the Bibliography.

—Walter J. Goralski

PREFACE TO THE FIRST EDITION

I wrote this SONET book for the same reason that I have written all of my other books. That is, I was fundamentally dissatisfied with the books I could find on the subject. When it came to SONET, there were books that were quite technically detailed at the byte level, but contained vendor or service provider information. Others provided details on the marketing of SONET, but almost nothing about how TL1 is currently used to manage SONET networks. Few mentioned things like wavelength division multiplexing.

I took good notes, talked to lots of people, and put everything into this SONET book. The bits and bytes are still here, of course, but not down to the last jot and tittle. Instead, the emphasis is on *working* with SONET. I hope this book appeals to SONET designers, installers, operators, maintainers, and (above all) users and customers. There is much in here for all.

—WALTER J. GORALSKI

ACKNOWLEDGMENTS TO THE FIRST EDITION

Medieval scribes would write out text in a manuscript, illuminate the book with figures as they pleased, and sometimes use paper or vellum they had made themselves. Publication often consisted of reading the work out in public, and distribution was carried out by the author passing the new book around. Advertising was done by the author as well, usually by word of mouth. Sales were limited in most cases to the transfer of ownership of the sole existing copy, but the proceeds went entirely and directly to the author. And some well-educated people could even read to themselves without moving their lips.

Book production has changed somewhat lately. The modern scribe calls upon a host of functionaries to review, edit, produce, distribute, and advertise a new book, all with the aid of computers and networks. This book could not have happened without the vision of Steve Elliot.

Information was provided on many essential matters by many corporations, in particular Nortel and Alcatel. In fact, without the numerous white papers available from Nortel, this book would have been quite different. Alcatel provided information with astonishing speed. The encouragement I receive from my colleagues and superiors at Hill Associates is invaluable.

I would also like to thank the reviewers. Gary Kessler saw the initial chapters and made several helpful suggestions. Michael Staufenberg put the whole book through the wringer and helped correct technical errors and many misconceptions that I harbored. Thanks to both of them, this is a better book. Of course, for any errors that remain, I must take full responsibility.

Some things never change with books, though. Some well-educated people can even read to themselves without moving their lips.

INTRODUCTION

[handwritten note: for WAN connection / LANs]

For the 1996 Olympics, the Olympic Stadium in Atlanta was wired with twelve miles of SONET (Synchronous Optical Network) fiber optic cable. Every Boeing 777 jet liner is wired with several miles of SONET-capable fiber. Either one can handle literally gigabits of information per second (a gigabit is 1 thousand megabits), which is the whole point of using fiber in the first place. Gigabits of generated information must be transported, processed, and presented to concerned users in adequate time to be useful. At the Olympics, the users were officials and judges and members of the press. In an aircraft, the users are pilots and, after the fact, potential investigators. Thankfully, most flight abnormalities do not result in disaster. But all of them must be investigated to avoid future risks. SONET can possibly be used to record not just cockpit sounds and instrument readings, but ultimately a video record of the cockpit operations and even the whole flight from a tail-mounted digital camera.

Admittedly, most users do not encounter SONET in competition or on an airplane today. However, the need for higher bandwidth is an unpleasant fact of life in all types of networks today. Until now, several newer technologies have helped to remedy this situation in a local area network (LAN) environment. Once limited to the choice of cabling (which kind of unshielded twisted pair?) for 10 Mbps Ethernet-type solutions, the LAN implementer today is faced with a wide range of choices for linking desktops together. Newer technologies, such as 100 Mbps Ethernet, Fibre Channel, switched Ethernets, and the much-anticipated Gigabit Ethernet, have started to appear or be planned for in many organizations needing faster connectivity between end systems.

[handwritten note: LAN transmission is getting faster use SONET for WAN to connect fast LANs.]

These faster networks are needed to link the more powerful—and less expensive and therefore more common—desktop systems, running client-server applications, and many with built-in multimedia capabilities. All of these reasons are explored more fully at the beginning of this book. For now, it is enough to point out that, for a variety of reasons, faster LANs are being built to address bandwidth limitations.

Although faster LANs have addressed to need for desktop bandwidth within a building or campus environment, LANs do nothing for increasing the bandwidth available for networking-dispersed buildings or campuses together over distances of more than a few miles. For this situation, a wide-area network (WAN) is needed. Until relatively recently, there was no easy way to link even 10 Mbps Ethernet LANs together with even a fraction of the bandwidth the LAN represented.

Fortunately, SONET provides a welcome means of relief from this growing bandwidth problem in the WAN environment. SONET is capable of linking LANs at separate sites not at a mere fraction of 10 Mbps, or even a full 10 Mbps. Rather, SONET can link several 10 Mbps Ethernets at a single site to other Ethernet LANs across the country. SONET links usually operate at speeds of 155 Mbps or better, and into the multi-Gbps range.

SONET Advantages

There is more to SONET than just a lot of bandwidth. SONET has numerous advantages in a number of areas that will be explored more fully in this book. However, this may be a good place to briefly outline some of SONET's more important and obvious advantages over other high-bandwidth networking schemes. These are summarized in Table I.1 below.

All of these reasons for the current level of interest in and popularity of SONET are fully explained later. For now, it is enough to note that there are many advantages that SONET enjoys over other networking methods.

This book explores SONET from top to bottom. It explains where SONET came from, what it is used for, and how it is used. The book explores all aspects of SONET in some technical detail, not to overwhelm the reader, but to point out how the SONET digital fiber standard addresses some of the problems that have plagued other WAN technologies in the past. All of the advantages of SONET for users and service providers alike are detailed, and the services deliverable on

TABLE I.1	Technology	Unprecedented speeds available on fiber optic cable
Advantages of SONET	Economics	Best interface for fiber optic networks
		Economical adding and dropping of channels
	Flexibility	Modular equipment design
		Adequate overhead for network management
	Compatibility	Works well with existing network hierarchies
		Allows multiple vendors' equipment to interoperate
		Worldwide standard (as Synchronous Digital Hierarchy)

SONET are fully described. Finally, the book examines specific service-provider SONET offerings and equipment providers' products. The goal is to make this work a comprehensive guide to SONET that is at the same time understandable to non-engineers and interesting to technicians.

Bandwidth

This book is about SONET. But just as important to the subject is the previous (and, of course, still existing) digital transmission system hierarchy in the United States, known as T-carrier. The rest of the world pretty much follows the E-carrier digital hierachy, which is different than T-carrier. Japan uses a variation of T-carrier usually called J-carrier. At this point, it is just necessary to note the associated bit rates of these schemes and not necessarily understand the designations and differences. These will be covered later.

This book constantly mentions terms like DS-1 and DS-3, and speeds of 1.5 Mbps (1.544 Mbps to purists) or 45 Mbps. In order to minimize the amount of flipping back and forth, and folding down of pages to mark spots, and the like, this may be an appropriate place to list several tables of T-carrier, E-carrier, and SONET speeds right at the front of the book where they would be somewhat easier to locate.

A Word on Terminology

The nice thing about words is that they can take on almost any meaning a writer wants. This can be a problem when one writer uses a word in one way or sense, and other writers do not immediately recognize that this meaning and sense is far superior to any other. Because it is difficult to write a technical book without using some common terms over and over, I thought it best to define exactly what I mean by these terms when they are used in this work. I admit each term and its meaning can be debated, but I am not so much concerned whose use is proper. It is more important that readers understand just what is meant by each of these key terms.

First, I use the term "organization" in preference to "company" or "business." In this book, the term "organization" is used in a generic sense to avoid limiting the discussion to a corporate environment. Corporations are all organizations, but so are federal, state, and local government departments, schools, hospitals, and private law firms, to name a

TABLE I.2

Digital carrier
hierarchy used
in North America,
Europe, and Japan

Digital Multiplexing Level	Number of Equivalent Voice Channels	Bit Rate (Mbps)		
		N. America	Europe	Japan
DS-0/E0/J0	1	0.064	0.064	0.064
DS-1/J1	24	1.544	—	1.544
E1	30	—	2.048	—
DS-1C/J1C	48*	3.152	—	3.152
DS-2/J2	96	6.312	—	6.312
E2	120	—	8.448	—
E3/J3	480	—	34.368	32.064
DS-3	672	44.736	—	—
DS-3C	1344*	91.053	—	—
J3C	1440*	—	—	97.728
E4	1920	—	139.264	—
DS-4	4032	274.176	—	—
J4	5760	—	—	397.200
5	7680	—	565.148	—

(*) Intermediate multiplexing rates.

few. An organization can even be a subsidiary or large department of a major corporation. All organizations need more bandwidth to link higher speed PCs and LANs together over a wide area.

Next, I use the term "service provider" in preference to "telco" (telephone company), "carrier," or anything else. In this book, the term "service provider" is used in a generic sense to avoid limiting the discussion to local exchange or interexchange telephone companies. All telephone companies are service providers, but so are cable TV companies, Internet service providers, power companies, and almost anyone who wants to be in today's deregulated telecommunications world. All service providers need more bandwidth to service more customers and provide each with the bandwidth they need over a wide area.

Also, I use the term "user" in preference to "customer" or "end-user." In this book, the term "user" is used in a generic sense to avoid limiting the discussion to bill-payers or people sitting in front of PCs. All customers

TABLE I.3

SONET Optical
Carrier (OC) and
SDH Synchronous
Transport Module
(STM) Levels

Optical Level	Electrical Level	Line Rate (Mbps)	Payload Rate (Mbps)	Overhead Rate (Mbps)	SDH Equivalent
OC-1	STS-1	51.840	50.112	1.728	—
OC-3	STS-3	155.520	150.336	5.184	STM-1
OC-9	STS-9	466.560	451.008	15.552	STM-3
OC-12	STS-12	622.080	601.344	20.736	STM-4
OC-18	STS-18	933.120	902.016	31.104	STM-6
OC-24	STS-24	1244.160	1202.688	41.472	STM-8
OC-36	STS-36	1866.240	1804.032	62.208	STM-13
OC-48	STS-48	2488.320	2405.376	82.944	STM-16
OC-96	STS-96	4976.640	4810.752	165.888	STM-32
OC-192	STS-192	9953.280	9621.504	331.776	STM-64

are users, but so are the members of a department on a corporate network, LAN administrators who rarely have time to sit down, let alone use a computer, and many others. All users need more bandwidth to accomplish what they set out to do with a network in the first place.

A minor point about the digital hierarchy itself should be mentioned here as well. Network engineers and educators will often insist on the use of the term "Digital Signal" (as in DS-3) when most network installers and customers would use the term "T" (as in T-3). As has been pointed out innumerable times, the terms are *not* technically interchangeable, but people often use them as if they were. This book will not attempt to decide whether any harm is done when calling a DS-1 a T-1, but will try to use the terms in the proper fashion consistently.

For some reason, many network folks will allow and feel comfortable with mentioning the speed of a T-3 or DS-3 as "45 Mbps." However, 45 Mbps is not the correct speed. The speed is actually 44.736 Mbps, and the rounding to 45 Mbps seems universal and acceptable. But any rounding of the T-1 or DS-1 speed to 1.5 Mbps is likely to be greeted with comment like "you mean 1.544 Mbps," which is indeed the actual bit rate of the link. Thus, in the interest of conserving 4s, this book will consistently round T-1 speeds to 1.5 Mbps, except where more accuracy is warranted to avoid confusion and misinterpretation, such as in mathematical expressions. I hope this practice offends no one.

Finally, there are many other terms defined along these same lines throughout the book. Where a term is confined to one particular aspect of a discussion, the term is defined at that point in the book (such as "client-server"). It is tempting to use these terms as if everyone had a common understanding of what they mean, but this is not the case. An undefined term is just *jargon,* and an indiscriminate use of jargon has probably caused more network problems than any other cause over the years.

ABOUT THE AUTHOR

Walter J. Goralski has more than 30 years in the data communications field, including 14 years with AT&T. He is currently a Senior Member of Technical Staff with Hill Associates, a technical training and consulting firm in Colchester, Vermont, and an adjunct professor of computer science at Pace University Graduate School in New York. He is also the author of several books on DSL, the Internet, TCP/IP, and SONET, as well as of articles on data communications and other technology issues.

ABOUT THE REVIEWERS

As the leading publisher of technical books for more than 100 years, McGraw-Hill prides itself on bringing you the most authoritative and up-to-date information available. To ensure that our books meet the highest standards of accuracy, we have asked a number of top professionals and experts to review the accuracy of the material you read.

We take great pleasure in thanking the following technical reviewers for their insights:

Gary C. Kessler is a senior network security analyst at SymQuest Group, a network integration consulting company with headquarters in South Burlington, Vermont. Gary's primary areas of interest are the Internet and TCP/IP applications and technologies, computer and network security, e-commerce, ISDN, and fast packet switching technologies. Gary is a frequent speaker at industry conferences, and has written two books and over 50 articles on a variety of topics. Gary holds a B.A. in mathematics and an M.S. in computer science. He is married and has two children. More information can be found at http://www.garykessler.net/.

Hang Lau is an adjunct professor at Concordia University in Montreal, Canada. He has worked as a systems engineer for Nortel Networks for the past 20 years.

Introduction to Fiber Optics and SONET

It sometimes seems that the pace of change in the world is always accelerating. The perception is especially common in the intimately related fields of computers and telecommunications. To a large extent, an accelerating pace of change is somewhat of an illusion. The fact is that sweeping technological change has been a characteristic of world culture, particularly in the United States since the beginnings of the Industrial Revolution. In spite of the vast changes that computerization and advances in telecommunications have wrought in the past 50 years or so, these changes are arguably less than the impact of the automobile or airplane on the world's population. After all, it was only fifty years from the first flight at Kitty Hawk to common commercial aviation.

There are significant differences between normal technological evolution and the evolution of the computer and telecommunications industries, however. Sometimes there is a deep affection for older technology, like classic cars. Every once in a while, a Model T appears on a superhighway. The attachment of a dedicated group to things like steam locomotives even seems somehow noble and uplifting. Never mind that these huffing giants required water every fifty miles and coal every hundred miles. Diesel locomotives do not seize the spirit like the hiss of steam.

With computers and telecommunications, however, the situation is different. No one treasures a classic IBM PC from the early 1980s and runs an old word processor just for fun. Groups do not form to explore the vanished world of 300 bit-per-second modems and marvel at the simple functional elegance of the teletype machine that once pounded away in the background of every news broadcast. Progress in computers and telecommunications has yielded not a treasure of memories but a mountain of worthless junk.

The telecommunications needs of businesses and residential users alike seem to be always one step ahead of the capability of technology to satisfy these needs. This is entirely the point. There would be a huge benefit to coming up with a technology that would be essentially bulletproof with regard to the future, especially the near future. It seems to be impossible with computers themselves. Advances in architectures, hardware, and software always appear to outstrip any one element in the threesome. For instance, a state-of-the-art hardware board will always succumb to a new backplane architecture that is incompatible with it, or a new software advance that makes the hardware function redundant and unnecessary.

The strange thing, and the whole intent of this book, is to describe an area of telecommunications technology where it actually seems that obsolescence may be decades away, instead of eighteen months (which is the average life-span of a major CPU chip architecture like the Pentium). The application of standards to fiber optic transmission offers a glimmer of hope to leapfrog end-system requirements at least for the immediately foreseeable future. The implications are large and the benefits many.

This section of the work describes the evolution of the SONET standard and how it came to overcome a variety of problems with older transmission methods. No knowledge of fiber optic transmission or digital communication is assumed or needed. The material is introductory and not intended to be challenging in nature; however, more advanced readers may still wish to read these early chapters. There is much to be learned about fiber optics and digital techniques that is not usually covered in standard texts on either topic. They are brought together here and examined with a historical perspective to prepare the reader for the changes that SONET will bring to the world of telecommunications (and perhaps even computing itself) in the next few years.

Chapter One describes how the concepts of SONET, Asynchronous Transfer Mode (ATM), and Broadband Integrated Services Digital Network (B-ISDN) were invented and intended to be used together for not only computer networks, but any network: from voice to cable TV. The chapter then explores how the related concepts got "de-coupled" and, in a sense, went their separate ways. It concludes with a consideration of whether it is too late to bring them all back together again, and even if it is wise to do so.

Chapter Two forms a tutorial of fiber optic communication. This chapter offers a fiber optic "survivor's guide" for readers who may not be as familiar as they would like to be with basic fiber optic transmission terminology and concepts. The key concepts of fiber cable characteristics, manufacturing, and usage are fully explained. The chapter includes a description of how typical fiber transmitters (i.e., lasers, LEDs) and receivers operate and surveys the current applications for fiber optic cable.

Chapter Three is new to this edition and introduces the optical concepts that make wavelength division multiplexing (WDM) possible. The chapter explores all of the ramifications and issues involved with WDM: scattering, other dispersion effects, fiber amplifiers, solitons, and so on. Then dense wavelength division multiplexing (DWDM) is investigated, along with all of the reasons that DWDM is one of the most exciting developments in the optical world today. State-of-the-art terabit systems are introduced here as well. This chapter is not SONET/SDH specific, but applies to optical networking in general.

Chapter Four is an introduction or review of the digital transmission hierarchy, not only in the United States, but around the world. This chapter explores the existing T-carrier and E-carrier digital hierarchies around the world, again for readers who may not be as familiar as they would like with these terms and concepts. Every effort is made not to limit discussion to North American methods (T-carrier), but the emphasis is here. The chapter includes an extensive discussion of the limitations of the current digital hierarchy (such as the lack of a true "mid-span meet") and how SONET addresses them.

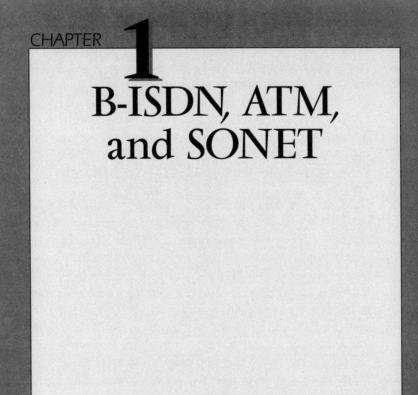

CHAPTER 1

B-ISDN, ATM, and SONET

Computer networks have been around since the mid-1960s. Telephone networks carrying voice and specialized applications, such as fax, are even older: since the 1870s in the case of the public switched telephone network (PSTN). Yet, the need to develop high-speed, low-delay networks has never been as intense in the past as it is today. The key to understanding the relationship between broadband-integrated service digital network (B-ISDN), asynchronous transfer mode (ATM), and synchronous optical network (SONET) is to understand how the standards-making bodies around the world predicted the direction of computer networking specifically and networking in general at the beginning of the 1980s. It was actually an impressive application of foresight, rare enough in any field.

Those in a position to do something about the future of networking looked mainly at three trends: the growth of powerful desktop systems linked at the endpoints of these networks; the rise of multimedia applications; and the trend toward what is usually called *distributed computing*. It is more common to encounter distributed computing used almost as a synonym for *client-server computing* (defined later). There is nothing wrong with this, in spite of definitions and "network speak." However, purists point out that distributed computing is a more encompassing term than is client-server; that is, client-server is only one way to accomplish distributed computing, although client-server architectures are certainly the most prevalent forms of distributed computing today.

This book begins by placing SONET in its proper context, which is no less than this: SONET is a high-bit-rate fiber-optic-based transport method that provides the foundation for linking high-speed ATM switches and multiplexers and providing users with B-ISDN-compliant services. There is a lot going on in this statement—thankfully; otherwise, this would be a very short book. In fact, the way SONET is deployed and marketed by service providers today has enhanced the position intended for SONET by international standards bodies and embodied in the above statement in many cases.

Before analyzing the statement about SONET's position, it may be a good idea to discuss details about the rising power of network end systems, multimedia issues, and trends in client-server computing. This detail is important because SONET includes a provision that tries to make networks based on SONET much more "future proof" than ever before. SONET is nearly unique in this ability to "scale" itself to networking needs and end-system technology.

Power on the Desktop

It is sometimes said that if you really want to get the best deal on the most powerful computer you can get, you will never buy anything. This argument applies to desktop and laptop personal computers (PCs) for the most part, but it is at least partially true today for UNIX-based workstations, minicomputers, and even mainframes (which are still around). This statement is often made because computer prices have been falling rapidly at the same rate computer capabilities have been growing. Why buy a computer when the price next month will be less—sometimes much less—and the performance will be enhanced as well?

Many good examples have been cited by authors Daniel Burstein and David Kline in their book, *Road Warriors*. For instance, a "singing" birthday greeting card has more processing power than existed in the entire world in 1950. A common video camcorder has more processing power than an IBM System/360 computer had in 1964. (An IBM engineer added this: "Well, the camcorder has an accurate clock. That means it's *already* more powerful than a System/360 in 1964." I will not comment further.) A simple $100 video game player found in many households today is more powerful than a Cray Supercomputer was in 1976; and the Cray cost about $4 million.

The accumulation of information that must be managed in organizations is staggering. Networks will form an increasingly crucial element as the reliance on information management technologies continues to increase. Organizations will seek to get more and better use from their capital investments and employee productivity (because, in many cases, fewer and fewer of them exist). Management technology is always finding better ways to integrate such previously diverse information technologies, such as optical storage, hypertext, neural networks, scientific visualization, and virtual reality.

In the area of optical storage, the emergence of inexpensive and abundant CD-ROM writers has addressed issues of inadequate tape and (in many cases) disk storage capabilities. It is almost inconceivable today to consider distributing compound documents (i.e., text and images merged with voice annotations), interactive graphics, and video files on diskettes. In the near future, such content will have to be distributed across a network to satisfy a number of application environments, such as computerized design applications and desktop publishing. These applications will require very high-capacity networks that bring together large amounts of information from diverse sources at a single, networked location.

Hypertext has been mainstreamed recently with the popularity of the World Wide Web on the Internet. The Web enables users to access truly staggering amounts of information in a nonlinear manner. This is the essence of hypertext. Anyone who has used a Windows-based "Help" file has experienced the power of hypertext. Instead of having to sequentially access the information stored on a computer or in a database, a user has the power to move through the information in a more intuitive and useful manner. Of course, the promise of hypertext, initially experimented as long ago as the 1960s, was fulfilled only with the development of powerful networks that allow the hypertext link to lead anywhere, not just to another portion of the same file on the same computer. Many organizations are increasingly relying on hypertext for online training and documentation, text management, and other forms of information distribution.

Neural networks are composed of a large number of microprocessors, all working together to solve a particular problem. Neural networks are particularly well-suited for information-processing-intensive industries, such as banking, securities trading, and insurance. All of these industries are distinguished by the fact that they seek to use information to gain a competitive advantage over other companies in the same field. The needed information must be quickly gathered and analyzed for trends and relationships. In these fast-moving fields, mere seconds can result in the gain (or loss) of several millions of dollars. In the world of companies characterized by downsized employee staffs, lack of middle-management personnel, and decentralized operations, these neural networks have become a vital business tool that allows the organization to be even more efficient under these lean conditions.

In the field of scientific visualization, increasing computer power has allowed organizations, such as oil and gas companies, to gather huge databases of seismic records, geological satellite images, and the like. Huge is probably an understatement in this context because the database can easily reach into the Terabyte (1,000 Megabytes) ranges. For example, data can be accessed so that a view of the earth 100 feet down can be imaged and color-coded in whatever way the viewer desires. In the chemical and pharmaceutical industries, molecular arrangements can be visualized, rotated, and folded, yielding new insights into the possibilities of new materials, drugs, and genetics.

Virtual reality is one of the latest examples of powerful computer applications that were only dreamed of a few short years ago. Images in movies and on television appear to be real, but are not: no real animals were used in *Jumanji*; Harrier jets were not flown over Miami in *True Lies*; and John Wayne did not resurrect himself to hawk beer. They all existed only in computers. But these are only films. Virtual reality extends this concept of fooling our senses to fooling ourselves into thinking that what we see and hear and sometimes even feel on the computer are as real as the world outside our window. Computer power today allows users to become as immersed in the virtual world as they wish, a computer-generated world of sight, sound, and mind. Virtual reality forms a real-time simulation that creates an artificial world where both objects and their environment have the illusion of being real.

This list could be extended much further, but the point is already clear: Processing power is growing, and fast, while prices continue to fall. Figure 1.1 illustrates the increase in millions of instructions per second (MIPS) that systems with Intel processors selling for about $2,400 were capable of from 1992 to 1996. Intel processors power about 85% of all desktop and laptop systems sold today. These figures, and all of the figures in this chapter, unless otherwise noted, come from advertisements appearing in *Windows* magazine, which published a retrospective on these trends in late 1996.

Rather than attempting to update all of these facts and figures and trying to hit a constantly changing target, this edition preserves the information through 1996. It is all still valid. Suffice to say that the "average" PC today boasts between 128 and 256 Megabytes of RAM, a 16 or 20 Gigabyte hard drive, and about a 750 MegaHertz processor, all for less than $1,500. It should be kept in mind that prices seem to be stabilizing somewhat, especially memory prices. However, monitors still remain relatively pricey, the result of a high video chip rejection rate and the fact

Figure 1.1
Intel MIPS for $2,400

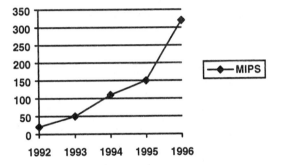

that monitors cannot shrink in size, but always grow. And new flat displays often cost more than the rest of the machine they are attached to.

Figures 1.2a and 1.2b chart the same $2,400 system against the amount of random access memory (RAM in megabytes) and storage (in megabytes) installed for the same years. Although not as steep as the processor curve, the trend is just as pronounced.

Systems based on the Intel processor continue to follow what is known as Moore's Law. Gordon Moore was one of Intel's founders who observed early on that doubling of semiconductor density and, therefore, processing power occurs about every eighteen months. Moore expected this trend to continue and, by 1996, it had resulted in Intel's introduction of five new Pentium versions and four new Pentium Pro chips within the previous twelve months. In spite of periodic negative statements about the limits of technology, such as semiconductors and masking techniques, the advance of integrated circuit and chip technology has continued without major difficulties for almost thirty years.

Not only is the hardware more powerful year by year, and sometimes month by month, but it is less expensive as well. Computer memory chips, in particular, have gone through a staggering decrease in

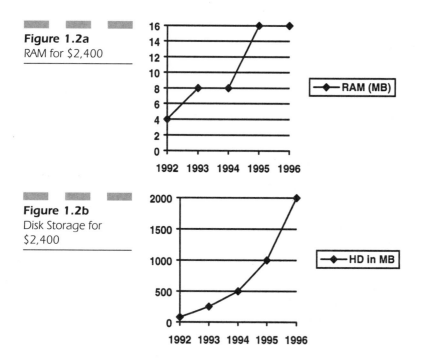

Figure 1.2a
RAM for $2,400

Figure 1.2b
Disk Storage for $2,400

TABLE 1.1

Memory prices

Size:	1992	1993	1994	1995	1996
1MB	$45	$65	$43	Not avail.	Not avail.
4MB	$143	$188	$156	$145	$37
16MB	Not avail.	Not avail.	$639	$514	$135*
32MB	Not avail.	Not avail.	$1,194	$1,066	$282

*Prices toward the end of 1996 for 16MB chips were in the $80 range.

price that has left users' heads spinning. Table 1.1 depicts the dramatic decrease in prices for PC dynamic RAM (DRAM) memory.

Most of the reduction in prices has come from the intense competition to supply memory-hungry Microsoft Windows 95 operating systems with enough memory to achieve the performance levels promised. When Windows 95 appeared, chip makers geared up production to supply the 16MB "recommended" (meaning "minimum") memory Microsoft mentioned frequently in documentation. DRAM pricing approached $5 per MB as manufacturers were accused of "dumping" (i.e., selling below manufacturing cost) to gain market share and reduce inventories.

The reduction in memory prices has been mirrored by the reduction in the prices of hard drives for storage. Most PCs use internal device enhanced (IDE) hard drives, and most prices reflect *bare drive* configurations. Bare drives are sold without the controller boards or even screws needed to mount the drives in a PC. This fact, however, makes the price a good reflection of hard drive technology alone.

It is not uncommon for most popular $2,400 PCs to include a 16 GB hard drive. And it's a good thing, too. Usually, about 1 GB will be taken up by the software bundled with the new system. Fortunately, vendors are happy to sell users as much hard disk storage as they need. In another stroke of luck, the price for hard disk storage is decreasing rapidly as well. Table 1.2 shows the decrease in IDE hard drive pricing through 1996.

Seasoned computer users, who were used to the more leisurely pace of technology development in the 1980s, are sometimes shocked to find that their PC hard drives, which were never backed up but served them well for five years, cannot be replaced when they finally fail. The faithful 120 MB hard drive is gone and a replacement is nowhere to be found, except perhaps at a backyard garage sale. The more pressing problem, however, is what to do about the lack of a backup!

TABLE 1.2

IDE hard drive prices

Size:	1992	1993	1994	1995	1996
200MB	$565	$301	$192	Not avail.	Not avail.
540MB	$1,589	$909	$369	$192	$160
1+GB	Not avail.	$1,208	$721	$320	$193
2+GB	Not avail.	$2,420	$1,519	$1,349	$285

Strangely, computer monitors have been an apparent exception to this declining price rule. Monitor vendors have struggled to lower prices at all, because monitors have little to gain directly from improvements in semiconductor densities, and the monitor manufacturing process suffers from a high rejection rate, which raises prices for even perfect monitors. Nevertheless, a high-quality 14-inch VGA monitor costing about $1,000 in 1988 went for about $300 at the end of 1996 and was of much better quality as well. New monitors have crisper images, higher refresh rates, and improved brightness.

This list could be extended to include CD-ROM drives and printers. Interested readers are urged to consult the pages of any PC magazine for additional information. However, one more example may be in order: In 1992, a 300dpi, 8ppm laser printer cost nearly $1,500. Five years later, the same device was selling for under $500.

What has all this to do with networks? Well, consider how much the end systems attached to these networks have changed in terms of power over the past five years. The obvious answer is radically. Now consider how much the wide area networks used to link these systems have changed in terms of speed over the past five years: hardly at all. Newer public network solutions exist, such as frame relay, but they often run at the same speeds as before.

A common guideline for networking is as follows: A computer can generate about 1 Mbps of network traffic for each MIPs that the computer CPU runs at. In the early 1980s, with only a couple of MIPS available on the most powerful desktops (and even minicomputers), it was not uncommon to see 200 to 300 PCs on a single Ethernet LAN. The 10 Mbps Ethernet handled it all with ease. In fact, the Ethernet design limit was 1,024 devices.

By 1987, the 386 chipset running at about 6 MIPS could push 6 Mbps out of the back of a PC onto a LAN. The only factor that kept Ethernet LANs from grinding to a halt was that traffic was extremely "bursty" (i.e., intermittent). As it was, the limit by the late 1980s became about 20 to 30 devices per Ethernet LAN. By the early 1990s, with speeds approaching 50 MIPS, a desktop could easily overwhelm a 10 Mbps Ethernet by itself. This need to give a server or even client a full 10 Mbps or faster led directly to the rise of switched Ethernet and 100 Mbps Ethernet on the LAN. However, as speeds go far beyond 250 MIPS, where is the bandwidth needed for the WAN?

Clearly, a need exists for new high-speed, wide-area networking methods to link these new desktops together. SONET fits this role admirably.

Client-Server Computing

The second development in network computing that became a driving force behind the bandwidth needs of networks is the popularity of client-server computing. The term *distributed computing* is closely linked to the concept of *client-server computing.* In fact, the use of client-server computing is sometimes seen as a reason for building the distributed network in the first place. If there seems to be no apparent reason why this should be so, it is undoubtedly true that until the rise of client-server computing there was little justification for an organization to go through the effort and expense of building a distributed network. This point needs further discussion.

One easy way to define a client is as follows: Any computing device on a desk with a person sitting in front of it doing work is a client. It makes no difference if the device is a terminal on an SNA network (SNA is IBM's System Network Architecture), a PC on a LAN, or a UNIX workstation on a TCP/IP network. These are all client devices.

Servers are easily definable too: Any computing device without a person sitting in front of it doing work is a server. Again, it makes no difference whether the device is an IBM mainframe, a DEC minicomputer, or a Novell Netware file server. These are all examples of servers. Servers do not have "workers" in front of them, but have at least one, and perhaps many, administrators who make the server available to a community of clients.

These definitions may seem simplistic, but they are extremely powerful in explaining the relationship of client-server computing and distributed networking, and the need for wide-area bandwidth as well. Once the basic definitions are understood, other terms fall into place.

For example, the terms *client-server computing* or *client-server architecture* describes this overall practice of viewing all computing devices as either clients or servers (or even both at once in some cases). *Client-server applications* are specifically designed to run in a client-server environment with a client portion and a server portion to the application, and so on. The term *distributed processing* is sometimes used synonymously with this *client-server model* to acknowledge the usual geographic separation of clients and servers.

The relationship of the client-server model of computing and distributed networks is now easy to understand. Clients attach to servers over networks. However, without a network, no easy way would exist for clients to reach the server they need. Therefore, the network is necessary to enable every client to access every server in the organization. The client-server concept is so important to networks that the key points are listed below:

1. Clients are desktop devices with workers

2. Servers are devices with administrators

3. Clients and servers are connected by networks

4. Any client should be able to attach to any server

5. Networks allow this to happen

Of course, all of this ties in with the growth of powerful desktop devices discussed previously. The process began in the 1980s with the introduction of the PC into many organizations to replace the "dumb" terminal devices workers had used before to access applications on remote mainframes and minicomputers.

By 1986, the top-of-the-line PC was as powerful as the IBM mainframe in 1979. Many organizations were quick to take advantage of some of this computing power. Instead of having two or three terminals at each employee location, organizations could have a single PC running SNA terminal emulation when attaching to the corporate mainframe, VT100 emulation when attaching to the departmental minicomputer, and also when attaching directly to the local file server. In fact, with the rise of multi-processing operating systems, such as Microsoft's Windows and IBM's OS/2, these PCs were capable of doing all three at the same time.

It soon became common enough to link all of these PCs with LANs, and all of the LANs with routers, and all of the routers with point-to-point leased, private lines. The bandwidth between the routers was only a small fraction of the bandwidth on the LAN (64 Kbps on the WAN compared to 10 Mbps Ethernet) in most cases, but the bursty and intermittent nature of most data traffic made this a viable solution, at least for the short term.

In the long term, this was a problem. As both clients and servers grew more numerous and powerful, the available bandwidth between the routers became more scarce and pushed the network to the limit. Organizations with many PGs put everything on LANs with routers, including SNA devices and specialized terminals. This led to increased pressure to install more bandwidth between the routers to avoid bottlenecks. However, few service providers had much above 1.544 Mbps to sell to users, and nothing above 45 Mbps, even for those who could afford it. A few examples of typical client-server applications used by organizations today should emphasize the fact that the WAN bandwidth is not adequate for many applications today.

Sticking with the definition of client as a desktop computer and the server as a "back office" computer, there are numerous examples of client-server applications in common use today. Most database applications are now used in a distributed, client-server fashion. The database files can be huge and represent, in some cases, the entire record of an organization's business activities past and present; these files may reside on a number of servers. These database servers do not have to be concentrated in a single location, of course, but may be distributed among many locations that are geographically dispersed. The database client software is used to access the information stored on these servers.

Database applications now need to be installed and used in this client-server fashion. Installing such a database package defaults to this configuration: client and server. In most cases, the software is not only shrink-wrapped separately, but comes in different packing cases and boxes. Some configuration and installation parameters must be changed not to run the software in a distributed, client-server fashion, but to install and run the database and "query engine" (the client portion) on the same computer.

Another application that has become common in many organizations and that has pushed the capabilities of networks to the limit is simple electronic mail (e-mail) and the increased use of electronic data exchange (EDI) technology. E-mail has evolved from simplistic text to messages including voice and video portions. It is not unusual today for some large corporations to generate nearly 300 e-mail messages *per day* for key employees and positions. All of these messages must be distributed over networks, of course, and because e-mail is essentially a connectionless service—the intended recipient does not have to be accessing the network before someone can send them e-mail—most of these messages must be stored somewhere on the network in large mail servers (or "post offices") until users access, read, and dispose of them.

EDI involves the transfer of computer-readable data between two separate organizations in a standard or agreed upon format. The partnered organizations may exchange EDI messages that represent an invoice, a packing list, or a purchase order. With EDI, instead of hand-processing multiple copy forms and mailing or faxing them to the proper personnel for processing, an organization can map the data into the correct electronic format and send it across a network. Upon arrival at the destination organization's computer system, the standard format can be translated into the organization's own internal format and processed automatically by the application software.

EDI, however, offers more than just a way to process batches of work orders. Graphics are increasingly being built into ordinary data transactions, and these graphics are often required in circumstances involving customized products. Merely ordering parts or materials from bland vendor listings is not sufficient to adequately convey the true nature of the products. Some graphic schematic or production drawing is usually necessary to complete the full business transaction. In some industries, notably the automobile industry in the United States, support for a host of standard EDI formats is required for all approved suppliers to the major automakers.

A common use for networks today in many organizations is *distance learning*. As training (and re-training, for that matter) becomes more important, especially in high-tech industries, the pressure to cut back on things such as travel expenses and time away from the immediate work environment is also becoming greater. Instead of traditional instructor-classroom courses and materials, many organizations have chosen to explore the use of networks, usually both data and video networks, to deliver training more effectively right to the work location.

Sometimes seen in a form called *asynchronous learning,* the use of the organization's network to deliver training and instruction almost anywhere and anytime puts additional pressure on the network. Some organizations have gone so far as to commit to a given level of distance learning education ("20% of all course materials") at the end of a certain time period. Distance learning will only increase in popularity and sophistication in the future.

An ever-increasing number of people are working at home with either home-based small businesses or as part of a corporate job. Both employers and employees are finding this to be a viable alternative to long commutes and are finding increased productivity and job satisfaction. In some cases, telecommuting is an alternative forced on employees so that large companies comply with national and state environmental regulations. For example, the Federal Clean Air Act mandates a work-at-home option for qualified employees and job descriptions. A home office usually includes a computer, modem, fax, and answering machine hardware. In many cases, some form of screen sharing software could allow the sharing of data, images, and documents with colleagues and customers. E-mail software and a World Wide Web browser provide access to e-mail networks and the Internet, allowing communications with colleagues around the world. Other software allows access to electronic "card file" databases of business colleagues which could automatically display information about a business caller, making the fact that an individual works at home transparent to customers.

What's wrong with this picture? Nothing, except the twin issues of bandwidth limitations and network delay. As attractive as many of these client-server applications and their variations are, the universal use of all of them is always limited by the amount of network bandwidth available and end-to-end network delay users encounter when accessing any of these client-server variations. In the vast majority of these cases, the network bottleneck is not on the LAN side(s) of the network, but on the part of the WAN employed between them.

Many organizations employ client-server applications to allow distributed computing to take place over the mix of LANs and WANs used at various sites. In the vast majority of cases, the LANs are 10 Mbps Ethernet-type; nearly 90% of all LANs built are one kind of Ethernet or another. The simple fact is that in spite of the availability for several years of technologies such as 100 Mbps Ethernet and 155 Mbps ATM employed to the desktop in essentially a LAN configuration, 10 Mbps Ethernet continues to be the LAN technology of choice for organizations.

Several reasons exist for the continued popularity of Ethernet LANs. First, the hardware and software components are extremely inexpensive and prices continue to drop. Next, several popular desktop architectures include built-in 10 Mbps Ethernet connectivity right on the computer motherboard. For example, a Sun Microsystems UNIX Workstation contains an integrated Ethernet connector. After all, what else would one do with a workstation besides connect it to an Ethernet LAN? Next, there is a vast pool of network managers, administrators, installers, and other miscellaneous personnel who know Ethernet well, and to the exclusion of familiarity with other LAN technologies. Colleges and universities around the world graduate more of these trained personnel every year. The pool is deep and wide and expertise for Ethernet is readily available. The increased use of high-speed Internet access to residences with multiple PCs has even led to the use of Ethernet right in the home.

The last reason for the popularity of Ethernet LANs is the one that is most relevant to this discussion. There is little incentive for organizations to increase the speed of their LANs, even for bandwidth-hungry client-server applications because the LAN is not the bottleneck in most cases. Although more powerful desktops have reduced the number of desktops linked by a single Ethernet from hundreds to tens to nearly one, the net result has been that the amount of traffic on a single Ethernet has remained fairly constant. The burden and bottleneck had now shifted to the interconnection devices between Ethernets and the bandwidth available on these links.

As common as the Ethernet is as a LAN type, so the router is as a LAN interconnection device. The roots of the popularity of the router go back to the days when LANs were almost universally linked by bridges. This is not really the place to debate the merits of bridging versus routing, but most industry observers agree that bridging makes sense in small, local LAN interconnection environments (such as when all of the LANs are in the same office park, or there are only a handful of dispersed LANs), while routing can handle anything bigger or more widely dispersed. Certainly the success of the Internet, which employs the router as the network node device of choice (currently), has also led to the increased popularity of the router itself as a network connectivity device.

The limiting factor in most client-server applications today, therefore, is the amount of bandwidth available between the organization's routers.

Multimedia

The third major development that has led to the need for more bandwidth is the increased use of multimedia in a number of different networked situations. Multimedia has already been mentioned in a variety of contexts previously in this chapter. Documents that run voice-overs while they display text, images that move, and graphics that evolve over time are all examples of multimedia. Multimedia is not just read, it is listened to and watched. Multimedia demands attention, it seems to many.

In spite of this almost obvious use of the term multimedia, multimedia remains a not-too-well-defined concept. Many people take the "I know it when I see it" approach and leave it at that. It seems clear that a "normal" database record will display on a computer screen while the user goes for coffee. But perhaps a multimedia database record will not. When the user returns refreshed, the multimedia aspect of the database record may have already "happened." The user has missed it. The event may be repeated, of course, but the point is the same. Multimedia happens: It is not static.

Multimedia has been variously described as either including, or even being characterized by, the inclusion of things such as video, three-dimensional graphics, animation, music, voice narration, voice recognition and synthesis, scanned documents, photographs, television clips, virtual reality, and even biofeedback-based input devices into otherwise data-only or text-only applications. Not all of these elements have the "evolving" characteristic emphasized above, but perhaps *multiple-media* use is just as good a definition. However it is defined, multimedia today seems to imply that networks must now support enough bandwidth to at least allow for the optional inclusion of voice, audio (quality sound), and video along with the data to be sent from server to client over the same physical network.

Standards such as MPEG and H.323 have been developed to define and support multimedia communications networks. Simple voice- and data-mixed applications could probably be adequately supported with existing WAN technologies; however, any applications involving mixed video, CD-quality stereo sound, and data will benefit by increased WAN bandwidths and lower delays. For example, experiments and trials are underway to allow the display of a Web site related to a sports event or even a night-time television drama right on the TV screen with the show.

PCs routinely play audio CDs in the same drive as that used for data-based CD-ROMs. Most PC systems come with better speakers (and features such as powered sub-woofers) than those found on stereo systems only a few years ago. PC users can easily listen to Mozart while writing a report for work or school. Cable TV connections can be hooked up to inexpensive PC boards to allow for the viewing of television shows in a monitor window. Strangely, this TV image can usually be iconized as would any other PC application window. But the TV icon contains a small, yet accurate, version of the television picture.

Multimedia is already pushing the limits of the capabilities of networks. The PC will probably become the focal point of home activity in the same way that the TV set is today. These combined TV/stereo/PC home units are already appearing and provide new multimedia interfaces between people and computers over networks. Instead of manipulating text and numbers, and later graphics and simple sound bites, a new breed of users will be manipulating moving images, changing the endings of movies or improvising a new climax to a popular piece of music. Photo-realistic graphics will blur the line between TV and PC.

A common thread runs through all of the above multimedia examples: The multimedia is either delivered locally to the user or accessed over a separate network built especially for this purpose. CD-ROMs with audio- and videolike multimedia encyclopedias are usually accessed directly from the local CD-ROM drive, not over a network, not even on a LAN. Audio CDs must be accessed and played in this fashion. The cable TV connection to the PC for video essentially makes a TV set out of the PC monitor. Any other network connectivity, for Internet access or telecommuting, must be maintained separately. The network, therefore, is a limiting factor here as well.

However, networks will have to handle a range of information: powerful PCs on LANs; client-server applications linking computers in diverse locations; and multimedia files and applications around the world. But what kind of network will be able to do it?

Broadband Networks

The term *broadband* is used at least as frequently and probably as imprecisely as the term multimedia. Sometimes a distinction seems to be made between a *broadband network* and a *broadband application,* but this is of little help in defining the term broadband more precisely because these terms usually are defined circularly. That is, a broadband network is needed to support a broadband application, and a broadband application is one that requires a broadband network to function. A better definition must be devised.

Actually, there is a standard, and internationally sanctioned, definition of a broadband network; as defined by the International Telecommunication Union (ITU), a broadband network is characterized by "speeds higher than the Primary Rate." The Primary Rate refers to the ISDN primary rate interface (PRI) speed of about 1.5 Mbps in the United States and about 2 Mbps in most of the rest of the world. This would seem to imply that any network running faster than a few Megabits per second would qualify as a "broadband" network. Many networks do—LANs and WANs alike. The problem is that this definition was established by the ITU well before the increase of PC power, the popularity of client-server, and the promise of multimedia put so much stress on network capabilities. Therefore, in spite of this definition, the term broadband has come to mean much more than just a network built out of links that run faster that a couple of Megabits per second.

Because one of the main goals of this book is to explore the position that SONET holds in relationship to broadband networks, perhaps there is a better definition of broadband, one more appropriate to what SONET is and does. For the purposes of this book, the term *broadband* is defined as a network capable of supporting *interactive multimedia* applications. This does not imply that all broadband networks necessarily are used for interactive multimedia, only that they are capable of providing such support if needed. Nothing in this definition prevents such networks, for example, from supporting only high-speed file transfer or rapid client-server database access.

This definition rather nicely gives a more intuitive feel to many of the terms that include the term broadband. For example, a *broadband network* is now an *interactive multimedia network*. A broadband application is now one that includes support for interactive multimedia in its software code. In a sense, however, the new, admittedly subjective, definition is just as circular as the old one. What is *interactive multimedia?*

No official definition of the term interactive multimedia exists. But perhaps that is not really a limitation, because official definitions from standards bodies, such as is the case with the term broadband itself, may be unenlightening. In this admittedly unofficial definition, the term interactive multimedia may be broken down into its two components: interactive and multimedia. Multimedia has already been defined as information delivered across a network that is not only read, but also viewed or heard.

Interactive may be defined as an application wherein user input action *now* affects application output *immediately.* User input action may be a click on a mouse button, the pressing of a function or enter key, or something else. Output is usually some monitor display, but could be some sound output or even printed images. Unfortunately, no absolute time intervals can be placed on the words *now* and *immediately* in this context. But again, this is not necessarily a bad thing. In fact, the only criteria are the ones users themselves apply to these terms. Not too long ago, an interactive (or real-time, or on-line, the terms varied) application was one in which user input resulted in output in about six seconds or less. Users today would never stand for such a sluggish system and network. Today, interactive users want sub-second response times from applications, systems, and networks.

The term, interactive multimedia, therefore, can be loosely defined as the delivery of multimedia content based on immediate user direction. When interactive multimedia is deployed according to the client-server model, the user is expected be seated at a client computer, and the multimedia content resides on a remote server, or even several remote servers, because multimedia implies multiple information streams. Each stream may derive from an independent source. The user can immediately alter what is seen and heard through the touch of the mouse, or push of a button. A broadband network can be defined as a network capable of interactive multimedia delivery in a client-server architecture.

Few networks can effectively deliver interactive multimedia today. Even LANs struggle with this task, to the extent that much interactive multimedia content must be accessed from the CD-ROM directly attached to the desktop system. This struggle occurs because interactive multimedia applications make two very different, and yet related, demands on a network.

Multimedia applications, whether interactive or not, are characterized by a need for great amounts of bandwidth. This is usually measured in bits per second. A text-only application, such as accessing a World Wide Web page on a remote Web server with a Web browser (client), can usually make do with about 30 kbps. Adding audio to the Web site will increase the loading time and make the higher bandwidth bonded ISDN channels, which run at 128 kbps, more desirable. Viewing acceptable video from a remote Web site may force users to seek bandwidths in the megabit ranges, usually 6 Mbps, but as low as 1.5 Mbps for some video.

Obviously, SONET is a way to provide all the bandwidth that a multimedia application needs. But there is more to the broadband network story than increased bandwidth. This is where the *interactive* portion enters into the broadband equation.

Interactive applications which, of course, do not always involve multimedia content, are characterized by low and stable delays across a network. Both low delays and stable delays are necessary. When the delay across the network varies too much, unacceptable jitter is the end result. Sound is distorted by jitter, and video suffers from annoying speed-ups and slow-downs, resulting in jerky presentation. Low delays are needed to make interactivity possible in the first place. User input must travel across the network to the remote source, affect the output, and then the altered output must travel back across the network to appear at the client system.

If SONET were to only affect the bandwidth available to users on the broadband network, what would cause delays on the broadband network? Actually, bandwidth has an affect on network delay in general. It is not an obvious one, however, and requires a few more words about the effects of bandwidth on delay.

Bandwidth and Delay

The relationship between bandwidth and delay on any network is seldom acknowledged by end users. Much of the confusion arises because both terms use the unit "seconds" (bits per seconds, seconds of delay) in their measurements. Consider the network link shown in Figures 1.3a and 1.3b. The upper figure illustrates how delay is measured in a network, while the lower figure illustrates how bandwidth is measured in a network.

In the upper figure, a source sends a frame across a network link. In actual practice, this measurement does not need to be done over a single link. The entire network consisting of network nodes (the switches or

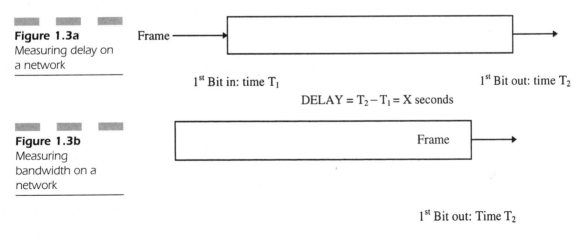

Figure 1.3a
Measuring delay on a network

Frame ⟶

1st Bit in: time T_1 1st Bit out: time T_2

DELAY = $T_2 - T_1$ = X seconds

Figure 1.3b
Measuring bandwidth on a network

Frame ⊢⟶

1st Bit out: Time T_2

BANDWIDTH = BITS in frame/ $T_3 - T_2$ = X bps

Last Bit out: Time T_3

routers) and trunks (links between network nodes) may occur between the end points of the measurements. The figure is only a illustration. The point is that the first bit of the frame leaves the source at some point in time (T_1) and this first bit arrives at the destination at some later point in time (T_2). The delay is just the time difference between these two time measurements of *first bit in* to *first bit out* of the network.

In the lower figure, the frame has arrived at the destination. There will also be a time interval for the entire frame to arrive off the network into the destination device. The number of bits in the frame (which varies) divided by the time interval between the first bit out (the same T_2 as before) of the network and the last bit of the frame out of the network (T_3) determines the *bandwidth* of the network available to the user.

Both the delay and the bandwidth will affect the *overall* response time that the user experiences from the network. To understand why this occurs, consider a network with *zero delay* (i.e., no delay at all) between end points. That is, the first bit of a frame appears at the destination as soon as it is sent ($T_1 = T_2$ in the figure). It is still possible for users to complain that "the network is too slow!" How can this be? Well, consider that the bandwidth available to users on this zero delay network is only 8 bps. Even a modestly sized 600 octet frame would take 600 seconds (10 minutes) to arrive. In the vast majority of cases, a computer cannot do anything with a frame until it is in memory completely because the error detection fields are usually located at the rear of the frame, and there is no sense in processing the beginning of a frame unless it is determined that no errors exist in the arriving frame.

It such cases, the network delay is fine (zero), but there is not enough bandwidth on the network for the users. In other cases, the bandwidth may be adequate, but the delay is too high for the user's tastes. It is possible to attempt to classify network applications as "bandwidth bound" when they are constrained by the amount of bandwidth available, or "delay bound" when these applications are sensitive to overall network delay. For example, file transfers are typically bandwidth-bound applications. File transfers can be speeded up by increasing bandwidth or lowering delay, but increasing bandwidth will probably give better results. Users do not usually care how long it takes pieces of a file to arrive, as long as the whole arrives within a certain amount of time. Conversely, voice is typically a delay-bound application. Giving 64 kbps digitized voice more bandwidth will not improve network performance. Only a suitable low and stable delay will make voice users happy.

It is important to appreciate the different effects that delay and bandwidth have on overall network speed. Users will always perceive both bandwidth-bound or delay-bound applications as caused by "slow" networks. When media, such as audio and data, are combined on multimedia networks, there must be both low and stable delays, as well as adequate bandwidth, to satisfy users. Interestingly, the International Telecommunication Union (ITU) defines the term *latency* on a network as the interval between *first bit in* and *last bit out* in several standards. This definition neatly combines the effects of bandwidth and delay. However, most network personnel loosely use the terms delay and latency interchangeably.

Bandwidth obviously has an effect on network delay, at least as perceived by the user; however, the more critical components of overall network delay are not dependent on bandwidth. There are two components of the network delays experienced by any user. These include the *propagation delay* of signals on the physical media and the *nodal processing delays* on each network node (switch or router) on the network. Only one of these elements can be speeded up in any realistic fashion.

The propagation delay on a network is a consequence of the physical fact that electricity, light, or any other form of electromagnetic signal can travel no faster than the speed of light from a source to a destination. In fact, in most networks, regardless of the physical media, the signals travel much slower, at about 2/3 of the speed of light. The speed of light is almost exactly 300,000 kilometers per second, or about 186,000 miles per second. This means that most signals take about 10 millionths of a second or 10 microseconds to travel a mile. One microsecond (abbreviated as μsec) is equal to 1/1000th of a millisecond. A circuit 2,000 miles long would thus have a propagation delay of $2,000 \times 10\ \mu sec = 20,000\ \mu sec = 20$ msec, as mentioned above.

Nothing on a network will decrease the propagation delay of a signal from the source to a destination, short of moving the destination closer. This is not a viable option in most cases; therefore, the only component of network delay that a network service provider can improve upon is the nodal processing delay.

Nodal processing delays in a network, in turn, are dependent on two factors: The first is the overall speed of the node, whether central office switch, router, hub, or other more exotic device. It takes so long to move bits from an input port to an output port, even when no other traffic occurs on the switch. The second factor is the load on the network node at any point. The more traffic there is, the slower the node will operate. Eventually, operation may slow noticeably, and the node is said to be *congested*. This basically means that the network node is operating outside of its design parameters in terms of delay. The presence or absence of

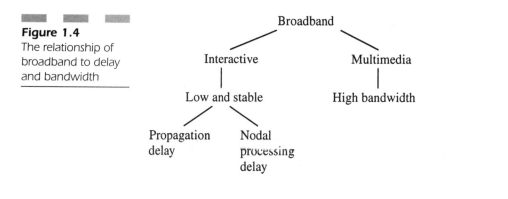

loads can make the end-to-end delay that an application manifests variable.

Figure 1.4 shows the relationship between all of the concepts discussed so far. Broadband is defined as support for interactive multimedia capability. The interactive portion depends on low and stable delays across the network. The delay, in turn, depends on the propagation delay and nodal processing speed. The bandwidth depends on the speed supported by the media available.

Creating Broadband Networks

Two ways exist for equipment designers to deal with the two factors of basic delay and congestion. The first way is to make the network node as fast as technology allows. The second way is to make sure that the capacity of the network node is adequate for even the most extreme network loading conditions usually by adding more trunks and memory.

In the public switched telephone network primarily used for voice, the network nodes are central office or toll office switches and the bandwidth is provided by unshielded twisted pair copper local loops to the switching office and digital trunks on a variety of media between the switching offices. There are exceptions, but this is the general architecture.

Voice services work well on this architecture for two reasons. First, the bandwidth needs for voice are relatively modest. Digitized voice consumes only 64 Kbps, and compression can lower this even further. Second, international standards require voice switches to operate quickly, in less than 1/2 a millisecond. This means that voice services enjoy adequate bandwidth and delays on the public switched telephone networks. Ten switches on a voice circuit add less than 10 milliseconds to the end-to-end delay.

Networks that support interactive multimedia and many other networks today benefit from the deployment of fiber optic cable and SONET. SONET can easily provide all the bandwidth needed for any application, from interactive multimedia and imaging to medical applications. The trickier part to address is the nodal processing delay.

Early attempts to combine voice and data applications go back to early days of X.25 packet-switched networks in the early 1980s. X.25 proved not to be adequate for the delivery of packetized voice services for two reasons. The older X.25 switch was not nearly as fast as a telephone central office switch, taking anywhere from 5 to 10 msec to process a packet. This was 10 to 20 times slower than the voice network. Ten switches on an X.25 network added from 50 to 100 msec of delay to the end-to-end delay experienced by the user or application. This was far above the propagation delay.

The second reason was that the delay in any particular X.25 switch was extremely dependent on the traffic load of the switch at any point as are all network devices.. This made the delay encountered by a particular packet containing digitized voice highly variable. The service was usually unacceptable because of great voice distortion. Vendors applied more inventive techniques to their X.25 products to compensate for these twin problems, but with modest success. (Of course, X.25 was never *designed* for voice—but some still tried.)

Of course, the answer to building network nodes with low and stable processing delays was simple to describe. The answer, however, was difficult to implement until computer architectures ran fast enough to perform the needed tasks. In a nutshell, the answer is to make the nodal processing delay, even in a worst-case scenario in terms of traffic load, an insignificant fraction of the propagation delay end-to-end through the network. All applications—voice, video, or whatever else—*had* to work within a given propagation delay, or they would not work at all.

For example, consider a circuit with an end-to-end propagation delay of 20 msec. Suppose that there are 10 network nodes along the circuit and that these contribute only 5 µsec ±1 µsec of delay regardless of network load. Then the total nodal processing delay would be about 50 µsec, or only about 1/400th of the propagation delay. The variability would be only ±10 µsec, or only 1/2000th of the overall propagation delay. These nodal processing delays of a fraction of a percent of the propagation delay would be indistinguishable to any end user, regardless of content.

Once this realization is made, the only question remaining is to decide which nodal architecture is the best one available to offer the low and stable processing delays required. Basically, any "fast-packet" technology will do. Fast-packet technology delivers packets of information with a low and stable enough delay to allow for multimedia operation across the network. The international standard for broadband services built on these fast packet networks is B-ISDN. The reference to ISDN is not accidental. B-ISDN, as conceived and standardized by the ITU in 1988 (the "Blue Books"), was intended as a logical extension of ISDN technology and services. The B-ISDN standard settled on the cell relaying ATM technology, for a variety of reasons. ATM switches provide the nodal processing delays that were low enough and stable enough for all services. SONET links provided the huge bandwidths needed to effectively combine voice and video and data networks on the same physical infrastructure.

In a sense, B-ISDN is a framework and an evolutionary standard. Work on SONET, ATM, and especially B-ISDN services was not completed in 1988. Indeed, the work had barely begun. The ITU is not the only standards organization working on completing the B-ISDN vision. Other organizations are working on various aspects of broadband networking. The Institute of Electrical and Electronics Engineers (IEEE) has extended broadband work into the LAN arena. In the United States especially, the American National Standards Institute (ANSI) and Bellcore (now Telcordia) have adapted ITU standards for use within the United States. Other organizations, such as the Electronic Industries Association (EIA) have contributed as well.

Figure 1.5 extends the previous picture to include the present broadband or fast-packet technologies. SONET is the choice to supply the

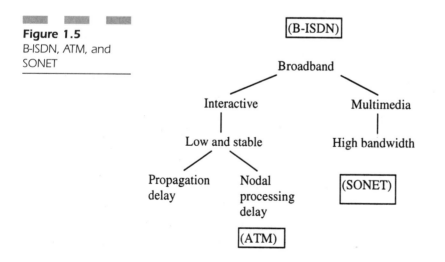

Figure 1.5
B-ISDN, ATM, and
SONET

bandwidth. ATM switches provide the network nodes. B-ISDN services may be offered based on this network architecture. Now, broadband networks may be built without SONET links and ATM switches. After all, customers buy services, not technology. But only a network based on SONET and ATM may offered ITU-compliant B-ISDN services to customers.

B-ISDN

To their enduring credit, the ITU recognized the limitations of ISDN in terms of raw speed and multimedia capabilities by the middle of the 1980s. The increased power of desktop computing devices and the widespread use of client-server architectures made it apparent that the 64 kbps to 2 Mbps (1.5 Mbps in United States) speeds of ISDN would not be able to keep pace with the rapid evolution of these devices for long. In 1988, the ITU produced a "blueprint" for taking telecommunications into the twenty-first century. This blueprint was B-ISDN.

B-ISDN was a vision of unified services, all delivered on the same unchannelized physical network with fast switching, state-of-the-art multiplexing, and very fast and yet low-bit-error-rate links tying all of the components together. Perhaps overly optimistically, some observers hailed B-ISDN as a network architecture for the next 30 years. No matter how long it takes to implement B-ISDN, or how long the B-ISDN vision remains viable, the whole concept remains an impressive achievement as an attempt to address network limitations.

The services that B-ISDN encompasses are all-inclusive. They amount to no more or less than nearly everything that has been done on a network before, everything that can be done now, and anything that can be done in the future. After all, once a network has all the bandwidth technology allows, operates fast enough for any interactive application, and virtually has no errors, what else could users possible need that a B-ISDN network could not deliver?

The fast switching and state-of-the-art multiplexing are provided in the B-ISDN scheme by ATM network nodes. The very fast and yet low-bit-error-rate links that tie everything together are SONET fiber links. The whole network has a public portion in the form of large, service provider-based ATM switches, which mostly switch and do some multiplexing, and a private component consisting of smaller (usually), customer-premises-based ATM switches, which mostly multiplex and do some switching within the customer site. This entire "hybrid" private/public ATM/SONET network delivers services defined by B-ISDN to the users.

The B-ISDN network concept is illustrated in Figure 1.6.

A partial list of B-ISDN services is listed in Table 1.3.

Network Connections

The list of services in Table 1.3 naturally tend to fall into one of a number of categories. There are voice and data services, of course, but also imaging (fax) and video services. Few of these depend on B-ISDN to allow them to be available to users who want them badly enough. Currently, for example, many organizations buy equipment to supply voice, fax, or data services for their users without a thought for B-ISDN; however, this means that all of these various types of user equipment and the network nodes that support the service must be linked together separately with some form of network connections to supply the needed connectivity.

In the United States especially, network connections between routers, LANs, and other network service delivery devices tend to take the form of leased private lines. Also called point-to-point links, dedicated circuits, or even some other terms, these leased private lines are basically the bandwidth on digital links bought from the telephone company (not always, but usually) on a long-term lease. They are considered "private" due to the fact that only the organization that pays the bill has the right to put bits on the link. Even when the link is idle for days, the service provider cannot reclaim the bandwidth for its own purposes. The bandwidth belongs to the customer. It is, however, still a leased or rental situation. If the link were to fail, the owner (the service provider) must fix it. If the lease were to expire or were not paid, the bandwidth would

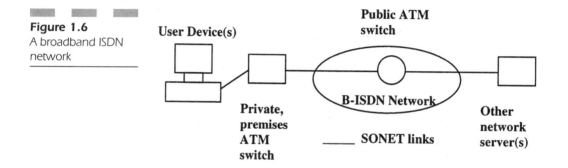

Figure 1.6

A broadband ISDN network

TABLE 1.3

B-ISDN services

Services		
Voice	CAD/CAM	Telemetry and security
LAN interconnectivity	Telex/Teletex	Home consumer services
Fast-packet service	E-mail	Electronic banking
High-definition TV	Fax	Teleshopping
Videoconferencing	Videotex	Telecommuting
Private line emulation	Electronic document interchange	

revert to the service provider, which would then be free to lease the bandwidth to another customer.

There are a number of reasons why the process of building network connections became one of leasing private lines in the United States. First, when they began building computer networks in the United States in the 1960s, organizations had plenty of money to spend on whatever they needed to make the networks possible. Second, service providers were only too happy to expand their facilities to meet the demand for private lines that the new data networks consumed.

Most organizations had enough money to afford their own network nodes. Whatever switches or other devices they needed could be purchased from the manufacturers. The leased private lines then became the method of choice for linking these network nodes together. This solution was neither inexpensive nor painless. Leased private lines were purchased by the mile, and more bandwidth cost more money. In an attempt to cut down on the increasing burden of leased line networking costs, many organizations built a bare minimum of private lines to link their sites. This also resulted in unintentional bottlenecks as traffic to and from remote sites funneled through intermediate sites. Because more bandwidth cost more money, organizations also attempted to provide connectivity with the absolute minimum amount of bandwidth required, usually the digitized voice bandwidth of 64 kbps.

This is not to say that organizations employed individual 64 kbps digital links on their own physical facilities. It was common for many organizations, especially larger ones, to lease "bundles" of 64 kbps private lines. These were known as channelized DS-1s or T-1s in the United States. The term, *channelized* refers to the fact that the bandwidth on these circuits was divided into 24 circuits, or channels, of 64 kbps each, called DS-0 channels. The whole practice became so common in the

1980s that the term *channelized networking* began to be applied to this configuration. This process was accomplished by time-division multiplexing, but it was the ends that was important, not the means. The whole idea of channelized T-1 networking is shown in Figure 1.7.

Organizations would divide the use of the channels on the T-1 according to their needs. It was not unusual to find an organization assigning six DS-0 channels to connect their corporate routers on various LANs, using six other DS-0s to link the corporate PBXs with private voice lines (called *tie-lines*), and using six more DS-0s to link videoconferencing equipment from one site to another. The other six DS-0s were usually reserved as spares for growth. This mixed data, voice, and video network achieved this support at the price of flexibility.

The bandwidth that the channels represented could not be easily reassigned or shared by other channels. For example, even if no one were on the phone at 3:00 A.M., the 384 kbps bandwidth associated with the tie-lines could not easily be used by the data channels. Each channel was a private line by itself, even though it was bundled and delivered on a T-1 (DS-1). In many cases, even the use of these T-1 circuits running at an aggregate bit rate of 1.544 Mbps did not help to alleviate the need for more bandwidth within most organizations due to the use of channelized networking.

In the rest of the world, the situation was different than in the United States. Other countries, instead of prospering in the 1950s, were attempting to recover from the devastation of World War II. Still other countries, although relatively unharmed by the nearly global conflict, nevertheless lacked the internal funds and resources to build the telecommunications infrastructure a leased-line market required. When computer networks began to be built around the world in the 1960s and

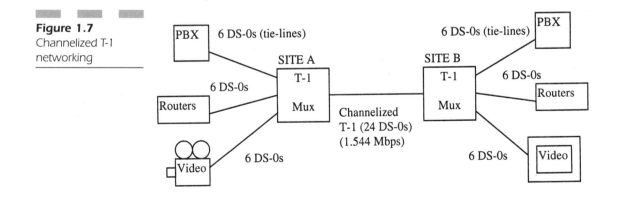

Figure 1.7
Channelized T-1
networking

1970s, the plentiful and inexpensive leased-line environment found in the United States was absent in the rest of the world, with very few exceptions.

To address this need for networks without private lines, other countries created public data networks. In a public network environment, the service provider, usually a government entity, such as the national telephone company, bore the expense of buying and installing the network nodes and the links connecting them. Users then bought the service that the node type delivered, but not the links or nodes themselves. It was a simple and elegant solution. Eventually, the ITU adapted the X.25 Packet Switch Data international standard to ensure compatibility and interoperability between these public data networks.

Just as the leased-line networks in the United States had limitations, so did the public data network have limits as well. In the case of public data networks, the problems were slightly different, but just as serious. The bandwidths available over X.25 public packet switched networks did not exceed 64 kbps, and in many cases did not exceed even 9.6 kbps. There was no technical reason for this. The X.25 standard simply peaked at 64 kbps. Delays were highly variable, just as in all packet-switched networks of the time, precluding the use of X.25 for packetized voice and video. Because much of the customer traffic was carried on the same public network facilities also meant that performance was highly variable, and security trickier to implement than on private lines.

None of these factors prevented X.25 from becoming the data network solution of choice in most countries around the world. X.25 was used for everything from SNA networks to the emerging client-server applications. Today, frame relay has extended X.25 capabilities into the multi-megabit ranges.

B-ISDN Splinters

By the beginning of the 1990s, all of the trends discussed to this point seemed to be coming to a crisis point. The power of computers threatened to overwhelm the networks they were increasingly dependent on to function within the client-server model. Demands for multimedia only worsened the situation. The practice of building leased-line networks with channelized private lines led to the splintering of the available bandwidth among many network connections, each with idle bandwidth. Even the use of public data networks did not improve matters greatly.

The need for increased networking bandwidth and faster network nodes eventually resulted in the creation of B-ISDN. B-ISDN used the incredible bandwidths available on fiber optic links and combined this with the unparalleled speed of ATM switching. A sophisticated multiplexing technique used with ATM made it possible to create unchannelized networks which allowed all of the available bandwidth to be used by whatever application (i.e., voice, video, or data) needed bandwidth. The invention of flexible (or dynamic) bandwidth allocation became popularized as *bandwidth on demand,* although this is a less precise term. Bandwidth on demand encourages the belief that bandwidth can be created out of nothing, while flexible bandwidth allocation correctly focuses on the dynamic reaction of the network to users' changing bandwidth needs.

However, even though B-ISDN offered a potential solution to all these problems, not all of the pieces of B-ISDN were equally mature and ready for implementation in hardware, software, and network protocols. The three components of B-ISDN (i.e., SONET, ATM, and B-ISDN services) were not pillars, but a tower. Supporting the whole network structure were the fiber SONET links, coincidentally the most mature part of the plan. On top of the SONET foundation were the ATM switches. Although ATM was emerging from laboratories and test networks by 1990, there is a lot more to networking than just sending bits from place to place, especially in a public network. Networks must be managed and controlled. It was in just these two areas that early ATM products proved most lacking. Without the firm support of stable and robust ATM products, the B-ISDN services, perched atop the B-ISDN tower, seemed ready to come tumbling down.

What actually happened in the 1990s was not as bad as that, but it was something that may not be fixed for some time to come. The different stages of the components of B-ISDN soon led to the breakdown of the whole structure, a splintering that may take years to put back together, if it happens at all. SONET was soon deployed independent of ATM and/or B-ISDN services; after all, bandwidth was just bandwidth. SONET neither required nor cared whether the bits carried over the link came from or went to ATM switches or equipment. SONET grew up in the 1990s as an orphan in some respects, but an orphan with a very bright future.

ATM suffered a somewhat different fate. ATM was effectively split in half by two related, but independent, camps. LAN equipment vendors seized the fast switching aspects of ATM and immediately made plans to use ATM to allow for unprecedented speeds to the desktop. Because network management and control are much simpler in a LAN environment, these ATM shortcomings were much less obvious in this application. At the same time, WAN equipment vendors embraced unchannelized support for voice, video, and data that was inherent in ATM and immediately made plans to use ATM internally with their network products. Service providers liked the idea of merging separate channelized facilities onto ATM networks, but needed the network management and control components desperately.

B-ISDN suffered the worst fate of all. All of the attention to SONET and ATM left little time or effort to devote to B-ISDN services. If SONET was just another private-line solution, and ATM was being buried inside LAN hubs or telephone company toll office switching networks, what was the sense of moving forward with B-ISDN service descriptions?

Each of the two lower pieces of B-ISDN—SONET and ATM—have taken on their own identities independent of B-ISDN. The question is now becoming one of whether SONET and ATM should be reunited with B-ISDN, no matter how far into the future. Of course, networks and the technologies that make them possible exist to serve users. If users are happy with SONET and ATM the way they are, why not leave well enough alone?

The answer to this question is complex. A full answer would point out the advantages of having a global standard for interoperable and compatible services. However, as time goes on, the success of the pieces of B-ISDN, especially SONET, makes it apparent that a reunification of SONET, ATM, and B-ISDN may not happen for years to come.

In the past few years, the rise of the Internet and the Web have all but laid to rest grand plans for a global, unified network based on B-ISDN applications, ATM switches, and SONET links. The brave new world of Web-based television and telephony, IP routers, and DWDM links has fundamentally changed the future of networking. Networks today grow from the ground up, like the Web, from instance to standard, rather than from standard to instance as outlined by the vision of a standards body, such as X.25.

Oddly, the only widespread use of H.323, the most mature piece of B-ISDN yet defined, is on the Internet in the form of IP telephony. But certainly people will not wait for B-ISDN applications while they buy and sell, work and play, and even live their lives on the Internet and the Web. The only role for ATM today appears to be as a high-speed backbone technology to connect IP routers, and even that role has also been challenged. Even SONET and SDH are challenged by leaner, more IP-friendly techniques within the DWDM framework.

A word here is appropriate about adding quality of service (QOS) to IP networks like the global public Internet. Like all technologies designed to meet mainly data needs, IP networks struggle to provide the low level of delays, stable delays, and reliability that are needed to support interactive multimedia applications such as voice and newer forms of video. These QOS parameters must be added either to the network or to the IP application, or to both, if the application is to become acceptable and common. There are currently more than a dozen schemes for adding this QOS to IP networks, including DiffServ, RSVP, and IntServ. It is not necessary to discuss these schemes further, only to point out that as long as there are multiple ways to add QOS to an IP network, the pressure to add bandwidth to an IP network will only increase.

Why should this be so? Because the benefits of adding bandwidth to a network are well known, studied, and accepted. The benefits of adding DiffServ, RSVP, or something else to a network to improve QOS are at the present unknown. The situation is like the early days of PCs: with limited cash on hand, and faced with a slow PC, the smartest thing to do was to add memory—as much as affordable. The benefits of the added memory were immediate and well documented. The reasons that bandwidth is the network equivalent of memory have been discussed already. So bandwidth will be the network QOS resource of choice for some time to come, until one IP QOS scheme or another emerges that is as reliable as adding bandwidth to improve overall network QOS.

SONET as Private-Line Service

In the United States, where facilities were plentiful and organizations had money to spend in relative abundance, most networks were built as private-line networks. Telephone companies were almost always the source of these private lines. Private lines amounted to the selling (actually, long-term leasing) of bandwidth on telephone facilities to customers. The telephone companies gratefully saw the increase in private-line sales as another revenue stream. It may seem strange that the selling of pieces of the telephone companies' lifeblood was considered a sound business practice, because the bandwidth leased to the customer could not be used internally by the service provider. The practice of leasing private lines, however, was embraced because a private line service is barely a service. The risk and exposure of the telephone company as a service provider (or *carrier* of the service) was minimal.

The risk and exposure were small because a private-line service is used to create a private network. In a private network, the service provider's main duty is to furnish the basic transport service, in which the service provider merely accepts bits from a piece of compatible customer premises equipment (CPE) and delivers them to a compatible CPE at the other end of the point-to-point private line. The service provider neither looks at the bits (voice, video, or data), stores the bits (e.g., messaging system), nor alters the bits (e.g., to ensure compatibility). In fact, telephone companies were forbidden from providing any more than basic transport services or selling the CPE to the customer. Anything beyond the basic transport of bits qualified as enhanced services. Enhanced services were strictly controlled by regulating agencies such as the Federal Communications Commission (FCC) before 1996.

Service providers may not be overjoyed at losing large chunks of bandwidth that could otherwise be used to deliver dial tone and other types of service; however, the tradeoff was still acceptable because the only thing the service provider was responsible for was the transport of bits from point A to point B. It made no difference whether the bits made sense or not because the CPE and not the service provider generated them. All that mattered was whether the bits arrived, and all that could really go wrong was that the link stopped relaying the supplied bits, which was fairly easy to fix. Coupled with the fact that the private line had to be paid for, month after month, whether it was used 100% of the time, 50% of the time, or not at all, these factors made private-line services quite popular in the United States.

What has all this to do with SONET? SONET started out as an international standard for fiber optic transmission and was quickly included in the B-ISDN plan by the late 1980s; however, B-ISDN is not a private networking solution. B-ISDN specifies the services on a public network. Yet, service providers in the United States quickly embraced SONET and deployed it as a private-line service independent of B-ISDN. Why did they do this?

There are two reasons that SONET is not tied to B-ISDN, especially in the United States. First, early SONET standards proposals predated the work on B-ISDN; therefore, SONET began independent of B-ISDN. Second, service providers (i.e., mainly telephone companies) in the United States needed SONET to keep up with the demands for private-line service. Each point will need a little more elaboration.

Although it is true that it forms the foundation of a B-ISDN and ATM network, SONET did not begin this way. The idea of creating a standard digital network scheme for fiber was around before the idea of combining SONET with high-speed switching and multiplexing in the form of ATM eventually created B-ISDN services and networks. Although SONET came to be *intended* for use on public B-ISDN networks, there is nothing to *prohibit* its use as a private-line solution.

Private-line services sell off "pieces of network" in the form of bandwidth to customers. Now, how much bandwidth any one customer can buy is limited only by the total amount of bandwidth available, which only makes sense. But as multimedia applications and powerful desktop computers began to proliferate in many organizations, the demand for increased bandwidth speeds could not easily be met with existing telephone company networks and facilities.

For example, early LANs and router networks could be linked productively at 64 kbps. To supply this bandwidth, the telephone company usually installed 24 channels of 64 kbps each, providing 1.5 Mbps combined. If the customer were really pressed for speed, the whole bandwidth of 1.5 Mbps could be linked to a single router. However, above 1.5 Mbps, things were expensive and difficult to provision. The only realistic solution above 1.5 Mbps was 45 Mbps, some 30 times faster. In many cases, this was overkill and, in most cases, the price was too high.

Several lower speed 1.5 Mbps would sometimes do for multimedia, but not always. It is the same in many other industries: If a customer were to need a race car that goes 200 miles per hour, selling the customer two that would go 100 miles per hour each would not be the same thing.

The highest standard speed used internally in the telephone company network was 45 Mbps. Ironically, higher speeds were possible, but were never standardized and remained proprietary. Equipment could only be purchased from the one vendor that made it. This clearly would not work in an environment where the CPE remained beyond the control of the service provider.

All of these factors combined to keep the cost of private-line bandwidth above 1.5 Mbps too high to be easily affordable by most organizations. Lack of facilities kept the cost high on the service provider side, and proprietary equipment kept the cost high on the CPE side. After all, if the piece of network being sold is 45 Mbps, there must be higher speed standard interfaces on the service provider's network backbone to deliver these speeds. Unfortunately, there really was not.

SONET private-line services will break the high bandwidth logjam. SONET runs at many times the 45 Mbps bandwidth customers frequently require for interactive multimedia services. SONET will effectively lower the price for previously expensive links running at 45 Mbps and even higher (of course, SONET may require changes to the wiring inside a building as well).

CHAPTER 2

A Fiber Optic
Tutorial

The synchronous optical network, or SONET, is the North American version of an international standard method for a family of high-speed transmission links on fiber optic cable. SONET takes advantage of the many attractive features available on today's fiber optic facilities. This chapter explores these features and advantages in more detail, and shows why SONET is built on fiber optic technology. This chapter assumes no special knowledge of fiber optic cable, or anything more than a passing knowledge of digital transmission systems. At the same time, this chapter attempts to detail the reasons why SONET and fiber optic transmission systems are invariably intertwined today.

The rapid deployment of fiber optic transmission systems allowed service providers to deploy equipment that could transmit at very high data rates. This process had been going on since the 1970s, and (in some areas like Chicago) even in the late 1960s. But, with a lack of standards above 45 Mbps (the DS-3 rate), this equipment utilized vendor-proprietary-multiplexing techniques to transmit data at rates of up to 1.7 Gbps. Before SONET came along, identical vendor equipment was needed at each end of a fiber span, and the addition or drop off (usually just called *add/drop*) of various data streams for switching (usually at the DS-1 level) necessitated demultiplexing the entire data stream. When identical equipment was not available at both ends of a fiber span, such as in a handoff between a local carrier (LEC) and an interexchange carrier (IXC), handoff occurred at the lower "standard" data rates of DS-3 and below.

The previous chapter explored the origins of private-line networks and why SONET forms such a key piece in these networks today. This chapter explores the reasons that SONET came to be based on fiber. Today, the answer seems obvious, because fiber optic networks have become common and accepted not only on the service provider's backbone network, but also on the desktop. But it was not too long ago that fiber optic networks seemed much too fragile and expensive to become the basis for a new generation of digital networking standards. Therefore, it is worthwhile to examine why fiber exploded in the networking arena.

It sometimes seems that fiber is the media of choice for an increasing number of cabling situations. It is not only the service providers that have installed miles and miles of fiber, about half of which is installed privately in a campus or building environment. Fiber optic cable installations will continue to increase both on the customer premises and on the service provider's network. Fiber is the media of choice in increasing telecommunications situations, from voice, to video, to data.

Fiber Optic Transmission

The transmission of digital signals on fiber optic cable is simple to explain and understand. It is really as simple as shining a light down a strand of fiber. Even a hand-held flashlight can do it, but not very well. It is quite common to simply use a brief flash of light (a *very* brief flash of light: the faster the link, the briefer the flash) to indicate a "1" bit in a time slot and the absence of a flash in a time slot, to indicate a "0" bit. This simple binary on/off technique is the easiest and cheapest to implement. Many will recognize this method as just another form of amplitude modulation. This method is assumed in this tutorial. However, it should be pointed out and always kept in mind that there are other modulation techniques used in many configurations. Newer fiber modulation techniques are very sophisticated and are nearly as elaborate as the complex modulation techniques used in high-speed modems. The idea is the same: increase the bit rate available on a given bandwidth.

Fiber optic cable is made of very pure glass. Actually, there are other constructions, including very pure plastic, but these fibers have more restricted applications. The manufacturing process is fascinating, but of limited interest here. Fiber optic cable is made by heating and drawing out a large chunk of very pure glass (called a "preform") into a strand of fiber thinner than a human hair (a typical fiber is 120 microns (millionths of a meter) in diameter, and a human hair is about 140 microns in diameter). Various layers of similar materials are added (called "buffers"), along with strengthening materials (usually Kevlar fibers). The composite fiber optic cable, which may contain numerous fiber strands, is jacketed, labeled, and taken up on reels. Early fibers had a reputation for fragility, and often broke due to bending stresses at installation time. Fiber is actually stronger than steel when it comes to pure tensile strength (pulling). Bending fiber causes stress fractures (cracks), however, and the strengthening material is present more to prevent kinking and resultant fracturing than to prevent excessive pulling forces.

For commercial applications, whether within a building or over a long distance, a laser or light-emitting diode (LED) is used, depending on the speed and distance the application must cover. LEDs have a more "spread out" light flash, or pulse. Lasers generate much sharper pulses, which make them useful for higher speeds and longer distances, as shown in Figure 2.1. Although an LED may only be able to generate a light pulse with useful power about 40 nanometers (nm, billionths of a

Figure 2.1
An LED and laser
light pulse compared

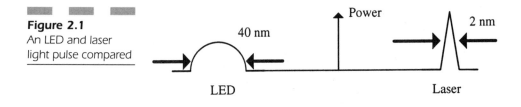

meter) in width, a laser pulse can be as short as 2 nm. Shorter pulses translate to higher bit rates in even the simplest modulation schemes.

Of course, many good safety rules exist that make the use of the lasers previously used for fiber optic communication difficult in general office space. Older lasers were usually not found on someone's desktop. The laser was usually down the hall in a locked communications closet with a warning sign on the door.

Perhaps a word should be said about lasers and the human eye, since safety is always a concern. Although it is true that the power of most commercial laser products have very low power rates (thousandths of a watt, or milliwatts), the human eye is a very effective lens. When looking directly at even the lowest power laser, the eye will focus all of the laser's power on the retina. In a relatively short amount of time, permanent damage could be done to a person's eyesight. This is mentioned not to frighten people, but only to remind them that a healthy respect is needed when lasers are in use nearby.

One of the major reasons that fiber optic cable is popular and effective is that the loss of signal strength over a given distance is much less than with other transmission media. That is not to say that fiber has no signal loss, or attenuation. Much of the loss of signal strength on a fiber link happens right at the interface between the light source and fiber. This is called the "injection loss" and is a function of what is known as the "numerical aperture" of the particular fiber. It is important to note that the interface between the light source and fiber does not have to be a bonded, soldered, or direct physical interface. If anything, the connection is essentially glued using a strong epoxy resin. Typically, the light source and fiber are just very close together. As a result, the light source can become misaligned with the fiber and needs periodic "tuning" to stay within specifications.

On the fiber optic cable itself, the light pulse travels as a coherent package of light waves. The light waves travel because of "total internal reflection" (it isn't really total) between the *core* and *cladding* structure of the fiber. The difference between the *index of refraction* (as a spoon appears bent in a cup of water) between the core and cladding confines the light-wave package to the core itself. The core is simply very pure glass, about 5-50 micrometers (millionths of a meter) in diameter. The cladding can be glass or plastic, and is about 120 micrometers in diameter. However, the fiber optic cable itself is typically heavy, due to the presence of multiple fiber strands, strengthening materials, and jacketing.

It is important to realize that fiber optic cable shares an important characteristic with other digital transmission systems. Fiber optic cables are almost universally deployed in pairs. One fiber is used for outbound digital signals and the other fiber is used for inbound digital signals between the same endpoint transmitters and receivers. Fiber can be made to run full-duplex on a single strand, but this tends to be more expensive and is not a common practice. Therefore, most diagrams of fiber networks in this book will show two strands of fiber connecting any two pieces of equipment.

At the receiver end of the link, which may be literally thousands of miles long, an avalanche photo detector (APD) or positive-intrinsic-negative (PIN) device detects the light. It is the nature of light to travel in a wave and be detected as particles called photons. The receiver counts the photons in the corresponding time slot. When the number of photons is above some threshold, a 1-bit was sent, anything else is a 0-bit. In theory, a 1-bit represented by 75 photons worth of light at the sender can be detected by 21 photons in the same time slot at the receiver. However, the number is usually about 38 (about half of those sent), which gives a characteristic bit error rate (BER) of 10^{-9} on the link. Again, this modulation technique is only intended for illustration purposes; actual techniques may be much more sophisticated.

Another aspect of the quantum physics "weirdness" of fiber and light transmission, in general, is that some of the photons will be detected *before* they are sent. A light pulse can "spread out" from the time slot in which it was actually sent, spreading forward as well as backwards. Fiber optic cable signals also do not travel at the speed of light through the fiber; light travels at 300,000 kilometers per second only in a vacuum. The index of refraction of the fiber slows the light down somewhat. Usually the propagation speed of light in a fiber optic cable is about 2/3 of the speed of light, or some 200,000 kilometers per second (about 125,000 miles per second).

Even from this simple description of fiber transmission systems, some major points are obvious: fiber links are sensitive to configuration differences and must be maintained; timing is obviously quite important on fiber links; and fiber cable is quite small. A simple picture of a fiber

Figure 2.2

A typical fiber optic
transmission link

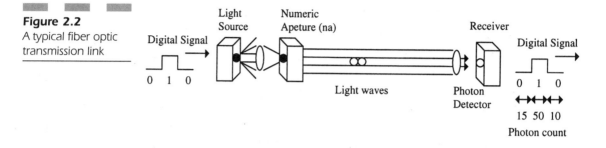

optic transmission system is shown in Figure 2.2. Although the numeric aperture (na) is represented as a "device," the actual na is just a theoretical construction of the fiber itself.

Fiber Optic Cable Types

A few words should be said about the various types of physical structure that fiber optic cables can have. Two basic classifications of optical fiber are defined here:

- Multimode fiber (mm fiber), which allows light to take many paths (modes) as it travels through the fiber.
- Single mode fiber (sm fiber), which has a core so small that only one path (mode) is available for light to travel.

Multimode Step-Index Fibers

The simplest type of fiber is the multimode step-index fiber. This fiber type typically has a core diameter ranging from 125—400+ microns (.005 inch to over .016 inch). This relatively large core allows many modes, or ways for light propagation down the fiber. The larger the core diameter, the more modes it has. Because light reflects at a different angle for each mode, some waves follow longer paths than others. A wave that travels straight down the core without any reflecting arrives at the other end sooner than other waves. Other waves arrive later in the receiving sample period, and the more times a wave is reflected, the later it arrives.

Therefore, light entering the fiber at the same time and representing the same bit or bits may arrive at the other end at slightly different times. Although it is not physically accurate, most people would say that the light has "spread out." This process of light-spreading is more accurately called *modal dispersion*. Modal dispersion limits both the speeds and distances that signals can be sent through a fiber.

Dispersion, along with some other complicating factors, makes multimode step-index fiber the least efficient of the fiber types. It is used for short runs, such as linking a computer mainframe to peripheral equipment and other LAN applications. Conversely, this type of fiber's relatively large size and simple construction do offer advantages.

This fiber gathers light well because of the large numerical aperture (the light-gathering cross-section of the fiber). Next, it is the easiest to install. Finally, it is quite inexpensive compared to other types. The disadvantages are the large dispersion of the light signals and the relatively low bandwidth, which limit both speed and distance.

Multimode Graded-Index Fiber

Obviously, modal dispersion can be a limiting factor in long-haul, fast-fiber networks, such as SONET. One way to reduce modal dispersion is to use multimode graded-index fiber, sometimes called GRINs. With this type of fiber, the index of refraction is highest at the center of the core and gradually tapers off toward the edges. Because light travels faster in a lower index of refraction medium, the light furthest from the center axis travels faster.

Because the waves following different paths travel at various speeds, the waves will reach the same point at roughly the same time, which is desired. Note that modal dispersion is still present, but now it is much lower than in step-index fiber. Light waves are now no longer sharply reflected, they are gently bent.

Multimode graded-index fibers are generally small, typically 125 microns. This type of fiber's size still makes it fairly easy to install, and the use of the graded index makes it more efficient. However, the more complex structure of the core makes it more expensive than the multimode step-index fibers, but it is very popular.

Overall, multimode graded-index fiber offers a good intermediate choice between the other two types (multimode step-index and single mode step-index), and the advantages and disadvantages of it are in between these two types. Multimode graded-index fibers are good for medium distances and medium speeds. Usually, this type of fiber is used for more than merely a single floor of a building.

Single-Mode Step-Index Fibers

Of course, the best way to remove the effects of modal dispersion is to reduce the fiber's core diameter until the fiber only propagates one mode. This method is used in the single-mode step-index fiber. This type of fiber has a core diameter of only 2—8 microns (.00008 to .0003 of an inch). These fibers are by far the most efficient, but their small size can make them difficult to handle for many applications. Use of this fiber is usually limited to very demanding, high-speed, long-distance applications, such as SONET. The advantages of single-mode fiber include minimal dispersion, higher efficiency, larger bandwidths, and thus longer distances and higher operating speeds. Conversely, these fibers are more expensive, more difficult to splice and connectorize, and require a laser as the light source because of their small numerical aperture.

Single-mode step-index fibers are used for long distances for telecommunications networks. Generally, spans can be run up to 30 miles without the need for repeaters. They can carry signals at high speeds, up to 10 Gbits. Naturally, most SONET networks use single-mode fiber.

Fiber Optic Advantages

What makes fiber so attractive for so many uses? Actually, there are eight distinct advantages that fiber has over almost any other medium. Some are simple and obvious, while others are more subtle. In fact, fiber offers more pluses and fewer minuses than any other medium in use today. The price of fiber optic links, which must include the cost of not only the fiber itself but also the transmitter, receivers, and any repeaters necessary, is now comparable to other, copper-based media, such as coaxial cable or unshielded twisted pair (UTP). Certainly, the installation cost per bit/second is the least of any cable or wireless medium. Beyond the fact that the bandwidth of fiber is practically unlimited, it is also true that no other cable medium is even being researched today.

Moreover, in the fiber industry, speeds achievable on fiber quadruple every two years, and that the distance achievable without digital repeaters also quadruples every two years. The eight fiber advantages are listed below:

- Lower BERs. Fiber offers lower bit error rates (BERs) than any other transmission media. In some cases, it can take three years of running a fiber link to generate the same number of errors as on a copper (coax, twisted pair) link (1,000 to 1 ratio).

- Higher bandwidths. Speeds achievable on fiber are unsurpassed. Unlike copper media, fiber speeds are increasing constantly, in some cases by orders of magnitude.

- Longer distances without repeaters. Digital repeaters, which detect and repeat the string of 0s and 1s onto another fiber link, require power and add potential points of failure. A given length of fiber will need fewer repeaters than the same length of cable composed of another medium.

- Immunity from interference. Most interference in the form of noise comes from the fact that copper cables are long antennas. Fiber will not pick up electromagnetic noise from the environment.

- Security. Fiber networks can be constantly monitored for signal loss increases, which may indicate the presence of taps.

- Less maintenance costs. Solid-state fiber optic components need much less maintenance than do other devices used in copper networks.

- Small size and weight. The same size crew can install much more fiber in one day than any other cable media.

- Bandwidth upgrades. When designed with this in mind, it is possible to increase the bandwidth available on the fiber link simply by changing the sender and receiver components.

Because the success of SONET is intimately linked with the advantages of fiber optic systems, let's take a closer look at each of these advantages.

Bit Error Time Intervals

It is not just the lower BERs that make fiber so attractive for transmission links and standards. Copper networks from T-1 to X.25 have always been engineered for BERs of about 10^{-6}. This indicates one bit in error for every million bits sent (10^6 is 1 million) which is not bad, but these links are used for very high bit rate transmission. For instance, with a BER of 10^{-6} at the DS-3 rate of 45 Mbps, a bit error will occur every 22.0 milliseconds on average, or about 45 bits in error every second. Actually, copper errors are usually in *bursts*, but the average BER is still a valid concept.

The significance of the BER for high-speed networks is twofold. First, at higher speeds, such as 135 Mbps, only fiber can give acceptable average bit error intervals. Second, at lower bit rates, fiber can give astonishingly low bit error intervals, leading to the common observation that "with fiber, bit errors just go away." This is not really an exaggeration. For example, some service providers offer 64 kbps links on fiber with a BER of 10^{-13} (1 per 10 *trillion* bits). At this rate, the average bit error interval at 64 kbps will be one bit error every five *years*. A decrease in the BER from 10^{-6} to 10^{-9} is a thousandfold decrease. In other words, the number of bit errors a network has in one day at 10^{-6} on copper is the same as the number of errors it has in three years at 10^{-9} on fiber!

Fiber links today run at 10^{-10} or better; therefore, the number of errors that a copper network has in one day, takes 30 years to occur on fiber. This fact fuels the perception that errors are eliminated with fiber. Table 2.1 shows some average bit-time intervals (i.e., time between bits in error) for a variety of line speeds.

Obviously, as networks' speeds climb, lower BERs become not a luxury, but an absolute necessity.

Typical Media Bandwidths

Nothing can match fiber for high bandwidth and speeds. In fact, no digital signal has ever been generated that fiber cannot carry. The bandwidth of a transmission medium is measured in terms of the analog frequency range (bandwidth is more properly an analog term) measured in Hertz (cycles per second, or Hz). The symbol for cycles per second is f, and the rule for determining the bandwidth of a medium is given by the formula $f = c/\lambda$, where c is the speed of light and λ is the characteristic wavelength of the medium.

For fiber, c is 186,000 miles per second or 300,000 kilometers per second, and λ is about 1 micron or 1,000 nanometers (billionths of a meter) for visible light. Using these numbers gives a characteristic bandwidth for fiber optic cable of about 300 TeraHertz, or 300,000 GigaHertz. This is only a naive computation, in the sense that it concerns only analog

TABLE 2.1

Average bit error time intervals

Bit Error Rate	64 kbps	256 kbps	1.5 Mbps	45 Mbps	135 Mbps
10^{-6} BER	16.0 seconds	3.9 seconds	0.7 seconds	22.0 millisec.	7.4 millisec.
10^{-9} BER	4.3 hours	65.0 minutes	11.0 minutes	22.0 seconds	7.4 seconds
10^{-12} BER	6.0 months	1.5 months	7.7 days	6.2 hours	2.1 hours

bandwidth and does not take into account digital modulation techniques for sending digital signals on the fiber. However, it is representative even assuming only 1 bit for each Hertz of bandwidth.

In contrast to fiber, Figure 2.3 shows the typical bandwidth of other common media. The scale is logarithmic, which means that each step on the vertical axis is actually 1,000 times the previous one.

Long Distances

Fiber offers the ability to traverse long distances without repeaters. Repeaters "clean up" and re-send the signal on long fiber spans. The distance between repeaters on a digital link is purely a function of the signal loss over the distance. Greater signal loss means increased numbers of repeaters are needed. Figure 2.4 illustrates the typical signal losses over one kilometer (about 5/8 of a mile) for various transmission media. In each case, the signal strength is rated at an arbitrary 1,000 "units."

The ratio of transmitted-to-received signal strength is known as attenuation. Engineers measure attenuation in decibels or dB, which are easier to work with mathematically than are simple ratios. The formula is dB = 10 log SNR, where SNR is the signal-to-noise ratio. By this scheme, a send/receive ratio of 1,000 to 1 causes a 30 dB signal loss.

Typical attenuations in terms of dB per kilometer are shown in Figure 2.4. However, it is more understandable to translate this into received signal strength. For example, coaxial cable has a typical attenuation of 20

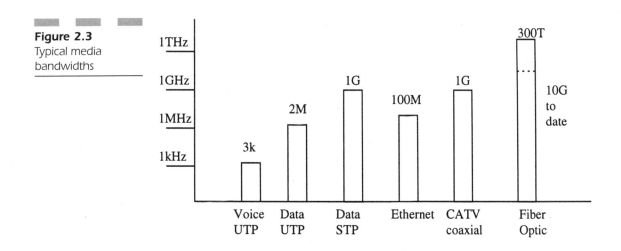

Figure 2.3
Typical media
bandwidths

Figure 2.4

Typical signal losses

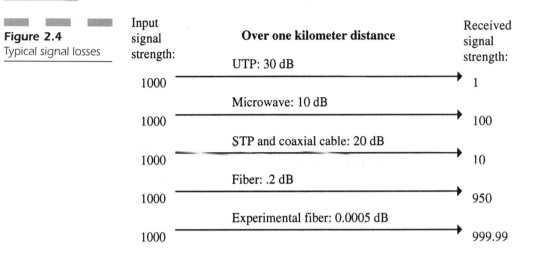

dB per kilometer and a received signal strength of 10 out of 1,000. Fiber comes in at an eye-catching 950 out of 1,000.

Incredibly, experimental fiber receives 999.99 of the 1,000 transmitted signal strength over a kilometer. Such "lossless" transmission has only previously been an engineer's dream.

Immunity from Interference

Copper media, from the twisted pair copper wires used in telephone company local loops to the coaxial cables used in cable TV systems, are basically long antennas. Because these all operate in the region of the electromagnetic spectrum used for wireless and other forms of transmission, copper wires are liable to pick up noise. Transmission lines can be shielded from noise, but this adds expense to the cable and must be maintained and installed properly to achieve maximum benefits.

Noise is just an unwanted signal that must be distinguished from the sent signal at the receiver. Noise interferes with the intended signal and must be addressed. Impulse noise can come from electrical motors and lighting strikes nearby. Other environmental noise can come from the normal operation of household appliances and TV sets. Even PCs can interfere with TV reception in another room or adjacent apartment.

On copper transport networks, interference is a nagging problem. Telephone companys' local loops share utility poles with power companies. These telephone cables must be carefully installed to avoid annoying hums from the power lines. When used for digital data, these hums are more than annoying, they are fatal to the bits, causing numerous errors.

Coaxial cable has a *self-shielding* property and, therefore, has a reputation for being more noise-resistant than unshielded wires. Although this is true, many sources of electromagnetic noise make their way onto coaxial cable as well. For instance, when cable TV companies began to experiment with two-way communication over the cable TV coaxial network, they quickly discovered that many common household devices generated noise that was inadvertently picked up by coaxial cables. These sources of noise and, thus, interference included microwave ovens, TV sets, and PCs. This is one reason why every PC has an FCC sticker. The data bus on every PC radiates electromagnetic signals into the surrounding area and may interfere with other electronic devices. Digital cell phones can even interfere, in some models, with heart pacemakers and hearing aids.

Fiber is not only noise resistant, but it is also basically impervious to noise because fiber operates in the optical range of the electromagnetic spectrum. Although stray electrical signals may be transmitted by copper media, no "stray light" effect occurs with fiber. Put more simply, although they overlap with those of copper media, "bandwidths" of storms, power supplies, and many wireless transmitters do not overlap with optical bandwidths.

Fiber optic cable has been used for years where noise is a factor in transmission systems. For example, many factories with heavy power equipment cannot even use coaxial cable for LANs on the factory floor. Many power companies that run communications and control lines along high-voltage power transmission lines cannot use it either. The 400,000 volts of alternating current will knock out the signal of many coaxial cables. Fiber, in many cases SONET fiber, is used for this purpose instead.

Security

One of the most persistent myths regarding fiber is the idea that fiber is more secure than copper because fiber is difficult to tap into. In fact, this is not true. Fiber used to have a reputation for being difficult to handle; this security myth was a consequence of that reputation. Today, all that is needed is a simple optical coupler. The fiber is then tapped and any signal that travels down the fiber is picked up by the eaves-dropping device.

Now, the fact that fiber is guided and not a broadcast medium (e.g., satellite, wireless, or microwave) affords it some measure of security over these media. However, any fiber link can be tapped by bending the fiber, scraping off the outer layers of cladding, and strapping a simple optical coupler device to the cable at that point.

Why does fiber have a reputation for being secure? Because when this coupler is inserted, it is very easy to detect. The inserted optical coupler will add at least 0.5 dB of signal loss (attenuation) to the fiber link. In normal situations, the attenuation on any link, even a fiber one, may increase with time—especially when the link is not maintained properly. However, a jump of 0.5 dB very quickly is not normal under any circumstances.

Security is handled on fiber networks by closely monitoring the end-to-end attenuation (signal loss) of a link. If this were to rapidly increase by a half dB (the dB is a standard measure of signal loss) or so, then some physical damage would have occurred. It could be innocent (e.g., environmental damage), or it could be an intruder. There are even ways to determine precisely how far along the fiber the loss was injected, focusing detection efforts at that area.

All of this indicates that fiber is attractive for secure and sensitive applications. Needless to say, the federal government and military prefer fiber to any other type of link in many cases. Even banks and many businesses have come to recognize the enhanced security that fiber optic networks can provide.

Less Maintenance Costs

Less maintenance costs means both fewer and lower in this case. All network links, whether fiber or not, need to be maintained. The cost factor is a function of the number of devices on the network that may fail, and the raw number of these devices throughout the entire network.

Fiber optic links and networks require fewer active devices that are less expensive to maintain. This translates to a cost-effective solution because there are lower costs involved in maintaining the fiber network than in any other media.

Even the fiber itself has lower maintenance costs. For instance, fiber does not corrode as copper media does. Furthermore, squirrels constantly chew aerial copper cables, especially in the spring, but have demonstrated less of an appetite for fiber. Fiber, however, does have its problems: It must be carefully protected from water. Even in the ground below the water table level, microscopic air bubbles in the fiber can fill with water and cause the fiber to fail. Ultraviolet radiation can destroy the glass in fiber, so fiber must be carefully deployed in aerial fashion. Fiber works best and lasts longest when buried underground, and this is how most fiber is deployed today.

The greatest risk to underground fiber today in North America is from gophers. There is actually a standard test placing fiber cable in a gopher enclosure to measure the risk (the gophers are not harmed). Squirrels will chew on anything, including aerial fiber. In other parts of the world, termites cause damage to fiber cables. However, reports of sharks chomping on undersea fiber are greatly exaggerated.

Low attenuation on fiber links means a smaller number of repeaters. Fiber links need much less time and energy devoted to maintenance because of this factor. The smaller number of active components needed on fiber links is a considerable cost savings in terms of installation costs. Solid-state electronics also draw less power and have a mean time to failure of about 20 years, just as a modern stereo system.

Once the fiber optic cable is in place, little can go wrong with it outside of simple cable cuts. Early fiber links were said to either work or not. Some never quite worked right, no matter how much post-installation tinkering occurred. However, if installed correctly, fiber links worked well. Some service providers even buried the fiber under poured concrete, which helped to prevent backhoes and other excavation equipment from damaging the fiber (known as "backhoe fade"). Other types of cable, however, needs constant access for maintenance purposes and cannot be buried under concrete.

Small Size and Weight

Fiber also offers numerous advantages in terms of size and weight. Even 10,000-foot fiber reels weigh only a couple hundred pounds. Smaller reels can be carried under a person's arm and weigh about 60 pounds. Fiber optic cable can be installed above hung ceilings without worry, and can be attached to a building slab without concern about an engineering report to determine if the building floor load factor may be exceeded. Fiber easily snakes its way through crowded conduits. In fact, one of the earliest incentives to run fiber optic telephone cable in metropolitan areas was to relieve the stress on crowded trunk cable conduits.

In contrast, 2,700 pair copper wire is 3.38 inches in diameter and weighs almost 7 pounds per foot. Installing such cable in hung ceilings or attached to floor slabs is usually out of the question due to the weight. The problems are just as bad outdoors. A 1,000-foot reel weighs close to 4 tons, and a four-person crew would struggle to install a fraction of a mile per day. Simply setting up the enormous trucks and pulleys consumes the better part of a day. Fiber crews work fast and lean, and the increased productivity is an added benefit not to be ignored today.

Fiber represents many times the bandwidth of the 2,700 pair cable. Is it any wonder that a service provider like a telephone company can install more bandwidth on fiber in a week than their entire network had on copper previously?

There is even an added benefit to fiber optic cable's small size and weight that is not obvious, but has come to be quite important: Aerial 2,700 pair (there is even 3,600 pair cable) is difficult to handle because much of it is buried below grade. When installed across bridges, the cable is particularly vulnerable, in spite of warning signs. A jackhammer can rip through the cable in a few seconds, seeming to the construction workers to be at first just a particularly tough patch of road.

This is just what happened in the summer of 1992 in White Plains, New York, when a bridge was being resurfaced. Immediately, a tent was erected on the site, traffic diverted, floodlights installed, and a crew worked around the clock for three days until all service was restored. It was not just a matter of "I found a purple and slate. Who has a purple and slate over there?" Documentation was delivered to the site and consulted, because service must be restored according to a particular customer hierarchy. First hospitals get service back, then police, then doctors, then pharmacies, and so on (details vary state by state), down to the ordinary residential customers. This task proved as time consuming as the actual splicing of the copper pairs themselves.

Now consider the same outage on fiber optic cable. A fiber optic cable may contain 12 to 36 strands of fiber. Each can be spliced in about 10 to 20 minutes, depending on the type of fiber and immediate conditions. Documentation is usually not needed, because thousands of users may be on the same fiber optic cable pair. Service is typically restored to all affected customers in a few hours, not a few days (although severe outages may take longer).

Bandwidth Upgrades

Another advantage of fiber is that when the link is configured with upgrading in mind, the same fiber link can be run at a higher speed just by exchanging the transmitter and receiver. This is a huge cost advantage for fiber networks.

Upgrading a digital link with copper networks has been difficult. For instance, converting from a 1.5 Mbps link (a DS-1 or T-1) to a 45 Mbps link (a DS-3 or T-3) requires the installation of new media in most cases. T-1 is defined on unshielded twisted pair (UTP) copper wire and T-3 is defined on coaxial cable, fiber, or microwave. These are pre-SONET definitions. Commercial buildings and even other types of buildings have plenty of UTP from the local telephone companies for supporting T-1s, but few have the media needed for T-3s.

Many organizations are asking for 45 Mbps links to support new interactive multimedia or other high bandwidth, broadband applications. The price to the customer for the T-3 must offset the cost of running the new media to the service provider in a reasonable payback period. This fact has kept the cost of T-3 links high, and forms yet another incentive for installing fiber optic cable initially.

Once installed, the fiber is not a limiting factor. In most designs, fiber installed to support 45 Mbps service can easily be upgraded to support 155 Mbps service, or even 622 Mbps service. The fiber link grows with customer expansion and is not obsolete in a short time frame (often before the payback period!). In these cases, it is only necessary to change the transmitters and receivers to make the fiber optic link function at the desired higher speed. Because most fiber is buried below grade—the perils faced by submersed and aerial fiber have already been mentioned—cost of burial is the most significant cost factor in fiber optic installations. In most parts of the country, it costs tens of thousands of dollars per mile simply to dig a hole in the ground. Obviously, the cost component is to be avoided when possible.

Again, it is important to note that this upgrade capability is not inherent to fiber optic networks. It must be carefully planned. But it can be done, and is a vast improvement over older digital media.

This chapter explored some of the operational characteristics and benefits of fiber optic cable. These are fiber benefits and not direct SONET benefits. But because SONET is an international standard for fiber optic transmission, SONET shares all of the advantages of fiber optic networks, with the added benefit of being an open standard tied to no single vendor or company.

This brief initiation into the world of fiber optic transmission has introduced the key concepts of fiber optic cable, transmitters, and receivers. But each of these three basic components of fiber optic networks such as SONET/SDH has undergone a revolution in the past few years. These developments have lead to the feasibility of advanced optical networks, independent of SONET/SDH, based on WDM and, more importantly, DWDM. This new world of optical networking deserves a chapter of its own.

3

Optical
Networking,
DWM, and
DWDM

The previous chapter introduced some of the basic concepts of fiber optic transmission. This chapter uses the basic terms and ideas examined in the previous chapter and builds on that foundation. This chapter explores some of the most interesting and vital ideas in networking today. There are three major areas: state-of-the-art optical networking, combining multiple wavelengths of light in the same fiber to produce wavelength division multiplexing (WDM), and, finally, literally jamming these wavelengths as close together as technically possible to produce a denser version of WDM known simply as DWDM. All three concepts are related and have implications for legacy SONET/SDH systems.

The very term *SONET* contains the key concept of an *optical network* as opposed to a network where 0s and 1s are represented by electricity. When applied to SONET, oddly enough, the term *optical network* is somewhat misleading. The fact of the matter is that SONET is *not* an optical network at all. The same is true of SDH, of course, although at least SDH makes no claims in its acronym to be any form of optical networking. SONET and SDH are still very much electrical transmission networks that just happen to consist of fiber optic links.

This just-optical-links aspect of SONET/SDH might not seem like much of a big deal, but it makes all the difference. In SONET/SDH the bits only shuttle back and forth between *network elements* (NEs) in the form of lightwaves on the optical links. In order to do anything at all interesting with these optical bits in a network element—switch them, route them, multiplex them, even simply check them for errors—it is first necessary to change the optical bits back into electrical bits. Naturally, this need for constant and frequent electro-optical (E/O) conversion is itself expensive and error-prone, and adds considerable complexity to the network. Even simple regenerators along a SONET/SDH link (traditionally about 40 km apart) must first convert the optical bits to electrical bits in order to send the bits over the next section of the line. But as long as the network elements are incapable of acting directly on the lightwaves, SONET/SDH has little choice but to deal with the electrical bits to do more than shuttle bits between two points over a fiber optic link.

The whole point of this chapter is that networks no longer have to function in this way. The term *optical networking* can be defined as *a network in which at least some of the network elements are capable of dealing with bits in an optical form*. This might seem like a broad definition, and it is. The network need not be all optical, and not all network elements need to be optically enabled. The key is that bits might be switched, routed, multiplexed, amplified, checked for errors, or all of the preceding, while still in their optical form without the need for E/O conversion and the reverse.

Once this definition of an optical network is established, more elaborate plans and variations of the basic concept are easily understood. For example, an *all-optical network* is an optical network in which all multiplexing, switching, routing, and so on, is done directly on the 0s and 1s while they are still in their optical form. A *passive optical network* is an optical network in which all of the components from sender to receiver do not require electrical power to operate. And so on.

Multiplexing has proven to be the simplest network function to make into an optical operation. This leads directly to the creation of WDM and DWDM. Optical cross-connects are right behind, and optical switches might follow soon. These more advanced, but related, network elements will be discussed in Chapter 16, where the impact of optical cross-connects and switches on SONET/SDH is assessed.

Optical Network Breakthroughs

Optical networks, and to a lesser extent WDM and DWDM, depend on the recent development of four key optical components:

- Special fibers with nonzero dispersion characteristics
- The in-fiber amplifier, such as the Erbium-Doped Fiber Amplifier (EDFA)
- The tunable laser diode operating in the "third window" at around 1550 nm
- The in-fiber Bragg grating to isolate individual wavelengths at the receiver

Figure 3.1 shows each one of these breakthroughs in use in a WDM system. A full discussion of WDM and DWDM appears in a later chapter. The emphasis here is on each of the optical networking breakthroughs that pose somewhat of a challenge to SONET/SDH and lead directly to WDM and DWDM systems.

Each one of these breakthroughs is important enough to deserve considerable discussion. In this chapter, there will be no specific mention of SONET/SDH itself. However, optical networking concepts apply equally to SONET/SDH and all other networking architectures that include fiber optic links, such as Gigabit Ethernet.

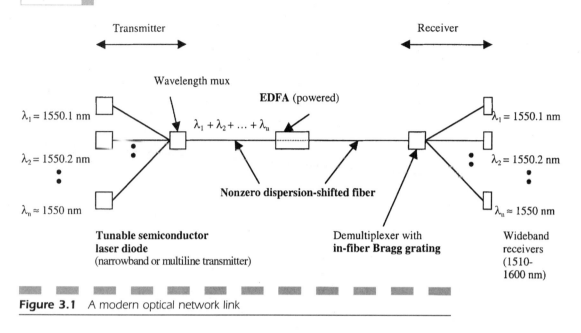

Figure 3.1   A modern optical network link

Special Fibers

The previous chapter outlined the main differences between multimode and single-mode fibers, the major forms of fiber optic cable. All of the advantages of fiber optic cable in general were detailed in that chapter, but mention should also be made of the fact that the connectivity in an optical network is optical, not electrical. This is not a trivial difference. Electrical communications systems always have the possibility of ground loops. Grounding is an important part of electrical networks and basically prevents static electricity from building up on the cable and preventing signals from being sent. But if there is an appreciable voltage potential difference between two grounding points, the resulting current flowing over the cable can easily swamp the actual signal sent. Furthermore, electrical connections can be hazardous since high voltages can harm people touching the wire, as anyone touching a "wet" T-1 can assert. Finally, electrical communications cables can be struck by lightning, even when buried underground. Lightning strikes constitute a hazard in many tropical areas around the world. Optical networks suffer from none of these drawbacks, except when electrical power is distributed to powered fiber components along the link. But this is just an argument for as many passive optical network elements as possible.

Other potential drawbacks of fiber optic networks include joining of the fiber cables, the risk of bending or kinking the cable, and the effects of gamma radiation and very high voltage electrical fields. The best way to join fiber is by fusion splicing: melting the glass so the ends fuse to one another. This is a problem outdoors during harsh weather. Connectors can be used, but add considerable loss to the link. The attenuation can be about 0.3 dB for multimode fiber and higher for single-mode, due to the smaller core diameter.

If fiber is bent too much, the light is no longer reflected into the core properly, and it escapes. The allowable bend radius depends on the actual difference between the refractive index (RI) between the cladding (lower RI) and core (higher RI). The RI is just the ratio of the speed of light in a vacuum to the speed of light in the material, which is glass in the case of fiber optic cable. Confusingly, in formulas relating to fiber optics, RI is represented by the letter n. If $n = 1.5$, for example, the speed of light in the fiber is about 200,000 km/sec, not 300,000 km/sec as in a vacuum (300,000/200,000 = 1.5).

The bigger the difference in core and cladding RI, the tighter the allowable bend radius. Unfortunately, the special fibers discussed in this section all tend to minimize the RI difference between core and cladding, leading to a larger allowable bend radius. Kinks in fiber are tight, permanent bends caused by physical damage during installation, such as dropping a heavy weight on a fiber cable. A more correct term for a kink is a microbend.

Finally, gamma radiation causes some types of glass to emit light (which interferes with the signal) and permanently darkens the glass as well. But this level of gamma radiation is normally found only inside a nuclear power plant. And when used in environments where voltages exceed 30,000 V (not all that uncommon today), fiberglass also emits light and discolors. Fiber needs to be shielded in high-voltage environments.

Light in Glass

The traditional picture of light traveling down a fiber optic cable as a little arrow is somewhat misleading. This arrow representation is known as the *ray theory of light* and was common right up until the last part of the nineteenth century. The current theory of light acknowledges that, in accordance with quantum physics, light is both a wave and a particle. Light normally travels as an electromagnetic wave and is detected as a particle. Both forms can be called *photons,* although it is more common to call a particle of light a photon and refer to light in its wave form as just a *light wave.*

This is important because light does not really travel down a fiber as a ray in the form of little arrows, bouncing off the walls between core and cladding from transmitter to receiver. The truth is much more complex. Light consists of fluctuating electric and magnetic waves, orthogonal (90 degrees apart) to each other (so they do not interfere). As light propagates, the whole structure normally rotates in a pattern known as *circular polarization.* When the light travels down a fiber, a whole array of effects takes over. Light may be elliptically polarized, as when the period of rotation of the fields is not the same as the wavelength of the light or when the propagation speed of the electric and magnetic fields are not quite the same, as in a fiber. The main point is that this complex light structure actually snakes its way down a fiber in a spiral known as a *helical pattern.*

Interference is another important issue when trying to understand how modern fibers function. Light interferes even with itself in ways that can be described perfectly with mathematics, but are hard to explain or understand in an everyday sense. Even a single photon of light can interfere in odd ways with itself when a system is set up just right. It is one thing to think of light as a wave phenomenon and to understand interference as similar to water waves on a pond interfering with each other. It is quite another to think that a single photon can somehow interfere with itself, when obviously there is only one of them to begin with.

Interference is important in all aspects of optical networking, even in the simple act of injecting light from air onto the fiber itself. Typically, about 4% of the incident light is reflected, and the higher the angle, the greater the reflection. Differences in RI always cause reflection—4% into the fiber at the sending end and 4% into the air at the other end of the fiber link. But when monochromatic light (one wavelength) shines on a thin sheet of glass, the total reflection is not 8%, as might be expected. There could be anywhere from 16% of the light reflected to no light reflected at all. The result depends on the thickness of the glass in terms of the wavelength of the light. The maximum of 16% reflection occurs when the round-trip thickness of the glass is exactly a multiple of the wavelength of the light. This is constructive interference. The minimum of zero reflection occurs when the round-trip thickness of the glass is an odd number of half wavelengths. This is an example of destructive interference in the fiber.

It is important to point out that interference does not destroy the energy present on the fiber. Energy is always conserved, so the lost energy just shows up in some other part of the optical system, probably as heat or noise. With this in mind, modern fibers include antireflection coatings of special thickness and RIs to keep reflections to a minimum. This is not to say that interference is always a negative. In optical networks, filters, gratings, and other related components all depend on light interference to function.

Wavelengths are always important in optical networking. But today many wavelengths and ranges of wavelengths are used for different purposes. This is a good place to take a quick look at the wavelengths that will be discussed in the rest of this chapter. Both of these are shown, and explained, in Figure 3.2.

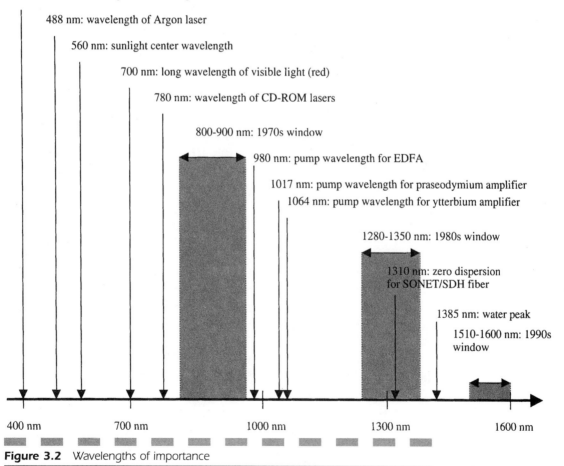

Figure 3.2 Wavelengths of importance

The best perspective on the wavelengths used for optical networks is linked to history. The range from 800 to 900 nm can be called the *1970s window* or *first window* (this is the term used by Harry Dutton of IBM). All early optical links in the 1970s used this range. The band around 1310 nm is the *1980s window* (*second window*), and this is where SONET/SDH, which matured in the early 1980s, operates. Today, the range from 1510 to 1600 nm is the *1990s window* (*third window*) and is used because it is at the lowest point of attenuation on the fiber. This window is used in all modern WDM and DWDM systems.

However, the opening of the third window was not easy. Light sources are expensive in this range, and there is more to the fiber story than simple attenuation. More was needed to enable fibers to function well in the 1990s window. This work concerned the degradation effects due mainly to *dispersion*.

There are many reasons why a pulse of light sent along a fiber will *degrade*, which is just another way of saying *change*. The light will be weakened, a pulse lengthened in time, and so on. There are five major reasons for the degrading of a pulse of light sent along a fiber optic cable.

ATTENUATION.   This is always the major concern. A light pulse will always weaken because the fiber absorbs light. The glass itself does not absorb the light, but any and all impurities in the glass will. And if the fiber is not manufactured with precise uniformity in diameter, and so on, these flaws will cause scattering of the light. Scattering effects are the biggest cause of light loss in fibers today.

DISPERSION.   This is exactly what it sounds like: a pulse of light becomes spread out (disperses) in time as it travels down a fiber. Over long enough distances, all of the pulses eventually merge together and the peaks needed at the receiver to distinguish 1s all disappear. There are many different types of dispersion, some of which can even be used to advantage in optical networking. Three types of dispersion are of primary concern. *Material dispersion* (also called *chromatic dispersion*) results from the fact that fibers present slightly different RIs at different wavelengths and so travel at slightly different speeds. Even the best lasers produce more than one wavelength of light (although in a very narrow range). *Waveguide dispersion* is caused by the shape of the core and RI profile across the core. Waveguide dispersion can actually cause the light pulse to bunch up in time and disappear. This also means that waveguide dispersion, if carefully planned, can counteract the effects of material dispersion. (It should be noted that in some references chromatic dispersion is defined as the sum of material dispersion and waveguide dispersion.) This will be explained more fully later on in this section. *Modal dispersion* occurs in multimode fibers because the distances traveled, and hence the arrival times, are slightly different for each mode.

POWER LIMITS.   One way of overcoming attenuation effects is just to pump up the input power. But the amount of optical power coupled onto a fiber is limited by nonlinear effects. This just means that doubling the input does not double the output and, in fact, might decrease the output to lower levels than before. Optical networks must function in the linear power region to be useful. Most single-mode fiber systems are limited to about a half a watt of input power to remain in the linear region. The nonlinear effects in a fiber are caused mainly by the intense electromagnetic fields in the core when light is present.

POLARIZATION EFFECTS.   The core of fiber optic cable should be perfectly round and symmetrical. But there are always imperfections, either in manufacture or during installation, which distort the core in places. As a result, light in the fiber is changed in polarization. Currently, few if any optical networks rely on polarization of the light to carry information. But in future systems, polarization effects could be a concern.

NOISE.   It might seem odd to speak of noise in an optical network, since immunity from electrical interference is one of the big advantages of fiber. It is true that optical networks will not pick up noise from *outside* the network. But noise is just an unwanted signal detected by the receiver. In optical networks, *modal noise* is a complex effect present on multimode fibers, and *mode partition noise* is a problem with single-mode fibers as power hops around (is partitioned) among many close wavelengths.

None of these optical impairments are fatal, fortunately, and most can be overcome by a variety of methods. Generally, it is desirable to make the fiber core as narrow as possible and allow only one mode of propagation at a particular wavelength. This is the idea behind single-mode fiber, of course. Next, the wavelength(s) chosen for operation must be selected based on the characteristics of the fiber itself. Cost is a factor as well, since the shorter the wavelength used in terms of nanometers (nm), the lower the cost. Also, dispersion effects that depend on wavelength can be minimized by narrowing the *spectral width* of the light source. The spectral width is a measure of the range of wave-lengths generated by the light source. Even lasers do not produce all of their power at one particular wavelength, although lasers are much better than LEDs. Modern semiconductor lasers can have spectral widths of around 1 to 5 nm, while some more sophisticated (and expensive) lasers can have spectral widths as low as 0.01 nm. LEDs on the other hand will have spectral widths between 30 and 150 nm. Finally, material and waveguide dispersions act to spread out and narrow waves, respectively, and so can be used together to counteract each other, to the benefit of the optical system.

All in all, modern optical network design depends on the interplay of two factors, with a third factor added in some cases. The two most important considerations are signal strength and dispersion effects. The third factor is noise, but noise is rarely a problem in simpler optical systems.

Fiber as Glass

Optical fibers are not made out of the same kind of glass as is used in windows. In scientific terms, a *glass* is any substance (mostly containing large portions of silicon) that does not transition to a solid state when cooled from a liquid. Glass acts as a solid but is still technically a liquid. However, even the oldest window panes are still just as thick at the top as they are at the bottom (reports of "flowing" medieval cathedral windows have proven to be greatly exaggerated).

Pure fused silica, or silicon dioxide, is the primary ingredient in optical fiber cables. But using silica requires high temperatures and the RI of silica is not particularly well suited for fiber optic systems. So *doping* is used to change the RI of the silica core or cladding, or both. Doping is just the process of intentionally adding impurities to a substance during fabrication. Even chandeliers can have lead oxide added in to produce the sparkle of "lead crystal" glass. In optical fiber, 4% to 10% of germanium oxide is added to the silica to increase the RI. This is a huge amount of dopant, considering that semiconductor fabrication processes dope materials with only 1 part in 10 million or so. Boron trioxide will decrease the RI. Other substances are also used, but they all tend to also increase the attenuation of the fiber.

When all is said and done, the typical attenuation characteristics of modern fibers are fairly consistent. It should be noted that the infrared range starts at about 730 nm, so none of these wavelengths are visible.

Also, the typical fiber attenuation in 1970 was about 20 dB/km, which is quite high for any medium. By the 1980s, this was reduced to about 1 dB/km, a figure widely used by vendors and writers. In the mid-1990s, attenuation on fiber of around 0.2 dB/km was common.

Fiber attenuation varies dramatically with wavelength. Figure 3.3 shows the variation in some detail. There is a "water peak" at around 1400 nm caused by water ions that are always present in the fiber.

So the battle against fiber attenuation can be fought and won by opening the 1990s window wavelengths for operation. However, dispersion effects have to be taken into consideration as well. One of the reasons that the 1970s window was not the best window was that the fiber *zero dispersion point*, where material and waveguide dispersions exactly canceled each other out, was around 1310 nm. So pulses at 1310 nm stayed nice and tight as they made their way along many kilometers of single-mode fiber, although they still weakened from attenuation, of course.

The standard single-mode fibers used in SONET/SDH had their zero dispersion point at 1310 nm. By using changed fiber fabrication meth-

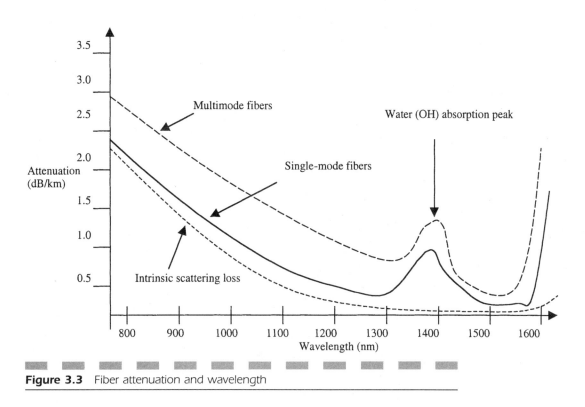

Figure 3.3 Fiber attenuation and wavelength

ods (doping) and manipulating the RI profiles across the fiber core, a whole generation of *dispersion-shifted fibers* appeared in the 1990s that made operation in the even lower attenuation 1990s window attractive. So in the 1990s newer fibers moved the zero dispersion point into the 1990s window, centered around 1550 nm. This interplay between material dispersion and waveguide dispersion is shown in Figure 3.4.

The story of special fibers for optical networks might stop here, if not for a few other important points. First, SONET/SDH still is defined to operate at 1310 nm, yet obviously the current way of building optical networks is firmly in the 1510- to 1600-nm range. Second, this is not to say that wavelengths in the 1310-nm range cannot be used with dispersion-shifted fibers (sometimes called DSF). However, the dispersion (and attenuation) will be higher at 1310 nm (the characteristic SONET/SDH wavelength) than at, say, 1550 nm. Third, optical networks today rely more on WDM and DWDM than a single serial stream of bits at a single wavelength. So a single zero dispersion point right in the middle of the transmission band is not always desirable. For WDM and DWDM, it would be better to have a range between 1510 and 1600 nm where dispersion is low, but still nonzero. Again, clever doping and RI profiles made the manufacture of such *nonzero dispersion-shifted fiber* (NZ-DSF) possible.

The RI profile of NZ-DSF is shown in Figure 3.5. These fibers are recommended for WDM and DWDM systems, although regular 1310-nm zero dispersion-shifted fibers can be used as well (as long as the higher dispersion is accounted for during system design).

There are many other interesting developments in the field of fiber optic cables. These can only be mentioned here. There are large effective-area fibers (LEAF) that will accept more optical power, dispersion flattened fibers with low dispersion in the 1990s window (but high attenuation),

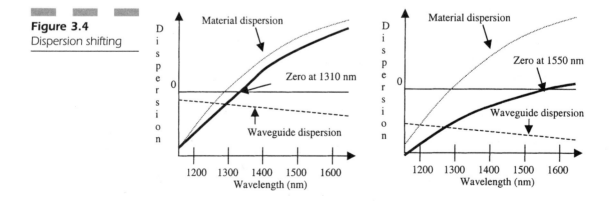

Figure 3.4
Dispersion shifting

Figure 3.5
Core RI profile of
NZ-DSF (dispersion
optimized) fiber

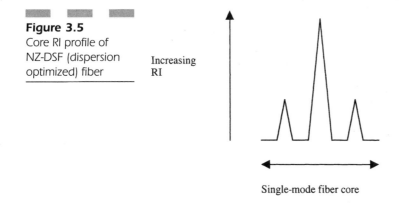

Increasing
RI

Single-mode fiber core

and dispersion compensating fibers that can be spliced in periodically to combat dispersion effects on long NZ-DSF runs.

The Erbium-Doped Fiber Amplifier (EDFA)

The second key piece of the optical networking puzzle, in addition to the creation of fibers with special characteristics in the 1500- to 1600-nm range suitable for WDM and DWDM, was the invention of the in-fiber optical amplifier. There are also semiconductor optical amplifiers (SOAs), but these are not considered here. The most well known and most widely deployed in-fiber optical amplifier is the optical amplifier based on the rare earth element erbium. When erbium is used, the amplifier is called an EDFA, or Erbium-Doped Fiber Amplifier. As might be imagined, an EDFA is created by doping an otherwise ordinary single-mode fiber with erbium during fabrication.

The use of EDFAs has become so widespread that the acronym EDFA is in danger of becoming a synonym for fiber amplifier in general. This is unfortunate, as other elements can be just as effective as erbium when it comes to amplifying optical signals. For example, praseodymim and ytterbium, which are also rare earth elements, can be used. Sometimes, this whole class of optical devices is called *rare earth doped optical amplifiers* (REDFAs), but this terminology is far from standard.

SONET/SDH does not—and right now cannot—use EDFAs or other forms of optical amplifiers. This is because SONET/SDH, as a firmly entrenched 1980s technology, uses the 1980s optical communications window, usually at 1310 nm. EDFAs, on the other hand, were developed in the 1990s and so function in the 1990s optical communications window from 1500 to 1600 nm. The essential characteristic of any optical amplifier is that these devices operate across a wide range of wavelengths, independently and effectively. The electro-optical repeaters used in SONET/SDH typically function only at the SONET/SDH wavelength of 1310 nm. If SONET/SDH links are to be mapped onto WDM or DWDM systems, it is necessary to use a transponder to shift the native SONET/SDH wavelength into the 1500- to 1600-nm range.

It is somewhat ironic that amplifiers were used in analog transmission systems but abandoned in favor of digital regeneration and repeater techniques, culminating in SONET/SDH eletro-optical repeaters, due to the cumulative noise effects when cascading amplifiers. In other words, in an analog system, noise is added by each amplifier, since analog noise is indistinguishable from the analog signal. In a digital system, noise can be removed at each regenerator as the signal is strengthened and cleaned up. Why then do optical networks emphasizing WDM and DWDM rely on in-fiber optical amplifiers instead of digital repeaters? Does not noise also accumulate in optical amplifiers? Of course it does! What saves optical amplifiers and allows them to be used in WDM and DWDM systems is the fact that the degree of noise in optical systems is very low to begin with. However, given enough EDFAs, low level of optical noise or not, eventually a long enough cascade will result in more noise than signal (this is the case with 80 or more EDFAs on a link several thousand kilometers long).

The attractions of EDFAs and all optical amplifiers in general are many. Electro-optical repeaters are complex and can fail. EDFAs are built into the fiber itself. Repeaters are tuned to a specific wavelength and bit rate. Amplifiers just amplify anything on the link. With WDM, each wavelength would have to be demultiplexed, converted to electricity, regenerated, converted back to optical signals, and then remultiplexed. In-fiber amplifiers make WDM technically feasible. Finally, repeaters cost much more than optical amplifiers. The differences between electro-optical repeaters and in-fiber optical amplifiers are shown in Figure 3.6.

The operation of an EDFA is simple to describe, in principle. A relatively high-powered (compared to the weakened input signal) laser source at 980 nm is mixed with the arriving input signal on a special section of erbium-doped fiber, typically about 10 m of fiber coiled to reduce the form factor of the device. This is called the pump signal (there is also a second pump window at 1490 nm). The erbium ions present in the EDFA fiber section are excited to a higher energy state by the pump laser. When photons from the input signal in the range from about 1530 to 1620 nm (the higher wavelengths are not as effectively amplified) strike the excited erbium ions, some of the erbium ions'

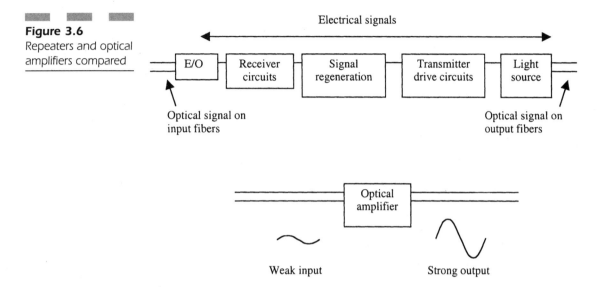

Figure 3.6
Repeaters and optical amplifiers compared

energy is transferred to the signal and the erbium ions return to a lower energy state (until they are pumped again). The nice thing is that the photons have exactly the same phase and direction as the incoming photons. There are just more of them, so the input signal is amplified. That is all there is to it.

This does not mean that there is no more to be said about EDFAs. For example, reflections at the end of the EDFA section can cause the EDFA fiber itself to become a laser, so an optical isolator is usually employed to prevent these reflections. Also, noise in an EDFA manifests itself as amplified spontaneous emissions (ASEs), which can emerge from both ends of the fiber link and must be dealt with. ASEs in the same direction as the signal become a major source of cumulative noise, while backward-propagating ASEs can harm source lasers if not filtered out.

Oddly, the pump signal can actually be injected into the EDFA section in the opposite direction from the input signal! In fact, there are firm technical reasons for doing so. The erbium ions do not go anywhere, of course. EFDAs can also be pumped from *both* directions. EDFAs can even be pumped remotely, from the transmitter or receiver side. In some configurations, an EDFA can be pumped *through* another EDFA section, as long as the attenuation at 980 nm is not too bad. Usually the pump laser is right at the EDFA section it pumps, however—although the electrical power needed to operate the pump laser can be delivered locally or remotely (sometimes in the metallic fiber protection sheath). Obviously, EDFA design is a very active area of research. EDFAs have been used not only as line amplifiers, but also as preamplifiers directly in front of a receiver and as power amplifiers placed directly off the transmitter. A more detailed schematic picture of EDFA operation is shown in Figure 3.7.

In summary, optical amplifiers have numerous advantages over repeaters. Optical amplifiers are simpler and smaller, and they cost much less than repeaters and yet they have much longer mean times to failure. They operate over a range of speeds and wavelengths at the same time and produce a gain in the weakened optical signal of anywhere from 25 to 50 dB. They amplify all forms of optical signals, whether digitally encoded or analog, and in fact the coding on the fiber can be changed without changing the in-fiber amplifier. They add no appreciable delay to the fiber run and, if designed properly, can allow the operation of diagnostic tools such as optical time domain reflectometers (OTDRs) right through the amplifier.

Naturally, there are drawbacks to optical amplifiers as well, especially since they are relatively new optical networking devices. The biggest problem is that they add noise to the system that can accumulate, as in any cascade of amplifiers, until the noise swamps the signal. They do not yet operate over the full WDM and DWDM wavelength window, and the gain produced is not the same at all wavelengths of interest (that is, the gain spectrum is not flat). They do not perform any clean up of the optical signal as regenerators do, such as reshaping pulses. This means that if an incoming optical signal is dispersed (distorted), the output of the optical amplifier will be stronger but still distorted. Finally, in many cases the optical amplifiers must be placed more closely than electro-optical repeaters.

Figure 3.7
Erbium-doped fiber amplifier (EDFA) schematic

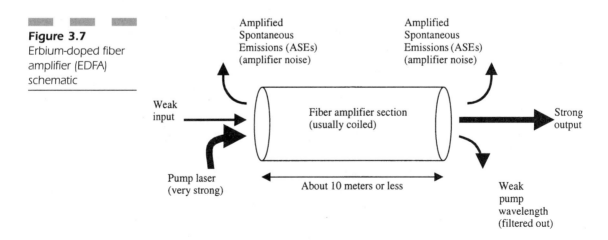

But when all is said and done, the in-fiber optical amplifier will be a key piece of any optical network.

The Tunable Laser Diode Operating at 1550 nm

The third key piece of the optical networking puzzle, in addition to the creation of special fibers and in-fiber amplifiers for WDM and DWDM, was the creation of the tunable laser diode operating at around 1550 nm, right in the middle of the 1990s optical communications window. There are actually three pieces to this invention. The first part is the *laser diode* itself. The second part is the *tunable* nature of the laser, which enables the same physical device to be used to generate different wavelengths of light. Finally, the third part is the operation around 1550 nm. All are equally important for optical networking today, and each aspect will be discussed in some detail.

In the early days of fiber optic systems, lasers were bulky affairs that were very expensive and hard to work with in spite of their numerous advantages over LED-driven fiber systems. Not only was the cost often prohibitive, but lasers often had to be locked in a closet with a warning sign on the door and anyone who entered had to put on protective eye-wear. Today, both LEDs and lasers are semiconductor devices mounted on a simple computer card assembly. LEDs remain simpler than lasers when implemented in a chipset, but they also have a lot of features in common. So it makes sense to start with a quick look at semiconductor LEDs.

An LED is just a *forward-biased p-n junction*, but this is not the place for a full discussion of semiconductor electronics. All that is needed is an appreciation of the fact that when the proper electric potential (voltage) is applied to a wafer of p-semiconductor material and n-semiconductor material, light will be emitted at the junction between the materials. That's the easy part. The hard part is getting the LED light at the desired wavelength out of the junction and onto the fiber. In practical LEDs, sometimes the light emerges from the surface of the device (SLEDs) or from the edges of the device (ELEDs). In most SLEDs, the n-semiconductor has a hole in the middle to allow the light to emerge from the junction. A very simple diagram of a SLED (sometimes called a "Burrus LED") is shown in Figure 3.8.

The characteristics of LEDs as light sources (transmitters) in optical networks compared to lasers are important. LEDs have been lower in cost than lasers, although this price differential has all but disappeared

Figure 3.8
Conceptual structure
(components not to
scale) of a surface-
emitting LED (SLED)

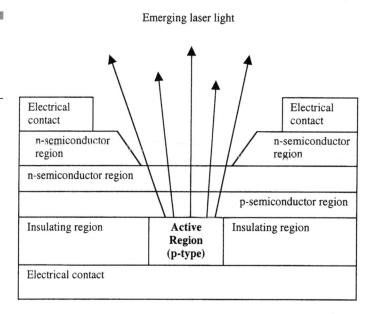

Emerging laser light

Electrical contact

n-semiconductor region

n-semiconductor region

n-semiconductor region

p-semiconductor region

Insulating region

Active Region (p-type)

Insulating region

Electrical contact

in recent years. Much of the price differential was based on production volume. Certainly the use of lasers in simple CD-ROM data and audio devices has leveled the cost playing field drastically. LEDs have had lower power outputs than lasers (100 microwatts or so), and produce light not at a single wavelength, but in a large band or spectral width of about 5% of the wavelength (50 to 100 nm). The spectral width can be reduced through the use of filters, but this also filters out part of the signal power, of course. LED light is not coherent as is a laser's light, and so the LED light must be focused with a lens onto the fiber (this is why LEDs are not used with single-mode fibers: it is too hard to get much light into the small core). LEDs cannot be turned on and off fast enough to drive gigabit-per-second systems as lasers can. Most LEDs top out at about 300 Mbps. On the other hand, LEDs can easily be used with analog modulation techniques. This can be done with lasers, but only with much more difficulty.

In contrast to LEDs, lasers are much better to use in optical networks as light sources. Laser light in theory is coherent and consists of a single wavelength. This is not strictly true in practice, as all lasers used for communications exhibit a characteristic *linewidth*. But lasers can be turned on and off very quickly, in some cases on the order of femtoseconds (one millionth of a nanosecond, or 10^{-15} seconds), and they produce higher powers than LEDs (about 20 milliwatts for communications systems, and up to 250 milliwatts for optical amplifier pump lasers). Laser light, which emerges highly parallel from the device, can inject 50% to 80% of its power onto a fiber.

But however close lasers and LEDs come in price, the fact remains that temperature control and power control are needed in lasers. These controls require electronic circuits to function. And lasers are generally nonlinear with respect to input power.

In spite of some claims to the contrary, most semiconductor lasers used for communications do *not* produce a single wavelength of light. Just like LEDs, a range of wavelengths is produced, and the spectral width is the measurement of this. This turns out to be important if a *tunable* laser is desired. Usually, the spectral width of the laser is about 8 nm, scattered across eight peaks or *modes* (not the same as fiber modes). Not all of these wavelength peaks are of equal strength. The laser first outputs power at a *dominant mode*, then at another mode and another, switching back and forth unpredictably. Thus power is distributed in a bell-shaped curve, with ill-defined edges. So it is common to quote spectral width as the distance between the points on the curve where the power drops to one-half the maximum mode power. This is called the *full width half maximum* (FWHM) and is a valid measure for LED spectral widths as well.

Spectral width is a crucial laser parameter for a number of reasons. First, the wider the spectrum, the greater the dispersion on the fiber. When using WDM or DWDM, the closer the wavelengths can be packed (i.e., the smaller the spectral width), the more channels can be used on a single fiber. Also, sophisticated modulation techniques such as phase modulation cannot be used unless the spectral width is very narrow. Finally, very narrow spectral widths, usually desirable, can actually cause unwanted nonlinear effects in the fiber, such as Stimulated Brillouin Scattering (SBS), discussed more fully later in this chapter.

In addition to spectral width, semiconductor laser output has a characteristic *linewidth*. So not only are the peaks spread across the spectral width, but each individual peak is slightly different in width. The relationship between spectral width and linewidth is shown in Figure 3.9.

There are many other characteristics of semiconductor lasers, such as coherence length and time, power, operational range in terms of wavelength, and wavelength stability. However, these are beyond the scope of this discussion.

Oddly enough, it is conceptually easy to make a semiconductor laser out of a semiconductor LED. Whenever light is confined in a proper material of the proper shape and size, the material will "lase." Sometimes

Figure 3.9
Spectral width and
linewidth

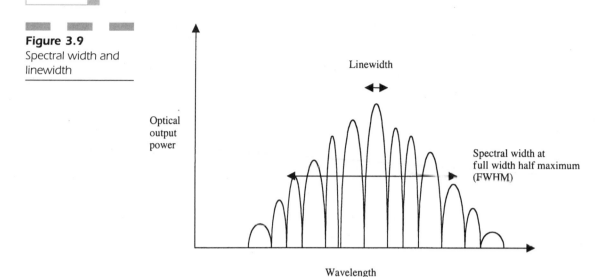

this can happen even when not desired, as with reflections in a fiber amplifier section. So to make a semiconductor laser out of a semiconductor LED, just add mirrors at the ends! In practice it is a little more complicated, but not by much. The end result is a Fabry-Perot laser, or FP laser. Essentially, the FP laser is an edge-emitting LED (ELED) with mirrors on the ends. The only problem is that simple FP lasers have very wide spectral widths, so a lot of time and effort are spent repackaging FP lasers to get just the right operational characteristics.

Usually, some form of diffraction (Bragg) grating is placed within the lasing cavity between the mirrors to narrow the spectral width, typically to the 0.2- to 0.3-nm range. A diffraction grating is just a series of etched lines (discussed more fully in the next section) which refracts (bends) the light in the laser cavity in a certain way. There are two major classes of this type of laser: the distributed feedback laser (DFB) and the distributed Bragg reflector laser (DBR). Of the two, the DFB laser is of the most interest in optical networking, because it leads directly to the tunable laser. A simple diagram of the active region of a DFB laser is shown in Figure 3.10.

The quarter wavelength phase shift in the middle of the grating is optional, but it narrows the linewidth of the laser output significantly. The main purpose of the grating is to reduce the spectral width of the basic FP laser. The grating is just a small variation in the RI of the material with a characteristic period (wavelength). In fact, since much of the action of the bending of the light below the lasing cavity

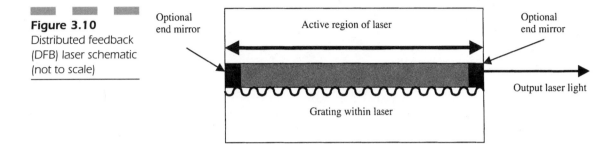

Figure 3.10
Distributed feedback
(DFB) laser schematic
(not to scale)

(where the grating is usually located today) is reflection, the DFB laser does not even need end mirrors, but they are usually present. The form and position of the grating can be manipulated to yield a tunable laser as well.

The major drawbacks of DFB lasers, besides expense, are that they are sensitive to reflections (light entering the device from the attached fiber will eventually destroy the DFB laser) and that they are sensitive to temperature variations (not only environmentally, but also due to the *internal* heat generated by sending a long string of 1 bits). So DFB laser bit streams are typically balanced to guarantee an equal number of 0 and 1 bits over time.

The inclusion of a Bragg grating within the structure of the DFB laser leads directly to consideration of tunable lasers. In WDM and DWDM systems, having physically identical light sources that can be tuned to slightly different wavelengths is highly desirable. In one sense, tunable lasers make a virtue of necessity. When attempting to construct simple WDM systems, it proved difficult to fix the wavelengths of the transmitters exactly. This was because slight variations in the semiconductor laser fabrication process meant that the lasers produced varied slightly in wavelength anyway. Whenever a precise wavelength was needed, many lasers had to be made, tested, and selected, and the rest discarded. This was wasteful and expensive. Also, the materials that make up the laser diode deteriorate over time and change the wavelength, leading to a risk of the wavelengths overlapping. If the laser could be tuned, this would correct both situations.

But the most important reason for using tunable lasers in optical networks is the interest in *optical wavelength routing* (more precisely, direct optical cross-connecting). Tuning lasers hooked into single strands of fiber allows optical network devices to quickly accept a signal arriving on one wavelength from an input fiber and route the signal onto an output fiber at another wavelength. The arriving and departing wavelengths can be changed quickly to establish new routes or paths through the optical network. It is even possible to set up wavelength paths, just like telephone calls, across such an optical network.

Typically, tunable lasers can be tuned directly over a range of about 10 nm, mostly centered around 1550 nm, which is right in the middle of the third optical communications window. This remains a very active area of research, naturally. To extend the tunable range even further, a *sampled grating* can be added to the basic DFB laser design. A sampled grating is just a short, but otherwise normal, refraction grating with periodic *blanked sections* between the grating sections. In the tunable laser, two sampled gratings are made to interact with each other so that a small, induced change in the RI of the device causes a larger change in the wavelength output.

The gratings produce reflection peaks at slightly different wavelengths. Most of the peaks from the two gratings will be at slightly different wavelengths, but one set of peaks can be made to coincide, reinforcing that wavelength and effectively tuning the laser to that wavelength in particular. When a tuning current is applied to one grating, the "comb" of reflection peaks shifts slightly, reinforcing another set of peaks at another wavelength. Tuning range can be extended in this fashion to 100 nm. The basic principle of operation of the tunable laser diode is shown in Figure 3.11.

The two gratings essentially produce two sets of output wavelengths with slightly different spectral widths. By "tuning" these sets of wavelengths, one pair of peaks can be made to coincide and reinforce each other, while the other peaks are made to interfere. This is somewhat of an oversimplification of what occurs, but it is close enough for this level of detail.

There are other forms of tunable laser currently under research, such as the external cavity DFB laser that employs a tunable grating on a crys-

Figure 3.11

Tunable distributed feedback (DFB) laser schematic (not to scale)

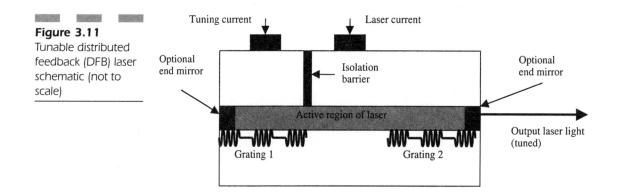

tal structure. There is more than enough interest in optical routing to make tunable lasers a highly active field for some time to come.

Recently, DWDM systems with hundreds of channels have appeared. Rather than use many lasers, *multiline* lasers have appeared to drive these systems. These encourage and equalize the multiple wavelengths for external modulation. However, the most common use of SLEDs remains in the 1970s' window for Gigabit Ethernet (GBE), ELEDs (FP) lasers for SONET/SDH around 1300 nm, and DFB lasers for DWDM.

In-Fiber Bragg Grating

The fourth and final key piece of the optical networking puzzle, in addition to the creation of special fibers, in-fiber amplifiers, and tunable lasers for WDM and DWDM, was the invention of the in-fiber Bragg grating. The in-fiber Bragg grating is used to isolate individual wavelengths at the receiver of the WDM or DWDM system. Some form of isolation is needed because the general-purpose receivers used in most optical systems today are wideband receivers, meaning they will register photons not only from one specific wavelength, but across a wide range of wavelengths. This problem is shown in Figure 3.12.

Of course, narrowband receivers tuned to a specific wavelength can be developed and used, but this approach is an expensive proposition and an increasingly difficult task for DWDM systems where wavelengths are

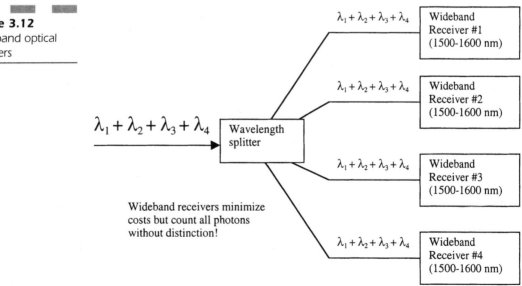

Figure 3.12
Wideband optical receivers

jammed closer and closer together all the time. There are other ways to filter out the unwanted wavelengths at the receiver besides the in-fiber grating, naturally. The attraction of the in-fiber Bragg grating, like the in-fiber optical amplifier, is that these optical devices are part of the actual fiber itself and allow the continued use of a standard, wideband optical receiver.

Some observers of the optical networking scene rank the invention of the in-fiber Bragg grating (sometimes called an FBG) as second only to the invention of the laser itself as a milestone in optical networking. The FBG is simple in the extreme to make, very inexpensive, and highly reliable. Not much more could be asked for. The FBG is just a short length (only a few centimeters) of ordinary single-mode fiber into which variations in the core RI have been etched using beams of ultraviolet radiation. When etched with just the right spacing, the in-fiber Bragg grating will reflect (select) one precise wavelength back in the direction from which the light entered the fiber and pass the other (nonselected) wavelengths through the fiber unchanged (there is some small attenuation effect on the other wavelengths). The general idea behind the FBG is shown in Figure 3.13.

The small changes in RI lengthwise in the fiber core of an FBG section will reflect a small amount of light at each discontinuity. If the wavelength of the light and the spacing (period) of the grating are the same, then there is positive reinforcement in the backward direction. In technical terms, the grating forms an *electromagnetic resonant circuit*.

There are many variations on this basic in-fiber Bragg grating design. Some of these can be used to deal with the issue that the wavelength selected is reflected *back* toward the source. It would be more useful to pass the desired wavelength *through* the fiber and eliminate the others, as a true filter should. And so it turns out that it is possible to etch many different FBGs into the same section of fiber. Then a *blazed grating* can be added which is built at an angle to the fiber core and reflects light out of the core. So it is possible to combine many FBGs and blazed gratings to filter out all but the desired wavelength.

Figure 3.13

General structure of an in-fiber Bragg grating (FBG)

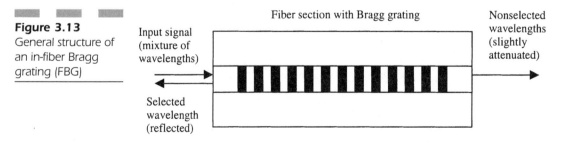

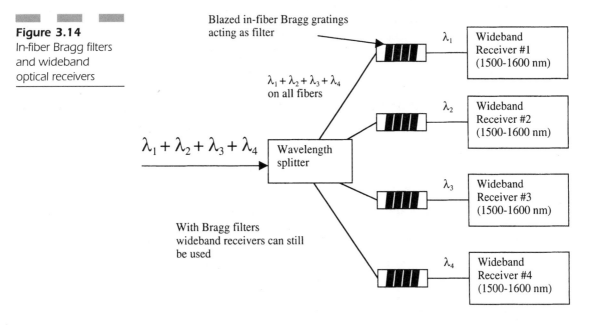

Figure 3.14
In-fiber Bragg filters
and wideband
optical receivers

The use of FBGs as filters at the receiving end of a simple WDM link allowing the use of identical wideband receivers is shown in Figure 3.14.

The only problems with FBGs have been some sensitivity to temperature variations and stretching of the fiber section with the gratings. Both effects have been relatively easy to control in practice. FBGs can be used not only as filters in WDM and DWDM systems, but also to control dispersion in the 1550 nm band and to stabilize the output of a laser, tunable or otherwise.

WDM

The first edition of this book did include a consideration of WDM systems. However, the WDM section appeared in a chapter on the future of SONET. So in just a few short years the practice of WDM has moved along to the point where the topic deserves to be treated right up front. This section basically reproduces the original WDM section, with minor updates along the way.

Strangely enough, there is really no such thing as WDM. Actually, it would be more accurate to say that the technology known as WDM is basically familiar analog frequency division multiplexing (FDM) applied to fiber optic transmission. This is because the frequency of any electromagnetic waveform (and light is just another form of electromagnetic wave phenomenon) is directly related to its wavelength by the formula $c = \lambda/f$. That is, the wavelength of any electromagnetic wave divided by its frequency is always equal to the speed of light (c).

Analog systems, which multiplex signals together in different frequency channels (e.g., broadcast TV channels set 6 MHz apart), use FDM. So did older L-carrier networks using coaxial cable. Each channel uses its assigned bandwidth all the time, but no channel gets the entire bandwidth. The formula $c = \lambda/f$ could easily be used to convert the megahertz or kilohertz channels to wavelengths, and then the method could technically be called a form of WDM. However, there is a little more to WDM than just FDM.

Digital transmission systems, in contrast to analog systems, usually use time division multiplexing (TDM) to combine signals onto a higher-speed medium. This is the multiplexing method used in both T-carrier and SONET, of course. In TDM, the entire bandwidth (frequency range) of the transmission medium is broken up into time slots, and each input uses the entire bandwidth, but only for its allotted time interval. For example, T-carrier divides the entire 1.544-Mbps bandwidth (less overhead) into 24 time slots of 64 Kbps each.

The key to understanding WDM is to realize that the enormous bandwidth of fiber optic cable, which spans the whole visible light spectrum and beyond, makes it possible to combine FDM and TDM on the same path through the same fiber. The differences between FDM, TDM, and WDM are shown in Figure 3.15.

Figure 3.15 shows that both FDM and TDM techniques can be combined on fiber optic links to create WDM. There is no reason that different "colors" of light cannot be used on the same fibers, as long as transmitter and receiver are tuned to function that way. In the same fashion, voice nursery intercoms can function over home electrical wiring without interference.

WDM and SONET

In conventional fiber optic transmission systems like SONET, information is transmitted in the form of light pulses. Pulses of light represent individual bits: zeros and ones. The electrical bits are converted to optical signals, transmitted over a fiber optic cable, and converted back into electrical signals at their destination.

Figure 3.15
FDM, TDM, and
WDM compared

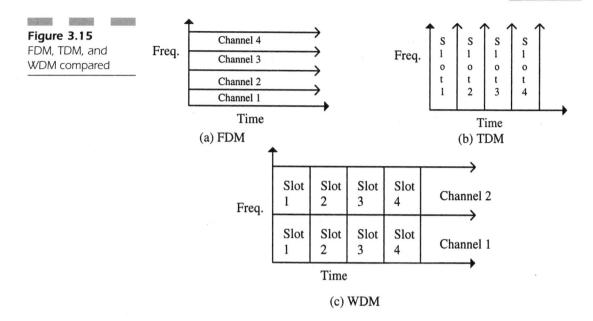

WDM involves the simultaneous transmission of light from multiple lasers with different wavelengths over a single fiber optic line. Sources send these wavelengths of light to a multiplexer, which consolidates them for transmission over a single WDM fiber link. WDM breaks the optical spectrum into channels, each with a different wavelength. Time slots, as in SONET, can be used with each of these channels.

Special optical amplifiers, usually spaced tens of kilometers apart, amplify all of the wavelengths simultaneously. Finally, at the far end of the WDM link, the signals arrive at a demultiplexer, where they are separated and sent to receivers at the destination points. WDM equipment also includes optical fiber filters, which selectively transmit or block a particular range of wavelengths, and transponders, which function as the transmit/receive systems.

WDM is useful whenever congestion occurs on fiber networks, SONET or otherwise. Greater capacity can be created without the need to install more fiber along the right-of-way. The trade-off occurs in the end-equipment expense, but electronics usually beat the backhoe anytime. WDM can be used to expand to greater capacities over the same number of fibers. WDM also gives a service provider one way of achieving 10 Gbps and even higher transmission rates.

Early WDM devices and links involved two-channel multiplexing. These used the 1310- and 1550-nm (nanometer) laser operating regions, or windows, in short-span applications, or two widely spaced channels in the 1550-nm region (e.g., 1538 and 1558 nm) in long-span applications.

WDM systems may be unidirectional or bidirectional. A bidirectional WDM device will pass a number of multiplexed frequencies in both directions on one fiber. For example, in a four-channel bidirectional WDM span, signals at 1549 and 1557 nm are multiplexed for transmission in one direction, and 1533- and 1541-nm signals are multiplexed for transmission in the other direction, all on the same working fiber. This is shown in Figure 3.16.

Conversely, a unidirectional WDM device multiplexes a number of different frequencies for transmission in one direction on one fiber. For example, in a four-channel *unidirectional* WDM span, signals at 1557, 1549, 1541, and 1533 nm are multiplexed for transmission in one direction on the fiber. This type of WDM is shown in Figure 3.17.

Both of these techniques allow for such fiber configurations as combining four OC-48s on a single fiber span for an aggregate signal speed of OC-192. This newer WDM technique can use a previously installed fiber base to support extremely narrowly spaced channels. This newer WDM system does have a number of requirements. For instance, using WDM to combine four OPC-48s requires maximum reuse of any existing OC-48 equipment: compatibility with current regenerator spacing, a minimum of 4-wavelength operation, the ability to upgrade to 8 or even 16 different wavelengths (or more) in the future, the capability to add or subtract wavelengths (no mean feat), and unidirectional and/or bidirectional operation.

Sometimes there are references to something called *dense WDM* (DWDM). There are varying interpretations of the difference between standard WDM and DWDM. Some would define DWDM as the multiplexing of two or more channels on a single fiber, with the channel spaced 2 nm or

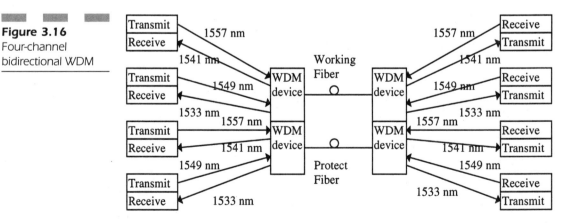

Figure 3.16
Four-channel
bidirectional WDM

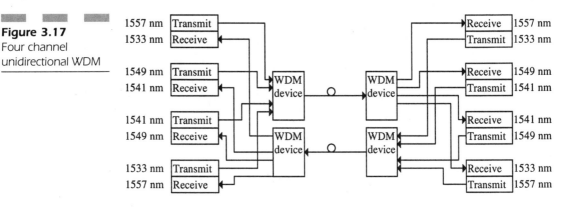

Figure 3.17
Four channel
unidirectional WDM

less from each other. Others basically consider anything involving four or more channels to be DWDM. The International Telecommunication Union (ITU) is busily working on a standard set of WDM guidelines.

The next generation of WDM is waiting in the wings. This type of WDM, which is known as *wide-spectrum multiwavelength multiplexing,* is ideal for SONET and all long-haul fiber applications because it allows for economical incremental growth. Such wide-spectrum WDM is needed to realize the full benefits of future fiber optic technologies, such as optical cross-connects.

In wide-spectrum multiwavelength multiplexing, the wavelengths needed for the transmission system can be produced by wavelength-specific transmitters, or *wavelength adapters.* Wavelength adapters allow transmission equipment to have any wavelength output at the transmitter. For example, a lower-cost 1310-nm laser could be used. This is good for multivendor environments or for backward-compatibility, but wavelength adapters are quite expensive and the high cost is accrued with every added wavelength.

Before implementing wide-spectrum WDM, however, a whole new set of challenges must be addressed. One such challenge is finding the proper spacing of transmission frequencies to allow a large number of channels without crosstalk or expensive filters. The evolving ITU-T standards have set the channel spacing for wide-spectrum WDM on 100 GHz intervals with 193 100-GHz as a reference frequency.

DWDM

Not only is WDM with wavelength spacings of about 10 nm an accepted optical technology today, but WDM has moved along to the point where it is *dense* WDM (DWDM) systems that are the major topics of discussion and consideration. The major difference between WDM systems and DWDM systems is that in DWDM the wavelengths used are packed within 1 nm or even 0.1 nm of each other, or about 10 to 100 times denser than in WDM. The current record for wavelength packing on a single fiber is 1,022 channels, although the total throughput is only about 40 Gbps. However, since the current serial bit transmission record on fiber is 160 Gbps, DWDM systems with thoughputs in excess of 160 Tbps (160,000 Gbps or 160,000,000 Mbps) are potentially possible.

So it would seem that there is very little real difference between WDM and DWDM. One has closer spacing of wavelength channels, true enough, but essentially WDM and DWDM are the same technologies. However, this is not really true. The close spacing of wavelengths in DWDM systems translates to a whole new set of concerns.

This is because in practice the closer spacing of wavelengths in DWDM makes a big difference. If the packing of wavelengths to within 0.1 nm (or less!) over a single fiber core were trivial, DWDM would merit no more than a footnote in WDM discussions. But there is a world of difference between 10-nm spacing and 0.1-nm spacing of light waves. For example, using 0.1-nm spacing allows 100 light sources to concentrate their energy over a 10-nm bandwidth into a single fiber core 9 microns or so in diameter. This is a huge energy flux concentrated in a very small space, although in terms of raw power the total amounts are still relatively small. But confining this electromagnetic energy in the fiber core for long distances without debilitating attenuation, dispersion, and other fiber effects is a challenge that deserves a section all its own.

Everything is magnified in DWDM, so that even small effects in WDM become crippling in DWDM unless steps are taken to compensate for these impairments. Fortunately, there are methods available that make DWDM systems not only practical, but attractive.

To take a simple example, in DWDM systems the center wavelength of a channel must be tuned carefully to make sure it is exactly where it is supposed to be. This is because even simple modulation of the carrier signal will essentially double the bandwidth required (in precise accordance with Nyquist) and might otherwise overlap and interfere with wavelengths still within their band, but not properly centered. Naturally, as the wavelengths move to within a fraction of a nanometer of each other, these *band edges* become more and more important.

Before getting into a full discussion of DWDM system impairments and what must be done to control their effects, this is the place to say a few words about two often neglected components of optical networks.

These are the *combiners* and *splitters* that mix wavelengths onto a single fiber at the sender and split the mixed wavelengths onto separate fibers at the receivers. These components have come a long way as well. In many cases, the same device can serve both functions. Used in one direction, it is a splitter. Turn it around and it is a combiner or coupler.

Simple couplers make use of a phenomenon called *resonant coupling*. This occurs when two fiber cores are brought into close proximity. If the lengths are just right, half of the input power ends up in each fiber core. This is called a *3-dB coupler*. The basic building block can be employed to make (for example) eight-way combiners or splitters. However, each stage causes some signal loss and attenuation. This is a physics phenomenon, and even two light streams directly coupled onto one fiber *must* lose half their power.

When operating at different wavelengths, DWDM couplers and splitters are more efficient. Modern systems can use optical technology such as *beamsplitter prisms* or *bifringent materials*. Many splitters make use of diffraction gratings of one form or another, including planar diffraction gratings.

At the source, some lasers are now *multiline lasers*. Instead of using the comb of spectral width wavelengths to tune the laser, mutliline lasers attempt to even out linewidths and equalize the power of the several wavelengths produced. This is probably a good idea in the long run. If DWDM systems evolve to 1,000 inputs or more, 1,000-input lasers might be needed. But if multiline lasers can produce 10 usable wavelengths, only 100 input lasers would be needed. Of course, the multiline laser must be modulated externally at each wavelength to send bits, since there is only one drive device, but this is simple enough to do.

DWDM Impairments

It has been established that jamming many wavelengths onto a single fiber core is difficult. This is due to a number of fiber impairments that are annoying in any optical system, but absolutely crippling to a DWDM system. Oddly, some effects can be controlled by jamming the wavelengths even closer together, and other impairments require keeping the wavelengths well apart. Some effects can be dealt with by boosting the transmitter power, while others require power limits. But just what are they, and why do they interact this way?

All fiber impairments can be roughly characterized as dispersion effects. In this book, they are further divided into modal dispersion

TABLE 3.1

Fiber impairments

Modal Effects	Cause	DWDM Workaround
Intermodal dispersion	Multiple modes interact	Single-mode fibers
Intramodal dispersion	Various:	Fiber construction methods:
Material (chromatic)	RI varies with wavelength	Effects cancel at 1300 nm
Waveguide	20% light travels in cladding	(Can be shifted)
Polarization mode	RI varies with polarity	Perfectly round core (difficult)
Scattering Effects		
Rayleigh	RI varies with fiber density	Perfectly uniform density
Raman (SRS)	Low wavelength pumps high, Produces "power tilt"	Lower power, limit span, pack wavelengths close together
Brillouin (SBS)	RI varies with acoustic waves	Lower power, limit number of channels, avoid two-way systems
Miscellaneous Effects		
Linear crosstalk	Mostly result of SRS	Use filters, separate wavelengths
Four-wave mixing (FWM)	Additive wave response	Watch channel wavelengths, avoid 10+ Gbps systems
Cross-phase modulation	RI varies with intensity (called *optical Kerr effect*)	Balance power in system
Self-phase modulation		Limit power in system

effects, scattering effects, and miscellaneous effects, but they all manifest themselves as dispersion. Table 3.1 lists the effect, the cause, and the workaround in DWDM systems. Keep in mind that this table is a distillation of dispersion effects that most impair DWDM systems. There are many other types of dispersion and fiber impairments.

It should be noted that many of these effects also vary with the serial bit speed used in the fiber channel of a DWDM system. As a general rule, doubling the speed from 5 to 10 Gbps will quadruple the dispersion on the same length of fiber. This only reinforces a truism in networking: to go farther, go slower. The following sections take a quick look at each effect in some detail.

Modal Effects

The first modal effect is *intermodal dispersion* and is included here really just for the sake of comparison. Intermodal dispersion is the result of the interplay between the various physical paths down which a beam of light can make its way down a fiber core. The workaround is to use single mode fiber (SMF), and SMFs are used in all practical DWDM systems. Even older single-mode fibers designed for 1300-nm use can be employed, but the attenuation is high and runs must be short.

But there is also *intramodal* dispersion that affects even a single propagated mode. The causes of this impairment are varied according to type, and all the workarounds concern fiber construction methods. *Material (chromatic) dispersion* is a result of the RI of the fiber varying with wavelength. *Waveguide dispersion,* which acts to counteract the effects of material dispersion, is the result of 20% of the light being carried not in the fiber core, but in the cladding. The difference in RI in the cladding causes dispersion. These effects usually cancel out at around 1300 nm, but dispersion-shifted fibers move this point to around 1550 nm. So most DWDM systems will use dispersion-shifted single mode fibers. The final modal effect is *polarization mode dispersion*. This is the result of the RI of the fiber varying with the polarity of the signal. It is generally only of concern at serial bit speeds of 10 Gbps and higher. The workaround is to use fiber with a perfectly round core. (Surprisingly, there are times that perfectly round cores are not desired and polarization mode dispersion is actually encouraged!) Now, most fiber fabrication techniques do their best to produce round cores. However, it is easy for the fiber to be stepped on or otherwise deformed during installation.

Scattering Effects

The *scattering effects* are an interesting group. These are named, curiously enough, after physicists and investigators into optics. *Rayleigh scattering,* common to all fibers, is unavoidable and results when light is scattered, or deflected, by particles in the fiber. It is a concern in DWDM systems because variations in the density of the fiber make the dispersion worse. Again, the only way to compensate is to construct the fiber to very exacting standards.

Of more interest is *stimulated Raman scattering* (SRS). SRS is caused when multiple wavelengths interact on one fiber through the crystal lattice of the fiber core and so became a concern only in WDM systems.

With SRS, lower wavelengths will "pump" higher wavelengths, resulting in a power loss at lower wavelengths and a power gain at higher wavelengths. This causes a pronounced *power tilt* in the received power of the wavelengths, even though all might be sent with the same power. SRS is not always bad: optical amplifiers are nothing more than fibers in which the pumping from the lower wavelength (the pump laser's wavelength) to all the higher ones is encouraged to the extreme. SRS can be controlled by lowering the input power, limiting the span distance, and packing the wavelengths (generally, the farther apart the wavelengths, the more pronounced the SRS between them).

Finally, there is *stimulated Brillouin scattering* (SBS). SBS occurs even at low power levels and is the result of acoustic vibrations in the fiber. That is, the atoms of the fiber literally move back and forth, creating an acoustic (pressure) wave in the fiber. The pressure change affects the RI, which scatters the light, producing SBS. Both SRS and SBS are *stimulated* in the sense that the light that is itself scattered produces the effect. Oddly, SBS is less severe at higher data rates as the bit time shrinks. However, SBS is much worse when light signals propagate in both directions at the same time in bidirectional WDM systems (because of the complex acoustic waves generated). In fact, SBS is one of the main reasons that two fibers are used, one for each direction, in most optical systems. But if runs are short enough, even DWDM can be used as long as the number of wavelengths used is limited. So the workarounds are limited power (which limits runs), limited wavelengths, and avoiding two-way transmission on a single fiber.

Miscellaneous Effects

The final category of *miscellaneous effects* is also an interesting group. The first member, *linear crosstalk,* is not really a separate impairment at all. Linear crosstalk is caused by SRS, but many sources treat linear crosstalk separately. Crosstalk is just the interaction or interference between two wavelengths. The workaround is to keep the wavelengths well separated and filter out extraneous wavelengths not used for information carrying.

The name *four-wave mixing* (FWM) is an unfortunate one. This impairment was first noticed in early WDM systems with four channels, hence the name. But it really should be *N-wave mixing*, since the interactions of any number of wavelengths can lead to this phenomenon. FWM is just an additive response to multiple wavelengths, the optical equivalent of *beats* in electrical systems. Three wavelengths will combine to yield a fourth wavelength at the receiver, according to a formula (so FWM is appropriate in this sense). The formula, in frequencies used, is $f_{new} = f_1 + f_2 - f_3$. Naturally, if the new wavelength is also used in the DWDM system, crosstalk results. The workaround is to watch the wavelengths used and space them out unequally (the ITU grid, discussed later, allows for this). FWM is also worse at 10 Gbps and higher.

Finally, there are two impairments known as *phase modulation* effects. As the result of the RI of the fiber varying with the intensity of the light, a phenomenon called the Kerr effect occurs. Cross-phase modulation (CPM) occurs between the wavelengths and self-phase modulation (SPM) occurs within even a single wavelength of light. CPM is compensated for by carefully balancing the power between the channels in a DWDM system, and SPM is dealt with by limiting the power used (which also limits distance, of course).

The ITU DWDM Grid

The ITU-T has been active in exploring international standards for DWDM. One specification is G.692, entitled "Optical Interfaces for Multichannel Systems with Optical Amplifiers," which contains a number of interesting points about DWDM. Unfortunately, DWDM developments have been proceeding so fast that the ITU-T and other standards organizations have been hard-pressed to keep up with system capabilities. These words might soon be obsolete, if they are not already.

The ITU-T specifications mainly address point-to-point DWDM systems using 4, 8, 16, or 32 channels. The maximum specified span without amplifiers for a DWDM link is 160 km (100 mi) or up to 640 km (400 mi) when amplifiers are used.

A standard wavelength grid is established based on multiples of 50-GHz spacing between the wavelengths used. There is also a center reference frequency ("wavelength") at 193.1 terahertz (THz: 1,000 GHz). The whole grid is sometimes called the "ITU 193 THz grid." In keeping with the tradition of citing frequency in ITU-T specifications, all of the wavelengths are specified in Hertz (frequency), not nanometers! But conversion is easy enough to do. The reference frequency of 193.1 THz equates to 1553.5 nm, right in the middle of the 1990s window, as might be expected. The 50-GHz spacing works out to about 0.4 nm between the wavelengths. The wavelengths used are in the range from 1528.77 nm to 1560.61 nm.

The G.692 draft proposal allows:

- 4 channels with 3.2-nm (400 GHz) spacing
- 8 channels with 1.6-nm (200 GHz) spacing
- 16 channels with 1.6-nm (200 GHz) spacing
- 32 channels with 0.8-nm (100 GHz) spacing

Supervisory wavelengths are allowed at 1310, 1480, 1510, and 1532 nm. In the draft, users are allowed to use any wavelength on the grid, but only those wavelengths on the grid. However, working DWDM systems have progressed to 128 channels or more, and spans in excess of 640 km are possible, challenging the ITU-T plans. Allowable speeds for each channel are 622.08 Mbps, 2.488 Gbps, and 9.9 Gbps. These correspond to SONET/SDH speeds of OC-12/STM-4, OC-48/STM-16, and OC-192/STM-64. Optical amplifiers of the EDFA type are expected to be used, operating in the 1550-nm region. In addition to the basic 50-GHz spacing, systems are allowed to use 100-GHz (8 or more channels), 200-GHz (4 or more channels), 400-GHz (4 or 8 channels only), 500-GHz (8 channels only), 600-GHz, (4 channels only), and even 1,000-GHz (4 channels only) spacings.

Potentially, the ITU-T grid spacing could be firmly allocated on a national basis, with each country limited to using the wavelengths assigned to it. This is actually a good idea, since the day may come (and many feel this day is closer than anyone thinks) when a single strand of fiber will literally circle the globe carrying hundreds or even thousands of wavelengths with in-fiber optical amplifiers. Each country would just attach to this global fiber ring and never fear interfering with some other countries' wavelengths. To communicate with any other country, just send on a channel allocated by the receiving country. To receive, just listen on the channel assigned for the return path.

It might seem odd to spend so much time in a book on SONET/SDH discussing optical networks, WDM, and DWDM. None of these concepts are dependent on SONET/SDH in and of itself. And SONET/SDH is defined to run at 1310 nm, in the 1980s window, reflecting the time frame for SONET/SDH origins. Nevertheless, it is quite common to find SONET/SDH running somewhere in even a DWDM system. This is because SONET/SDH is a much more mature technology and includes consideration in a standard manner for essential network operational features such as network management and protection switching. In networking, bit transfer is really the easy part. Controlling all aspects of the process is much more difficult.

There are a number of ways to use SONET/SDH as a *management channel* in a DWDM system. One way is just to couple in a SONET/SDH channel at 1310 nm with the DWDM channels at 1550 nm. After all, there is only one fiber. If the SONET/SDH management channel indicates a problem, there is obviously a problem in the whole fiber. Alternatively, a transponder can be used to shift the 1310-nm management channel into the band centered around 1550 nm. Sometimes the SONET/SDH management channel is used to carry additional network management information about the other DWDM channels, such as error-correcting codes. Additional issues concerning the future of SONET/SDH in a DWDM world will be considered in a later chapter.

CHAPTER **4**

The Digital
Hierarchy

SONET is an international standard for very high-speed digital transmission over fiber optic cable. SONET is an abbreviation for Synchronous Optical NETwork and was first considered by Bell Telephone Laboratories (Bell Labs, now Lucent Technologies) in the early 1980s to solve a series of nagging problems with existing digital transmission systems. Digital transmission systems are characterized by the fact that these communication links only carry information in the form of binary digits (universally known as bits). Binary digits can only represent a "0" or a "1." Strings of 0s and 1s can be constructed to represent almost anything from computer-based data to digitized voice to stereo audio on a music CD (compact disk) to the soundtrack of a movie (in several varieties), even potentially to the movie itself.

It may seem that we are now living in a "digital world" where nearly everything from music to movies comes out of a computer. This impression is not far from true. Music can be generated much more precisely and accurately (not to mention much less expensively) from a computer than from a human. Movies are so laden with digital special effects that it sometimes seems silly to even generate the images on film for projection onto a screen. Why not just watch a big computer monitor?

How has this transition to the digital representation of almost everything come about? What has all this to do with SONET? In order to understand the position of SONET in this digital world, it is necessary to understand the problems with digital transmission that SONET was designed to overcome.

Analog and Digital

The alternative method of transmitting voice, video, or data from one place to another is to employ analog transmission systems. Analog signals can take on more values than merely 0 or 1—for instance, a signal of "0.5" is perfectly fine in an analog system, but meaningless in a digital transmission system. In fact, analog systems allow signals to have any value between some maximum and minimum value. Typically, analog signals will vary smoothly over time in an unpredictable pattern, wandering between maximum and minimum as the input signal (voice or otherwise) varies. In voice transmission, the input signal is an electrical representation of the human voice's acoustic pressure waves (sound). This analog signal is sent over the telephone wire, which has a number of physical forms, as an analog electrical signal and reconverted to an acoustic pressure wave at the receiver.

These analog systems evolved to deliver voice over the national telephone network, or public switched telephone network (PSTN) as it is more properly called. Because people cannot talk in 0s and 1s, the builders of the PSTN had little choice but to initially deploy the national voice network as a purely analog network. Sometimes it is said that the PSTN is "optimized for voice," meaning that nothing can really be sent over it unless it is an analog sound. Even office and home computer users with personal computers (PCs) on their desks are usually familiar with the use of modems (which is a contraction of MOdulator/DEModulator) to communicate over analog voice telephone lines. A modem changes the 0 and 1 digital signals that computers generate into "sounds" (actually, the electrical representation of that "sound") which can be carried over the PSTN. Simply put, a "0" bit "sounds" differently than a "1" bit to the receiver.

The differences between analog and digital signals are shown in Figure 4.1. Note that the signals vary by amplitude (signal strength), but other signal parameters, such as frequency, are possible and even quite common. Also note that digital signals are not limited to basic 0s and 1s, but may include more possible values. Even when this is the case, however, all signal values are interpreted as strings of 0s and 1s, and the signal is considered to be undefined when not expressed as one of this limited set of values. For example, four level digital signals would express the bit strings 00, 01, 10, and 11.

It is somewhat ironic that the analog voice telephone network largely replaced a perfectly functional digital network that was some thirty years older than the telephone, which was invented in 1876. This was the national telegraph network. After all, Morse Code is easily understood as

Figure 4.1
Analog signals and
digital signals

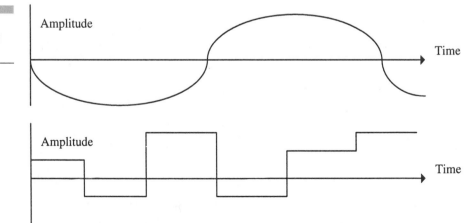

representing information as a string of 0s ("dots") and 1s ("dashes"). The letter "a" is just a dot followed by a dash in Morse Code (01), the letter "b" is dash-dot-dot-dot (1000), and so on. Morse—it was really his assistant, Theodore Vail, who invented "Morse" Code—gave common letters shorter strings and less common letters longer strings out of consideration that Morse Code was keyed by a human being.

Alexander Graham Bell invented the telephone while trying to develop a way to *multiplex* (combine) telegraph signals on a single wire. The concept of multiplexing has enormous significance in SONET, but it obviously is an idea that has been common for many years. Without multiplexing, many wires must be installed to increase the total message-carrying capacity of the network. Multiplexing makes more efficient use of existing facilities, minimizes expense in terms of physical wiring, and is mostly limited only by the degree of ingenuity of electrical engineers operating within the limits of a few basic physical laws, such as the fact that signals cannot travel faster than the speed of light.

Bell's background in education of the hearing impaired uniquely qualified him to hear in the tones of multiplexed telegraph signals not the mixture of 0s and 1s or dots and dashes, but the sound of the human voice. Time was needed to work out the details, but the telephone was the first major network technology that was analog instead of digital. All previous methods, from smoke signals to war drums to "one if by land and two if by sea" were inherently digital in nature. The analog telephone required no special knowledge or training to use.

It soon became apparent to early telephone users, however, that the blessing of analog had a dark side as well: Analog signals are much more susceptible to *noise* than are digital signals. In any transmission system, noise is simply a spurious and unwanted signal that interferes with the signal sent into the system. To understand why, consider the fact that in Morse Code, a dash was simply defined as being at least three times longer than a dot. Early telegraph operators all had a distinctive "hand" because their basic dots and dashes all varied somewhat in duration (the phrase "heavy-handed" comes from this era). When a noise on the telegraph line (usually a nearby lightning strike) induced a spurious electrical signal into the line that somehow stretched a dot to twice its normal duration, the receiving operators were still likely to say to themselves "that wasn't a dash, so it must have been a dot."

Analog signals can vary smoothly from maximum to minimum. A signal twice as long as a dot was just as likely as one three times as long, or one and a half times as long, or nearly any duration at all. In analog systems, it quickly becomes impossible for receivers to distinguish noise (the unwanted part) from the signal (the wanted part). In the telephone network, this noise was simply *noise* which swamped the sender's speech in a constant background hiss of clicks, pops, and buzzes, mostly due to adjacent digital telegraph lines.

Early telephone pioneers quickly invented ways to minimize the analog noise on telephone lines. But it was never eliminated and limited the quality of voice that could be delivered, especially over longer and longer distances, to a barely acceptable level. Clearly, there was a reason to try to invent a more effective way to minimize noise effects on the analog network. The answer, when it came, was to digitize analog voice signals.

The Digitization of Voice

Many people in the telecommunications field today associate the digitization of voice with such modern developments as integrated services digital network (ISDN) and digital T-carrier networks. However, efforts to digitize voice began around World War II mainly as an attempt to eliminate nagging noise problems on analog voice circuits. These early efforts were interrupted by the war, but work continued in the United States after the war. By the 1960s, the phenomenal growth of the United States economy after World War II was the added incentive needed to complete this work.

Before the war, only about 40% of households in the United States found it economically feasible and socially desirable to have a telephone. Letters, backyard fences, and other social activities filled the hours people spent not otherwise engaged in work and related activities. But after the war, when the United States was essentially the only developed nation not devastated by the conflict, the economy expanded rapidly as other countries understandably turned to the United States to purchase goods and raw materials to rebuild their shattered cities. The United States was only too glad to oblige, and the immediate result was a kind of "Golden Age" in the 1950s, when companies rapidly expanded to meet the growing demand for goods not only overseas, but for the newly prosperous citizens in the United States.

One of the luxury items at the top of nearly everyone's list was a telephone. The telephone companies, especially the AT&T Bell System (as the self-declared "nation's telephone company"), were hard pressed to keep up with the demand for local loops, switching offices, and trunking networks that the swelling system now required. Local loops were needed to provide a pair of wires for every new telephone. Switching offices (or central offices, as they were frequently called) were needed to switch the calls to their destinations. Trunking networks were needed to link these offices together; trunks are telephone lines that link not individual telephones to switching offices, but switching offices to each other. These trunks became the focal point of digitization efforts in the 1960s.

Many switching offices employed an analog trunking system that multiplexed, or combined, 12 voice channels carrying telephone calls onto two pairs of copper wires. Each voice channel was assigned a separate frequency band for this purpose. This technique was known as *frequency division multiplexing* (FDM) because each voice channel had its own frequency band assigned to it regardless of whether it was in use or not; therefore, idle channels existed. When someone made a telephone call that terminated at a different switch than the one that the telephone's local loop was attached to, an idle trunk channel frequency band was assigned to that person's conversation for the duration of the call. A typical switching office could have up to 10,000 local loops, while trunks between switching offices typically numbered only about hundred. This basic telephone network structure of analog loop, switch, and trunk is illustrated in Figure 4.2.

As network use grew due to the number of new telephones and telephone calls, this trunking network threatened to be overwhelmed. If no trunks were available between switches, no more calls could be completed, and the telephone company would lose potential revenue. Obviously, adequate trunk capacity was a key parameter in telephone network design and operation. The problem was that adding trunk capacity was neither easy nor quick. Funds had to be allocated, routes planned, materials purchased, and the work done. The task was made even more difficult by the fact that the switching offices usually were many miles apart.

Figure 4.2
Local loops,
switches, and trunks

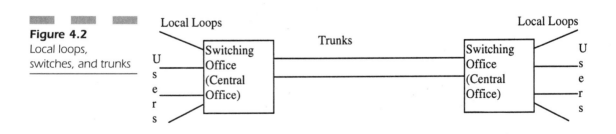

The search for a more efficient way to quickly add capacity to a trunking network led several groups of engineers back to digitization. It was already understood that digitized voice was less noisy than analog circuits. However, the end equipment (transmitters and receivers) needed to convert analog voice to digital voice and back again was quite expensive. An added economic motive was sought to justify the expense of the equipment: Additional trunking capacity provided that added motive.

A digital system was designed that could double the capacity of 12-channel analog voice trunks to 24-voice channels. Multiplexing trunk networks were known as *carrier* systems to the telephone companies. This allowed telephone companies to cost-justify the new digital equipment needed by doubling capacity on the trunking network without having to go through the expensive process of laying new cable and had the additional benefit of improving voice quality by reducing noise. The new digital multiplexing trunk network was designated by the engineers at Bell Telephone Laboratories as *T-carrier*. The "T" had no significance other than a letter designation. There were also N- and L-carrier systems, among others, mostly analog methods. But it was the digital T-carrier that fueled a revolution in digital transmission which led directly to SONET.

T-Carrier

With the advent of T-carrier trunking networks in the early 1960s, telephone companies addressed one of a number of problems brought on by the unprecedented growth demands of the rapidly expanding economy. At about the same time, electronic components and processors were introduced into the switching offices. These looked similar to 1960s-style computers, except that the Bell System was reluctant to call them what they were, because an agreement in 1955 (the "Final Judgment" of an anti-trust suit) with the federal government essentially banned AT&T from developing and marketing anything that could be construed as a computer. Therefore, under the name "Electronic Switching Systems" (ESS) computers slowly made their way into the AT&T network. The introduction of area codes (called "Numbering Plan Areas," or NPAs by AT&T) helped to complete long-distance calls without requiring the wholesale hiring of more long-distance operators. Understanding these related developments helps to place the T-carrier within the proper deployment context.

T-carrier was the basis of a whole family of digital trunking methods. The basic unit was the digitized voice channel, which produced a stream of bits at the constant rate of 64,000 bits per second, or 64 kbps. The 64 kbps digitized voice channel was designated Digital Signal-0 (DS-0) in the T-carrier system. The "0" indicated the lowest level of what was to become the T-carrier hierarchy. There was no carrier involved in DS-0, because this basic signal was not multiplexed with anything else, but merely served as the lowest level T-carrier input and output.

Twenty-four DS-0 signals were combined to yield a DS-1. The "1" indicated the first level of the T-carrier hierarchy. The DS-1 signal was transmitted on a T-1 physical network. Although DS-1 specified a certain structure of 0s and 1s from 24 input DS-0s, T-1 specified the transmitters, receivers, and wiring requirements for transporting the DS-1. Most people use the terms "DS-1" and "T-1" interchangeably, although this is technically imprecise.

A DS-1 performed its multiplexing task in a fundamentally different fashion than did analog carrier systems using FDM. The DS-0 produced 64 kbps by generating 8 bits 8,000 times per second. There were many technical reasons that 8 bits and 8,000 analog "samples" were taken, but these are beyond the scope of this discussion. It is only important to realize that each digitized voice channel (DS-0) produced 8 bits every 1/8,000th of a second, or 125 microseconds (μsec). Obviously, the combined output of 24 DS-0s would require a way of sending at least 24×8 bits = 192 bits every 125 microseconds. To accomplish this, a DS-1 was organized into *frames*. Each frame was sent and received 8,000 times per second. An additional bit was added to help the receivers distinguish the beginning of one DS-1 frame from the end of the previous DS-1 frame. So the aggregate bit rate of a DS-1 was 193 bits/frame $\times$ 8,000 frame/second = 1,544,000 bits per second or 1.544 Mbps. The structure of a DS-1 frame is shown in Figure 4.3.

Within the DS-1 frame, a DS-0 input channel always occupied the same position in the frame. For instance, a particular DS-0 channel may be the first 8 bits after the framing bit, or the next 8 bits after that, and so on. Because the DS-1 frame was thus divided by *time* and not by frequency, this method was referred to as time division multiplexing

Figure 4.3
The DS-1 frame structure

←	193 bits (125 microseconds)	→

| F | Time Slot #1 | Time Slot #2 | Time Slot #3 | | | Time Slot #24 |

(TDM). Although it is not impossible to FDM digital signals, or to TDM analog signals, the equipment needed to do so is complex and expensive. It is easier and cheaper to multiplex analog signals with FDM and multiplex digital signals with TDM, and this approach is nearly universally observed today.

It is important to note that there is some *overhead* associated the DS-1 frame structure. Overhead are bits that are needed, but do not represent the information sent. Because the 24 DS-0s only generate 24 × 8 bits/frame × 8,000 frames/second = 1,536,000 bits/second, a difference of 1.544 Mbps – 1.536 Mbps = .008 Mbps = 8 kbps occurs between the aggregate DS-1 bit rate and the sum of the inputs. This "extra" 8,000 bits per second of course is due to the presence of the framing bit in every DS-1 frame. Telephone companies use this overhead for a variety of purposes today, such as error control,but 8,000 bits per second is not much in a world where even 30,000 bits per second or so is considered barely adequate for checking e-mail from a home computer.

Other information needed to be conveyed over a T-carrier link as well. For example, both ends needed to know which trunks were idle, which were in use, which were now being seized from one end or the other for a voice channel. No place existed in the T-1 frame structure to carry this signaling and supervisory information; therefore, the T-carrier equipment routinely "robbed" bits from each individual voice channel for signaling and supervisory and maintenance purposes. This *robbed bit signaling* became a distinguishing characteristic of the T-1 frame structure.

It quickly became apparent that robbing a bit from each voice channel 8,000 times a second was overkill for signaling purposes. It was simply not necessary to know to the nearest 125 microseconds when an idle trunk channel was about to be seized. When not used for signaling purposes, the robbed bit in each voice channel (DS-0) came to be used for other things as well. Because T-1s indicated 0 bits by a lack of voltage on the line, too many consecutive 0s in a row caused the receivers to go out of synchronization with the senders. It was then impossible to determine whether 20 or 21 consecutive 0s were actually sent. Initially, T-1 multiplexers set the maximum number of consecutive 0s sent to 15. If a user were to generate more than 15 consecutive 0 bits on adjacent DS-0 channels, the robbed bit would be used to enforce this minimal "1s density" requirement.

The robbed bit was also used to convey certain alarm conditions (specifically, *yellow alarm*) and to carry special maintenance messages between the T-1 multiplexers. These four uses of the robbed bit—signaling, 1s density, yellow alarm, and maintenance—effectively limited the useable bandwidth of a DS-0 channel in a T-carrier network to 7 bits/frame × 8,000 frames/second = 56 kbps, especially in data applications. The quality of normal voice conversations were unaffected by this wholesale bit robbery.

T-carrier grew to include many hierarchical levels. That is, DS-1s were combined in a variety of ways to yield higher and higher capacity trunks. The goal was to aggregate as much traffic as possible on a single trunk T-carrier, and it was enormously successful in doing this in a cost-effective fashion.

Other Digital Carrier Schemes

While all of this development was going on in the United States, the rest of the world, mainly meaning Europe, was not idle. The Europeans were busily completing their own hierarchy of digital carrier systems, which differed from the United States version in a number of significant ways. First, even though the European systems also produced 8 bits 8,000 times per second, the 8 bits were coded differently than in the United States version, and actually produced even less noise than the United States digital encoding method; however, the differences were great enough to prevent a digital voice channel from being directly connected to an international circuit in most cases. Confusingly, both are often designated DS-0, which means the exact coding method must be determined by context.

Next, the Europeans also improved on the United States DS-1 frame structure. The 8 kbps DS-1 frame overhead that was barely adequate for management purposes, and totally inadequate for signaling in switched environments, such as normal voice operation, was replaced by a more advanced structure. In the United States, many functions had to be done by "robbing" a bit from each voice channel, limiting many applications to 56 kbps instead of 64 kbps at the DS-0 level.

The Europeans eliminated bit robbing by adding a full 16 bits of overhead to each frame, rather than just a single bit, as in the United States T-carrier system. A group of 8 bits usually is referred to as an *octet*, but in the United States this group of 8 bits is more commonly, but less correctly, called a *byte*. There were, and perhaps still are, bytes that are not always 8 bits long. But all octets are exactly 8 bits long, by definition.

The 16 bits of overhead in the European digital carrier standard was organized into 2 octets. The first octet was located at the beginning of the frame and allowed sending and receiving equipment to stay in timing synchronization for the entire duration of the frame, regardless of how many consecutive 0s were encountered in adjacent DS-0s. This framing "word" (yet another term for a group of 8 bits) was followed by the octets from 15 DS-0 digital inputs. Then a group of 8 bits used for signaling, alarms, and maintenance was inserted in the frame. The frame concluded with another series of octets from another 15 DS-0 digital inputs. The whole frame structure yielded 30 DS-0 64 kbps channels with 2 overhead channels, also functioning at 64 kbps. But this allowed the European digital carrier system to offer the full 64 kbps channel to any application.

This frame structure gave an aggregate bit rate of $32 \times 64,000$ bits/second = 2,048,000 bps = 2048 Mbps. This became the Conference of European Postal and Telecommunications (CEPT) administrations standard. This was designated E-1 to distinguish it from the T-1 frame structure of 1.544 Mbps; the structure of the E-1 frame is shown in Figure 4.4.

Digital Hierarchies

T-1 was just the beginning of a whole hierarchy of multiplexing schemes. It made sense (and saved money) to aggregate larger and larger groups of digital channels onto a single trunk or transmission channel. After all, the channels in the trunk were all going between the same two points anyway. This hierarchical structure of course was not unique to digital multiplexing. Analog multiplexing hierarchies based on FDM (such as L-carrier) had existed, and continued to exist, side by side with digital multiplexing based on TDM. Unfortunately, the differences between T-1 and E-1 frame structures translated to differences in their higher levels of multiplexing as well. Even more confusingly, other nations, which had adopted the basic T-1 frame structure (e.g., Japan), organized higher levels differently.

To sort all this out, it became common to refer to the first-level digital carrier hierarchy as T-1 in the United States (technically North America), E-1 in Europe, and J-1 in Japan. (There are some other differences between North American and Japanese carrier systems beyond the

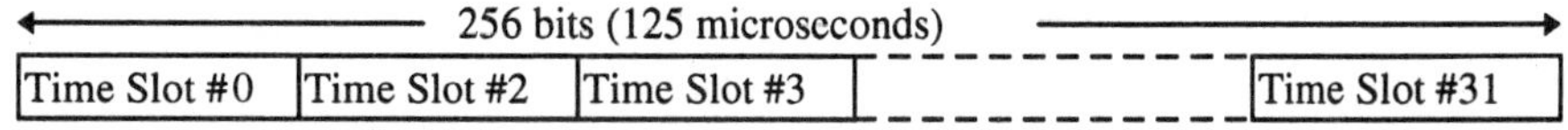

Figure 4.4
The E-1 frame structure

	256 bits (125 microseconds)				
Time Slot #0	Time Slot #2	Time Slot #3			Time Slot #31

scope of the present discussion.) This terminology extended to all levels of the hierarchy as well. Table 4.1 shows all of the defined levels of digital TDM multiplexing in use around the world. (This table also appears at the beginning of this book for ease of reference.)

The use of the asterisk to indicate "intermediate multiplexing rates" simply means that the indicated level of the hierarchy was formed primarily to act as a intermediate stage on the way to some other level. For instance, the North American DS-1C was formed primarily to act as an intermediate stage when going from a DS-1 to a DS-2. Until relatively recently, these intermediate levels were not seen as user interfaces or even on the transmission network itself.

Of course, it is nice to have all of the world's digital hierarchies in one place for reference purposes. However, SONET was designed and developed primarily to replace the existing T-carrier hierarchy in North America. This is not to say that SONET is not an important part of the

TABLE 4.1

Digital TDM hierarchy used in North America, Europe, and Japan

Digital Multiplexing Level	Number of Equivalent Voice Channels	Bit Rate (Mbps)		
		N. America	Europe	Japan
DS-0/E0/J0	1	0.064	0.064	0.064
DS-1/J1	24	1.544	—	1.544
E1	30	—	2.048	—
DS-1C/J1C	48*	3.152	—	3.152
DS-2/J2	96	6.312	—	6.312
E2	120	—	8.448	—
E3/J3	480	—	34.368	32.064
DS-3	672	44.736	—	—
DS-3C	1344*	91.053	—	—
J3C	1440*	—	—	97.728
E4	1920	—	139.264	—
DS-4	4032	274.176	—	—
J4	5760	—	—	397.200
5	7680	—	565.148	—

(*) Intermediate multiplexing rates.

full international standard. The fact that SONET is primarily intended for use in North America merely acknowledges SONET's origin and main sphere of interest.

It may be instructive to describe the levels of the T-carrier portion of the hierarchy used in North America. These are all digital multiplexing techniques using time-division multiplexing (TDM). The T-carrier hierarchy is detailed in Table 4.2.

The term *intermediate multiplexing* rates again refers to the same practice as before. DS-3Cs were primarily formed to carry digital signals on digitized microwave links and to carry North American television signals on older fiber optic networks. For many years, events like Monday Night Football carried the video portion of their signal on a nationwide DS-3C network from interexchange carriers such as AT&T or MCI.

The hierarchy begins with the basic building block of a single voice channel digitized at the international standard rate of 64 kbps (which is 0.064 Mbps). No transmission facility is associated with this rate, which is known as DS-0 (digital signal level 0). Twenty-four of these DS-0s are multiplexed together with the TDM method to become a DS-1. In this case, the transmission facility is known as the T-1 (T-carrier Level 1). T-carrier describes certain factors, such as the copper wire pairs or coax cable characteristics and the transmitter and receiver specifications, for the whole hierarchy. The DS part describes the signal format on these physical links.

Most private networks in organizations today are built from leased T-1s and T-3s, or DS-1s and DS-3s. In either case, thousands of organiza-

TABLE 4.2

Digital TDM hierarchy used in North America

Digital Multiplexing Level	Number of Equivalent Voice Channels	Bit Rate (Mbps)
DS-0	1	0.064
DS-1	24	1.544
DS-1C	48*	3.152
DS-2	96	6.312
DS-3	672	44.736
DS-3C	1344*	91.053
DS-4	4032	274.176

(*) Intermediate multiplexing rates.

tions now have at least one T-1, and many have 10 or more. T-3s are less common, but are selling well today.

T-2 has begun to be sold, mostly for LAN connectivity. T-3C exists partly because it fits nicely into a microwave tower. Note that the "C" is always capitalized and stands for "concatenation." A "concatenated T-1" (T-1C) is two T-1s "pasted" together for transmission purposes. No one has ever seen a standard DS-4 or T-4. The standard was never developed and most service providers have tried to fill in the gap with various proprietary ways to aggregate multiple DS-3 signals onto a fiber optic link.

T-carrier links do not have to be channelized into a given number of voice channels. It is possible to have an *unchannelized* T-1, which offers an unstructured, raw bit rate of 1.536 Mbps (and, in a few cases, even the full unframed 1.544 Mbps).

Proprietary T-3 and Above

Because leased line T-carrier networks give a user "all of the bandwidth all of the time," the philosophy of service providers has always been to sell "pieces" of bandwidth delivered on higher-bandwidth links farther up the digital multiplexing hierarchy. For instance, in the early 1980s, when T-1 was the most common digital link used internally between switching offices on the service provider network, the highest commonly available rate was the basic DS-0 line running at 64 kbps.

By the mid 1980s, service providers (then almost exclusively the local and long-distance telephone companies) began to offer full T-1 rates transmitting at 1.5 Mbps. These T-1 connections could either be provided channelized into twenty-four 64 kbps DS-0s or unchannelized as an unstructured data steam transmitting at 1.5 Mbps. It made little difference to the service providers because the entire 1.5 Mbps could no longer be used by the service provider for anything other than the transport of the customer's bits.

Of course, by the late 1980s, users began to experience pressure to be able to go even faster. Multimedia, networked applications, imaging, and the raw horsepower of LAN-attached desktop computers began to overwhelm even a full T-1 in many cases. The next logical step beyond T-1 in the digital hierarchy was T-3; T-2 equipment existed, but mainly as an intermediate step to T-3. However, if a service provider were going to offer T-3 speeds to customers, for technical and economic reasons the service would need to be based on something that operated even faster than T-3. The problem was that no standard really existed beyond T-3, in spite of the existence of T-4 on paper, and a standard fiber optic interface for T-3 was never fully completed and standardized.

This did not stop many service providers from more or less inventing their own ways of carrying and delivering bit rates above T-3 over their networks. All of these methods were nonstandard and effectively proprietary, meaning that a piece of sending equipment from one vendor had to be matched by a piece of receiving gear from the same vendor at the other end of the link. Because the user was not buying the "T-4," but rather the T-3s (or multiple T-1s) that it carried, this was not a bad solution.

Even the T-3s became the target of much activity. Because no standard fiber optic interface existed for T-3, many service providers became creative with how a T-3 signal operating at 45 Mbps was picked up from a customer site or switching office and carried internally on the service provider's network. There was nothing wrong with this approach. After all, users did not care much about how their bits got from place to place on the networks, only that they got there.

Many proprietary T-3 and above digital multiplexing schemes were in use by 1990 in North America; these are listed in Table 4.3.

Most of the levels listed in Table 4.3 were implemented on fiber optic cable, and all are tagged with a nonstandard designation. The terms applied to the levels in the digital hierarchy are in quotes because of the lack of a standard with these carrier systems.

TABLE 4.3

Proprietary T-3 and above links in North America

Carrier Multiplexing Level	Number of Equivalent T-3s	Bit Rate (Mbps)
N.S. T-3	1	48-50
N.S. T-3C	2	90
N.S. T-3C	3	135
N.S. "T-4"	9	405 or 432
N.S. "T-4"	12	560 or 565
N.S. "T-4C"	18	810
N.S. "T-4C"	24	1,100 (1.1 Gbps)
N.S. "T-4C"	36	1,700 (1.7 Gbps)

(N.S. = Nonstandard)

As long as these digital systems were used internally, no real problem with the proprietary nature of the technology existed. However, if any of these systems were to be offered to customers as leased private lines (and expensive ones at that), the equipment located at the customer site would be considered as customer premises equipment (CPE). As such, the service provider had no control over the customer's vendor selection for this equipment. However, if the proper vendor were not chosen—and in many cases, a low bid process or other open supplier procedure had to be followed—then the link would not work. Thus, these proprietary levels of the digital hierarchy were never offered for customer leasing.

The lack of a fully standardized T-3 rate and above method for digital multiplexing and transporting high-bit rate data streams was one of the problems that ultimately led to the development of SONET. Nevertheless, the use of all levels of the T-carrier hierarchy has become the accepted way to build private networks.

T-Carrier and Private Networks

As strange as it may seem at first, most major corporations in the United States today (and in many others around the world) are little telephone companies. People do not often think of the corporate networking department in this way, but corporations in the last 20 years have not looked to public telephone companies for networking solutions to corporate networking problems. Rather, they have built their own private networks to provide custom networking solutions.

The difference between a "public" and "private" network is an important one, especially when it comes to a discussion of SONET. All networks consist of a number of simple elements. One of these elements is the "end system" itself: the source and/or destination of the information traveling across the network. Another element includes the "network nodes" that reside within the network and forward the traffic from an input port to an output port. Information makes its way across the network in a series of "hops" from one network node to another from the source end system to the destination end system.

In a public network, the service provider owns and operates the network nodes. These network nodes may be switches or routers or even some more exotic network device. The key is that the user/customer need not own or operate any of the network nodes. The service provider takes care of all of this on behalf of the customers who are usually called subscribers on a public network (such as a frame relay network).

In a private network, conversely, the customer owns and operates the network nodes. If any role exists for the public service provider, it would be to supply the bandwidth needed to link the customer-owned-and-operated network nodes together. These "private leased lines" are really just rented pieces of the telephone company network that supply the bandwidth needed for this private network node connectivity.

Telephone carriers in the United States have more or less accepted their role as simple bandwidth providers for their corporate customers, at least for the present. Of course, corporate customers have had to assume a larger role in building, operating, and managing private networks. In this sense, corporations have had little choice but to become small telephone companies on behalf of their employees.

The origins of this trend are no mystery. Even before the rise of large corporate data networks, corporate voice networks evolved to the point where it made sense to have a small voice switch on the corporate premises. This was the origin of the "private branch exchange" (PBX). The term "private" accurately reflected the migration of the "exchange" (voice switching) function from being the responsibility of the public telephone company to being the responsibility of the private corporation. The resulting voice network was small (a "branch" office of the larger public telephone switch to which it was still attached), but the message was clear: A service that was usually provided by a public service provider was now supplied to customers (i.e., employees of the corporation) by a private network. Another arrangement, called *outsourcing*, treats another's private network similarly to a public one.

When data networks became common in the corporate environment, which usually meant the deployment of an IBM SNA network, the trend continued. There was no longer any question of obtaining network services for SNA networks from a public network service provider because IBM sold front end processors (FEPs) directly to corporate customers. These FEPs formed the network nodes of the IBM SNA network. Corporate data managers liked the control they enjoyed over the corporate network, and because the bandwidth had to be purchased (i.e., leased) from the telephone company anyway, the public telephone company profited without needing to understand SNA data networking.

Outside of the United States, there were more determined efforts to provide public data networking services. The X.25 international standard for public packet-switched data networks, explained in a little more detail later in this work, was a major step. Bandwidth in the form of leased private lines was much more expensive and facilities scarce outside of the United States for economic and policy reasons, and X.25

enjoyed mild success, especially in Europe. Within the United States, the practice of building private corporate networks had become such a part of the corporate SNA networking environment that efforts on the part of telephone companies to promote public X.25 services largely failed. IBM helped by pricing X.25 support on SNA network so high that it was much cheaper to lease point-to-point private lines to link FEPs together.

Linking LANs with Private Lines

Before the mid 1980s, corporate networks usually consisted of a hierarchical SNA network which employed low-speed, analog- and digital-leased private lines. In many cases, the SNA network spanned the entire organization, coast-to-coast, and the cost of leasing these private lines by the mile from telephone companies was controlled by using a *multidrop* arrangement. With a multidrop circuit, many remote SNA cluster controllers could share a single port on the mainframe's FEP. Because the remote cluster controllers in an SNA network could not communicate directly, all traffic to and from a particular site had to pass through the mainframe location. The use of multidrop lines was an elegant solution for the modest amounts of text-based information sent over an SNA network from terminal device to host application and back again.

By the mid 1980s, many companies were replacing desktop terminals with PCs, which could do more for an individual employee than provide access to the corporate mainframe. To be sure, mainframe access was still an important part of the employee's job function. In fact, most workers simply used the desktop PC to access the corporate mainframe applications they needed to perform their jobs. The PC did not magically make applications and data jump off of the mainframe onto the desktop PC. A whole class of PC applications known as *terminal emulation programs* evolved to meet this continued need for central computer access. With terminal emulation, a PC user could still be wired to the cluster controller for mainframe access.

The corporate networking environment situation changed in the 1980s, quickly and dramatically. Most private corporate networks had been built to accommodate the IBM SNA environment. Even when other computer vendors' networking products were used (e.g., DECnet and HP networking products), the resulting networks looked much the same as SNA networks: hierarchical, star-shaped, wide area networks (WANs).

These networks typically linked a remote corporate site to a central location where the IBM mainframe or other vendor's minicomputer essentially ran the whole network. In this sense, the networks were hierarchical, with the central computer forming the top of a pyramid and the remote sites only communicating, if at all, through the central computer. Building private networks for this environment was easy and relatively inexpensive. Private lines were leased from the telephone companies to link all the remote sites to the central site. Although private lines were leased by the mile (i.e., a 1,000-mile link cost more than a 100-mile link), there were various ways around letting this variable expense limitation impose too much of a burden on the private network.

This situation is illustrated in Figure 4.5; five sites are networked together with point-to-point leased lines. The connectivity needed is simple: hook all four of the remote sites to the remaining central site where the main corporate computer was located.

A quick look at the figure makes it easy to see how many links are needed to create this network. Four private leased lines are depicted; fewer cannot be used to link each remote site directly to the central location. These links are the main expense when building and operating such private corporate networks. These links, however, are not the only expense; a communications port must be maintained on each computer in the network. This expense is minimal at each remote site. A glance at the figure shows that each remote site needs only one communications port and related hardware. The situation is different at the central computer site where the communications port and related hardware needs to link to each of the remote sites. Of course, the central computer must be configured with the total number of ports needed for remote site connectivity. This last requirement usually was not a problem, especially in the IBM mainframe environment.

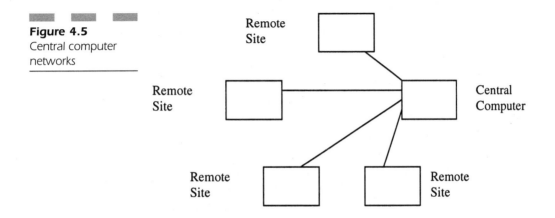

Figure 4.5
Central computer networks

However, what if the number of remote sites grew to ten? Twenty? One hundred? How many links and ports would then be needed to deploy such a hierarchical network? As corporations—and corporate networks—merged, expanded, and otherwise grew throughout the 1970s and into the 1980s, this issue became a problem for corporate data network designers.

Fortunately, it is not necessary to draw detailed pictures of these larger networks to determine how many links and ports are needed. A simple mathematical relationship can be used to determine how many links and ports are needed to link any number of sites into a hierarchical, star-shaped network. When the number of sites is designated by the letter "N" (including the central site), then the number of links needed would be N–1. For instance, in the above figure, N = 5 and the number of links needed is N–1 = 4. The number of communication ports needed throughout the network is given by 2(N–1) (read as "2 times N–1"). When N = 5, the number of ports is 2(N–1) = 8. Fully half of these ports (given by N–1 = 4) would be needed at the central site.

It is now easy to figure out that when the number of sites is 20 (N = 20), the number of links would be 19 (N–1 = 19) and the number of ports would be 38 (2(N–1) = 38), with 19 of them (N–1) at the central site. If N = 100 (100 locations on the corporate network, a figure easily approached in larger corporations and governmental agencies as well), the number of links would be 99 and the number of communications ports would be 198, with 99 at the central site. These networks were large and expensive, but not prohibitive.

What has all of this to do with T-carrier? Simply this: The rise of corporate LANs and client-server computing in the 1980s has meant that building private corporate networks in hierarchical stars is no longer adequate for private corporate networks. This is not the place to discuss the evolution of LANs and client-server in detail. It is enough to understand that LANs and PCs running client-server software (e.g., database client package to a database server or even corporate e-mail applications) are best served by networks with more than just a single link to some central location.

In a client-server LAN environment, it can no longer be assumed that all communications occur between a remote site and a central location, as hierarchical networks assumed. In a client-server environment with LANs connected by WANs, any client may need to access any server, no matter where the client- or server-PC happened to be located. Client-server LANs at corporate sites that need to be connected are better served by peer, mesh-connected networks.

This need for a different type of private corporate network created problems. The number of T-carrier links and ports needed for peer, mesh-connected private LAN networks were much higher than the modest link and port needs in the older hierarchical, star environment.

To see why this is true, Figure 4.6 illustrates the problem. A mesh network consists of direct point-to-point links between every pair of locations (LANs) on the corporate network. This way, no client and server is separated by more than one direct link; however, the associated numbers for the required links and ports have exploded.

The figure demonstrates that the peer network requires ten links and four communications ports at *each* location to establish the required mesh connectivity. The total number of ports needed is now twenty. The formulas, again based on well-understood mathematical principles, are now $N(N–1)/2 = (5 \cdot 4)/2 = 10$ for the number of point-to-point links needed to mesh connect $N = 5$ sites and $N(N–1) = (5 \cdot 4) = 20$ for the total number of communications ports needed (four at each site).

For twenty sites ($N = 20$), the numbers would include $(20 \cdot 19)/2 = 190$ for the links and $20 \cdot 19 = 380$ for the ports (19 at each site). Although it would not be impossible to configure 19 ports for each site, the hardware expense alone would be enormously high, even prohibitive. Most network managers would balk at providing 190 WAN leased lines paid for by the mile to link the sites together. For 100 sites ($N = 100$), the numbers would require an astonishing 4,950 links $((100 \cdot 99)/2)$ and 9,900 communications ports $(100 \cdot 99)$. Each site would need hardware to support 99 links to other sites, an impossible task for any common communications device architecture.

Again, various strategies were employed by corporations to keep LAN connectivity networking expenses in line. Partial meshes were deployed

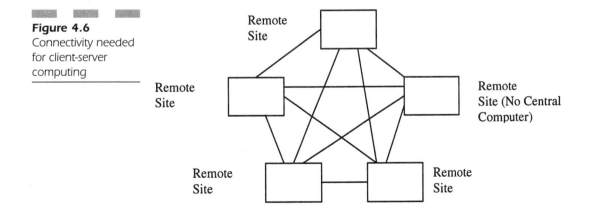

Figure 4.6
Connectivity needed
for client-server
computing

Remote
Site

Remote
Site

Remote
Site (No Central
Computer)

Remote
Site

Remote
Site

in varying "backbone" and "access network" configurations. These measures were quite successful, in the sense that these arrangements satisfied connectivity requirements at a minimal cost.

T-Carrier and Network Applications

The above section mentioned some of the "workarounds" that corporate data communications managers have employed to try to avoid the expense involved with creating mesh-connected LAN internetworks. These strategies all involved paring down the number of links and ports needed so that even if all the sites were not point-to-point directly connected, at least each site was reachable from any other site. In most cases, this meant that any client could still attach to and access data on any server, but not always directly. This put more of a traffic load on some network links and network nodes. Higher speed T-carrier links offered relief from this situation. Instead of a few 64 bkps links between sites, a full, unchannelized T-1 running 1.5 Mbps was normally employed.

There is another reason that T-carriers, especially full T-1 links operating at 1.5 Mbps, have been employed for many of the applications that users run on LANs in a client-server environment. The reason is referred to here informally as "application overload," which means that LAN speeds and application data exchange needs have made the WAN portion of the corporate data network an automatic bottleneck that slows these networks to a crawl when the link bandwidth is not adequate.

Initially, the new and powerful PCs in an organization were used mainly to access applications and data on the mainframe running terminal emulation. This worked, but hardly took advantage of the computing power now available on the desktop. Fortunately, this computing power did not go untapped for long. Organizations began to develop—or even purchase off-the-shelf—applications that operated and accessed data available locally, within the department in which the employee was located. This movement toward *departmental computing* led to a separate need to network these PCs not throughout the organization, but within individual departments. The LAN was a perfect solution. LANs provided networks for many intelligent devices (PCs) at a single location, rather than the WAN connectivity provided by the SNA network.

By about 1990, a whole new way of looking at corporate applications and the networks they ran on had evolved. The need to share information and resources among users led to the development of the idea of distributed processing, most often along the lines of the client-server model. Although some of these terms have no precise definitions, it is generally agreed that sharing "information" refers to the need for PC-based applications to sometimes access data stored on another computer, whether local or not. Shared "resources" refers to the expensive hardware that PC users sometimes need to complete their jobs, hardware that could not effectively or affordably be provided to each individual PC user. Not long ago, both modems and laser printers fell into this "shared resource" category. Falling hardware prices have served to make both of these pieces of computing equipment common on most desktops, but other devices that are new and expensive fit into the sharing category, such as color laser printers and DVD-ROM writing devices.

Distributed processing acknowledges the fact that desktop PCs are powerful enough to perform work-related functions that were previously reserved for the central computing site. There are various forms that any movement toward distributed processing or computing can take. One of the most common is the employment of the client-server model for distributed processing. In the client-server model, most desktop PCs are classified as clients, meaning these PCs run some form a LAN Network Operating System (NOS) software. The NOS (for example, Novell's highly successful NetWare product) allows the client PC to attach to another kind of PC device known as a server over the LAN to which both devices must be attached. The server runs the bulk of the NOS software, and clients can only communicate among themselves by first sending data or messages to the server, which forwards the data or messages across the LAN to the correct destination client PC.

The whole scheme is quite elegant and takes advantage of the tremendous processing power available on each desktop in a modern corporation. A problem arose with this idea when departments with similar functions had a need to share information and resources not over an individual LAN, but with other remote sites in the same company. For example, a national company often had a need to share information between a client on the west coast and a server on the east coast. This commonly occurred when a corporate customer located on the east coast was traveling on business to the west coast. An order placed to the west coast sales office meant that the sales representative had to access the customer's sales account information on the east coast server.

The most efficient and effective way of doing this was to allow the west coast client to access the east coast server. This meant that the LANs in both locations had to be networked together over a WAN. Because the applications that had been developed in the client-server environment typically used a Windows-based graphical user interface (GUI) rather than the older text-based terminal interface, the amounts of data exchanged by a relatively simple sales transaction were much larger than anything that the corporate SNA network had been designed to handle.

As a result, in most cases separate WAN links were leased and used to connect special LAN internetworking devices together at the various locations needing such connectivity. These special network devices could be LAN bridges or routers, but as time progressed the LAN internetworking device of choice became the router, for a variety of reasons beyond the scope of the present discussion. A new philosophy and terminology of corporate networking arose in this environment, which is most commonly known as "router-based networking."

The most critical networking consideration in router-based networking is the need for efficient router connectivity across the corporate WAN. Only then could the logical conclusion of the client-server model ("any client can access any server") be realized. But this led to a new series of challenges for the builder and managers of this new breed of networking.

The number of PCs attached to corporate LANs increased rapidly. The number of LANs needing connectivity and the associated corporate support across a WAN increased rapidly as well. It quickly became apparent that client-server LAN traffic was not the same as SNA traffic. LAN traffic is much more diverse, of a much higher volume, and much harder to predict in terms of traffic patterns than is the traffic on SNA networks.

As the day-to-day operational applications in corporations became more dependent on the LAN-based client-server applications, LAN internetwork traffic grew dramatically. The level of network complexity grew rapidly as well because client-server router-based networks are not well served by either low-speed multidrop lines or hierarchical, star-shaped WANs. The needs of LAN applications led to attempts to connect each router with a direct link to every other router. As the router network complexity grew, it became impossible to keep up with client-server demands, especially as more SNA devices came to be attached to LANs as well.

A newer 1990s corporate trend developed Internet-based corporate services. These newer applications often needed the services of a more powerful UNIX-based workstation to provide the company with a "Web site" presence on the logical subset of the public Internet known as the World Wide Web. Other companies relied on UNIX-powered workstations for more processor-intensive applications, such as computer-aided design/computer-aided manufacturing (CAD/CAM) that remained beyond the abilities of the desktop PC. New multimedia applications requiring a network handling not only data, but audio and video were developed for both PC and UNIX computing platforms. The TCP/IP protocol played a large part in this as well.

In this environment, routers were best served by "mesh connectivity" with direct point-to-point leased lines to every other router site. The numbers of communication ports needed on each device grew as well. A compromise solution deployed "partial mesh" and "router backbone" networks, but the amount of LAN traffic quickly exhausted the capabilities of these compromises to deliver the speed and bandwidth that LAN-based client-server user applications demanded.

Older LANs built in the early 1980s operated at the familiar Ethernet and Token Ring speeds of 10 megabits (millions of bits) and 4 megabits per second (Mbps), respectively. By the mid 1980s, Token Ring LANs operated at 16 Mbps. Of course, these LANs had to be connected over a private corporate WAN with leased lines from the telephone companies. The most common affordable speed for these links was 64 kilobits per second (Kbps, thousands of bits per second). One million is 1,000 times larger than one thousand; therefore, trying to let LANs running at millions of bits per second communicate over WAN links running at only 64 kbps created a severe congestion problem on the corporate network, with the associated delays that exasperated users.

Even when a corporation employed a full T-1 operating at 1.5 Mbps between LAN sites, the bottleneck created trying to link two Ethernet LANs running at 10 Mbps over the T-1 was still about 6 to 1 (6 times 1.5 Mbps = 9.0 Mbps). There was only one sure way to make this type of network functioned even marginally acceptably: reduce the amount of information that needed to be sent over the WAN portion of the network.

Unfortunately, this philosophy went against the trend of the applications the users were running that led to the need for LAN internetworking in the first place. By the early 1990s, these applications were no longer just sending modest amounts of data bits from client to server and back across the WAN. Company employees were looking at including multimedia in their LAN applications. Multimedia combined images, graphics, audio, and video with the traditional data into new and exciting applications for sales and marketing, research and development, and even information systems departments.

The distinguishing characteristics of multimedia applications in a client-server environment are two-fold. First, multimedia applications need plenty of horsepower on every client desktop to deliver realistic audio and full-motion video to the user. The rapidly falling cost of PC hardware guaranteed that this would not be an insurmountable problem for most organizations, either for the clients or the servers. Second, multimedia applications need plenty of bandwidth to deliver the much greater amount of bits these applications required, and within a smaller network delay time frame. Users who would wait patiently for about 10 seconds for a database record or Web page would not stand for 10 "freezes" of video or 10 seconds of silence during an audio playback.

Furthermore, the whole philosophy of distributed, client-server applications prevented the organization from simply duplicating information at every LAN site in the organization. The resulting complexity of propagating updates everywhere, and the sheer volume of the updates sent over the network in this case, easily offset any gain realized from cutting back on transporting the multimedia information across the overburdened WAN links. By the mid 1990s, the problem had only gotten worse.

Even LANs running at 10 and 16 Mbps were being swamped by many users all running the newer applications over shared-media LANs which were, after all, based on early 1980s technology. Even when multimedia applications were not an issue, newer PCs were so fast that they could easily fill up a 10 or 16 Mbps LANs with traffic from only a handful of users.

LAN network managers responded with one of two solutions, and sometimes even employed both in many cases. First, so-called "switch LANs" appeared, giving each user attached to the LAN essentially a dedicated 10 or 16 Mbps right to each PC. The result was that there were even more bits waiting to travel across the WAN link to remote locations, in some cases 10 times as many bits or more. Second, newer LAN technologies appeared that ran at 100 Mbps, turning a barely acceptable LAN-to-WAN 6 to 1 bottleneck into an unbearable 60 to 1 bottleneck at the LAN-WAN interface.

These network devices employed at the LAN-WAN interface were usually routers, which were capable of linking two separate LANs over a WAN link. Sometimes bridges were used for the same purpose, and although the two devices behaved differently, both enabled clients on one LAN to access a server or servers on another LAN at a remote location.

A much better fit for 10 Mbps Ethernet and 16 Mbps Token Ring LAN connectivity would be T-3 running at 45 Mbps. Many organizations have begun to use these higher speeds for this purpose. Until recently, T-3 were relatively rare and expensive (in part because of their scarcity) and so were not easily used to link LANs together.

Many modern applications need more bandwidth than is easily affordable or available to build the types of LAN-WAN-LAN connectivity required for distributed, client-server environments, especially when faster LANs are factored in. Additional problems arose with the T-carrier structure which became more apparent in the late 1980s.

The Trouble with T-Carrier

By the 1980s it became obvious that drawbacks existed with the digital hierarchy that were neither debilitating nor trivial, yet serious enough to require action. T-carrier worked well within certain limitations, but given the accelerated pace of desktop device development and LAN speeds and capabilities, problems needed to be addressed quickly. Accentuating the problem was the fact that more often, T-carrier was used to connect LANs in an organization.

T-carrier was basically a single vendor world at the time. The Western Electric Company (WECO), the manufacturing arm of the Bell System, made T-carrier equipment for all components of the Bell System. It was difficult to obtain accurate performance statistics for these T-carrier links, and they were awkward to troubleshoot .

The Mid-Span Meet and Missing Pieces

The strange fact about the T-carrier hierarchy is that it was never completed. Annoying gaps in the original Bell Labs specifications existed, some of which have been mentioned above. For instance, no DS-3 method of sending a T-3 signal over a fiber optic link was fully specified and deployed. Therefore, even a simple thing, such as getting a DS-3 from New York to New Jersey, was often an adventure and costly. DS-4 was in even poorer shape, but even less of it existed at the time. For these reasons, service providers that wanted to deploy the T-carrier hierarchy—which was all of them—had to fill in the blanks in the specifications with their own proprietary methods. As long as the Bell System remained intact and all of these carriers bought WECO, a problem did not exist. After all, how many T-1s would start at Illinois Bell and end up at GTE? Not very many. If worse came to worse, the signal could be "re-analoged" for the interface. This lack of ability to have one vendor's equipment or telco on one side of the link and another vendor's equipment on the other side of the link is known as the lack of the *mid-span meet*. In a mid-span meet, one carrier is responsible for one end of the link and another carrier is responsible for the other side of the link.

However, after 1984, there were a lot of situations where the two ends of a T-carrier link were under the control of two different organizations. With the "equal access" rules of divestiture, local callers had the right to use their long-distance carrier of choice, on a per call basis. This required the use of digital T-carrier trunks from the local exchange carrier's (LEC) central offices to the long-distance carrier's points of presence (POP). "No problem," said the new AT&T (which still included WECO), "Everybody just buy WECO." This philosophy was not well received by anyone.

The question then became one of completing the T-carrier standard or trying to invent something new in the world of divestiture.

The Need for Operational Procedures

Network management, troubleshooting, and other administrative networking functions are needed on T-carrier networks as well as any other type of network. Digital T-carrier links need to be monitored, problems detected and fixed, and so forth. Because the customer has basically leased only the bandwidth that the T-carrier link represents, and not the transmission facilities themselves, this network management function remains the responsibility of the service provider. The customer may initiate the process (usually the customer is the first to know that a problem exists on the link), but the service provider carries out the tasks needed to maintain the network link up to the standard set for the leased line service (errors, availability, and so on).

On a T-carrier network, this network management function is known as Operations, Administration, Maintenance, and Provisioning (OAM&P); outside of North America, OAM&P is usually known as OAM, especially in Europe. The process of OAM&P on a T-carrier network is rather primitive. All OAM&P measures in T-carrier, from alarms to maintenance signals, robbed bits from the user channels. The service providers really had no choice because the overhead in a T-carrier transmission frame (1 bit out of the total 193 bits in the transmission frame, or 0.518%) was not adequate for even the simplest OAM&P tasks.

One of the side effects of this was to limit the effective bandwidth of a DS-0 channel to 56 kbps because one in eight bits in every channel had to be "robbed" for OAM&P (the term "robbed" may be unfortunate, but this is what it was usually called). There were other functions for this "robbed bit," such as signaling and maintaining something called "1s density," but only OAM&P is important in this limited discussion of the robbed bit. Even the simple act of measuring the Bit Error Rate (BER) on a T-1 link (usually because the customer had complained about it) involved taking the link out of service, performing a BER test (BERT), and then returning it to service (many times with a "no trouble found" designation because the problem was frequently in the CPE).

The overhead in the T-carrier network for OAM&P, even at the higher levels of the hierarchy, was totally inadequate for both the customer and service provider needs.

Byte and Bit Interleaved Multiplexing

The problem of bit interleaved multiplexing is fairly complex. However, because it is an important topic in T-carrier, especially when contrasted with SONET, some time will be spent explaining the related concepts of bit and byte-interleaved multiplexing.

Whenever digital data streams are combined to form a higher-rate carrier system (so called because they "carry" many low-speed links in one higher-speed package), the source digital data streams are almost invariably organized as bytes. Bytes are groups of 8 bits (0s and/or 1s), and officially called octets in international standards. This once was not universally true, but in a world where almost everything from data, to voice samples, to video conferencing comes from a computer, the organization of digital information into bytes (or octets) of exactly 8 bits is a standard. Whether the bits come from a desktop PC, or a digital PBX, or a digitized video stream, a computer chip pumps out those bits somewhere.

Now combining these individual source streams involves the actual multiplexing of them onto a higher rate link and of course de-multiplexing them into the composite signals (if desired) at the other end of the link. There are two main ways to accomplish this multiplexing of bits, and both of them are used in the T-carrier hierarchy. Both involve *interleaving,* or merging, because it would hardly be possible to take more than a few bits from one input port and forget about the others, even for a short period of time. The receiver would notice a "gap" in the bits arriving immediately and fail. All inputs must be interleaved, even idle ones, because absence of bits always is interpreted as a network failure by the T-carrier receivers. The question is whether to take the input data stream and interleave by *bytes* or by *bits.*

In byte-interleaved multiplexing, input digital streams are multiplexed by interleaving 8 bits (the byte) from each input port with the bytes from the other input ports. The result from one round of servicing all input ports (active or not) is usually organized into a *transmission frame* that is sent on the link. The term *transmission frame* is preferred because there is less confusion when this term is used with other types of frames generally seen on networks, such as Ethernet frames or Token Ring frames.

In a T-1, the multiplexing is byte-interleaved multiplexing. That is, 8 bits are taken from the 24 input ports on the *T-1 mux* (short for multiplexer) CPE. A T-1 mux is also called a *channel bank* when voice is the primary focus of the equipment. The T-1 mux/channel bank terminology came about when channel banks were positioned as voice-only T-1 muxes and the term *T-1 mux* was reserved for data muxes, a distinction based on technology. It is a frustrating fact of life in both the computer and networking industries that distinctions based on technology never last, but terminology based on technology never dies. So the term channel bank is still heard occasionally. Whatever the device, these input bits are taken from each input port and combined to form a DS-1 frame which is sent over the link. This process is shown in Figure 4.7. The transmission frame has the familiar time slot arrangement characteristic of all time-division multiplexing techniques.

Many nice things happen when byte-interleaved multiplexing is used. First, it is simple and efficient because the processor chips used in the T-1 mux itself move bits around 8 bits at a time, in byte form the same way as a computer would. Second, it is efficient to store bytes in the memory of the T-1 mux and form the frames with a minimum of effort and does not require much memory (only a byte or so on the input side, only a frame or so on the output side). Third, it is easy for equipment

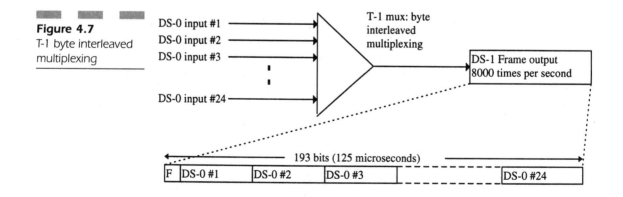

Figure 4.7
T-1 byte interleaved multiplexing

DS-0 input #1
DS-0 input #2
DS-0 input #3

DS-0 input #24

T-1 mux: byte
interleaved
multiplexing

DS-1 Frame output
8000 times per second

193 bits (125 microseconds)

| F | DS-0 #1 | DS-0 #2 | DS-0 #3 | | | DS-0 #24 |

located within the network itself to find an individual DS-0 inside the DS-1 frame. There are several reasons to do this, such as in equipment known as an add-drop multiplexer (ADM).

ADMs are common pieces of equipment in T-carrier networks. Of course, they are not really called ADMs; the most common term is to call these devices digital cross-connect systems (DCS) because these devices do essentially the same thing as a cross-connect panel does in an analog network. A cross-connect panel used to be a piece of plywood nailed to the wall that enabled technicians to manually connect an input link to an output link, usually by "punching down" the wire pairs associated with the signal. Today cross-connects are rack-mounted, high-tech units in many cases, but manual cross-connect panels still exist. A DCS is just a digitized, computerized, manageable, high-tech version of the manual cross-connect that does electronically what technicians ("craft" people in the telephone industry) had to do manually in the past.

A DCS enables a service provider—or customer when the DCS is configured as a CPE device—to *drop* one or more DS-0 data streams and *add* one or more DS-0 data streams to replace them. Another name for this process is *drop and insert*, but the process is the same. This gives more flexibility to the user of the link.

For example, consider an organization with offices in San Francisco, Atlanta, and New York. Like most organizations, the network linking these sites is a T-carrier network made up of leased private lines. Some DS-0 channels handle voice channels from PBXs, others handle data between routers at these locations, but all are multiplexed onto a T-1 (DS-1). The question is how can these sites be linked most efficiently?

The answer varies based on actual user needs; however, it is not unusual to find the organization employing T-1 muxes in San Francisco and New York, and an ADM in Atlanta. Here is why: When a single T-1 is used between San Francisco and New York, then obviously all 24 channels that start on the west coast end up on the east coast. But what about Atlanta? The organization could lease another two T-1s to link both San Francisco and New York to Atlanta, which gives the network the appearance shown in Figure 4.8. This configuration requires three links and, because these links are paid by the mile, this could be an expensive solution, especially when only 10 or so of the DS-0 links are needed in each location. The usual pricing for T-1s from most service providers means that a T-1 is typically less expensive than as few as six DS-0s.

A better solution does exist. Again, the details would depend on the actual network and usage patterns. The organization would need to

Figure 4.8
A network with three
T-1s for connectivity

employ an ADM as a CPE device in Atlanta. The ADM could *drop* 12 (for example) DS-0s from San Francisco in Atlanta and *insert* (or *add*) 12 DS-0 in Atlanta on the way to New York. The same thing would happen to digital signals sent in the other direction. As a result, 12 DS-0s occur between all three locations, but are delivered over only two T-1s. The basic operation of an ADM is shown in Figure 4.9 showing the adding and dropping of only two DS-0s for simplicity.

It is important to note that in the figure, the DCS shares many of the features of the basic T-1 mux. The DCS both demultiplexes and multiplexes, creates DS-1 transmission frames, has memory buffers, and so on. A buffer is merely a special memory area dedicated to an input or output port on a network device. The input and output DS-1s do not need

Figure 4.9
Basic operation of
a DCS

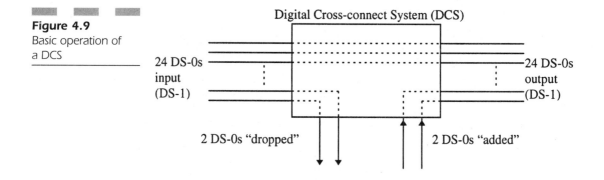

the DCS. It makes no difference to the links whether two of the DS-0 channels going into the device are not the same as the two input channels. "Twenty-four-in" is still equal to "twenty-four-out." Of course, the input frame must be processed to yield the 24 time slots, and rebuilt on the way out of the DCS; however, this happens quickly. Although the remaining 22 DS-0s in the figure are shown passively shuttled through the DCS, a few configuration commands could allow any of these other channels to be cross-connected as well.

Technically, the term "add-drop" or "drop and insert" should be used only when DS-0s are terminated at a site. The term "cross-connect" should be reserved for the process of rearranging or aggregating DS-0s (a process sometimes called "grooming"). However, the terms are used loosely in many instances today.

The network now looks as it does in Figure 4.10. The network is still private, the links are still leased, but the cost to the organization is much less. DCS and its associated adding and dropping of constituent channels is much easier when byte-interleaved multiplexing is used.

Byte-interleaved multiplexing is often called *synchronous* multiplexing, but the term is not universal. The use of the term *synchronous* is unfortunate because it also appears in the acronym SONET. The unpleasant fact is that words such as *synchronous* and *asynchronous* are overused in the telecommunications and networking industry, and often the precise meaning of the term is unclear. In the context of byte-interleaved mul-

Figure 4.10
A network with a
DCS in operation

tiplexing, the term, synchronous refers to the fact that the input clocks must be running at the same speed as the output clock.

Not all multiplexing in the T-carrier hierarchy is byte-interleaved multiplexing. In fact, byte-interleaved multiplexing is the exception rather than the rule. With the exception of a special kind of T-3, explored more fully in Chapter 5, the remaining digital T-carrier hierarchy forms higher bit rate frames, employs *bit*-interleaved multiplexing rather than byte-interleaved multiplexing used when going from 24 DS-0s at 64 kbps to 1 DS-1 frame at 1.5 Mbps. As may be imagined, the term, *asynchronous multiplexing* is often used as a synonym for bit-interleaved multiplexing.

In bit-interleaved multiplexing, the input digital data streams (i.e., voice, video, or data) are not taken one byte (8 bit octets) at a time to form a transmission frame, but one bit at a time. This leads to some interesting—that is, complex and error-prone—situations on the network link. With byte-interleaved multiplexing, bits are accumulated into a buffer (usually only two bytes or so long, but not necessarily) on the input side. When a bit arrives a microsecond of so "late" or "early" (due to network timing inaccuracies) the input buffer can mask these timing differences from the actual multiplexing portion of the equipment that builds the output transmission frame. It makes no difference if 9 bits or 10 bits were actually in the input buffer, as long as 8 bits (the byte) were ready to go into the output frame. T-1 muxes are specially designed so that this will always occur.

Because bit-interleaved multiplexing relies on smaller buffers and feeds only one bit at a time from each input buffer to be sent out with the output transmission frame, there may be times, for example, when a bit is needed from input port 22 and there is no bit in the buffer. Obviously, the position must be occupied by something, or two things will happen, neither of them particularly good from a networking standpoint. First, the receiving multiplexer will not be able to find the bit in the arriving transmission frame to be demultiplexing to port 22 on the receiving side of the link. Second, the total number of bits sent per unit time would fall below the accepted line rate of the link (such as 44.735 Mbps instead of 44.736 Mbps). In either case, the equipment and link will fail. This makes bit-interleaved multiplexing more awkward to handle than is byte-interleaved multiplexing. The reasons for this will be explored in more detail a little later in this chapter. The concept of bit-interleaved multiplexing is illustrated in Figure 4.11, which may be compared to Figure 4.7.

Although Figure 4.11 is accurate in concept, some details are worth noting. First, the framing and overhead bits for a DS-3 frame are much

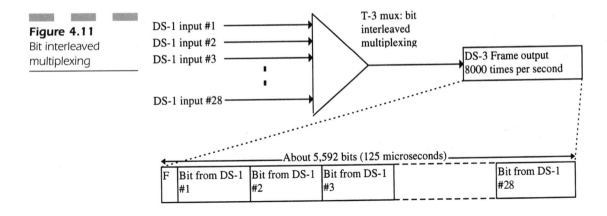

Figure 4.11
Bit interleaved
multiplexing

more complex than is the simple framing that the byte-interleaved DS-1 requires. If a bit were not ready to be sent from an input buffer (i.e., the input DS-1 clock is running slower than the output DS-3 clock), then a bit must be "invented" by the T-3 mux in order to maintain proper frame structure and line speed. These "stuffed bits" must be removed by the receiving equipment because this is not a "live" data bit on one of the T-1s. Much of the added overhead bits in the full DS-3 frame structure are dedicated to allow receiver equipment to find and eliminate these stuffed bits, if employed. Another important point is that the output data rate is about 5,592 bits every 125 microseconds (8,000[th] of a second) because when the whole T-carrier structure was set up in the 1960s and 1970s, clocks in processors had difficulty in dealing with such small time slots. After all, 5,592 bits in an 8,000[th] of a second is only about 0.022 microseconds per bit. At these speeds, the multiplexing equipment had to settle for approximations. There may have been 5,596 bits actually sent in the 8,000[th] of a second, or 5,588. This variation in timing on a digital link and, thus, the number of bits actually sent or received per unit time, is known as *jitter.* Another term for the same effect is *wander.*

Byte-interleaved multiplexing leads to easy byte (or 8 bit octet) handling in equipment memory and results in easy add/drop and cross-connect equipment arrangements. However, the use of bit-interleaved multiplexing is more complex, error-prone, and awkward both for memory buffer handling and the adding and dropping of both DS-1s and DS-0s—how do you easily find the 8-bit bytes when spread around the arriving frame as individual bits? With this known, then why does the rest of the digital T-carrier hierarchy employ bit-interleaved multiplexing?

The answer is that, at the time, no other way existed to do multiplexing at these speeds except using bit-interleaved multiplexing. Also, the cost of the memory used in the buffer was much more expensive than memory costs today. (In 1961, a Meg of memory cost about $4 million.) A T-1 must only buffer 1.5 Megabits to buffer a complete frame. A T-3 mux must buffer 45 Megabits to accomplish the same task, some 30 times more. Also, the simple fact is that jitter is easier to deal with when bit-interleaved multiplexing is used, and dealt with more quickly with fewer buffers. This lack of fast processors and inexpensive memory led to the use of bit-interleaved multiplexing at levels above T-1 in the North American digital hierarchy.

DS-1 to DS-3 Multiplexing

A good deal of the initial work done that led to the development of SONET was done to address issues of DS-1 to DS-3 multiplexing. Therefore, a more detailed look at this process is in order, above and beyond the brief conceptual introduction above.

The biggest liability in the T-carrier hierarchy had to do with the way that TDM worked in T-carrier. The device used in a T-carrier network that went from 28 DS-1s (each with 24 DS-0s in most cases) to a single DS-3 is known as an M13 multiplexer (pronounced "M-one-three"). The M13 multiplexing process involves the combination of 28 DS-1 (1.544 Mbps) signals into a single DS-3 signal (44.736 Mbps). This is done in a two-stage process, the first stage of which is diagrammed in Figure 4.12. Two stages were needed because of the relatively rudimentary electronics used at the time.

The construction of the DS-1 signal is performed in a synchronous (or *isochronous*[1]) fashion by byte-interleaving single-byte voice samples or other input bits from 24 DS-0 (64 kbps) signals into an aggregate signal of 1.536 Mbps (without the 8 kbps framing overhead). In this process, bytes are grouped into transmission frames of 192 bits each (24 channels x 1 byte per channel). These frames are repeated 8,000 times per second. At the beginning of each 192-bit frame, a single framing bit is added bringing the aggregate DS-1 rate up to 1.544 Mbps.

In Figure 4.12, Stage 1 of the M13 multiplexing process involves the combination of four DS-1 (1.544 Mbps) signals into a single DS-2 signal

---

[1]From the Greek "iso" (same) and "chronous" (time). The meaning is then "uniform in time; having equal duration; recurring at regular intervals."

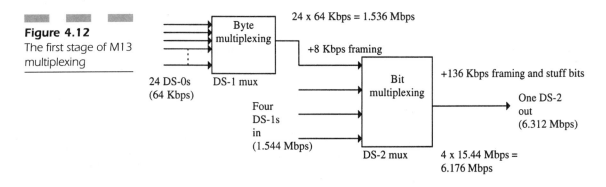

Figure 4.12

The first stage of M13 multiplexing

(6.312 Mbps). In this process, 136 kbps of overhead is added to the DS-1 signals to produce the DS-2 signal. Part of that overhead is for DS-2 framing purposes. The remainder consists of "stuff bits" that are added to the DS-1 data to make up for differences in the timing rate (jitter) of the incoming DS-1 signals. Jitter also affects the "phase" of the arriving frame. Technically, the term for two or more signaling events that occur at *nominally* the same rate (within certain defined limits), but that may occur with any phase difference, is *plesiochronous*[2]. Thus, the DS-1 to DS-2 multiplexing process is said to be plesiochronous. In fact, because the only isochronous piece of the T-carrier digital hierarchy is DS-1, the whole T-carrier hierarchy is often called the plesiochronous digital hierarchy (PDH). This term is much more often used outside of North America, where E-carrier is used.

The important point is that the step from 24 DS-0s to one DS-1 is *byte*-interleaved, while the step from four DS-1s to a DS-2 is *bit*-interleaved. It is easy to find a DS-0 in a DS-1 because all of the bits are grouped together. But it is next to impossible to find a DS-0 (or even a whole DS-1) inside a DS-2 because the bits are scattered throughout the frame and include stuff bits as well.

The second stage of the M13 multiplexing process is shown in Figure 4.13. This involves the combination of seven DS-2 (6.312 Mbps) signals into a single DS-3 signal (44.736 Mbps). This is also a plesiochronous process. Of the 552 kbps of overhead that is added to the DS-2 signals to produce the DS-3 aggregate, much of it consists of "stuff bits," which ensure rate and phase alignment of the incoming DS-2s.

When the M13 multiplexing is complete, T-3 data streams are often multiplexed with additional T-3s and transported over a single fiber

---

[2]Literally "nearly the same." Two bit streams having nominally the same bit rate, but controlled by different clocks, result in bit rates that are nearly the same, but not quite.

Figure 4.13

The second stage of M13 multiplexing and E/O conversion

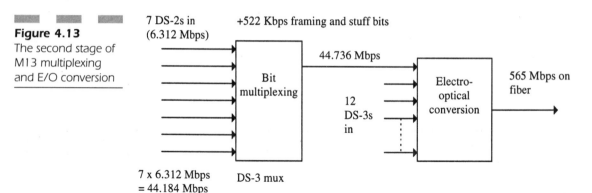

7 DS-2s in
(6.312 Mbps)

+522 Kbps framing and stuff bits

44.736 Mbps

Bit multiplexing

12 DS-3s in

Electro-optical conversion

565 Mbps on fiber

7 x 6.312 Mbps
= 44.184 Mbps

DS-3 mux

optic link, but this is a nonstandard process. Therefore, this process is proprietary (i.e., different vendors' equipment complete the process in various fashions) and necessitates the need for matching equipment at both ends of the fiber span.

Figure 4.13 also shows an electro-optical conversion process that combines 12 DS-3 signals (metallic) into a single 565 Mbps optical carrier signal. Although the example shown on the visual is not hypothetical (i.e., equipment that performs this process exists), many other variants of this electro-optical conversion process also exist. Unfortunately, the example shown, along with the aforementioned variants, is proprietary to each particular equipment manufacturer.

Problems with M13 Multiplexing

The M13 multiplexing process has several inherent disadvantages. From a user perspective, the most important of these disadvantages is that no fiber optic carrier standard existed at rates above DS-3. In today's environment of bandwidth-hungry, distributed applications (e.g., video, imaging), 44.736 Mbps simply falls short of user requirements and provides a strong impetus for development of standards above the DS-3 rate, such as SONET.

From a carrier perspective, the lack of electro-optical standards is equally problematic, especially in the United States. With no optical transmission standard, carriers that purchase equipment from different vendors are unable to interoperate at interface points. This lack of a "mid-span meet" standard was one of the factors that prompted the development of the SONET standard.

Due to the plesiochronous nature of the M13 multiplexing process, full demultiplexing is required to access an individual DS-1 stream within the DS-3 aggregate. Bits that were stuffed in the multiplexing process must be destuffed in the demultiplexing process. In the M13 process, two stages of destuffing/demultiplexing are required. Provision of enhanced services, such as DS-1 cross-connection and DS-1 add/drop, are thus rendered difficult and expensive in the M13 environment.

Finally, OAM&P interfaces and systems in the M13 environment are typically vendor-specific. Craftspeople need to be trained on a number of such systems to become proficient in fault diagnosis and repair operations. In addition, even when equipment is purchased from a single vendor, OAM&P systems may not be fully integrated; that is, a craftsperson performing a diagnostic operation on an M13 multiplexer may not be able to "see through" its associated electro-optical converter to another M13 multiplexer.

What a Concept!

This chapter concludes by tying the advantages of fiber optic network links discussed in the previous chapter to the nagging and persistent shortcomings of T-carrier links. Why shouldn't service providers struggling with the limitations of T-carrier use the advantages of fiber optics to offset the drawbacks of T-carrier networks?

The advantages of fiber could be employed as a basis for a family of scaleable, standard transmission systems and equipment. It would be more flexible, and easier and cheaper to install, run, and maintain. Such a network would use optical transmission with synchronous multiplexing in a fully standardized fashion. This network would be a synchronous optical network—SONET!

Importance of SONET

The drive toward higher capacity transmission systems with low-deployment costs helped lead to the development of the SONET standard. This standard has several key features. First, the SONET standard defines a transmission hierarchy that begins where the current digital hierarchy ends; in T-carrier systems (in North America) that rate is 45 Mbps, in ETSI systems (in Europe) that rate is 139 Mbps. The second feature of the SONET standard is the synchronous nature of its multiplexing system.

Lower bit-rate channels (called "subrate channels" in SONET) can be identified, extracted, or inserted by a single device without the need to demultiplex the higher rate SONET signal. This allows simplified add/drop capability. The third feature is SONET's support for OAM&P systems. Bandwidth is allocated for use as a common OAM&P transmission path. This allows each element in the transmission system to communicate with a centralized management system, or to distribute the management responsibility throughout the elements. This extensive OAM&P support could conceivably redefine the way communication networks are managed and the way these networks provide services to customers.

How SONET
Works

P art One of this book introduced how fiber optic transmission links and the T-carrier digital hierarchy used in North America function in a private line network. These chapters concluded with the thought that the standardization of fiber optic transmission would both take advantage of all of the positive features of fiber optic systems and overcome all of the drawbacks that were becoming more obvious in the T-carrier world. However, an idea is not reality. Much hard work remained to be done before SONET was created as a standard fiber optic transmission system.

This section of the book explores how the standardized fiber optic transmission system eventually gave birth to SONET, and details some of SONET's operation. SONET has a very detailed architecture and series of protocols to implement this architecture. Key aspects of this architecture include considerations of network timing and synchronization. Some of this information was introduced briefly during the survey of T-carrier in Part One of this book. This is the time to explore this architecture and timing structure in more detail, at the risk of introducing more technical terms and relatively advanced concepts. These terms and concepts are discussed on a nonmathematical basis, however, because details about things like network synchronization—while crucial to network operation (especially SONET operation)—remain somewhat arcane to users and even technicians.

This part of the book is organized into four chapters. Chapter 5 forms the basis for a high-level description of SONET functioning. It also investigates the formation of the SONET standard from important forerunners, such as SYNTRAN. SYNTRAN is discussed and contrasted with T-3 (really M13) multiplexing as a first step on the road to SONET. An important section of this chapter explains why differences exist between North American SONET and synchronous digital hierarchy (SDH) intended for use in the rest of the world. An important point is that both are standard, both are compatible in the vast majority of cases, yet both will remain different (for very good reasons).

Chapter 6, which is of necessity quite long, forms a complete exposition of the SONET architecture and operational protocols. All aspects of SONET are explored in this chapter, from the basic SONET STS-1 frame structure, to sub-rate payloads maintained for backward compatibility with existing T-carrier services, to super-rate payloads, such as STS-3c (OC-3c to some), which are intended for use with asynchronous transfer mode (ATM) and B-ISDN networks.

Chapter 7 is new to the second edition. This chapter explores the details of four possible SONET payloads. There are other payload types defined, but these four constitute the bulk of all SONET (and SDH) uses today. The chapter begins by considering SONET for 64 kbps DS-0 voice, which is still the main content of many SONET networks. Then the chapter moves on to investigate asynchronous DS-3 support, another key SONET application. Next the chapter explores one of the original goals for SONET/SDH transport: the filling of payloads with ATM cells. The chapter closes with a discussion of probably the most interesting aspect of SONET (and SDH) today: using SONET to transport IP packets directly instead of inside ATM cells.

Chapter 8 contains details about the operation of two key elements and features of SONET: synchronization and timing. The chapter describes how SONET performs all aspects of network synchronization and timing. The emphasis is not a theoretical discussion of these concepts, but rather describes their real world application in SONET.

CHAPTER 5

The Evolution of SONET

The first part of this book detailed the need for higher bandwidth links for private-line networks and the lack of adequate standards or even technologies to provide them. Higher-speed equipment like desktop computers needs higher-speed links to connect them over a wide area. These links were traditional private lines, but no such private-line standard or technology existed in a service provider's network to accommodate the customer's needs.

A real force behind this drive to create SONET was a significant lack of hierarchical transmission standards greater than 45 Mbps. However, prior to SONET, a technology known as SYNTRAN (SYNchronous TRANsmission), operating at the DS-3 rate of 45 Mbps represented the upper end of the bandwidth. Only highly specialized, and generally proprietary, applications operated at speeds in excess of 45 Mbps. As a result, little standards work had been done for rates greater than DS-3.

The importance of SYNTRAN to the ultimate evolution of SONET cannot be overemphasized. Because SYNTRAN shared (and still shares) many of the basic architectural features of SONET, especially with regard to synchronization and timing (as the name implies), a brief look at SYNTRAN is in order.

The second driving force behind the efforts to develop the SONET standard was the need for what was called the *mid-span meet*. This was an open and standard interface so that different service providers, such as local and long-distance telephone companies, could establish links between their switching offices. These twin forces—high-bandwidth synchronous needs and mid-span meets—combined to produce SONET.

However, the time has also come to introduce the main players in the creation of the SONET standard, namely the international and national standards organizations themselves. Because these organizations figure prominently in the development of SYNTRAN and mid-span meets, this chapter will explore them first.

Meet the Players

Many organizations play a role when it comes to developing and overseeing SONET standards. Most of the main players are shown in Figure 5.1. The role of each is described below, with emphasis on their contributions to SONET standardization.

Figure 5.1
Major SONET standards groups

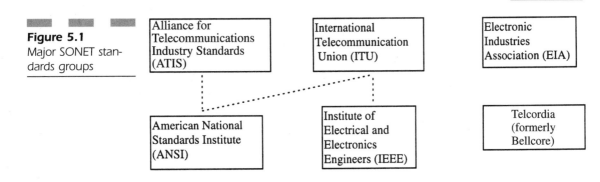

American National Standards Institute

American National Standards Institute (ANSI) was created in 1918, making it one of the first standards bodies to be formed in the United States. Oddly, ANSI develops no standards of its own. Rather, ANSI's role is to play an important part in standards coordination and approval. ANSI's main function is procedural; that is, ANSI coordinates and blends private sector standards development. ANSI is supposed to be an impartial mediator, resolving conflicts and duplications of effort among various private groups working on technology implementation. ANSI is also the United States representative to the International Standards Organization (ISO).

ANSI has over 300 different standards committees, as well as associated groups like the ATIS. Specific procedures must be followed for standards development. Under the various groupings, such as ATIS, are various technical committees and task groups. These subcommittees and working groups perform the actual standards research in a variety of computer and communications areas.

Most relevant here are the T1 subcommittees:

- T1E1—Carrier to customer installation interfaces
- T1M1—Internetwork operations, administration, maintenance, and provisioning (OAM&P)
- T1S1—Services, architecture, and signaling
- T1Q1—Performance
- T1X1—Digital hierarchy and synchronization
- T1Y1—Special subjects, such as video and data

The T1 subcommittees that are especially active in SONET standards include the T1X1 and T1M1 subcommittees. The T1X1 group played a key role in organizing the effort that led to the specification of the SONET rates and format, as well as the optical interface for the single-mode fiber. The TlMl group has responsibility for guiding the development of the SONET standard for OAM&P. These efforts, crucial in any networking scheme, focus on such things as alarm surveillance, performance monitoring, fault-location, and link management.

OAM&P will be a matter of ongoing standards activity for years to come. Even after they are implemented, the initial standards for OAM&P will not cover all issues for all SONET products operating under every circumstance. This fact can be frustrating to users and service providers, but it is a fact of life.

Alliance for Telecommunications Industry Standards

ATIS was organized after divestiture (discussed in more detail later) to represent the interests of interexchange carriers. After the breakup of the Bell System, some degree of cooperation between these competitors was still needed. Formed in February, 1984 as the Exchange Carriers Standards Association (ECSA), ATIS's Committee Tl-Telecommunications is a direct result of the divestiture process. ATIS sponsors the operation of and provides administrative assistance to this standards-making group. The Committee T1 has more than 100 member companies and operates under the formal rules and procedures of ANSI.

More specifically, the Committee T1 is focused on the functionality and characteristics associated with the interconnection and interoperability of telecommunications networks. This, in turn, concerns various interfaces with end-user systems, other carriers, Internet service providers (ISPs), and other types of enhanced service providers (ESPs).

The Committee T1 does not have wide-ranging responsibility for its own set of standards. However, through the combined processes of direct liaison, cross-membership, and coordination with other national standards bodies (much of it through ANSI's Joint Telecommunications Standards Coordinating Committee), ATIS still has much to say about standards. The Committee T1 standardization process is becoming more efficient at addressing a wide range connectivity issues, including those related to the development of SONET.

ATIS standards usually make their way to other standards bodies in the United States and eventually become ANSI standards. It was ANSI that adopted the SONET Phase 1 specification for SONET implementation (discussed in greater detail later) in September of 1988; in November of that same year, the International Telecommunication Union (ITU; then the CCITT) also approved it. Phase 1 of SONET contains over 90% of the SONET standard.

Telcordia (Bellcore)

Bell Communications Research (Bellcore), now Telcordia, was founded as the result of the divestiture process splitting off the operating companies from American Telephone and Telegraph (AT&T) (the Bell System) in 1984. Once the former Bell Operating Companies (or Regional BOCs, the "RBOCs") were separated from AT&T, the new localized telephone companies did not have easy and convenient access to AT&T Bell Laboratories, which was retained by AT&T under the divestiture decree. The seven Regional Bell Holding Companies (RBHCs, the "owners" of the RBOCs) established and supported (by way of "contributions") Telcordia as their research and development arm. Telcordia has played a key role in developing industry standards, including those for SONET, mainly because of the need for equal access capability, which is detailed later. Bellcore is now known as Telcordia and is totally independent of the RBOCs. Although many SONET documents now say "Telcordia," the work embodied in them is mostly Bellcore's!

Telcordia issues a number of documents that pertain to telecommunications in general and SONET specifically. Telcordia issues technical advisories (TAs) which inform the industry about standards issues. These roughly correspond to draft requests for comments (RFCs) in the Internet world. Interested parties can provide input to Telcordia before the TAs become finalized into technical requirements (TR) documents. The TRs provide the information necessary for vendors to begin developing equipment that complies with the technical requirements of the RBOCs, which naturally constitute a huge market for these vendors.

The major ANSI standards for SONET are largely the result of a close cooperative effort between ATIS and Bellcore. The main impact of the Telcordia TRs was on equipment, while the interface definitions were the primary contribution of ATIS. For example, the fiber optical line signal in the ATIS interface standard specifies the use of bit-interleaved parity bytes to perform error checking. The Bellcore TRs specify that the receiving equipment monitor the parity bytes. If a disagreement were to occur between the transmitted and received parity byte values, a transmission error would have resulted, of course. SONET equipment determines the number of error-free seconds, erred seconds, severely erred seconds, and unavailable seconds. This information is stored and may be relayed to the network management monitoring station in the form of instantaneous alarms, performance statistics, or both. Thanks to the presence of the relevant Bellcore TRs, all vendors can develop SONET-compliant equipment in this respect.

Telcordia also issues standards in the area of SONET test equipment. Test equipment standards allow service providers to implement sophisticated diagnostic routines to identify problems with regard to such SONET features as virtual tributaries, fiber optic regenerators, and other network equipment. The test procedures developed by Telcordia will probably penetrate to the user community through the Electronic Industries Association (EIA).

Electronic Industries Association

The EIA was founded in 1924 as the Radio Manufacturers Association. The EIA is a trade group, meaning a group whose members all engage in a like activity or sell a similar product. EIA members represent manufacturers of electronics devices in the United States. Several thousand industry representatives participate in over 200 EIA technical committees to develop mainly hardware standards. An EIA committee known as TR-30 developed the famous RS-232 (which is now officially the EIA-232) and RS-449 physical interface standards. Through the TR-30 committee, EIA interacts with both ITU and ANSI. In fact, EIA sponsors many of the ANSI committees.

With SONET, the EIA role is to assist in the development of optical interfaces for customer premises equipment (CPE). Another EIA committee, the FO2.1.4 Committee, along with Telcordia, will continue to develop testing procedures. Why are both Telcordia and EIA involved? Because Telcordia develops SONET test procedures for service providers, while EIA develops SONET test procedures that will be used by end-users employing SONET CPE and buyers of private SONET equipment. There will be a lot of joint activity between Telcordia and the EIA to achieve the highest degree of compatibility in this key area.

Institute of Electrical and Electronics Engineers

The Institute of Electrical and Electronics Engineers (IEEE) is an international professional society for electrical engineers. The major contribution of the IEEE to telecommunications has been the ongoing development of Local Area Network (LAN) standards. The IEEE has been only peripherally involved with the SONET standardization process, but the IEEE Technical Activities Board will have a great impact on SONET, due to the IEEE's LAN involvement.

The IEEE 802.x series of standards specify the characteristics of Ethernet (IEEE 802.3), Token Ring (IEEE 802.5), and various other LAN architectures. The LAN work of the IEEE 802 committee is currently organized into the following subcommittees:

802.1 High Level Interface

802.2 Logical Link Control

802.3 CSMA/CD Networks

802.4 Token Bus Networks

802.5 Token Ring Networks

802.6 Metropolitan Area Networks—Distributed Queue Dual Bus (DQDB)

802.7 Broadband Technical Advisory Group

802.8 Fiber Optic Technical Advisory Group

802.9 Integrated Data and Voice Networks

802.10 Network Security

802.11 Wireless Networks

802.12 Demand Priority Networks

802.14 Residential Networks

The IEEE works in cooperation with other standards bodies. IEEE committees routinely pass their recommendations to ANSI for approval as national standards, and on to the ITU for approval as international standards.

International Telecommunication Union

Technically, it is the International Telecommunication Union-Telecommunication Standardization Sector (ITU-T) that is concerned with SONET. The ITU-T was formerly known as the CCITT. The ITU is supported by the United Nations and is the body that makes recommendations and coordinates the development of telecommunications standards for the entire world, including North America. The official United States representative to the ITU is the Department of State. The ATIS Committee T1 provides much input to the United States ITU-T

Study Groups, such as Study Group C. They, in turn, support the Department of State in establishing positions of the United States at ITU meetings, which is a curious blend of technical talk and political infighting.

Study groups in the ITU have specific technical responsibilities for developing international standards, in much the same way that technical subcommittees of the ATIS Committee T1 have specific responsibilities in national standards development. The ITU is comprised of 15 study groups, each of which has specific responsibilities in the field of data communications. The study group that has made the most contributions toward the development of an international SONET standard is Study Group 15—formerly Study Group XVIII, but the study groups lost their Roman numerals (and were renumbered) when the CCITT changed its name. This study group focuses on digital networks, including B-ISDN.

The study groups work on technical issues assigned to them by the Plenary Assembly of the ITU, which meets every four years at various sites around the world to approve the work of the study groups and to issue new assignments for the next development cycle. The study groups hold meetings six to nine months apart. It is important to note that although they are only "recommended," the standards approved by the ITU do carry the weight of law in many European countries. The ITU has clout: Service providers can be sanctioned for violating ITU "recommendations" in many countries. (The ITU specifications are called "Recommendations" only because the ITU has no enforcement agency. The ITU "recommends" that each nation enforce ITU rules on its own.)

SONET Interoperability Forum

Fairly new on the SONET scene is the SONET Interoperability Forum (SIF), which was formed in 1994 to address the interoperability issues in SONET. Particular attention was focused on the methods needed to ensure that equipment from different vendors worked together. The forum emerged mainly because of the shortcomings in the original SONET standards, as defined by the ANSI T1X1 Committee. These standards, helpful as they were, gave vendors far too much latitude for developing variations of messages contained in the built-in SONET data communications channel. This made interoperability difficult, if not impossible. SIF will consider interoperability proposals from members and, after a consensus is reached, issue guidelines or specifications. SIF activities should be monitored closely by interested parties.

Forging Standards

The organizations introduced above do not follow explicit rules when developing standards. All things considered, developing communication standards remains an art rather than a science. Although many industry observers treat standards as explicit rules, standards are most often open to considerable interpretation by vendors, software developers, and others. Although mostly perceived to be driven by advances in technology, standards are often the result of severe political infighting between corporations, governments, standards developers, and even influential individuals. Therefore, in the real world, all standards—SONET included—are documents that are continually evolving.

Standards are not static and do not spring fully grown from standards bodies like Athena from the brow of Zeus. Even a firmly established standard like X.25, for instance, which defines the facilities offered by the worldwide public packet networks, has changed over the years. X.25 was developed in stages and has remained open to amendment since the first draft was issued in 1971. This was the document known as the "gray book." The X.25 gray book was followed by a series of amendment documents in 1976, 1980, 1984, 1988, 1992, and 1996, all at four-year intervals. All of these were issued by the International Telegraph and Telephone Consultative Committee (CCITT, now called the ITU-T), the international standards body; even though X.25-based public packet networks enjoy worldwide acceptance, the ITU will continue this four-year cycle of improvements to X.25 into the future.

Generally, the main function of all standards bodies is to coordinate activities or directly engage in activities that lead to a consensus on how best to implement a particular technology. This process can eliminate needless duplication of effort, waste of often limited resources, and alleviate the confusion that inevitably results when too many organizations produce their own (proprietary) solutions.

A good example of the standardization process is from the early days of telephony. Each telephone company—hundreds had emerged after the original Bell telephone patents expired about the turn of the century—set up its network differently from all the others. As a result, users could only communicate with each other if they were served by the same telephone company. If a subscriber wanted to call someone served by another telephone company, connections to both networks were required, which often meant having more than one telephone.

Although the eventual solution to this problem perhaps should not be repeated (i.e., AT&T systematically bought up all the independent companies and put them on the "Bell System Standard"), this standardization process did lead to the explosion in popularity of this new communications medium. Therefore, AT&T came to occupy its position in the telephony standards arena that it sustained through the breakup of the Bell System.

SYNTRAN

A full recapitulation of how M13 multiplexing operates in the T-carrier environment is unnecessary here. But it is important to note that the M13 multiplexing process that formed one DS-3 out of 28 input DS-1s was a plesiochronous process that involves the bit-interleaved multiplexing of the input DS-1s. As a result of this plesiochronous operation, the formation of the DS-3 transmission frame also involved using many overhead bits to determine the presence (or absence) of "stuffing bits" based on the differences between input timing (called *clocking*, in most cases), and output timing (clocking).

Assuring the same rate of ticking (which also included phase of ticking because ticks could occur at the same rate, but not at the same instant) between input clock and output clock on the network would allow the multiplexing equipment to dispense with the need to stuff bits into the DS-3 transmission frame. This would, in turn, allow byte-interleaved multiplexing in an isochronous fashion to occur at the T-3 level of 45 Mbps. As has been shown, a real incentive exists to perform byte-interleaved multiplexing at any level. Byte-interleaved multiplexing allows for the easy *recovery* of constituent DS-1s from a DS-3, and the recovery of any DS-0s directly from the DS-3 frame.

This recovery of a DS-0 (64 kbps) from a DS-3 frame (45 Mbps), of course, could be done without SYNTRAN or SONET. However, it was tedious and messy because the DS-1s with their embedded DS-0s were scattered bit-wise throughout the DS-3 frame. The DS-0 channels were needed in many cases to add and drop individual 64 kbps channels to make more efficient and cost-effective use of the links. In T-carrier networks operating at the T-3 rate, it was necessary to fully recover each of the twenty-eight DS-1s in order to determine the proper DS-0 channel needed. This resulted in what came to be known as *back-to-back multiplexing* in networks using M13 equipment.

With back-to-back multiplexing, an incoming DS-3 frame was broken down into its constituent DS-1s, the relevant DS-0s were cross-connected (or dropped and inserted) as required, and then the entire DS-3 frame was reassembled for the outbound link. Naturally, if the T-3 were delivered on fiber optic links, the proper fiber optic termination system (FOTS) equipment would also be needed. This whole operation is shown in somewhat simplified form in Figure 5.2.

In actual practice, the process of back-to-back multiplexing was much more complex than is shown in Figure 5.2. The short lines in the figure labeled as "DSX-3" were typically coaxial cables that snaked around the site to a cross-connect panel, where they extended back to the M13 mux itself. Of course, the same thing happened on the other side. Naturally, the lines labeled "DSX-1" were typically masses of twisted pair copper wires extending across or under the floor as well. The appearance of the site was usually a nightmare of four telephone-booth-sized equipment boxes (the FOTS and M13 muxes) and tangled masses of coaxial and copper cables. All of the various DSX cross-connect electrical interfaces had maximum cable distances associated with them. Otherwise timing problems and their related alarms occurred. These distances for T-carrier DSX interfaces are shown in Table 5.1. Plus, all of these devices had problems: They needed power supplies, generated heat, and were prone to failures.

One of the main reasons that SYNTRAN was developed was to try to reduce the complexity of this arrangement needed to get at a poor 64 kbps DS-0. Work on SYNTRAN began in the mid 1980s and was approved as an ANSI standard in 1987 as ANSI T1.103. SYNTRAN's intent was to "augment the asynchronous, bit-stuffing multiplexing design currently in use." SYNTRAN allowed "direct" DS-0 and DS-1 "observability" in the 45 Mbps bit stream. It would be more common today to use the terms, *plesiochronous* and *recovery,* instead of *asynchronous*

Figure 5.2

Back-to-back multiplexing with M13s

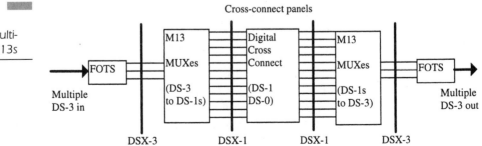

TABLE 5.1

Cross-connect
interface distances

Cross-connect Type	Maximum Cable Footage
DSX-1	655 Feet
DSX-1C	655 Feet
DSX-2	1000 Feet
DSX-3	450 Feet
DSX-4 (nonstandard)	225 Feet

and *direct observability*, but the meaning is identical. The document also described the DS-1s inside the DS-3 as *tributaries*, a term that SONET eventually used in the same context.

SYNTRAN allowed the new, "synchronous" DS-3 frame format to be transmitted over the same link with the same transmitters and receivers as before. After all, 45 Mbps is 45 Mbps. Only the format of the bits changed from bit-stuffed to byte interleaved. The equipment changed, the transmission frame structure changed, but not the bit rate.

The fact that much of the DS-3 frame overhead bits were no longer needed in SYNTRAN to determine whether stuff bits were employed allowed these bits to be used for other purposes. In an obvious moment of triumph, the ANSI standard announced that "the stuffing and control bits, which have been liberated by the synchronous format, are assigned new functions."

These new functions included an embedded 64 kbps data link just for network OAM&P, an identification field, and so on. One new function was the inclusion of some error checking on the entire transmission frame. This was not for retransmission purposes as with data frames, which would have been impossible for voice anyway, but rather to allow for the gathering of error statistics either on the link or even end to end. Such error statistics could be used to verify performance levels or to detect problems.

There is no need to discuss the details of the new SYNTRAN frame format any more than there was a need to discuss the older DS-3 frame in a SONET book. It should only be noted that SYNTRAN still struggled with timing variations (i.e., jitter). SYNTRAN introduced a process called *controlled slips* to prevent the accumulation of timing variations on the link which would throw the framing process off.

SYNTRAN was good, as far as it went. In many cases, however, SYNTRAN was too little, too late; events of the 1980s in the telecommunications industry in the United States overshadowed the accomplishments of SYNTRAN, as formidable as they had seemed. The main event was *divestiture.*

Quest for the Mid-Span Meet

The SONET concept was born in the early 1980s, before the signal event in the telecommunications field in the United States known as *divestiture.* Before 1984, the bulk of the telephone system in the United States was owned and operated by AT&T. AT&T Bell Laboratories was the research and development arm of the organization, and had invented T-carrier and SYNTRAN. AT&T Long Lines handled the vast majority of "long distance" telephone calls, those made beyond a certain regional area. AT&T also handled local telephone calls through wholly owned subsidiaries called the Bell Operating Companies (BOCs). In fact, because people made mostly local calls, the whole conglomerate was usually referred to as the Bell System.

Other independent local telephone companies existed mostly in rural areas at first. Some grew quite large in their own right, such as GTE around the country and Southern New England Telephone (SNET) in Connecticut. Others remained quite small, such as the Sylvan Lake Telephone Company in rural New York, with only a few hundred subscribers. All of these local companies more or less were forced to use AT&T Long Lines so that their subscribers could make long distance calls. In order to do so, a link (trunk) was needed to connect a local switching office to the AT&T Long Lines switching office. In these cases, one end of the trunk used was owned and operated by the local company and the other end was owned and operated by AT&T. Because the responsibility for the link was assumed to change from one company to the other in the middle of the link, this arrangement was known as the *mid-span meet.* Naturally, in order to work, the equipment at both ends had to be compatible.

This was not much of a problem as long as AT&T was the only game in town. AT&T also had a manufacturing branch, called Western Electric (WECO), that made equipment used throughout the Bell System. As long as AT&T Long Lines was the only one needed, any local company could just buy WECO equipment and be sure that the long-distance trunks to the AT&T Long Lines office would work.

In 1984, the process of divestiture, however, broke up the Bell System into several large, regional local telephone companies (Regional Bell Operating Companies or RBOCs). The other pieces of AT&T, Long

Lines, WECO, and Bell Labs, remained more or less intact, although the new RBOCs founded their own research and development company, Telcordia, with a lot of personnel from Bell Labs. A key piece of the divestiture process was called *equal access.* Equal access meant that a local company, and any subscriber to any local company's telephone service, was free to use other long-distance carriers besides AT&T. These carriers, called Interexchange Carriers (IXCs), were represented by a new breed of competitive long-distance service providers. The most notable were MCI (which originally meant Microwave Communications Incorporated), whose competitive long distance efforts had largely brought about divestiture in the first place, and Sprint (originally part of the Southern Pacific Railroad network).

Under the new divestiture rules, a local company could just as easily connect to an MCI or Sprint switching office (called *point of presence* or POP) for long-distance service as to AT&T. In fact, in most cases local companies had installed trunks to MCI and Sprint POPs so that their customers could enjoy the equal access that divestiture promised. The question was, what type of link should these be?

It seemed obvious that the equipment at each end still had to be compatible. But because neither end of the link was AT&T in many cases, it was no longer a given that WECO T-carrier equipment would be used in all cases. Certainly, neither MCI, Sprint, nor any other IXC was interested in using WECO equipment. However, the T-carrier hierarchy was never fully completed at the T-3 level and above. Mid-span meets at these levels required both end companies to purchase the same vendor's equipment. It hardly seemed possible (or probably violated the spirit of the equal access provisions) to require the newly competitive telephone systems to all buy the same equipment. Therefore, how could the anticipated flood of calls be handled to the new long-distance companies with a variety of vendor's equipment when no universal, scaleable, fiber-based digital trunking standard existed for the mid-span meet?

SONET and SDH

MCI issued a request to the ICCF, set up for this very purpose, that they go to ANSI's T1 Committee and ask them to develop standards for interconnection of multivendor and multiple-owner fiber optic transmission systems. This request, also known as the mid-span meet capability request, was one of several that the ICCF was working on with ANSI, and one of two submitted to the T1 Committee.

Early in 1985, Bellcore proposed to the ANSI T1X1 group a network solution to fiber system standardization. Their proposal was designed to allow for the interconnection of multivendor, multi-owner fiber optic terminals. It also offered interconnection of functionally different network hardware. For example, it would allow, for the first time, line-terminating multiplexers from different vendors to be connected simultaneously to a digital cross-connect system. Further, Bellcore proposed a digital signal/transmission rate hierarchy, wherein transmission speeds were integer multiples of some predefined base rate. Finally, the proposal offered a bit-interleaved multiplexing scheme that would allow for optimal use of available bandwidth.

Both of these offerings, SYNTRAN and the mid-span meet initiative taken together, resulted in the creation of the Synchronous Optical Network, or SONET standard.

This was not an easy task. Until 1988, no universal fiber-optic system standards existed. Digital standards had developed differently around the world, to the point that the multiplex hierarchies that govern the public networks of major commerce centers, such as North America, Europe, and Japan, supported different transmission rates.

Therefore, at the T-3 equivalent rate and above, each manufacturer's product was designed according to a proprietary set of specifications. As a result, no two manufacturer's fiber optic products were compatible with each other. This meant that an installation at one end of a connection between networks had to be exactly matched at the other end with a corresponding product from the same manufacturer.

The situation led to a strong tendency among carriers to choose a single supplier of fiber systems, which resulted in a loss of some potential economic and technical benefits of a competitive multi-vendor market. To ensure interconnectivity across network boundaries, either the network operators had to agree to use the exact same fiber products, or they had to confine themselves to electrical interfaces that limited them to transmission rates far below the actual bandwidth capacity of the fiber optic systems being used. This was usually done using multiple T-1s instead of the higher capacity T-3s.

Moreover, based on the technological promise of the synchronous standard, many manufacturers were already looking at synchronous multiplexing, and a network approach to bandwidth offerings. Some of them were already planning product rollouts based on minimal knowledge of what was coming. Naturally, this complicated the standards-development process. Manufacturers were so anxious to get on the SONET bandwagon that they did not wait for all of the standards to appear. Telephone companies needed and wanted SONET badly. As a result, the momentum for its design and standardization grew quickly. The SONET standard originated from a cooperative effort between various standards bodies and private industry.

As mentioned above, in 1984, MCI proposed an interconnecting fabric of multiple carrier and manufacturer fiber optic transmission connections to the ICCF. The ICCF, in turn, requested Committee T1, the carrier-to-carrier interfaces subcommittee of ECSA, working under ANSI T1X1, to develop standards for an optical interface.

By February, 1985, Bellcore had proposed to T1X1 the SONET concept, which went beyond the MCI request to allow the interconnection of all network elements with fiber terminations. The Bellcore proposal was based on the following assumptions about the future direction of telecommunications networks:

- Fiber optic media is the transmission medium of choice
- Customers will require high-bandwidth services (broadband services)
- Remote switches connected by fiber optic media are highly desirable
- Switching equipment will eventually accommodate a direct optical interface

Using this proposal from Bellcore as a foundation, the T1X1 committee's efforts eventually involved more than 400 detailed technical proposals from 120 individuals representing more than 50 corporations. That year, a Bellcore Technical Advisory (TA) was issued at the Technology Requirements Industry Forum (TRIF).

In August of 1985, the T1X1 Committee rolled out a proposal document on the SONET principle. Of course, the design was not strictly theirs. It was based on input from a variety of manufacturers, local and long-distance carriers, ad hoc user groups, and others. One of the greatest points of contention existed between Bellcore and AT&T: Bellcore wanted the baseline speed to be 50.688 Mbps; AT&T wanted it to be 146.432 Mbps. After much debate, a compromise rate of 49.94 Mbps was reached and the "virtual tributary" concept was introduced as a way to transport DS-1 based services (backward compatibility).

By January of 1987, widespread agreement had been reached on SONET design, and a draft document was prepared for a vote by the T1X1 Committee. Just as the committee was to vote in session, the ITU-T (then the CCITT) threw a wrench into the works. In 1986, the international community had taken notice of the standardization effort in the United States, but these initiatives addressed the T-carrier hierarchy in the United States, not the one used in Europe. Therefore, ITU-T began an effort of its own that became known as the Synchronous Digital Hierarchy (SDH).

Throughout the SONET design process, the needs of domestic consumers in the United States had driven the efforts of T1X1, with little regard for the needs of those who may want to use SONET as a standard outside of the United States. T1X1 saw SONET as the answer to many problems, most notably a way to stop the virtually uncontrolled growth of incompatible fiber optic terminals in the manufacturing sector to make mid-span meets easy. In Europe, however, where the manufacturing arm of the telecommunications business was still completely regulated, no such urgent need was perceived. As a result, the ITU was less willing to assign SONET the priority that delegates from the United States wanted. Little incentive existed in completing ANSI work when the ITU was dallying.

Just before things came to a complete halt, the problem was somewhat resolved when delegates from Japan and Great Britain began to attend the United States T1X1 Committee meetings. This resulted in two benefits. First, European interests began to be fairly represented in T1X1 and, second, a communication channel of sorts was created between the European standards group CEPT (the acronym is not easily rendered in English), and the T1X1 Committee.

In February of 1987, after a great deal of development and deliberation, the United States formally proposed SONET to the ITU as an answer to the network node interface (NNI) study, begun in 1986. NNI was a study established to find a non-media-specific interface, different from that specified for B-ISDN. In the eyes of T1X1, SONET seemed an ideal NNI solution; however, it was not ideal for those outside the United States. The proposal specified a base rate of 50 Mbps. European transmission systems, which have no benchmark level anywhere near 50 Mbps, asked that the new hierarchy be based on 150 Mbps, because this would be closer to their 139.264 Mbps basic signal rate. This was a substantial request. They executed the necessary modifications and, within three months, had an acceptable design, based on what is called the STS-3 frame structure.

In July of 1987, the ITU asked all its subscribing administrations to consider two proposals for a 150 Mbps NNI specification, one from CEPT, and one from the United States. For technical reasons, the United

States proposal fell out of favor. Instead, CEPT asked the United States to make a few simple modifications to their proposal that would bring it in line with European needs. These involved changing the basic bit rate from 49.92 Mbps to 51.84 Mbps and employing a byte-interleaved multiplexed frame structure, as SYNTRAN did. Oddly, initial SONET proposals still used bit-interleaved multiplexing. In any case, the United States agreed to the changes. Following some intense work, a draft proposal was finally created.

In early 1988, T1X1 accepted the changes, and SONET was on its way to becoming a true international standard. In February, 1988, ANSI T1X1.4 submitted for ballot two documents: T1X1.4/87-505R4, specifying fiber optic rates and format; and T1X1.4/87-014R4, specifying the optical interface for a single-mode fiber.

All of the comments about the ballot were resolved by a new Optical Hierarchical Interfaces working group T1X1.5. Final approval for SONET was achieved in June, 1988, in the "blue books." The SDH specifications are known as G.707, G.708, and G.709.

In the United States, equivalent specifications, ANSI T1.105-1988 and ANSI T1.106-1988, published later that year are now commonly referred to as the SONET Phase 1 specifications. This provided a North American Standard. The ITU-T also recommended an international standard based on OC-3 which would allow Europeans to multiplex a 34 Mbps signal as readily as Americans could multiplex a DS-3 signal.

Refinements to all of these resulting documents continued into 1989. Most of the activity centering around the details for full product internetworking and network management. Because it is an internationally recognized standard, SONET has been the target of considerable interest among organizations that operate international networks. With SONET deployed around the world, users will not only realize cost-savings in transport networks and be able to expand the range of useful applications, but enjoy seamless interconnectivity across international boundaries.

It is important to state that SONET *is* in many ways an international standard, although the SDH rules take precedence on international links. All SONET specifications are in compliance (despite some differences in implementation) with the ITU's Synchronous Digital Hierarchy (SDH) recommendations. In the United States, service providers and equipment vendors will follow the ANSI and Telcordia specifications. For international links and SONET use, the ITU specifications are followed.

SONET is both a United States and, as SDH, an international standard. SONET is being implemented in phases due to its complexity, worldwide scope, and required collaboration on the part of implementers. Because of this phased SONET standardization, vendors are all adopting different strategies to design equipment that can be readily upgraded to conform to emerging standards as they become available.

Because all vendors have followed the phased approach to SONET rather closely, it may be helpful to explore the SONET phases in more detail.

SONET Standards: Three Phases

The standards development work was divided into three separate efforts.

The first phase (Phase I) of this activity set the stage for interoperability between different vendors' equipment, regardless of transmission rate. This stage defined things such as signal construction (i.e., attributes that define the frame structure and the assignment of overhead versus user payload), multiplexing parameters (i.e., definition of the method used to achieve higher rates in the SONET hierarchy), optical requirement (i.e., specification of the pulse characteristics and optical parameters necessary for light-based transmission systems), and payload mapping (i.e., methodologies used to carry user data in the SONET).

Phase II addressed the problems associated with OAM&P functions. Because SONET has its own embedded operations channel, called the data communications channel, a new protocol would have to be designed to allow operations messages to be swapped between dissimilar network elements (i.e., hardware components). Phase II activities also defined an electrical interface specification (for optical-to-electrical conversions). The work associated with Phase II took considerably longer than expected. A benchmark model, based on the OSI Reference Model, was still not ready to serve as a basis for OAM&P design work when the T1X1 Committee met in February of 1990.

The solution to the overburdened Phase II took the form of a third phase, which further defined specific message formats and message sets for performing the OAM&P functions. Phase III was completed with the adoption of the OSI network management protocol message set.

With the completion of Phase III, the SONET standards became effectively complete. Three standards organizations now have specifications for defining SONET. The three sets of specifications—ITU-T's G.707-G.708, ANSI's T1.105, and Telcordia's TR-NWT-000253—define a common architecture and common interfaces, and employ a common implementation for SONET. Equipment designed to one of the three specifications will interoperate with equipment designed to another.

Although the presence of a stable Phase III means that SONET Phase I and Phase II will increasingly be of only historical interest, it may still be informative to detail the major parts of Phase I and Phase II. The emphasis today is on SONET Phase III equipment capabilities.

Phase I

Phase I of SONET standards was issued in 1988 as the initial ANSI and related Telcordia documents. Phase I has two components, namely T1.105 and T1.106. The T1.105 defines the following:

- Byte-interleaved multiplexing format
- Line rates for STS-1, 3, 9, 12, 18, 24, 36, and 48
- Mappings for DS0, DS1, DS2, DS3, and SDS3 (SYNTRAN)
- Monitoring mechanisms for Section, Line 1E Path structures
- A 192 kbps and 576 kbps Data Communications Channels (DCC)

All of these concepts are discussed in detail in Chapter 6. T1.106 specifies the optical parameters for the longer distances with single-mode fiber cable systems.

Phase II

Phase II standards, released during 1990/1991, had three components. These included T1.105R1, T1.117, and T1.102-199X. T1.105R1 is a revision of T1.105 issued earlier and specifies a number of key SONET points.

1. SONET format clarifications and enhancements. Timing and synchronization enhancements are addressed in Section 7. The synchronization issues are still pending completion as is the jitter caused by the SONET pointer actions (described in more detail later). Due to its serious nature, jitter is addressed in Phase III standards as well.

2. Automatic Protection Switching or APS (Section 13).

3. The full seven-layer protocol stack for DCC and Embedded Operations channel (Section 14).

4. Mapping of North American DS-4 (139 Mb/s) signal into STS-3c.

These concepts are fully discussed in Chapters 7 and 13. T1.117 specifies the optical parameters for short haul (less than 2 kilometers), multimode fiber cable systems. T1.102-199X gives electrical specifications for STS-1 and STS-3 signals.

Phase III

Phase III of SONET addresses the crucial, but still controversial, issue of communication over the embedded operations channel for OAM&P. The controversy involves continued use of non-standard features in operations packages. Without this aspect of the SONET standards, equipment of different vendors cannot exchange control information and be managed by other than a specific vendor's network management software packaging. As recently as 1990, traffic could be successfully passed at the OC-1 and OC-3 rates in a multi-vendor SONET environment, but network management commands could not be exchanged.

Without this resolution of the OAM&P message format, existing RBOC operational support systems (OSS) cannot work with SONET's network management schemes. The SONET OSS is still being defined for operations, administration, maintenance, and provisioning of network services. Present RBOC OSS's are embedded in an ASCII-based TL1 or Transaction Language 1. This works fairly well for smaller SONET links and networks, but TL1 is still a "man-to-machine" language. This means that a human being must issue the TL1 commands at one end of the link. At SONET speeds, automation is preferred.

SONET network applications with elaborate meshed and multi-vendor links will require a machine-to-machine language, such as one based on the OSI-RM's network management standards. Resolution of this OSS issue is one of the sticky items in the completion of the SONET standards. These issues will be discussed in more detail later in Chapter 13.

Phase III consists of four components. These include T1.105.01, T1.105.05, T1.105.03 and Tl.1 19.

ANSI T1.105.01-1994 provides the requirements for two-fiber and four-fiber bidirectional, line-switched SONET rings. These rings allow for failure restoration within 50 milliseconds and are an important part of many network recovery plans.

ANSI T1.105.05-1994 describes the Tandem Connection Overhead layer for SONET. This allows carriers to maintain series of STS-1 signals as a single unit and offers users improved monitoring and service as well.

ANSI T1.105.03-1994 provides the jitter requirements at SONET interfaces and at interfaces between SONET and asynchronous (or plesiochronous) networks. The Phase III SONET jitter specifications are important for efficient SONET performance.

ANSI T1.119-1994 allows for the management of SONET network equipment using an OSI-Reference-Model-based (OSI-RM) interface. The interface is based on the OSI-RM Common Management Information Protocol (CMIP). CMIP, in turn, is based on the OSI Common Management Information Service Element (CMISE), which provides a standard set of application program interfaces (APIs) for OSS, and the Abstract Syntax Notation version 1 (ASN.1) standard, which provides a common format for internal data representation. The use of CMIP will provide a standard for SONET OAM&P.

The significance of these SONET phases is as follows: Phase I enabled SONET equipment to pass bits back and forth; Phase II enables SONET equipment from different vendors to pass bits back and forth; and Phase III will enable SONET equipment from different vendors to pass bits back and forth and to be managed by service provider OSS software regardless of vendor.

SONET has been deployed long before the complete resolution of the OSS issues with Phase III. This requires implementers to live with OSS operational inefficiencies. However, a service provider or builder of private SONET network links should always require their vendors to upgrade to the appropriate Phase III (or Phase II for older equipment) at a minimal cost to the customer. This protects the customer's investment and guards against instant obsolescence.

CHAPTER 6

SONET
Architecture
and Protocols

SONET is a set of coordinated ITU, ANSI, and Bellcore (now Telcordia) standards that defines a hierarchical set of transmission rates and transmission formats. These are to be used by equipment vendors and service providers at or above a basic optical transmission rate of 51.840 Mbps, which forms the base transmission rate for SONET). Today, SONET addresses transmission speeds as high as 10 Gbps.

This chapter describes the fundamental architectural layers of SONET and the protocols used by SONET at these various layers. The basic SONET frame formats are introduced, as well as the important concepts of virtual tributaries and super-rate frame structures. Virtual tributaries are used for backward compatibility with existing digital hierarchies, not only the North American T-carrier hierarchy, but also others. Super-rate frame structures are used for the transport of B-ISDN traffic, most notably the cells used in Asynchronous Transfer Mode (ATM) networks. asynchronous transfer mode (ATM) networks.

Because this chapter mentions ATM cell transport inside SONET, an important distinction must be made. The term *asynchronous* is part of the ATM acronym and the term *synchronous* is part of the SONET acronym. So how can asynchronous ATM cells be transported inside synchronous SONET transmission frames? The answer lies in the overuse of the terms asynchronous and synchronous in the telecommunications industry. Each of the terms may have a variety of meanings depending on context. In the case of ATM, asynchronous refers to the way ATM performs multiplexing in an asynchronous, or non-time-division-multiplexed fashion. In the case of SONET, synchronous refers to the way the SONET performs multiplexing in a synchronous, or purely time-division-multiplexed fashion, with input clocks synchronized to output clocks. ATM and SONET can still work together because they function at different levels of the overall B-ISDN architecture that brought them together.

A more complete discussion of key SONET concepts, such as clocking and timing, jitter, and network synchronization, is deferred until the next chapter. Some of the mechanisms for dealing with the effects of these timing variations, however, are examined in this chapter. This should not be a great inconvenience because the variations are examined in this chapter, however. This should not be a great inconvenience, however, since the exact details of factors like jitter will not prevent a basic understanding of SONET itself.

SONET Speeds

Table 6.1 shows the basic speeds of the SONET hierarchy. The SONET hierarchy, which consists of a few basic building blocks of terms and speeds. Each level has an optical carrier (OC) level and an electrical level transmission frame structure to go with it, called the synchronous transport signal (STS). It makes sense that an STS-3 frame is sent on an OC-3 fiber optic link, just as a T-3 frame is sent on a T-3 link in the T-carrier hierarchy.

Table 6.1 also details the physical line bit rate, the payload bit rate, and the overhead bit rate for each SONET level shown. The payload is merely the bit rate remaining after the overhead bits, which cannot be used for customer data, are subtracted. The payload is carried inside an envelope in SONET. Finally, the table gives the synchronous digital hierarchy (SDH) designations for each of the SONET levels. Obviously, the STS level divided by three gives the synchronous transport module (STM) level. In other words, SDH counts by threes, while SONET counts by ones. The counting units are the basic 51.84 Mbps bit rate of OC-1. STM levels increase by units of 155.52 Mbps (3 x 51.84 Mbps).

It should be noted that all SONET/SDH levels and speeds listed are not currently supported by equipment. Not all levels were needed, some

TABLE 6.1

SONET digital hierarchy

Optical Level	Electrical Level	Line Rate (Mbps)	Payload Rate (Mbps)	Overhead Rate (Mbps)	SDH Equivalent
OC-1	STS-1	51.840	50.112	1.728	—
OC-3	STS-3	155.520	150.336	5.184	STM-1
OC-9	STS-9	466.560	451.008	15.552	
OC-12	STS-12	622.080	601.344	20.736	STM-4
OC-18	STS-18	933.120	902.016	31.104	
OC-24	STS-24	1244.160	1202.688	41.472	
OC-36	STS-36	1866.240	1804.032	62.208	
OC-48	STS-48	2488.320	2405.376	82.944	STM-16
OC-96	STS-96	4976.640	4810.752	165.888	
OC-192	STS-192	9953.280	9621.504	331.776	STM-64

were not economically viable (in the sense that, for a little more money, a higher line rate could be used), and this limitation made life much easier for equipment vendors and service providers. Today, only the levels shown in **bold** in the table are supported. SONET speeds are OC-3, OC-12, OC-48, and OC-192. The corresponding SDH speeds are STM-1, STM-4, STM-16, and STM-64.

It is absolutely crucial to remember that unless other arrangements are made in equipment configurations, *SONET will remain as channelized as a T-carrier link*. That is, an STS-3 contains three STS-1s. An STS-12 contains twelve STS-1s, and so on. All of the STS-1s inside an STS-48 on a OC-48 fiber optic link run at 51.84 Mbps, the STS-1 rate. In the same fashion, a DS-3 frame contains twenty-eight T-1s, all operating at 1.544 Mbps. Ironically, SONET, despite the incredible speeds available at the higher ends of the hierarchy, would seem to limit networking to channels operating at 51.84 Mbps from any individual user.

However, just as DS-3 (or DS-1 for that matter) is available as an *unchannelized* transport without any structure (save the DS-3 frame overhead bits), so is SONET. An unchannelized DS-3 merely offers a raw bit rate (no more 28 DS-1s) at 45 Mbps. In the same way, an unchannelized STS-12 would offer the user a raw bit rate (no more 12 STS-1s) at about 622 Mbps. More details on "unchannelized" SONET levels — they are not called unchannelized STS frames in SONET, but concatenated—are detailed later in this chapter and in the next one.

A nice trick for converting a given OC level to Gbps is to divide the OC number by two and then again by ten: The result is the line rate in Gbps. Of course, this works by dividing by ten and then two, or even directly by twenty. The outcome is the same. For example, the speed of an OC-48 is 48 / 2 = 24 / 10 = about 2.4 Gbps. The speed of an OC-12 is 12 / 2 = 6 / 10 = about .6 Gbps or 600 Mbps.

The terms used in Table 6.1 appear frequently in the rest of this book. Because these form the basis of the SONET architecture, a review of these terms will aid in the understanding of the architecture.

OC-N: This notation refers to the SONET transmission characteristics of an Nth level transmission link. N can take on values between 1 and 255 (N is represented by an 8-bit field), but not all values are supported. This notation is analogous to the T-1 notation in today's T-carrier transmission network. Within SONET, an OC-3 transmission link is assumed to have an STS-3 frame structure.

STS-N: In this notation, STS refers to the SONET frame structure of an Nth level transmission link. N can take on values between 1 and 255 (an 8-bit field), as before. This notation is analogous to the DS-1 notation in today's transmission network. Although most SONET links should properly be referred to as STS-N, it is much more common (and just as correct) to speak of OC-N. T-carrier labors under the same burden, but has not suffered for it because most people say "T-1" when they really mean "DS-1".

Payload: The term used to indicate user data within a SONET frame. Payloads can be sub-STS-1 level (e.g., T-1s, E-1s, DS-2s), STS-1 level (e.g., DS-3s), and super rate (e.g., SMDS cells, ATM cells).

Envelope: The portion of the STS-1 frame used to carry payload and end system overhead.

Overhead: The portion of the STS-N frame used to carry management data for OAM&P.

Concatenation: A term that refers to the linking together of multiple STS-1 frames to form an envelope capable of carrying super rate payloads, such as when SONET is used to carry ATM cells. A concatenated SONET link is referred to as an STS-3c or an OC-3c, or whatever the SONET level happens to be. Note that in T-carrier, a form of concatenation existed as well and was also referred to by adding a letter "C," as in DS-3C. But also note that in T-carrier the concatenation indicator was always a capital "C", while in SONET the concatenation indicator is always a lower case "c". This is not an example of change for the sake of change. The concatenation in SONET is fundamentally different thatn the concatenation done in T-carrier. In SONET, the "c" really stands for "unchannelized," as will become obvious later on in this chapter.

Note that not all possible values of N are represented in SONET. For example, no OC-5 or OC-7 exists. Although the value of N can technically take on any value from 1 to 255, SONET standards define only a few of the levels. Otherwise, there would potentially be 255 different types of SONET equipment, making a joke of interoperability in spite of the presence of standards. At N = 255, SONET links would runoperate at 13.2192 Gbps, a speed that current optical equipment has already far exceeded.

Table 6.1 shows that not all speeds of SONET are supported. It seems strange to think of standards in this light, but the fact is that customers and equipment vendors have favored some speeds of SONET more than others. Sometimes there is a technical reason behind this preference. For instance, early equipment vendors found that it was actually more cost effective to build SONET gear operating at the OC-12 level than at the

OC-9 level because of laser chip costs and economies of scale. Little OC-9 SONET equipment has been built, while the OC-12 equipment market has flourished.

A unique aspect of the SONET hierarchy, as opposed to the T-carrier hierarchy, is the relationship between the speeds of the different levels within the hierarchy. In SONET, a simple multiplication by 51.84 Mbps yields the correct speed at any level of the hierarchy. STS-3 operates three times faster than an STS-1, and N x 51.84 Mbps = 155.52 Mbps. This simple multiplication is an aspect of SONET's synchronous, byte-interleaved multiplexing, and cannot be done with other carrier systems like T-carrier.

For example, a DS-3 frame contains the combined bits of 28 DS-1 frames, scattered throughout the DS-3 frame bit-by-bit. There is a lot of additional overhead in the DS-3 frame needed to locate these scattered bits properly at the receiving end of the T-3 link. The line rate of 28 combined DS-1 frames is just 28 x 1.544 Mbps = 43.232 Mbps; however, a T-3 operates at 44.736 Mbps. The extra bits are the added overhead from the DS-3 frame structure (e.g., C-bits, M-bits). This amounts to 44.736 Mbps - 43.232 Mbps = 1.504 Mbps, which is almost an added DS-1 all by itself.

SONET/SDH, conversely, adds no additional overhead at all at higher levels of the multiplexing hierarchy. All the overhead needed for synchronous, byte-interleaved multiplexing is present even at the lowest level of the hierarchy, at the STS-1 level. Of course, the trade-off is higher levels of overhead at the lower levels for no additional overhead at the higher levels. But this is not an insignificant advantage, especially when the need for higher and higher rate transports is considered. The overhead percentage in SONET/SDH is fixed, while the overhead percentage continually rises in other, asynchronous multiplexing schemes. In SONET/SDH, the ratio of overhead to payload (1.728 Mbps to 50.112 Mbps) remains constant at 3.45%, regardless of the value of N.

Figure 6.1 shows the percentage of overhead-to-payloads in SONET/SDH and several common T-carrier-based line signaling rates. The burden of asynchronous, plesiochronous multiplexing becomes obvious at higher line rates. The advantage of SONET/SDH becomes more pronounced at these levels, the very levels needed for modern network links.

STS/OC Distinction

A few final words are in order about the distinction between the STS and OC designations. The remainder of the SONET frame after the

Figure 6.1
SONET/SDH
overhead

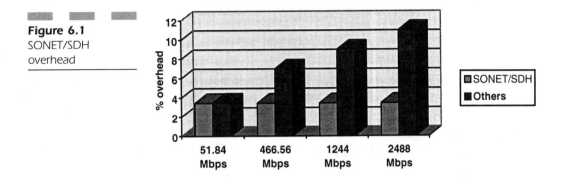

overhead bits is the SONET payload itself, more commonly (but less accurately) known as *user data*. This payload may be information from the existing T-carrier hierarchy (e.g., DS-1, and so on), ATM cells (if the SONET link were concatenated, like an OC-3c), or even something entirely different (e.g., MPEG video).

The designation STS refers to the electrical side of SONET. This seems odd at first, because SONET is above all an optical network. Although modern network equipment excels at the sending and receiving of serial bits over a fiber optic link, the same is not true of any other network task. All SONET frames must be created, multiplexed, managed, and so on. This must be done in the electrical world, rather than in the optical world. The outbound user to the network side of this process is illustrated in Figure 6.2.

In Figure 6.2, user data is sent to a SONET STS-1 multiplexing device. This device adds the Path Overhead (POH) to yield a synchronous payload envelope (SPE), and then adds the Transport Overhead (TOH) to yield the entire SONET STS-1 frame. Multiple user data steams can be combined as well. The SONET multiplexing device next sends the STS-1 frame to a SONET transmission device, which can be up to 450 feet away (an STSX-1 interface on copper for interequipment links for an STS-1) or

Figure 6.2
The STS/OC
distinction

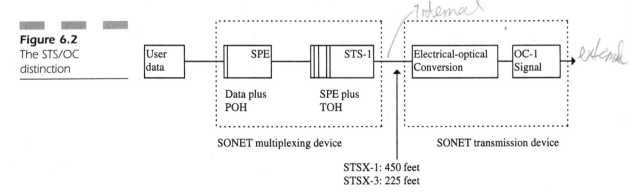

225 feet away (an STSX-3 interface on copper for interequipment links for STS-3). Multiple SONET multiplexers cannot share a transmitter, although several user data streams can share a SONET multiplexer, as in the case when three STS-1s are using a single STS-3.

Therefore, the multiplexing hierarchy, framing formats, overhead functions, and so on are properly referenced by STS terminology. An STS-1 frame can carry a clear-channel DS-3 signal (44.736 Mbps) or a combination of sub-DS-3-rate signals, such as DS-1 or DS-0. Other possibilities are also mentioned above. All of these signals are electrical in origin.

Note that distance limits are strictly imposed by various standards on the distance that SONET signals can travel electrically. The serial bit durations must be kept long enough so that electrical noise and high-frequency attenuation remain at a minimum. Otherwise, timing is lost and the frame structure degrades to the point where it can no longer be found on the receiving equipment. The distances are comparable to DSX-3 and DSX-4 distance limits and should not be a burden to users or equipment vendors.

For transmission on fiber optic links, the STS-1 structure is converted and an optical carrier level 1 (OC-1) signal results. The OC signal is sent after scrambling to avoid long strings of zeros and ones, and to enable receiver clock recovery from transitions in the incoming data stream. The OC terminology is typically used when the carrier system itself is discussed, but even customers seldom use the STS terminology. Most users and service providers alike typically refer to SONET speeds by OC references.

Purists still argue that since an OC-3, for example, is just an unstructured serial stream of bits, the concatenation term should properly be applied only to the STS-3c frame that rides the OC-3. Yet most service providers and equipment vendors happily speak and write about "OC-3c" or even "OC-3C" just because the terminology is easier. (One SONET vendor admitted labeling equipment "OC-3C" only because the die printer had no lowercase letters!) Users of SDH tend to look at this SONET controversy and shrug. This is because SDH tends to use the term STM for *both* electrical and optical signals. If ever any confusion were to arise, the terms STME-1 and STMO-1 might be used to indicate the electrical and optical aspects of STM, respectively.

STS-1 Frame Structure

The basic building block of the SONET digital transmission hierarchy is called the STS level one, or STS-1 frame. The SONET frame is a transmission frame used for the transport of a package of bits over a physical link; therefore, a SONET frame exists at a lower level on a network than the more well-known Ethernet or Token Ring LAN frames. The SONET frame is the exact analogy of the T-carrier level transmission frames in function, such as the T-1 frame structure.

SDH contains no STS-1 equivalent frame. But since an *STM-1* frame is quite complex, it is common even in SDH environments to speak of a "conceptual STM-0" frame identical to an STS-1 purely for educational purposes.

The basic STS-1 SONET frame consists of 810 bytes, transmitted 8,000 times per second (or once every 125 microseconds) to form a 51.840 Mbps signal rate. This basic rate, the fundamental building block of SONET, is derived as follows:

810 bytes/frame x 8000 frames/sec x 8 bits/byte = 51.840 Mbps line rate

In other words, the 810 bytes of the basic SONET frame structure are sent 8,000 times per second, and because each byte consists of 8 bits, the signaling rate on the link is 51.84 Mbps.

Figure 6.3 shows the basic structure of the SONET frame in visual format. The STS-1 frame is 9 rows of 90 columns; it is always shown in this format, so that the overhead bytes will line up properly at the beginning of the frame. The STS-1 frame is transmitted one row at a time, from top to bottom, and from left to right within each row. Therefore, the byte in row 1, column 1 is sent first, and the byte in row 9, column 90 is sent last. After the 90th byte is sent at the end of row 1, the next byte sent is the first byte in row 2, the byte in column 1.

One frame is sent every 125 microseconds; therefore, SONET can maintain time-slot synchronization that is required for delivery of nor-

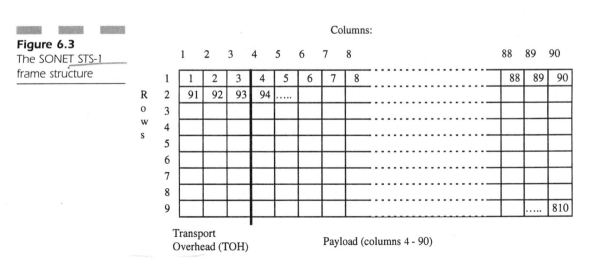

Figure 6.3
The SONET STS-1 frame structure

mal, uncompressed PCM voice (8 bits 8,000 times per second or 64 kbps). SONET also adheres to frame synchronization timing with existing asynchronous network standards (e.g., DS-1, E-1, and DS-3). More details on this timing synchronization are examined in the next chapter.

An STS frame is composed of two main sections, each with their own structures. The first three columns of the STS-1 frame form the TOH for the entire frame. All of the SONET overhead information (divided into Section, Line, and Path) that is used to manage defined parts of the SONET network and transported data (called a Payload), is in the first three columns of the frame. This overhead section, therefore, consists of 27 bytes (9 rows x 3 bytes/row) sent as part of each and every SONET frame. This overhead cannot be eliminated or converted for user data.

It is important to note that this means that the overhead in SONET is sent three bytes at a time throughout the entire SONET frame. That is, when a SONET frame is sent, three bytes of overhead begin the frame, and then the payload follows with 87 bytes of user data. These payload bytes comprise the remainder of the first row of the frame. The 91st through 93rd bytes are again overhead bytes, in this case the first three bytes of row 2 of the SONET frame. Then 87 bytes of payload follow, and so on throughout the entire 125 microsecond frame time.

The SONET payload is carried in the synchronous payload envelope (SPE). The capacity of the SPE is 9 rows of 87 columns (since the first 3 columns in each row are for overhead). This comes to 783 bytes (9 rows x 87 columns) of payload in each frame, giving a total user data rate of:

783 bytes/frame x 8000 frames/sec x 8 bits/byte = 50.112 Mbps payload rate

Actually, there is a little more to the SPE structure than just this. For example, more overhead, called POH, is contained in the SPE but considered part of the user data. However, this should be enough to understand the basic SONET STS-1 frame structure. The SPE, just as the entire SONET frame that contains it, is sent 8,000 times per second. Finer points about the SPE and POH will be explored a little later.

The basic structures of the STS-1 frame (and conceptual STM-0 frame) are described below.

Section and Line Overhead

The SONET STS-1 frame transports a considerable amount of overhead associated with operational functions within the SONET network. The first three columns of the STS-1 frame are dedicated to Section Overhead (SOH) and Line Overhead (LOH), performing a wide variety of functions. Together they are considered the TOH. The first three rows of the TOH is the section overhead (SOH; nine bytes). The last six rows of the TOH is the lone overhead (eighteen bytes).

Synchronous Payload Envelope Capacity (Information Payload)

The remaining 87 columns of the STS constitute the SPE capacity. It is more precise to say that these 87 columns represent the SONET Information Payload and that the SPE is the structure of the Information Payload, but usually only the SPE itself is singled out for attention.

Synchronous Payload Envelope ← POH + payload

move

The SPE contains both overhead information, called POH, and the actual user data. The POH, combined with the user data (Payload) that follows, constitutes an SPE. The POH can begin in any byte position within the SPE capacity. This typically results in the SPE overlapping into the next frame.

It should be noted that the extensive amount of overhead in SONET is useful for surveillance and network management, as well as other activities. Although criticized for excessive overhead, the fact that SONET overhead does not increase proportionately at higher levels of the hierarchy is a real advantage.

It is easy to construct the structure of a frame at any level of the SONET hierarchy once the basic STS-1 format is understood. All SONET frames are sent 8,000 times per second. All SONET frames are exactly 9 rows. The only variable is the number of columns. This may sound simplistic, but it is true and results in all of the different speeds at which SONET operates.

For example, an STS-3 frame (which is the same as an STM-1 frame) consists of 9 rows and is sent 8,000 times per second; however, an STS-3 frame is not 90 columns wide. The STS-3 frame is three times wider (N=3, after all). Therefore, the STS-3 frame is 270 columns wide, and that is not all. The STS-3 overhead columns are multiplied by three as well, as are the SPE capacity columns. An STS-3 frame then is 270 columns wide (3 x 90), of which the first 9 columns are TOH (3 x 3) and the remaining 261 (3 x 87) are payload capacity. The whole STS-3 frame is 2,430 bytes (9 rows x 270 columns). The line rate for an OC-3, therefore, must be:

2430bytes/frame x 8000 frames/sec x 8 bits/byte = 155.52 Mbps line rate

The same result can be derived simply by multiplying the OC-1 line rate of 51.84 Mbps by 3. Thus, the "trick" of SONET frame structures and line rates is that each frame is N times larger in terms of columns than an STS-1 frame, so the line must signal N times faster than the OC-1 line rate to eliminate the same 8,000 frames per second. After all, the STS-3 frames may be carrying hundreds of digital voice samples, all of which must arrive every 8,000[th] of second.

Figure 6.4 shows some basic structures of other common SONET frame types.

SPE Mapping into STS—1 Frame

It may seem obvious that the first byte of the SPE should be the 4th byte of the SONET STS-1 frame. After all, if the SPE contains the user data and the SPE is sent 8,000 time per second, why shouldn't the SPE be aligned with the overall SONET frame structure? In a perfect world, there would be no answer to this question. But this is not a perfect world, and the clocks used in networks to time bit streams do not always cooperate and remain synchronized. The whole question of timing and network synchronization is explored more fully in Chapter Eight. For now, it is enough to point out that jitter and phase differences make it difficult to fix the SPE inside the SONET frame in all cases and at all times. In fact, an insistence on fixing the SPE may actually be foolish and ultimately counter-productive.

The SPE, therefore, does not have to be aligned with the first row and fourth column of the SONET frame. It all depends on the clock source and SONET network configuration. Figure 6.5 depicts how an SPE would be mapped into a SONET frame when there are phase differences occur between the incoming and outgoing payloads. In plain English, this determines how the customer's data is carried on the SONET system.

Study a few items in Figure 6.5. First, the SPE is not necessarily frame-aligned with the beginning of the SONET frame. This is due to the fact that the SPE must be frame synchronous with the *customer's* payload (which may be a DS-3 CSU/DSU, or other CPE device) and not the service provider's SONET system. This only makes sense. A SONET link does not generate any payload of its own to transport, but delivers customer payloads from site to site over the SONET link. Phase differences also occur in the multiplexing of various customer payloads onto a higher speed SONET link as well, such as when three STS-1 payloads are combined into a single STS-3 link.

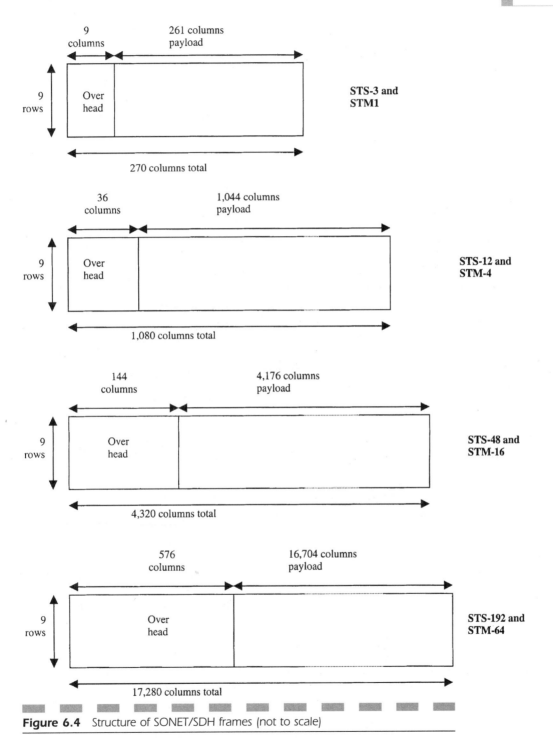

Figure 6.4 Structure of SONET/SDH frames (not to scale)

The customer's payload then is mapped into a payload envelope and the POH is added, forming an SPE. This is done by the CPE device at the customer's site. The SPE is then mapped into the SONET frame. A good analogy is to compare the SPE construction to the functioning of a T-1 CSU (a portion of the T-carrier CPE) and the final STS-1 frame construction to the functioning of a T-1 DSU (another portion of the T-carrier CPE).

The second item to notice in Figure 6.5 is that a mechanism must exist to find the SPE at the receiving side of the SONET link because the SPE is separated from the TOH by its position within the frame. That is, the SPE is no longer fixed in the fourth column of the SONET frame but can now begin almost anywhere within the Information Payload section of the SONET frame. How can the SPE be found by the receiving SONET equipment?

The answer is that components of the TOH (namely "pointer bytes") identify the position of the start of the SPE within the SONET frame. This allows the receiving SONET transmission equipment to align with the incoming transmission stream and to identify the start of the SPE. Otherwise, the buffers would be needed to delay the sending of the SPE until the start of the next SONET frame. This does not sound like much of a penalty, but the buffers would be quite large at SONET speeds. If an SPE were ready to be sent halfway through a SONET frame on average, then

Figure 6.5

Frame structures of some common SONET levels

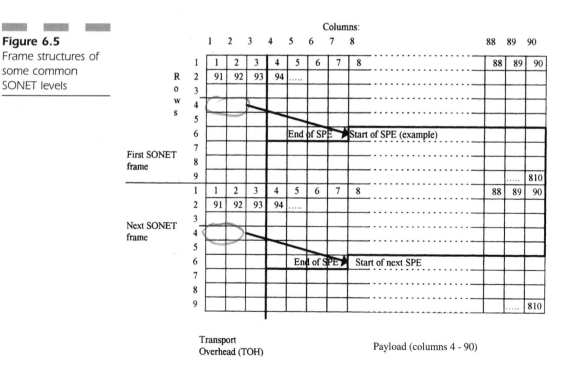

the buffer needed would be half the size of the SONET frame. This is about 1,200 bytes at the OC-3 level, but a huge 80,000 bytes or so at OC-192. STS-192 frames are 9 x 90 x 192 = 155,520 bytes long. Larger buffers add expense and complexity in the form of buffer management to SONET equipment. Today, buffers (memory) are quite inexpensive, but this was not the case when SONET/SDH was defined.

In practice, the delay needed to align the incoming SPE with the outgoing SONET frame is usually achieved once at the ingress point of the SONET network. Then for the rest of the links until the SPE emerges at the other customer site, the SPE is fixed at column 4 of the SONET frame on the service provider's SONET network. This "floating SPE" mechanism also allows the SONET equipment to identify sub-rate elements even at the native, much faster SONET speeds.

The last point in the Figure 6.5 is the concept that the SONET frames and the SPEs within them are continuous. Therefore, a realistic drawing of the SONET transmission systems would show SONET frames and SPEs before and after the sequence shown in the figure. That is, the last byte of an SPE is immediately followed by the first byte of the next SPE, regardless of where the SPE may appear in the SONET frame.

SONET Architecture Layers

SONET, like almost everything else in computers and telecommunications today, has a layered architecture. Layers help to break up the communications tasks into more manageable pieces, and can furnish a certain amount of flexibility for implementers by making the layers more or less functionally independent of each other. Thus, vendors have a fair amount of latitude when implementing layer functions internally, as long as the interface between layers is maintained in a standard fashion.

Although SONET has a layered architecture, it is important to note that SONET mainly deals with the Physical Layer of the OSI Reference Model (OSI-RM), similar to T-1 or T-3 services. In other words, no relationship exists between the SONET layers and the layers of the OSI reference model, except for the existence of SONET at the Physical Layer of the OSI-RM. A "Layer 3" in SONET has no relationship to the Layer 3 (Network Layer) of the OSI-RM.

There are four layers in the SONET architecture. These are shown in Figure 6.6. The Photonic Layer refers to the optical properties of the transmission path, which involves the sending of the serial 0s and 1s from a sender to a receiver. SONET lasers and LEDs and their paired receiving devices operate at this layer. The Photonic Layer deals with the transport

Figure 6.6
SONET architecture layers

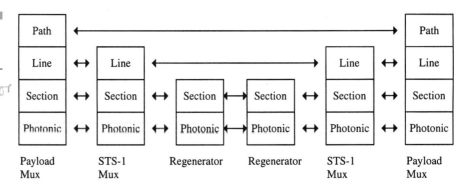

[handwritten annotations in left margin: "within SPE of STS bit layer (sender or receiver properties)", "Path", "Line", "Section", "Photonic", "regenerator for signal (framing + scrambling)", "creating & maintaining SONET payload (SPE) in STS level (multiplexing + synchronization)"]

of bits across this physical medium. No overhead is associated with this layer—it is only a stream of 0s and 1s. The main function of this layer is the conversion of STS electrical frames into optical OC bit signals.

Associated with each of the remaining levels is a set of overhead bytes that govern the function of each layer. These overhead bytes are logically known as SOH, Line Overhead (LOH), and POH. The SOH and LOH make up the TOH located in the first columns of the STS frame. The POH bits are included as the first column of the SPE.

The Section Layer refers to the *regenerator section* of the transmission link. In fiber optic transmission, as in most digital schemes, 0s and 1s must be regenerated at fairly regular intervals. Of course, the signal does not necessarily terminate at this point; therefore, a single SONET link may have regenerators spaced every 10 kilometers or so on a 100 kilometer link. In this case, every segment between regenerators forms a section in SONET.

The Section Layer manages the transport of STS frames across the physical path, using the Photonic Layer. The functions of this layer include section error monitoring, framing, signal scrambling, and transport of Section Layer overhead. Section Layer overhead consists of nine bytes of information that contain the information required for the Section Layer to perform its functions. This overhead is created or used by what is known in SONET as section terminating equipment (STE).

The Line Layer refers to the *maintenance span*. A maintenance span forms a segment between two SONET devices, excluding the lower layer regenerators. A single SONET link from one user site to another may consist of many such spans. In fact, much of the terminology in this chapter that refers to SONET "links" would be more properly expressed in terms of SONET "spans." However, outside of SONET network maintenance and engineering groups, the term "span" is rarely heard. To most, a SONET network consists of many SONET "links" between sites. However, the term mid-span meet is a constant reminder of the importance of the span concept in SONET networks, especially when maintenance and protection is considered.

The Line Layer manages the transport of entire SONET payloads, which are embedded in a sequence of STS frames, across the physical medium. The Line Layer functions include multiplexing and synchronization, both required for creating and maintaining SONET payloads. Line Layer overhead consists of 18 bytes of information (twice as much as the SOH) that provide the Line Layer with its ability to perform its functions, its ability to communicate with the layers that surround it, and to provide certain protection and maintenance features. This overhead is used and created by SONET line-terminating equipment (LTE).

Finally, the Path Layer covers end-to-end transmission. In this case, end-to-end refers to customer-to-customer transmission. One end of the Path is always where the bits in the SONET payload (i.e., SPE) originate and the other end of the Path is always where the bits in the SPE are terminated. This may not be the actual end user device (such as a desktop workstation or server), but usually refers to some kind of premises SONET multiplexing device. Such devices can combine the bit streams from many end user devices and send them on to another site, where the data streams are broken out in the companion SONET CPE. The POH associated with the SONET Path is considered to be part of the SPE for this very reason. The POH is passed unchanged through the SONET Line, Section, and Photonic Layers and is thus indistinguishable from user data to the SONET equipment (but the POH still must be there).

The Path Layer transports actual network services between SONET multiplexing equipment. These services would include the transport of customer DS-1s, DS-3s, ATM cells, and so on. The Path Layer maps these service components into a format that the Line Layer requires and then communicates end-to-end, using functions made possible by the POH bytes (nine in all) to ensure overall transmission integrity. The POH remains with the data (payload) until it reaches the other end of the SONET link.

Conceptually, each SONET layer requires the services of the layers below it to perform its functions properly. If two Path Layer processes were exchanging a DS-3 data steam from one customer site to another over an OC-3, the three layers would complement each other in the following fashion: The DS-3 signal and the associated POH bits would be encapsulated into a SONET payload (SPE) and handed off to the Line Layer. The Line Layer would then multiplex several signals (but only the link is larger than an OC-1, which can only fit one DS-3) from the Path Layer and add Line Overhead (LOH) to the mixture. It would then hand the signal element off to the Section Layer, which would add its own overhead, perform framing and scrambling, and then hand the whole frame off to the Photonic Layer.

The overhead in SONET will be totally unfamiliar to those experienced with T-carrier. SONET overheads obey a strict hierarchy in SONET equipment types; therefore, it may be a good idea to explore these relationships before examining the actual overhead byte functions in detail.

Many types of SONET equipment and the ways that they may be combined to provide SONET services are explored later in this book. For now, it is enough to realize that some types of SONET equipment are designed for CPE use, and so must generate the proper POH bits and SPE packaging. These equipment types are Path Terminating Equipment (PTE) in SONET talk. In is important also to point out that the SONET CPE must also form an termination point for a SONET span (LOH) and even the section (SOH). It is common in SONET, as in T-carrier, that the CPE performs a regenerator function as well. Obviously, the SONET CPE must form and process *all* of the overhead bytes of whatever function.

Other pieces of SONET equipment have more specialized functions with regard to overhead. For example, a user's SONET signal at the OC-1 level may be combined with other users, perhaps from other sites, to form a single OC-3 in the same way that 28 T-1s from various users may be combined by the service provider to form a single T-3. The individual user's POH will not change, but this LTE will generate its own LOH and SOH from the new STS-3 frame structure. So an STE performs both LOH and SOH processing on the maintenance span.

At the lowest level, each optical segment between any sender and receiver at all in SONET forms a Section. Regenerators all examine, process, and change the SOH, but do not touch the LOH or POH. This SONET overhead hierarchy, with different pieces of equipment performing different overhead functions and terminating sections, lines, and paths as required, is shown in Figure 6.7.

In Figure 6.7, two PTE devices communicate by exchanging SONET payloads over a SONET link. Not only do these CPE devices terminate the Path, but the Line and Section as well. Once the user's payload arrives at the service provider's network, the payload is typically combined with the payloads of other users in the same geographical area. Therefore, this piece of SONET LTE terminates the line (maintenance span) and the Section, but not the users' Paths. Finally, each regenerator along the way terminates the Section, but neither the Line nor the Path.

Between each component, down to the regenerator, which is no longer the relatively passive device it once was in T-carrier, SONET SOH is ter-

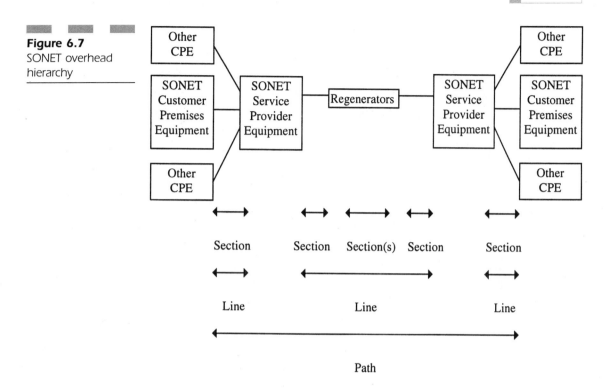

Figure 6.7
SONET overhead
hierarchy

minated, examined, and modified. The LOH passes unchanged through the regenerator, but is terminated, examined, and modified by what SONET calls the *major components* of the link, which generally translates to everything else but the regenerators. The SOH is processed by these major components as well.

Of course, the POH passes unchanged across the entire link, from CPE to CPE. All POH generated by one CPE arrives unprocessed and unchanged at the other CPE. Thus, the Line and POH is just "data" or more bits to the Sections and POH is just "data" or more bits to the Lines (i.e., maintenance spans).

Due to the hierarchical nature of the SONET overhead(s), each Line endpoint is also a Section endpoint and each Path endpoint is also a Line and Section endpoint. However, the converse is not true: Each Section endpoint is not necessarily a Line or Path endpoint and each Line endpoint is not necessarily a Path endpoint. Therefore, the required overhead processing varies at each component of a SONET network. It all depends on exact function.

A simple way to refer to SONET devices without reference to LTE or PTE terminology is just to refer to them as *SONET network elements* (NEs).

SONET Overhead

The overhead in SONET is both a distinctive feature of SONET as well as a much needed change from T-carrier overhead methods. This reflects both the more ambitious uses for SONET and the byte-oriented nature of SONET itself. Because overhead bytes in SONET play such an important role in many of the chapters that follow, this is a good place to outline the functions of the SOH, LOH, and POH bytes in SONET. Many of these functions overlap because there are separate Section, Line, and Path components on a SONET link that all need overhead benefits like error monitoring and failure protection.

Mention is made during this discussion of *composite* SONET signals. When several SONET data streams are multiplexing into one higher rate data streams (such as from three STS-1s to a single STS-3), the resulting SONET frame structure is called a *composite*. Every frame except a basic STS-1 frame is a composite frame of one form or another. Composites still retain *all* of the overhead bytes from each signal source. Obviously, an STS-3 frame that has nine columns of TOH, instead of just three from each of the STS-1s, still forms a unit and not just three frames traveling together. Thus, many of the "repeated" overhead bytes are essentially ignored ("undefined" in SONET talk) in composite frames. The full set of 27 TOH bytes is retained only in the first STS-1 of any composite. In the other STS-1s, all of the TOH bytes may be present, but many of them are neither examined nor processed.

SECTION OVERHEAD The Section Layer contains overhead information used by all SONET equipment along a network path, including signal regenerators. The SOH bytes are shown in Figure 6.8.

Figure 6.8
Section overhead

A1 Framing	A2 Framing	J0/Z0 (STS-ID) Trace/Growth
B1/undefined BIP-8	E1/undefined Orderwire	F1/undefined User
D1/undefined Data Com	D2/undefined Data Com	D3/undefined Data Com

The SOH is contained in the top three bytes of the first three columns in the basic STS-1 frame structure. The SOH's nine constituent bytes function as follows.

Framing Bytes (A1, A2): This two-byte code (1111-0110-0010-1000, or F628 in hexadecimal notation) is used for frame alignment. These bytes uniquely identify the start of each STS-1 frame and are not scrambled during the transmission process. Lack of scrambling makes their detection that much easier, and there is little chance of these bytes containing long strings of 0s. When multiple STS-Ns are sent in a higher rate STS-N, framing bytes must still appear in every STS-1 of the composite signal.

Section Trace (J0): This byte is used to trace the origin of an STS-1 frame as it travels across the SONET network. As SONET frames make their way across a series of links, the frames are often multiplexed to higher and higher levels, and then most likely back down again as they near their destination. Many senders process the SONET frames, and create new packages for the SPEs inside. It is obviously a plus to be able to trace the origin of any particular STS-1 frame to a particular piece of equipment from any point on the network. The exact format of this field is for future study. One possibility is that the source will send the SONET network address of the originating device in this field one byte at a time repeatedly; however, no standard has been defined. In the case of multiple STS-1s in an STS-N, the J0 byte is defined only in the first STS-1 because all of the frames must come from the same device. In other STS-1s in an STS-N, the use of this byte is just for Growth (Z0).

This overhead byte used to be known as the STS-1 ID (C1) byte. This single byte was used to uniquely identify the STS in question. The intent was to keep track of which frame is being transmitted at any time. The STS-1 ID was a binary number that corresponds closely to the number of STS-1 frames being sent as a composite STS-N. For example, if the signal being sent were a single STS-1, then the octet would be 0000-0001 (01 in hex). If an STS-N were being sent, the first STS-1 inside would be numbered as 0000-0001, the second as 0000-0010 (02 in hex), the third as 0000-0011 (03 in hex), and so on, up to N. Naturally, the STS-1 ID appeared in each STS-1 of an STS-N frame. However, as SONET speeds approach the limit of 13 Gbps and beyond, this 8-bit number threatens to limit the speeds at which SONET links can be deployed. Thus, a Section Trace is a much better use of this byte. However, since the full standard for the use of the Section Trace is not yet complete, the current use of the J0 bute is the same as that of the C1 byte.

BIP-8 (B1): A single, interleaved parity byte is used to provide STS-N error monitoring. BIP is an abbreviation for bit-interleaved parity. It performs a routine even-parity check on the previous STS-1 frame, after scrambling. The parity is then inserted in the BIP-8 field of the current frame before scrambling. SONET relies on even parity for its error-checking calculations. During parity checking, the first bit of the BIP-8 field is set so the total number of ones in the first positions of all octets in the previously scrambled STS-1 frame is always an even number. The second bit of the BIP-8 field is used in exactly the same way, except it performs a check on the second bits of each octet, and so on. This octet is only defined in the first STS-1 frame of a composite STS-N signal.

Orderwire (E1): This channel was a leftover from the days of copper wire carrier systems and historically was available to craft personnel for maintenance communications. In SONET, the channel is a 64 kbps voice path used for communication between remote terminals and regenerators. A telephone can be "plugged in" at the repeater location into a more or less standard voice jack, allowing craft personnel instant voice access on the fiber span. This octet is only defined in the first STS-1 of an STS-N frame. It is unlikely that anyone ever uses the Orderwire overhead in a SONET network, except in some rural areas, since craft personnel all carry cell phones today, but it seemed like a good idea at the time.

User (F1): The Section User Channel is reserved for use by the network services provider for its own application management activities. Its design is unspecified because it can be tailored by the network provider. F1 is only defined in the first STS-1 of an STS-N frame. Currently, many vendors of SONET equipment use the F1 byte, which operates at 64 kbps, for configuration and maintenance communications. A standard implementation of the F1 field is undefined, giving vendors a lot of flexibility, but limiting interoperability.

Data Communications Channel (DCC; D1-D3): These three bytes are used as a 192 kbps data channel for operations functions, such as OAM&P. These capabilities constitute a (if not *the*) primary advantage offered by SONET technology. The DCCs ship such things as alarms, administration data, signal control information, and maintenance messages. D1 through D3 are only defined in the first STS-1 of an STS-N frame. The use of this channel should comply with Telecommunicaitons Management Network (TMN) standards; it will be a while before specifications are widely used. Several interim methods are currently used. The line DCC is often called the Embedded Operations Channel (EOC) in SONET and the Embedded Communications (or Control) Channel (ECC) in SDH.

LINE OVERHEAD   LOH is contained in the bottom six bytes of the first three columns. This overhead is processed by all pieces of SONET network equipment, except for the regenerators. The LOH bytes are shown in Figure 6.9.

Pointer (H1, H2): The H1 and H2 Pointer is comprised of two bytes used to indicate the offset between the pointer bytes themselves and the beginning of the STS SPE. It allows for the dynamic alignment of the SPE within the allowable capacity of the envelope itself. In other words, because the SPE can begin anywhere within the Information Payload of a SONET frame (i.e., anywhere but the overhead fields), it would be nice to give the receiver a hint as to exactly where this may be. This is not to say that SPEs jump around haphazardly. After all, SONET is supposed to be synchronous. But once in a while, due to

Figure 6.9
SONET line overhead

H1 Pointer	H2 Pointer	H3 Pointer Action
B2/undefined BIP-8	K1/undefined APS	K2/undefined APS
D4/undefined Data Com	D5/undefined Data Com	D6/undefined Data Com
D7/undefined Data Com	D8/undefined Data Com	D9/undefined Data Com
D10/undefined Data Com	D11/undefined Data Com	D12/undefined Data Com
S1/Z1 Sync Status/ Growth	M0 or M1/Z2 REI-L/ Growth	E2/undefined Orderwire

(Handwritten annotations: "offset % itself & beginning of STS SPE (because of jitter & timing difference)"; "point for negative timing justification"; "Automatic Protection Switching")

jitter and timing differences, the start of the SPE may move one byte forward or backward within the Information Payload. The Pointer bytes provide a mechanism for the sender to inform the receiver of the correct position of the SPE at all times.

The actual inner workings of the H-bytes are straightforward. The first four bits of the H1 byte are called the new data flag (NDF). The NDF allows the pointer position to change when a major change occurs in the SPE position in the payload. When a new pointer value is introduced, the NDF is inverted (in this case 0110 changes to 1001), but the remaining fields stay the same. In this way, the H1 and H2 pointer bytes always point to the first byte of the SPE, which is the start of the POH bytes as well.

The next two bits, which are zeros during normal operation, have no significance: they are simply "place keepers."

The following ten bits, which overlap the H1 and H2 octets, constitute the actual pointer. In normal operation, this pointer field can have any value between 0 and 782 (the STS-1 frame size minus the overhead bytes). Values greater than 782 are undefined. When the pointer is all zeros (00 0000 0000), this indicates that the SPE begins immediately to the right of the H3 octet. For example, a value of 87 indicates that the SPE begins immediately to the right of the K2 octet, one row down. More details on pointer bytes will be examined in Chapter 8.

Pointer Action (H3): The pointer action byte is allocated to compensate for the SPE timing variations mentioned above. In the course of otherwise normal operation, the arriving SPE data rate will exceed the frame capacity. This means that within a 125-microsecond period, more than 783 bytes may be ready to be sent out in an SPE. If this excess were to be less than 8 bits, the extra bits would be just buffered sequentially and sent out as the first bits of the next frame; however, when a full byte (8 bits) has accumulated in the buffer, the pointer action byte is used to carry the "extra" 784th byte. This is called a *negative timing justification* and is governed by strict rules of SONET equipment operation. Naturally, a companion *positive timing justification* is used when 783 full bytes are not in the CPE buffer to fill the SPE exactly.

The H3 byte must be provided in all STS-1 signals within an STS-N signal. The value contained in this byte, when it is not used to carry the SPE data from a positive justification, is not defined and ignored. The H1 and H2 pointer bytes tell the receiver when the H3 byte is used for useful information. This byte is necessary due to the variations in timing across different service provider's networks or when CPE clocking feeds a SONET link.

The whole mechanism of "floating SPEs" with the H1, H2, and H3 bytes is quite complex and many service providers seek to minimize its use. In many cases, then, the value of the H1 and H2 bytes is locked to 522 (20A in hex), which points to row 1, column 4 of the next SONET frame. This method of "locking" the SPE to a fixed position in the STS-1 frame minimizes pointer justifications, but increases buffer requirements and management in the face of continued timing jitter.

BIP-8 (B2): Similar to the B1 byte, this BIP byte is used for STS-1 error monitoring. It calculates the parity check based on all bits of the LOH and the STS-1 frame capacity of the previous frame prior to scrambling. Note that the Section Overhead is specifically excluded, reflecting the hierarchical nature of the SONET overhead bytes. Then, the LOH places the BIP-8 into the B2 field of the current frame before the current frame is scrambled. This field must be present and used in all STS-1s of an STS-N signal.

APS (K1, K2): These two bytes communicate Automatic Protection Switching commands and error conditions between pieces of LTE. They are specifically used for link recovery following a network failure. These two bytes are only used in the first STS-1 of an STS-N signal. The full purpose and use of the K1 and K2 bytes are somewhat flexible, but are used in many pieces of SONET equipment to switch lines on SONET *rings* (more in Chapter 14). The K2 byte is also used to detect other types of line-level alarms on SONET links.

DCC (D4—D12): These bytes represent a 576 Kbps message-based channel used for shipping OAM&P messages between SONET line-level network equipment. This channel uses the same protocols as those used in the D1—D3 components of SOH. Typical messages may be for maintenance, administration, or alarms, and can be either internally generated or obtained from an outside source. In some cases, they may be manufacturer-specific. The DCC bytes are only defined in the first STS-1 of an STS-N signal. As with Section Overhead DCC bytes, TMN is supposed to be used. However, interim methods such as Transaction Language 1 (TL1) are often still used today.

Synchronization Status (S1): This byte is used to carry the synchronization status of the SONET network element. The use of this byte is defined for the first STS-1 in an STS-N. Only bits 5—8 are currently defined, leaving bits 1—4 for future functions. The purpose is to allow SONET equipment to actually choose the best clocking source from among several potential timing sources. This helps to avoid the creation of disastrous timing loops within the network. The S1 byte cannot prevent timing loops all but itself, however. In other STS-1s present in an STS-N, this byte is defined as a Growth (Z1) byte. Growth bytes are used for future functions as yet undefined by SONET standards.

STS-1 REI-L (M0) or STS-N REI-L (M1): The use of this byte is complex, and requires some background discussion. SONET frames may be a basic STS-1, or have a channelized STS-N structure with N STS-1s inside; other structures are possible, but do not affect the present discussion. Because two major frame structure types exist, two purposes for this overhead byte are necessary. It all depends on whether the STS-1 stands alone or sits inside an STS-N.

When the SONET link is an OC-1, which can only contain an STS-1 or an electrical STS-1, then this byte is the M0 byte. The M0 byte is used for a Line Level Remote Error function (REI-L), formerly called the line Far End Block Error. This conveys the Line BIP-8 (B2) error count back to the source. Bits 5—8 of the M0 byte are used for this purpose, and bits 1—4 are undefined.

Conversely, for an STS-N frame structure, this becomes the M1 byte. Not all STS-1s in an STS-N have the M1 byte. Only the "third" (as defined by SONET documentation) STS-1 in the STS-N will carry the M1 byte. Obviously, the value of N must be three or greater. The M1 byte has the same purpose and function as the M0 byte, with minor differences above the STS-48 level.

Growth (Z1, Z2): When the first two bytes of the last row of the SONET frame are not used for the S1 and M0 or M1 functions, these overhead bytes are defined as Growth bytes. Growth bytes are reserved for future functions as yet undefined by SONET standards.

Orderwire (E2): Similar to the E1 byte, this is a 64 kbps voice channel. It is only defined in the first STS-1 of an STS-N signal.

PATH OVERHEAD In addition to user data, the SPE contains POH bytes. These are processed at SONET STS-1 terminating equipment (usually CPE, but not always) because they travel as part of the payload envelope and are processed everywhere it is processed. The SPE contains nine bytes of POH. These bytes form a "column" in the SPE portion of the SONET frame, meaning the POH bytes are always in a column. However, because the position of the SPE can "float" within the STS-1 frame Information Payload area, the position of the POH bytes can float as well.

The POH bytes are shown in Figure 6.10.

STS Path Trace (J1): This field transmits a repeated, 64-byte, fixed-length string that enables the receiving Path Terminating Equipment to verify its connection to the device sending the SPE. This is a user programmable field. If no message has been loaded by the user, then a string of 64 null characters is sent. This field can be something as simple as the IP address or E.164 address (i.e., telephone number) of the CPE device. Because few customers have SONET CPE-retaining T-carrier equipment that just feeds SONET links, this field is typically used by the service provider.

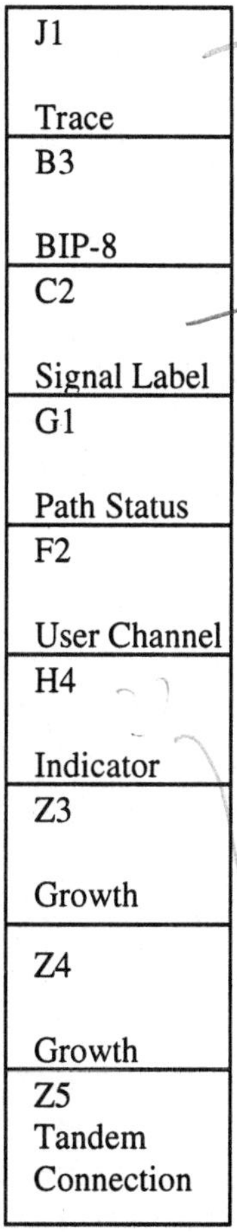

Figure 6.10
SONET path overhead

	J1
	Trace
	B3
	BIP-8
	C2
	Signal Label
	G1
	Path Status
	F2
	User Channel
	H4
	Indicator
	Z3
	Growth
	Z4
	Growth
	Z5 Tandem Connection

[handwritten: user programmable, identify sending device (e.g. IP addr)]

[handwritten: what SPE signal it is (e.g. VT, ATM cells, IP....etc.)]

Code in Hex	STS SPE Content
00	Unequipped (not used for live information)
01	Equipped – nonspecific payload
02	Virtual Tributaries (VTs) inside ("default")
03	VTs in locked mode (no longer supported)
04	Asynchronous DS-3 mapping
12	Asynchronous DS-4NA mapping
13	ATM cell mapping
14	DQDB cell mapping
15	Asynchronous FDDI mapping
16	HDLC-over-SONET mapping (used for IP)
CF	"Experimental" value for IP inside PPP
FE	Test signal mapping (see ITU Rec. G.707)

[handwritten: STS1 # for VT superframe = 4×STS1 frame]

When used by the service provider, in most cases the local telephone company, the J1 byte contains the CLLI™ code (pronounced "silly code") which stands for Common Language Location Identifier. The CLLI code is 8 to 11 bytes long and identifies a particular End Office (Central Office) within the national telephone system. Because the CLLI code is much less than 64 bytes long, the remaining J1 bytes are padded with nulls and terminated with ASCII carriage return (CR) and line feed (LF).

Path BIP-8 (B3): This field's function is analogous to that of Line and Section BIP-8 fields. It uses even parity, and is derived from the parity of the previous SPE prior to scrambling. Note that the Line and SOHs are specifically excluded, again reflecting the hierarchical nature of SONET overhead.

STS Path Signal Label (C2): The Signal Label tells network equipment what is contained in the SPE (i.e., how it is constructed). This allows for transport of multiple services simultaneously. This means that a single SONET device can actually interleave SONET frames (technically the SPEs) containing a DS-3 with frames containing ATM cells, or Fiber Distributed Data Interface (FDDI) data frames and so on.

The C2 Signal Label byte has some interesting history and carries a kind of warning to SONET vendors and implementers. SONET is essentially a point-to-point, trunking technology. That is, SONET does no switching based on SPE content: the SPEs can be cross-connected, but that is all. So it came as a surprise to equipment vendors when a label was included in the POH to identify SPE content. After all, end equipment is usually configured to send and receive one type of traffic and no other. Otherwise, it just won't work. If one end is set up to send IP packets, and the other end expects ATM cells, that is easy enough to find and correct.

So why bother labeling the SPE contents? The end equipment still has to be configured independently of the label. The answer is that given the huge bandwidths SONET/SDH can offer, it was seen as entirely possible that, on one SONET/SDH link, equipment could carry both IP packets and ATM cells and much else at the same time. But then the sender had to tell the receiver what was inside this particular SPE. That is what the signal label was invented for.

But none of the early implementers used SONET/SDH for much other than traditional voice, which was carried inside SPEs as *Virtual Tributaries* (VTs). The signal label value for the most common way of carrying VTs was 02 (0000 0010), so this became effectively the default value for the C2 byte in a lot of SONET gear. It could not even be changed in many cases, but the equipment hummed merrily along because it was configured to handle IP or ATM cells and did not bother to check the C2 byte. However, once SONET (and SDH) specifications matured to the point where mismatching the signal label to the traffic type generated an alarm, new SONET NEs went into alarm condition (or even took themselves out of service) if the C2 byte value did not match the traffic type in the SPE! Many vendors had to upgrade or patch their software and firmware quickly.

The moral of the story is that ignoring a specification just because it is not currently enforced can be a risky proposition. This is especially true of international standards. Currently defined values of the C2 byte are included in Figure 6.10. Some of these will be discussed more fully in the next chapter.

Path Status (G1): This field notifies the originating end of the Path performance and status of the entire duplex Path. It carries two maintenance signals, the B3 error count known as the Path Remote Error Indicator (REI-P), formerly called Path FEBE, in bits 1—4, and a Path Remote Defect Indicator (RDI-P) in bits 5—7. This allows the entire Path to be monitored from either end. The remaining bit is undefined.

Path User Channel (F2): This field, like its line/section cousin, can be used by the network provider for internal network communications. The F2 byte is also used in one odd case to carry Layer Management information generated by Distributed Queue Dual Bus (DQDB) networks.

Indicator (H4): ~~for VT superframe which contains 4 STS1~~ This field is used when a frame is organized into various mappings. Mappings determine the actual structure of the portion of the SPE carrying user data. One of the most common mappings is virtual tributaries. Several types of virtual tributaries are defined, but all exist to provide some degree of backward compatibility for SONET links. With virtual tributaries, it is possible for users to have a T-1 at each end of a private line, but SONET in the middle. SONET transports virtual tributaries as easily as ATM cells or anything else, using different mappings.

In ATM networks, the Indicator byte was used to denote cell boundaries, but this use is now officially "not recommended" (ITU talk for "totally obsolete"). This feature allowed SONET to transport such cell relay services as B-ISDN (ATM).

In DQDB networks, the Indicator byte is used to carry Link Status information.

Growth (Z3–Z5): These bytes are reserved for future use. However, in DQDB the Z3 byte is used to carry more DQDB Layer Management information.

Tandem Connection (Z5): Although still considered a Growth byte by Bellcore/Telcordia, ANSI has defined the Z5 byte for use as a Tandem Connection Maintenance Channel and a Path DCC. A tandem is a switching office that switches between trunks and, therefore, has no CPE devices on the end of the links. This can be a problem with SPE handling because no direct contact occurs with the originator of the SPE. Finally, the Path DCC performs the same functions as the Line and Section DCCs.

Figure 6.11 shows all SONET overhead bytes together. More details on SONET overhead values are given in Appendix A.

SONET Payload Pointers

Surely the most confusing part of any SONET overhead discussion is the Payload pointers. Payload pointers are one of the great innovations, and curses, of SONET. They have two primary functions. First, they are used for multiplexing synchronization in a plesiochronous environment. The multiplexing synchronization requirement has to do with the byte-interleaved nature of SONET multiplexing. Because bit-stuffing is not used in SONET, only full bytes can be multiplexed and sent in a SONET frame. The presence of extra bytes, or a shortfall of bytes at a CPE, will force SONET to change the position of the SPE within the STS frame. Payload pointers are used to help receivers figure all of this out.

What is a plesiochronous environment? Plesiochronous is a word coined by the ITU to mean just about everything except SONET, which is part of the ITU SDH. To the ITU, T-carrier, E-carrier, and anything else are part of the plesiochronous digital hierarchy (PDH). Officially, signals are considered to be plesiochronous if their "significant instants occur at nominally the same rate", with any variations constrained to specified limits. This is much the same as saying that two identical and equally accurate watches both tick 59, 60, or 61 times per minute (variations constrained to specified limits). But even when both count off exactly 60 ticks in the same minute (significant instants occur at nominally the same rate), there may be *phase differences* between the two. Within a given one-second interval, watch A may tick at the beginning of the interval, while watch B may tick toward the end of the interval. There are still 60

Figure 6.11
SONET overhead

A1 Framing	A2 Framing	J0/Z0 (STS-ID) Trace/Growth
B1/undefined BIP-8	E1/undefined Orderwire	F1/undefined User
D1/undefined Data Com	D2/undefined Data Com	D3/undefined Data Com
H1 Pointer	H2 Pointer	H3 Pointer Action
B2/undefined BIP-8	K1/undefined APS	K2/undefined APS
D4/undefined Data Com	D5/undefined Data Com	D6/undefined Data Com
D7/undefined Data Com	D8/undefined Data Com	D9/undefined Data Com
D10/undefined Data Com	D11/undefined Data Com	D12/undefined Data Com
S1/Z1 Sync Status/ Growth	M0 or M1/Z2 REI-L/ Growth	E2/undefined Orderwire

Section and Line Overhead

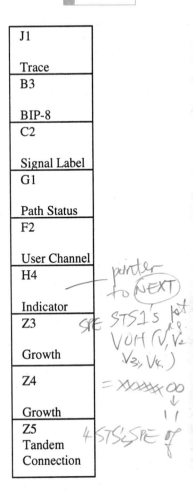

J1 Trace
B3 BIP-8
C2 Signal Label
G1 Path Status
F2 User Channel
H4 Indicator
Z3 Growth
Z4 Growth
Z5 Tandem Connection

Path Overhead

[handwritten annotations: "pointer to NEXT", "SPE STS1's 1st VOH (V₁, V₂, V₃, V₄) = XXXXXX 00", "4 STS's SPE of"]

ticks per minute, but the two watches are not guaranteed to tick at the same "instant." This concept is illustrated in Figure 6.12.

In other words, the network timing signal has jitters and wanders, but for the most part stays within predefined limits, although the phase can still vary. The PDH is "sort of" synchronous; however, even SONET is not

Figure 6.12
Two equally accurate
clocks with different
phases

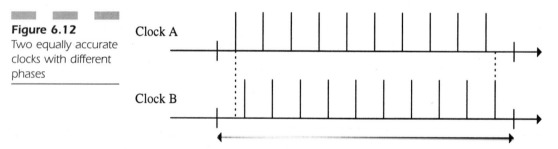

Figure 6.12
Two equally accurate
clocks with different
phases

Both clocks "tick" 10 times in one second

totally synchronous, especially when SONET payload must be filled and
emptied by plesiochronous T-carrier equipment. Of course, SONET is
much more stable, timing-wise, than is the plesiochronous T-carrier environment. But even SONET is not truly synchronous in the sense of
guaranteeing that all network clocks are timed from the same source.

SONET can be deployed in a fashion offering much better timing
characteristics than PDH. Payload pointers, needless to say, are superfluous in a totally synchronous network. However, all SONET equipment
must support payload pointer adjustments. This means that when
SONET links are deployed with T-carrier sources and destinations, or
when two service providers operate mid-span meets with SONET, payload pointers become critical in keeping the receivers informed about
the SPE's location in a SONET frame.

Another function of the payload pointers is to allow timing slips to
be handled in a SONET environment. If an incoming payload's timing
were to drift with respect to the SONET system—as would occur if the
timing network were not mesochronous—the pointers would also drift;
the pointers then must be recalculated for the payload periodically.

It should be noted that the use of pointers in SONET's synchronous,
byte-interleaved multiplexing is what makes direct add/drop multiplexing without complete de-multiplexing possible, even when the network
timing jitter is pretty awful. All of the constituent parts of SONET
payloads can easily be found, even when the source is totally plesiochronous.

Super Rate Payloads in SONET

This chapter has stated several times that a SONET STS-N frame contains N STS-1s. This is just another way of saying that SONET is as chan-

nelized as T-carrier. Thus, a DS-1 contains 24 DS-0 operating at 64 Kbps each. However, because a single DS-1 operates at 1.5 Mbps, perhaps the user's needs would be better addressed with a single channel operating at the full 1.5 Mbps, instead of 24 channels operating at only 64 Kbps each. This is, of course, what an unchannelized T-1 does for a user. Technically, the term should be *nonchannelized,* but everyone just says unchannelized.

This need may be even more critical in SONET. The unprecedented speeds available in SONET are capable of solving a number of user problems, however, not when an OC-3 operating at 155.52 Mbps is only useful as three STS-1s running at 51.84 Mbps. An unchannelized SONET link is needed. This would give a user just a raw bit rate at 155.52 Mbps (minus the STS-3 overhead). Of course, such an unchannelized frame structure exists in SONET but, in SONET, these are known as *concatenated* SONET frames that give a user a *super rate payload* because the rate is now greater than the usual 51.84 Mbps.

Using boxcars to represent STS-1 frames, a comparison can be drawn between channelized multiplexing of SONET frames to achieve higher transmission rates, and the unchannelized concatenation of SONET frames, which leads to the creation of larger payload envelopes that operate at the same higher rates. This is shown in Figure 6.13.

At the top of Figure 6.13, STS-1 frames are transmitted at a rate of 8,000 frames per second. This results in a data rate of 51.84 Mbps. In

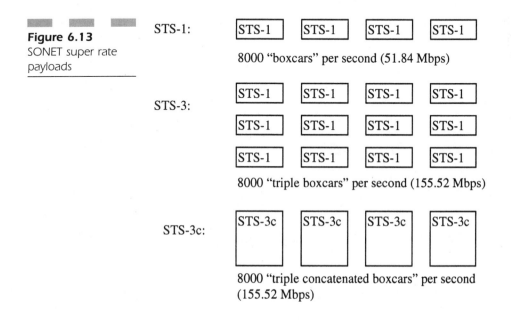

Figure 6.13
SONET super rate payloads

STS-1:

| STS-1 | STS-1 | STS-1 | STS-1 |

8000 "boxcars" per second (51.84 Mbps)

STS-3:

STS-1	STS-1	STS-1	STS-1
STS-1	STS-1	STS-1	STS-1
STS-1	STS-1	STS-1	STS-1

8000 "triple boxcars" per second (155.52 Mbps)

STS-3c:

| STS-3c | STS-3c | STS-3c | STS-3c |

8000 "triple concatenated boxcars" per second
(155.52 Mbps)

order to build up STS-N frames structures that operate faster, STS-1 frames can be multiplexed (by byte interleaving them) and these new larger frames are still transmitted at a rate of 8,000 per second. The middle of the figure shows three STS-1s multiplexed together to create an STS-3. Note that although the STS-1s are multiplexed in a single STS-3 sent on an OC-3, the STS-1s are still considered three independent frame streams and each has its own set of payload pointers. All of the user information must fit within an STS-1 frame SPE, and the frames are simply multiplexed prior to transmission so that the high speeds of fiber can be used more efficiently. This is just like multiplexing 28 DS-1s into a DS-3. The original user bit stream is still less than or equal to the DS-1 rate. The DS-3 allows greater speeds on the transmission media, but that is all.

At the bottom of Figure 6.13, SONET concatenation is illustrated. Here, three STS-1 frames are "pasted together" (concatenated) to create a single, large frame structure which, in this case, is called an STS-3c. Note the presence of the little "c." This lets people know that they are dealing not with three STS-1 frames inside the STS-3, but a single STS frame payload structure.

Because the individual frames are concatenated, one SPE is present and this only has one set of payload pointers. Concatenation is used when data is too fast to fit within an STS-1 frame. Larger frames need to be created for this situation. This is similar to providing a customer with a DS-3 that is not channelized—it is one big, 45 Mbps pipe. Note that an unchannelized DS-3 looks just like the DS-3 containing the 28 DS-0s, but the internal structure of the frame is not quite the same.

STS-3 SPEs

Because channelized STS-3 are the rule rather than the exception, the STS-3 frame structure contains a number of features designed expressly for combining three STS-1s into a single payload. First, the overall frame structure is derived by simply multiplexing the three input STS-1s a single byte at a time into the outgoing STS-3 frame structure. Of course, this overall STS-3 frame is still 9 rows, but has 270 columns (90 x 3). The effect of this byte-wise multiplexing is to effectively interleave the *columns* of the input STS-1s.

Thus, the 1st, 4th, 7th, and so on up to the 268th column of the STS-3 frame is totally derived from the bytes of the first STS-1. Likewise, the

2nd, 5th, 8th, and so on up to the 269th column of the STS-3 frame is totally derived from the bytes of the second STS-1. Naturally, the 3rd, 6th, 9th, and so on up to the 270th column of the STS-3 frame is totally derived from the bytes of the third STS-1. Because the first three columns of the STS-1s are the SONET overhead bytes, the net result is that the first 9 columns of the STS-3 frame are SONET overhead, while the remaining 261 columns are Information Payloads derived from the three input STS-1s.

There are three sets of overhead bytes, and three column interleaved payloads in each STS-3 frame. Because the error checking and many other SONET overhead functions only have to be done once for the whole STS-3 frame, overhead bytes as the B1 Section BIP-8 K1/K2 APS bytes are undefined and ignored except in the "first" STS-1 of the STS-3 (columns 1, 4, and 7). Also, what should be done when two error fields do not agree, such as the B1 Section BIP-8s? Their undefined status neatly solves the problem. The general structure of the STS-3 frame is shown in Figure 6.14.

Figure 6.14 shows a normal, completely channelized STS-3 with three STS-1s inside. Note that there is one frame (the STS-3 frame), but three

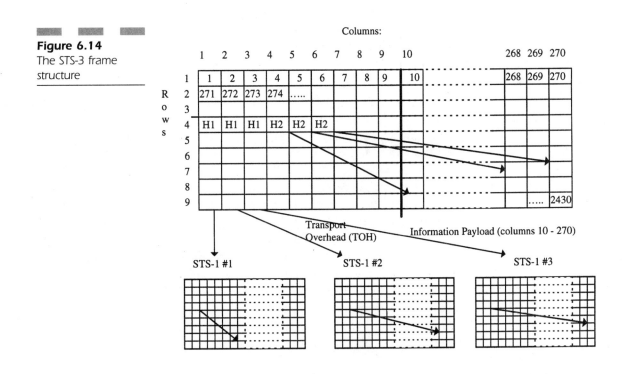

Figure 6.14
The STS-3 frame structure

sets of overheads and payloads (the three STS-1 payloads). Therefore, there are three sets of live payload pointers and three completely separate POHs. Of course, it is still possible to lock the three STS-1 payloads to the beginning of the STS-3 frame, as with the individual STS-1s; however, this introduces a great delay and then requires buffering.

Multiplexing occurs by columns, as usual, and the three sets of columns are "exploded out" of the STS-3 frame in Figure 6.14.

STS-3c (OC-3c) SPE

As nice as it is to combine three STS-1s into a single STS-3 for transmission over a single OC-3 fiber link (which is the whole point), this provides little advantage to the customer. Many customers do not need three STS-1s, but rather a single STS-3 operating at 155.52 Mbps (including overhead) to satisfy their bandwidth needs. Fortunately, SONET allows for the creation of *unchannelized* STS-3s, as well as other speeds. In SONET, this process is known as *concatenation;* thus, a concatenated STS-3 is an unchannelized STS-3 and contains not three, but one SPE. The designation for a concatenated STS-N is to add a lower case "c" to it. So a concatenated STS-3 becomes an STS-3c.

It is important to distinguish the lower case SONET "c" in terms of function from the upper case "C" used in T-carrier. Both mean "concatenation," but each uses the term in a different sense. An FT-3C, for instance, is two T-3s transmitting on fiber "pasted" together. But there are still two distinct T-3s, each operating at 45 Mbps. It makes no difference whether either is channelized into 28 T-1s or not. Two bit steams still operate at 45 Mbps. However, in SONET, an STS-3c is strictly unchannelized into a single bit stream operating at 155.52 Mbps. There are 9 columns of overhead (the overhead stays in SONET, no matter what), but only one SPE that spans the entire 261 columns of the Information Payload.

It is technically proper only to speak of an OC-3 as the physical transport for either an STS-3 or STS-3c. After all, senders and receivers care nothing about presence or lack of structure in the frames they transfer. However, just as T-carrier people came to speak of a "T-1" as an overall term for "DS-1" and much else, SONET people have come to speak of "OC-3c" to indicate that the STS-3 frames sent on the link are STS-3c frames; whether referred to as STS-3c or OC-3c, the structure is the same.

Figure 6.15 shows the structure of an STS-3c or OC-3c frame. The concatenation is indicated by putting special values in the H1/H2 pointers

Figure 6.15

STS-3c (OC-3c) frame structure

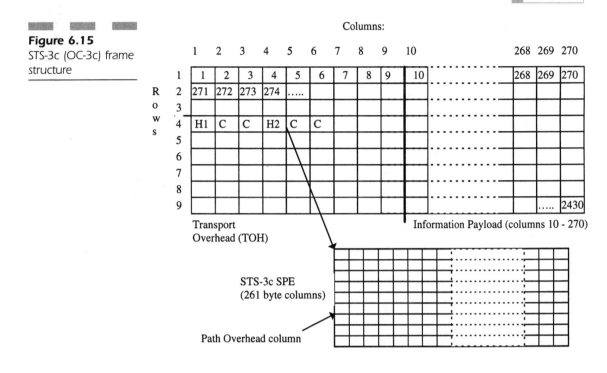

(called the *concatenation indicator*) where the pointers for the other STS-1s would ordinarily be positioned. The concatenation value is not a valid pointer value (i.e., not between 0 and 782); thus, no confusion over receivers exists. Obviously, there is only one SPE, so one set of H1/H2 pointers will do. Because only one SPE exists, only one set of POH bytes is needed. But it is still the first column of the SPE. As with STS-1 SPEs, the H1/H2 pointers must point to the first byte of the POH.

The SPE may still be set and locked into column 4, but this is another matter. There is, however, an important point regarding the pointer behavior when the SPE is allowed to roam within the STS-3c frame. Many people quickly observe that the single STS-3c SPE can start anywhere within the 261 columns of the Information Payload; the SPE can never occupy the overhead column positions. This gives 9 rows x 261 columns = 2,349 or 2,349 potential starting positions; however, the H1/H2 pointer is a single 10-bit field with valid values only up to 782. The other H1/H2 fields are lost to the concatenation indicator(s). How then can only the values 0 to 782 indicate starting positions from 0 to 2,348?

The answer is actually a pretty neat trick. SONET receiving equipment cannot distinguish between an incoming STS-3 frame or an STS-3c frame, except for the presence of concatenation indicators. When these indicators are detected, the concatenation indicator basically tells the receiver to increase the values of the H1/H2 pointers in the first overhead positions (columns 1 and 4) by a factor of 3. Thus, an H1/H2 offset of 1 really means 3 and an offset of 782 really means 2,346. The problem is neatly solved. (It should be noted that SDH pointers can operate in a totally different fashion—and usually do. SDH pointers are discussed in Chapter 18.)

Of course, the sending SONET equipment has a role to play in this scheme as well. The sender must buffer not 8 bits before moving the payload pointer value as before, but 24 bits (3 bytes) instead. Buffering also increases by a factor of 3. When SONET STS-3c SPE pointers adjust themselves, whether positive or negative, they do so in 3-byte units. For instance, a value of 2,346 points to the last 3-byte unit of the SONET frame. The whole idea is simple and effective.

As has been stated, an STS-3c has only a single SPE, which cannot be used for transporting existing DS-1s or DS-3s (that is what STS-1s are for). So what good is an STS-3c? The STS-3c was created expressly for the transport of ATM cells as part of a B-ISDN network. This corresponds to the STM-1 in the SDH hierarchy. In fact, the initial ITU B-ISDN specification for ATM cell transport had no provisions for the transport of ATM cells at speeds lower than 155.52 Mbps. Broadband needs speed above all.

Lower Rate Payloads

One of the advantages of the SONET standard is that it will transport all of the existing higher rate signals, such as DS-3. An STS-3c can carry a stream of ATM cells, but what about the lower rate signals of the T-carrier hierarchy? It seems wasteful to use a 51.84 Mbps pipe just to transport a 1.5 Mbps payload.

The SONET standard handles this situation by mapping the "lower" rate—which indicates rates lower than DS-3 in SONET talk—signals into "sections" of an STS-1 frame. These sections are each called a *virtual tributary.* Apparently the thought was that each lower rate payload forms a "tributary" flowing into the main data "stream." Whatever river-based image was intended, the terminology has stuck. Each of the virtual tributary sections are independent of each other and can carry different types of *sub-rate payloads.* A sub-rate payload is just a payload that does not map directly to an STS-1, but requires virtual tributaries.

This concept can be visualized by extending the boxcar example introduced to illustrate super-rate payloads. For sub-rate payloads, each boxcar can be divided into sections, or virtual tributaries. Each virtual tributary can contain any type of lower rate data; therefore, they all do not need to contain the same type of information. DS-1s, CEPT-1s, and DS-2s could all be in one "boxcar," which is the basic STS-1 frame.

In SONET, each STS-1 frame is divided into exactly seven virtual tributary groups (VTGs), and it is all or nothing. A single STS-1 frame cannot have, for example, only four VT groups and use the rest of the payload for something else (like ATM cell transport). Either the STS-1 frame is chopped up and sectioned off into exactly seven VT groups, or it is not.

Figure 6.16 shows this VTG process. For illustration purposes, the figure shows only three VTGs: a single boxcar can be divided into sections for the transport of different types of things.

Virtual Tributaries

There are four virtual tributary sizes currently defined in SONET. These are listed below:

- *VT 1.5:* Carries enough bandwidth to transport a DS-1 signal of 24 DS-0s at 64 Kbps. Each is contained in three 9-byte columns (27 bytes). A single VTG can carry 4 VT-1.5s.

- *VT 2:* Carries enough bandwidth to transport an E-1, 2.048 Mbps signal. It is contained in four 9-byte columns (36 bytes).

- *VT 3:* Carries enough bandwidth to transport a DS-1C signal. It is contained in six 9-byte columns (54 bytes).

- *VT 6:* Carries enough bandwidth to transport a DS-2 signal. It is contained in twelve 9-byte columns (108 bytes).

The seven VTGs in an STS-1 frame consist of 108 bytes each. Because each column in a SONET frame has 9 rows, each VTG occupies 12

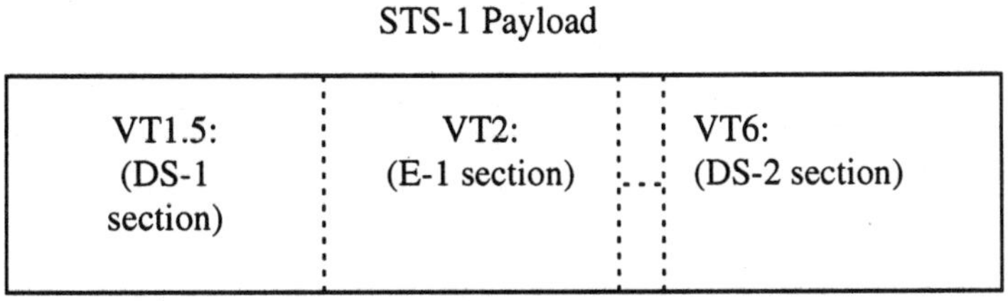

Figure 6.16
Lower rate payloads

STS-1 Payload

VT1.5: (DS-1 section)	VT2: (E-1 section)	VT6: (DS-2 section)

columns. Now, 7 VTGs of 12 columns each take up 84 of the 87 columns of the SPE. One of these "leftover" 3 columns is the POH, which must be present. The other two columns are reserved and have no currently defined function in a VTG divided SPE.

The mapping of various lower rate payloads into a Virtual Tributary is shown in Figure 6.17.

There is a pleasing symmetry in the VT mappings. Because every VTG consists of 12 columns of SPE payload, all of the four VT mappings can be carried in a VTG simply by varying the number of columns used. For instance, a DS-1 signal at 1.544 Mbps can be carried by 3 columns of an SPE. Three columns have 3 x 9 or 27 bytes. Because each DS-1 frame has 24 bytes plus a framing bit, there is plenty of room in 3 columns for the VT1.5, with a few bytes left over. Not surprisingly, these bytes are used for VT overhead. Thus, four VT1.5s can be carried in a single VTG (12/3). This is especially good because the 28 DS-1s inside a single DS-3 can be carried by an STS-1 efficiently (four DS-1s per VTG, and 7 VTGs per STS-1). It should be noted that a DS-3 can be mapped directly into an STS-1 without concern for any DS-1 inside. This is called an *asynchronous DS-3*, and is not covered by VT mapping, as shown in Figure 6.17.

Similar logic to that employed above reveals that an E-1 operating at 2.048 Mbps can be carried by 4 columns of a VTG, a DS-1C can be carried by 6 columns, and a DS-2 can be carried by all 12 columns. Happily, 3, 4, 6, and 12 are all factors of the column VTG, so things fit quite nicely.

The maximum capacity of an STS-1 used for Virtual Tributaries depends on the structure of the VTGs. Any VTG can technically carry any of the sub-rate VT types (e.g., VT1.5, VT2), but it will be much more

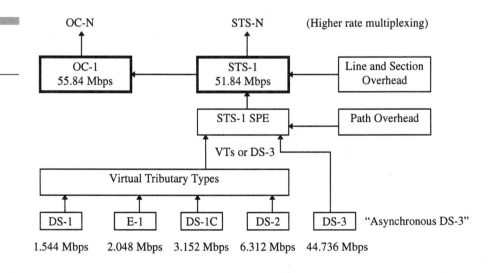

Figure 6.17
SONET with VT payloads

common to see STS-1s used for all one type of Virtual Tributary, especially DS-1, of course.

Table 6.2 shows the STS-1 capacity when used with various VT types. The table lists each VT type and which PDH signal type and rate it represents. The table then shows the number of bytes (always multiples of 9 row columns) needed to transport the sub-rate signal. Next, the number of this VT type in each 108 byte (12 column) VTG is given. Finally, the total number of VT type signals in an STS-1 frame is given, since each STS-1 can carry exactly 7 VTGs.

A single STS-1 employed to carry Virtual Tributaries can carry 28 DS-1s, 21 E-1s, 14 DS-1Cs, or 7 DS-2. Of course, more sophisticated mappings are possible. Several service providers have planned to carry a couple of DS-2 in some VTGs and use the rest of VT1.5s. For example, 3 VTGs could carry 3 DS-2s, while the other 4 VTGs could carry a total of 16 DS-1s (4 DS-1s per VTG). This flexibility is a nice feature of SONET Virtual Tributaries.

It is important to note that each of the VT groups total 108 bytes each. Because there are exactly 7 VT groups in each sub-rate SONET frame, the difference in the virtual tributary capacity is the total number of VT types in each 108-byte VT group. A VT1.5 has 4 DS-1s, but a VT3 has 2 DS-1Cs, both in a 108-byte package. Of course, VT types cannot be mixed *within* a single VTG.

An important characteristic of virtual tributaries is that they can be added to or dropped from the SPE without demultiplexing. This capability results from having pointers in the virtual tributary overhead.

VT Superframe

A couple of potential problems exist with VT mappings that have not been described yet. For instance, a DS-1 consists of more than just series of DS-1 frames one after the other. DS-1 frames are further organized into a series of 12 frames or 24 frames known as a *superframe*. When a

TABLE 6.2

Virtual tributary capacity

VT Type	PDH Signal	Rate	Bytes/VTG	#/VTG	#/STS-1
VT1.5	DS-1	1.544 Mbps	27	4 (4 × 27 = 108)	28
VT2	E-1	2.048 Mbps	36	3 (3 × 36 = 108)	21
VT3	DS-1C	3.152 Mbps	54	2 (2 × 54 = 108)	14
VT6	DS-2	6.312 Mbps	108	1 (1 × 108 = 108)	7

*VT superframe =
4 × STS 1 SPE*

DS-1 is transported inside an STS-1 VTG, it is important not only to preserve the DS-1 frame structure, but the entire superframe structure as well. There are other considerations involving timing differences, but the end result is that when an STS-1 is used for the transport of VTGs, a sequence of four STS-1 frames (actually the SPE sequence) is considered to be a VT superframe.

One immediate question involves the identification of the VT superframe sequence number by the receiver. Obviously, this sequence number must be put into a SONET overhead field somewhere. The solution was to basically reinterpret some of the SONET overhead bytes when the link was used to transport VTGs. Therefore, the SONET H4 POH pointer now takes on a structure that includes a 2-bit field indicating which frame of a 4-frame VT superframe sequence is being sent.

The VT superframe structure is embedded within a sequence of four SONET SPEs, naturally. VT overhead bytes are associated with the VT superframe as well. The structure of the VT superframe essentially mimics the overall operation of the STS-1 frame and SPE. That is, just as SPEs are allowed to float within an STS frame, so the VTs inside a VTG are allowed to float within a VT superframe. Just as with an SPE, the question is how to allow a receiver to find these floating VTs.

The answer is to create additional VT overhead bytes to serve as pointers to the VT payload starting position. When used to transport VTGs, the first byte following the POH J1 byte in each SPE becomes V1, V2, V3, and V4, following the repeating 0-1-2-3 counter in the H4 VT superframe sequence number field. The V1, V2, V3, and V4 bytes serve as payload pointers and mimic the behavior of the H1, H2, H3, and H4 pointer bytes in the STS-1 frame.

Consider as an example the transport of DS-1s inside VTGs on an STS-1. Because the DS-1 frame can begin anywhere in the 108 bytes sequence, it is important to let the receiver know just where this position is. So the V1 and V2 bytes form pointers at the beginning of the DS-1 superframe. Another VT overhead byte marks this position, the V5 VT POH byte. The V5 byte contains a scaled-down version of the error monitoring available in the STS-1 TOH, but is applied to VTs in this case.

There are even three more VT overhead bytes. Early documentation just referred to these bytes as V6, V7, and V8. Current SONET documents call these the J2, Z6, and Z7 bytes. Other documents even call them the J2, N2, and K4 bytes. In this text, the J2, Z6, Z7 terminology is used.

Just to make SONET overhead more confusing, there exists older documentation, mostly from ANSI, assigning functions to the Telcordia specification Z3, Z4, and Z5 POH bytes. This text mainly follows Telcordia specifications, but mention should be made of the alternative uses of these POH bytes.

The Z3 byte is called the F3 byte in some documents. This forms another vendor (or "user") path communications channel, just as with the F2 byte. The Z4 byte is also called the K3 byte, with full functionality not yet worked out. But by extension, some form of protection switching is intended, by analogy to the K1 and K2 bytes in the line overhead.

Only the use of the Z5 POH byte is unique to current ANSI specifications. This is called the N1 byte, or tandem connection byte. A *tandem* is a trunk-to-trunk switching office that is also often used for long-distance interexchange carriers (IXCs) to link up with the local SONET networks of the local exchange carriers (LECs). The problem on the path level is that SONET links passing from an IXC directly to a customer site do not allow status and error information from the CPE to reach the IXC through the line and section overhead. So the NI byte in the POH neatly solves this problem, since path overhead passes through the LEC network unchanged. There is an error count and an embedded communications channel to allow direct access to the CPE inside the N1 byte. Telcordia, as Bellcore, representing only LECs, rightly left the defining of this function up to ANSI.

A comparison of the STS-1 overhead bytes and their VT superframe byte counterparts are shown in Table 6.3. In the table, the functions of the STS-1 pointer bytes are listed first. H1, H2, and H3 exactly correspond to V1, V2, and V3 functions. V4 is reserved for future functions. In the POH byte area, V5 has no exact counterpart in the STS-1 POH world; however, J2 is the J1 counterpart, and if N2 and K4 become fully standardized, they will be the counterparts of the K3 (Z4) and N1 (Z5) STS-1 POH bytes.

VT Floating and Locked Mode

A quick understanding of Virtual Tributaries is not easy to accomplish. Coming on the heels of the new terminology of STS-1 itself, the vocab-

TABLE 6.3

STS-1 and VT overhead comparison

Frames	Pointer Bytes	Path Overhead Bytes
STS-1	H1 H2 H3 H4	J1, B3, C2, G1, F2, H4, Z3, Z4, Z5 (N1)
VT	V1 V2 V3 V4	V5, J2, Z6 (N2), Z7 (K4)

ulary of VTs is not readily added. Nevertheless, VTs remain an important consideration of SONET compatibility with existing digital links.

One other important concept relating to VTs must be discussed in this chapter. In addition to VT mappings, SONET defines two different VT modes of operation. They are called *floating* and *locked* mode.

The easiest way to understand the two is to recall that, inside an STS-1 frame, an SPE can float and begin almost anywhere within the STS-1 payload, except in the TOH columns. The value of the H1 and H2 pointer bytes indicate this position; however, it is also possible to lock the SPE position to always begin in Row 1, Column 1 of the STS-1 frame. In this case, the value of the H1 and H2 pointers is always 522. Of course, more buffering must be done on the payloads to ensure that the SPE always begins in this position.

The same may be done with VTs. Floating VTs minimize the delay and buffer size needed to load VTs into an STS-1 payload. But locked mode is obviously attractive for simplicity and less processor-intense activity. The trick is to make locked mode easy to implement; this has not turned out to be the case. Locking things such as DS-1s into a fixed position within a VT1.5 has proved to be extremely difficult, mainly due to the timing variations in the T-carrier digital hierarchy. These timing differences are discussed more fully in Chapter 8. For now, it is enough to point out that locked VT mode is not currently supported (i.e., "available") in SONET standards, although it continues to be defined. In locked VT mode, the V3 pointer byte is not used because the VT payload location cannot move.

To make things even more complex, along with the simple VT mapping types (e.g., VT1.5), SONET documentation defines three different ways that each VT type can be mapped into the VTGs of an SPE inside an STS-1 frame. These are *asynchronous, bit-synchronous,* and *byte-synchronous.* More details will be discussed in the next chapter.

All three are distinguished by important timing considerations, as may be inferred by their designations. A full discussion of the details will be given in Chapter 8. For now, all that needs to be understood is that each one has specific benefits to implementers and customers alike. Locking modes are attractive, but expensive and not currently supported. Floating modes are all supported, but while asynchronous floating mode is easiest, this mode makes it hardest to find VT payload contents, such as DS-0s. Byte-synchronous floating mode makes it easy to find any DS-0 anywhere, anytime, but is the most demanding from a timing standpoint because each DS-1 inside a VT1.5 must start a frame at exactly the same time. Bit-synchronous floating mode is a little less demanding from a timing standpoint because each DS-1 inside a VT1.5 can start a frame at different *instants.* Bit-synchronous equipment, however, needs more buffers than byte-synchronous equipment to handle the differences in frame arrival times.

All in all, the most common SONET equipment for VT transport should be byte-synchronous, floating VT equipment.

CHAPTER **7**

SONET Payloads

It is apparent that SONET (and SDH as well) is much more complex than either T-carrier or E-carrier when it comes to information-carrying capabilities. In the simpler T-carrier world, the structure and information content of an individual DS-0 or even an unchannelized DS-1 was totally up to the user or customer. Anything that fit into a 64-kbps DS-0 channel (in some cases, the DS-0 was limited to 56 kbps) popped out at the other end of the T-1 link. T-carrier expected voice in the DS-0, but anything else that "looked like voice" went right through the link. However, care was required in setting up T-1 links that carried computer data and not voice. A service provider could cross-connect voice, but it made no sense at all to try to reach inside a DS-0 carrying IP packets 8 bits at a time and "cross-connect" a piece of an IP packet.

However, since T-carrier and E-carrier expected to find voice inside their individual channels, there was nothing at all in the frame formats of either T-carrier or E-carrier to specifically identify the content of a T-1 or E-1 frame. Things are very different in SONET and SDH. This chapter considers the more popular SONET payload contents and formats. A later chapter considers SDH payload contents.

This is not to say that SONET and SDH do not *expect* to find voice inside their payloads most of the time. In spite of the popularity of the Internet and the Web, the majority of SONET and SDH links still happily carry channelized, digitized voice at 64 kbps inside their payload areas. As shown in Chapter 6, equipment vendors and service providers alike got used to the "default" Signal Label of the POH indicating voice channels inside the SONET SPE. But at least SONET and SDH allow for the possibility that payloads might not be digitized voice. And that makes all the difference.

This chapter begins by considering the structure of the SONET SPE when used for voice channels. These "voice" channels might not actually be carrying true digitized voice. The DS-0 might easily be carrying data from computer to computer. The point is that this DS-0 bit stream *looks exactly like voice* to the service provider and can be cross-connected or otherwise processed without regard to the details of the channel bit stream structure. Then the chapter considers three other popular SPE contents and provides some details on how each is carried inside the SONET SPE.

The four SPE contents or payload types considered in this chapter are:

■ Virtual Tributaries (VTs; channelized, digitized 64-kbps DS-0s)

■ Asynchronous DS-3 (unstructured DS-3 frames)

■ ATM cells (SONET as a B-ISDN transport)

■ Packet over SONET (POS; any Layer 3 packet, but almost always IP)

It is safe to say that the vast majority of SONET links in North America carry one of these four types of traffic in the SPE. As the values of the Signal Label (C2) byte listed in the previous chapter show, there are more than just these four possibilities for SPE content. But most service providers have never seen anything else inside an SPE but one of the four types of traffic listed here. And many service providers have seen nothing but 64-kbps voice inside their SPEs.

All of these payload types apply to the basic STS-1 level of the SONET hierarchy. In addition, the ATM cell and Packet over SONET payloads can be carried inside the larger SPEs available on STS-3c (OC-3c) and STS-12c (OC-12c) frames and links.

Virtual Tributaries

As important as Virtual Tributaries are as SONET SPE contents, there is a tendency to over-emphasize the details of VT operation in SONET, so discussions of SONET VTs tend to take over a chapter or even a whole book. This section will therefore necessarily be somewhat limited in the amount of detail offered. All the essentials are here, but the operational details of things like VT alarms are not given exhaustive treatment. Most SONET equipment vendors are more than willing to supply details regarding how their particular SONET NE handles these issues. There are also slight differences from vendor to vendor. The interested reader is referred to the individual vendors in such cases.

This emphasis and high-level discussion makes the topic manageable in a chapter investigating three other payload types. All that is really necessary to fill in the blanks, so to speak, that this section might leave is to realize that the intention of SONET VTs is to allow existing voice cross-connects and SONET-aware cross-connects to treat the voice channels in a SONET SPE exactly the same way they would a voice channel riding a T-1.

The figures in this chapter will offer slightly different representations of the SONET frame seen in previous chapters. Since the topic in this chapter is the form and function of the SONET SPE, the figures will always represent only the SPE itself. Keep in mind that the SPE would still be offset or locked in position in the overall SONET frame structure and found by the value of the H1 and H2 pointer bytes. But this chapter will deal with a 9-row, 87-column SPE structure, not the full frame. The first column will always be the Path Overhead column. So the total size of the SPE is 783 bytes (9 × 87), but the area available for information is 9 bytes less (the POH column), or 774 bytes. This is shown in Figure 7.1.

$C2 = 0 \times 02$
$= VT$

When the Signal Label POH byte (C2) is set to a value of 02x (02 in hex, or 0000 0010), then the SONET receiver knows that the arriving SPE must be carrying VTs. The first thing to realize is that the number of VTs present is not arbitrary. Each SPE must be divided into exactly seven *Virtual Tributary Groups* (VTGs). This is the first major division of the SPE for VTs.

The columns assigned to the seven VTGs in the SPE mimic the pattern established for multiplexing three STS-1s into a single STS-3 frame. That is, multiplexing proceeds byte by byte from each source until an entire row of the frame is filled. The result is that the multiplexing process creates a column-by-column structure where each column belongs to a separate multiplexing source. This frame creation was detailed in Chapter 6. An SPE consisting of seven multiplexed VTGs follows the same pattern. The intent is the same: to minimize the impact of delay on voice by interleaving sources instead of employing buffers.

The only complication is that 7 does not divide 86 columns equally: there are two columns left over. So 86 columns/7 VTGs = 12 columns for each VTG, with a remainder of 2 columns. The solution here is to "skip over" two columns when multiplexing the seven VTGs into an SPE. The two columns "reserved" (really, "ignored": their content is defined as "fixed stuff" bytes that must be ignored by the receiver) are columns 30 and 59 of the SPE, counting the POH column as column 1. Once this interleaving process is defined, the structure of an STS-1 SPE containing seven VTGs can be represented as in Figure 7.2.

In fact, now that this structure for the VTGs has been defined, it is a simple matter to generate a table showing the columns occupied by each

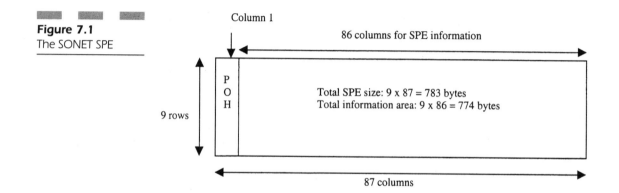

Figure 7.1
The SONET SPE

Column 1

86 columns for SPE information

P O H

9 rows

Total SPE size: 9 x 87 = 783 bytes
Total information area: 9 x 86 = 774 bytes

87 columns

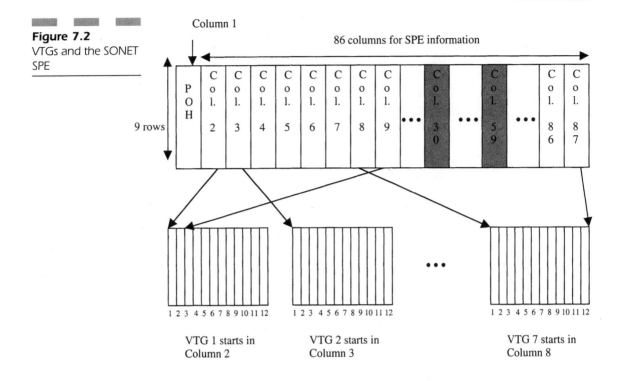

Figure 7.2
VTGs and the SONET SPE

VTG 1 starts in
Column 2

VTG 2 starts in
Column 3

VTG 7 starts in
Column 8

of the VTGs, which are always numbered VTG 1 through VTG 7. This is shown in Table 7.1.

Several features of the table are noteworthy. First, the table starts with column 2 of the SPE. This just reflects the position of the POH column always as column 1 of the SPE. Next, the VTG column assignments skip

TABLE 7.1

SPE columns occupied by the VTGs

Virtual Tributary Group	SPE Columns Used by the VTG
1	2 9 16 23 31 38 45 52 60 67 74 81
2	3 10 17 24 32 39 46 53 61 68 75 82
3	4 11 18 25 33 40 47 54 62 69 76 83
4	5 12 19 26 34 41 48 55 63 70 77 84
5	6 13 20 27 35 42 49 56 64 71 78 85
6	7 14 21 28 36 43 50 57 65 72 79 86
7	8 15 22 29 37 44 51 58 66 73 80 87

over columns 30 and 59, columns that would normally be between VTG 7 and VTG 1.

It is important to realize that when an SPE is divided into seven VTGs, the SPE cannot carry information organized into anything *but* VTGs. Once the VTG columns are established inside the SPE, that is all that the SPE can be used for. This does *not* mean that the Signal Label cannot change with the arrival of the *next* SPE: Signal Labels can actually change, SPE by SPE. However, this feature of SONET was controversial when introduced and has never been used on a large scale. Certainly it makes no sense to flip-flop Signal Labels on an STS-1 link when the intention is to carry DS-0 voice. There is just no room for anything else on that link. The 8000 SONET frames (SPEs) generated each second will carry 8-bit voice samples in each column and row byte, yielding 64 kbps, but no more.

Although this discussion of VTGs is limited to SONET in this chapter, it is worth mentioning that the SONET VTG structure is exactly the same in SDH. Of course, the name and acronym must be different. So in SDH, a VTG is called a *Tributary Unit Group-Level 2*, or TUG-2.

So far, the process of dividing the SPE into VTGs has produced seven identical structures consisting of 12 columns each, spread across 84 columns of the SPE. So each VTG has exactly 108 bytes (9 bytes in a column, and 12 columns). The next step is to look inside each of the VTGs and see the structure of this basic 108-byte unit.

Inside a VTG

Just because all VTGs form identical 108-byte units at different places in the SPE does not mean that the content of all VTGs is the same. There are four different forms that individual VTGs can take on. Although VTGs are optimized for the transport of DS-0 channels, not all DS-0s arrive at a SONET NE inside the same level of the T-carrier hierarchy. So the four possible VTG formats are:

- VT1.5, used for DS-1 transport (1.544 Mbps)
- VT2, used for E-1 transport (2.048 Mbps)
- VT3, used for DS-1C transport (3.152 Mbps)
- VT6, used for DS-2 transport (6.312 Mbps)

The only real surprise on the list is the definition of the VT2, used for E-1 transport. E-1 support is provided in SONET because whenever links cross a border beyond the United States, international standards apply. With a VT2, a service provider in another country can easily hand off an E-1 to a SONET NE and all the DS-0 channels inside will be carried to the end-user equipment. As an aside, a real problem before SONET was that 30 voice channels on an E-1 arriving in the United States had to be handed off to a DS-1 with only 24 voice channels! Either 6 channels could not be used (but were paid for) or a second DS-1 had to be purchased (but only 6 channels were used). SONET VT2s are a much better solution.

Of the four possible VTG structures, the most important is the VT1.5. Since there are many more DS-1s in service in North America than DS-1Cs or DS-2s, SONET links will most likely encounter DS-0s inside a DS-1 than in any other form. Even when T-carrier links arrive at a SONET NE in the form of a DS-3, chances are that the DS-3 is carrying exactly 28 DS-1s, each with 24 DS-0s. As it turns out, when an STS-1 is divided into seven VTGs and each VTG is structured as a VT1-5, the carrying capacity of the STS-1 is exactly the same as the capacity of a DS-3: 672 DS-0 channels running at 64 kbps.

So a SONET SPE consisting of VT1.5s can accept and carry and cross-connect 28 separate DS-1s or a single DS-3 containing 28 DS-1s. This is no accident. As has been pointed out, the whole reason that the STS-1 level exists in SONET is to be backward-compatible with existing DS-3 links (and the DS-1s most likely inside them). This allows an easy, gradual migration to an all-SONET network with flash cutovers and the like.

VT Structures

Each VTG is 108 bytes long (9 rows, 12 columns). However, the structure and carrying capacity of the VTG varies with VT type. In other words, a VTG containing DS-1s is fundamentally different from a VTG containing E-1s, and so on, for all four VT types. The carrying capacity of each VT type is determined by the frame structure of the T-carrier or E-carrier link feeding the SONET NE. It is probably easier to look at each VT type in turn by example rather than attempting to describe the structure in the abstract.

Consider VT1.5. It is designed to carry DS-1s. DS-1s consist of frames generated 8,000 times per second with a structure of 193 bits. The 24 DS-0s, each 8 bits, make 192 bits. The 193rd bit is the framing bit. The framing bit is important, but look at the 24 DS-0s by themselves. The 24 DS-0s do not divide the 108-byte VTG evenly. There would be 12 bytes left over (108/24 = 4, with 12 bytes remainder). So four DS-1s would fit nicely inside a single VTG. But what about the leftover 12 bytes? Well, there are four DS-1s. The 12 bytes are distributed between the four DS-1s inside the VT1.5 so that each DS-1 gets 3 bytes of *VT overhead* (VTOH). What is this overhead for? The same types of things as SONET overhead: alarms, performance information, and other OAM&P functions.

So each DS-1 now requires 24 bytes of the VT1.5 for the DS-0 and gets another 3 bytes for overhead. That makes 27 bytes total. And since SPEs (and SONET frames) are 9 rows, the DS-1 fits nicely in 3 columns of the 12-column VTG! Further, 4 DS-1s of 3 columns each precisely fill the 12-column VTG. So the structure of a VTG according to VT1.5 looks like Figure 7.3.

In the figure, each column of the VTG is labeled A, B, C, D. These represent the four DS-1s inside the VTG. The multiplexing of the DS-1s inside the VTG is done column by column, as expected and in keeping with the general SONET/SDH columnwise multiplexing approach. Above the VTG, the columns that comprise each VT1.5 are given under their respective representations. The first VT1.5 (labeled VT1.5 #1 [A]), has arrows pointing to the A columns of the VTG. The other three VT1.5s do not have arrows, but only to keep the clutter in the figure to a minimum.

The key point is that each VTG structured according to VT1.5 carries four interleaved DS-1s. If each VTG is structured according to VT1.5, then the total DS-1 capacity of the SONET STS-1 SPE is 28 DS-1s (4 x 7).

Figure 7.3.
A VTG structured according to VT1.5.

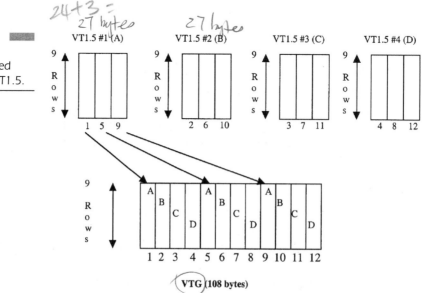

It is no accident that this is also the number of DS-1s inside a DS-3. Notice that the DS-3 *frame structure* is not carried directly inside the SONET SPE. (Recall that DS-3 uses bitwise multiplexing, not byte-multiplexing, as does SONET/SDH.) Rather, the 28 DS-1s inside the DS-3 frame are extracted by the SONET NE and then repackaged as 28 DS-1s inside seven VTGs.

The other three VT types' carrying capacities can be determined in the same way by considering the size of the frame each VTG must carry. For VT2, the E-1 frame has 30 DS-0s, plus 1 byte for synchronization and 1 byte for signaling. These 32 bytes do not divide the 108-byte VTG evenly. There can be three E-1s inside the VTG and, once again, 12 bytes left over ($3 \times 32 = 96$). These 12 bytes form VT overhead as before, but in the case of VT2s, the 12 VTOH bytes are distributed among three E-1s, not four DS-1s. So the number of overhead bytes allotted to each E-1 is 4, not 3. When all seven VTGs are carrying VT2s, the carrying capacity of the SONET SPE is 21 E-1s (3×7).

A VT3 is designed to carry a DS-1C. A DS-1C is basically two DS-1s pasted together with some additional overhead bits. So the DS-1C frame structure is twice as large as the DS-1 frame and carries 48 DS-0s. So a VTG can carry only two DS-1Cs, again with 12 bytes left over for VTOH ($48 \times 2 = 96$). The number of overhead bytes allotted to each DS-1C is now 6. When all seven VTGs are carrying VT3s, the carrying capacity of the SONET SPE is 14 DS-1Cs (2×7). Note that this is half the DS-1 capacity, but each DS-1C carries twice as many DS-0s, so the overall channel capacity is still the same (672 DS-0 channels).

A VT6 is for DS-2. The frame structure of a DS-2 is rather complex, but the bit rate of a DS-2 is 6.312 Mbps and a DS-2 frame carries 96 DS-0s. These 96 bytes fit inside the 108-byte VTG nicely, with the same 12 bytes left over for VTOH. This time, all the VTOH goes for the single DS-2 inside the VTG. When all seven VTGs are carrying VT6s, the carrying capacity of the SONET SPE is 7 DS-2s (1×7).

All of this is most easily expressed in a table. Table 7.2 lists all of the important parameters of the four VT types.

In the table, the numbers in the "Columns per VT" and "Number in VTG" must multiply out to 12, which is the number of columns in each VTG. The "Maximum Number in SPE" column applies when all seven VTGs carry that type of VT, as does the "Maximum DS-0s" entry.

The VT structures of VT2, VT3, and VT6 are shown in Figure 7.4. This figure follows the same conventions used in Figure 7.3.

Before looking at each VT type in a little more detail, there are a few rules that apply to VTGs and the VT types they contain that should be

TABLE 7.2

SONET Virtual
Tributaries

VT Type	Tributary	Columns per VT	Number in VTG	Maximum number in SPE	Maximum DS-0s
VT1.5	DS-1	3	4	28	672
VT2	E-1	4	3	21	630
VT3	DS-1C	6	2	14	672
VT6	DS-2	12	1	7	672

examined first. The first rule has already been stated: when a SONET
SPE is divided into VTGs, there are exactly seven VTGs, as detailed in
Table 7.1. The two rules that apply to VTs themselves are given as follows.

Each VTG can have a different VT type. That is, VTG 1 can be structured
according to VT1,5, VTG 2 structured according to VT6, and so on
through all seven VTGs. Although this mixing of VT types is allowed,
in practice this is seldom done.

Each individual VTG can contain only one VT type (VT1.5, VT2, VT3, or VT6).
That is, if a given VTG is structured according to VT1.5, then no other
type of VT can be mixed into that particular VTG, either in that
frame or in subsequent SPEs. VT types are an all-or-nothing
proposition. VT types are configured at each end of the SONET link
to be one type. There is no real justification for this limitation, besides
a wish not to make life too hard for SONET NEs. There *is* a VT Signal
Label in each VTG, but this Signal Label is not used for the same
purpose as the POH Signal Label (C2). A further problem is that the
basic unit of the VTG is *not* a single SPE. The basic unit of the VTG is
a sequence of four SONET SPEs, called a VT superframe. Allowing
changes of VT types would require careful handling of VT superframes. So each individual VTG must contain only one VT type, as
identified by the configuration of the SONET NEs.

It is tempting to look at the 108-byte VT1.5 structure, for instance, and
decide that each and every SPE should carry a single DS-1 frame locked
into position so that each DS-0 always appears at the same place in the
VT1.5. This would make cross-connecting the DS-0s very easy—as easy as
finding a DS-0 inside a DS-1. This is what *payload visibility* in SONET is
all about.

This *locked mode* method of operation was allowed in SONET, but is
no longer supported. VTs must be allowed to float within the SPE, in
the same way that SPEs float inside their SONET frames and are found

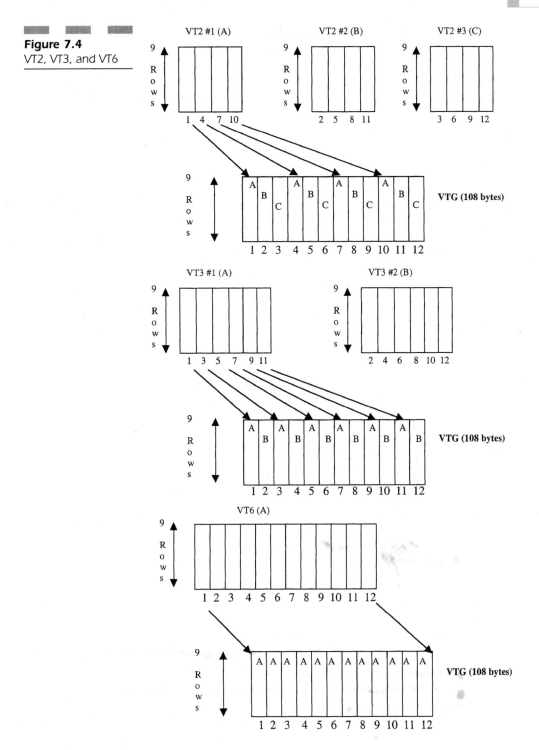

Figure 7.4
VT2, VT3, and VT6

with the H1/H2 pointer bytes. The issue with locked VT mode was that the arriving DS-1 clocks varied so much that processing, buffering, and loading locked VTs proved to be very troublesome and more effort than could be easily justified. So *floating mode* is the rule today. (The POH Signal Label [C2] value of 03x [03 in hex, or 0000 0011 in binary] is still defined, but only for backward-compatibility to prevent future mappings from using the 03x value and thus conflicting with locked mode legacy equipment.)

However, if VTs are allowed to float inside their respective VTGs, a whole system of pointers and other forms of VTOH had to be specified. This is exactly what was done, especially for the key VT1.5 type. Note that locked mode offered little advantage over floating mode and required a lot of extra work, since the floating mode VTOH had to be supported in SONET NEs handling VTs anyway.

So each VT type was given one or more special *mappings* to determine the form and function of the VTOH in each case. The key component of the VT mappings is the unit in SONET SPEs known as the *VT superframe*.

VT1.5 Superframes

In this section, the basic unit under discussion is the VT itself. In other words, each of the VT1.5s (for example) in Figure 7.3 will be presented as if it were a unit in and of itself. There is nothing wrong with thinking of the VT1.5 like this, but it should be kept in mind that things are really a lot more complicated. For instance, consider the VT1.5 in Figure 7.3 labeled "VT1.5 #3 (C)." Note that the three columns of this VT1.5 actually occupy columns 3, 7, and 11 of the VTG that the VT1.5 is configured in. And suppose that this is the 5th VTG of the SPE. A look a Table 7.1 will show that this VTG occupies columns 6, 13, 20, and so on, of the SPE itself. Not only are the VT1.5 columns spread out, but so are the VTG columns! Where in the overall 87-column SPE is VT1.5 #3 of VTG #5?

As it turns out, once all of this reverse mapping is done, VT1.5 #3 of VTG #5 occupies columns 20, 49, and 78 of the overall SPE. This is a good place to present a table for VT1.5, detailing the location of all 28 DS-1s inside the SPE (Table 7.3). Note that this table is completely valid only when all seven VTGs are VT1.5s. The only complexity in constructing such a table is that the DS-1s are organized and numbered by VTG order, not VT order. So *all* of the first VT1.5s (the A's) in all seven VTGs load first into the SPE. The VT1.5 #2s (B) follow, and so forth. In other words, the DS-1 numbering is consecutive by SPE position, not VT position. To

TABLE 7.3

DS-1s inside a
VT1.5 SPE

DS-1 No.	VTG No.	VT No. (letter)	VTG Columns	SPE Columns
1	1	1 (A)	1, 5, 9	2, 31, 60
2	2	1 (A)	1, 5, 9	3, 32, 61
3	3	1 (A)	1, 5, 9	4, 33, 62
4	4	1 (A)	1, 5, 9	5, 34, 63
5	5	1 (A)	1, 5, 9	6, 35, 64
6	6	1 (A)	1, 5, 9	7, 36, 65
7	7	1 (A)	1, 5, 9	8, 37, 66
8	1	2 (B)	2, 6, 10	9, 38, 67
9	2	2 (B)	2, 6, 10	10, 39, 68
10	3	2 (B)	2, 6, 10	11, 40, 69
11	4	2 (B)	2, 6, 10	12, 41, 70
12	5	2 (B)	2, 6, 10	13, 42, 71
13	6	2 (B)	2, 6, 10	14, 43, 72
14	7	2 (B)	2, 6, 10	15, 44, 73
15	1	3 (C)	3, 7, 11	16, 45, 74
16	2	3 (C)	3, 7, 11	17, 46, 75
17	3	3 (C)	3, 7, 11	18, 47, 76
18	4	3 (C)	3, 7, 11	19, 48, 77
19	5	3 (C)	3, 7, 11	20, 49, 78
20	6	3 (C)	3, 7, 11	21, 50, 79
21	7	3 (C)	3, 7, 11	22, 51, 80
22	1	4 (D)	4, 8, 12	23, 52, 81
23	2	4 (D)	4, 8, 12	24, 53, 82
24	3	4 (D)	4, 8, 12	25, 54, 83
25	4	4 (D)	4, 8, 12	26, 55, 84
26	5	4 (D)	4, 8, 12	27, 56, 85
27	6	4 (D)	4, 8, 12	28, 57, 86
28	7	4 (D)	4, 8, 12	29, 58, 87

find the 15th DS-1 inside a VT1.5-structured SPE (perhaps in order to cross-connect the whole DS-1 or a DS-0 inside), all a SONET NE need do is extract the 16th, 45th, and 74th columns of the SPE. Naturally, other VTG structures locate other VTs in the SPE using other rules.

The whole point of the exercise is to help you realize that even when a VT1.5 superframe is treated as a unit in this section, 12 columns of the SPE are actually involved.

The VT superframe is defined as four consecutive SPEs. Since each SPE arrives in 125 μseconds, the entire VT superframe repeats every 500 μseconds. The SPEs are numbered 00, 01, 10, and 11 in binary to help receivers determine the VT superframe boundaries. All VT superframes begin with SPE 00 and end with SPE 11. The two bits involved are in the H4 byte of the SPE POH and are called the *multiframe pointer* (some SONET documentation calls the VT superframe the *VT multiframe*).

When structured according to VT1.5, each of the four VT1.5s in the VTG occupies three columns and is therefore composed of 27 bytes. This is just another way of saying that there are four DS-1s in the VT1.5, and each VT1.5 gets 3 VTOH bytes per SPE. The first VTOH byte in each of the four SPEs making up the VT superframe is special. These are the V1, V2, V3, and V4 VTOH bytes. They are found by the value of the H4 multiframe pointer in the POH of the *previous* SPE. So a 00 value in the H4 POH byte of an SPE says, essentially, "get ready for the V1 VTOH byte in the *next* SPE."

Subtracting the V1, V2, V3, or V4 VTOH bytes from the 27-byte VT1.5 leaves 26 bytes per SPE. Now, 24 of these bytes are for the DS-0 channels inside the VT1.5, so this leaves 2 other bytes per SPE for other functions. Since there are four SPEs in a VT superframe, a total of 8 bytes in the VT1.5 form additional overhead. One of these is the V5 VTOH byte. The position of the V5 byte can be anywhere in the VT1.5 superframe, except for the initial byte position reserved for the V1, V2, V3, or V4 bytes in each SPE.

What are all these V overhead bytes introduced so far for? The V1 and V2 bytes perform many of the same tasks for floating VTs as the H1, H2, and H3 bytes in the LOH do for floating SPEs. For example, some of the bits in the V1 and V2 bytes form a pointer to the V5 byte, which is the VT path overhead byte. The V5 byte always determines the beginning of a sequence of four DS-1 frames spread over the VT1.5 superframe. The V3 and V4 bytes help the receiver to determine when the location of the V5 VT path overhead byte must move, most likely because of timing variations between the arriving DS-1 clock and the SONET clocks. More details about these timing considerations will be explored in the next chapter. For now, it is enough to realize that the V1/V2 bytes point to the V5 byte in the VT superframe.

The relationship between the SPE POH H4 multiframe pointer and the VT1.5 superframe V1 through V5 VTOH bytes is shown in Figure 7.5. The position of the V5 byte in the figure is totally arbitrary and for illustration only. The byte immediately following the V3 byte, marked with an *, is the *positive stuff VT byte* and performs in ways similar to the H3 byte in the SONET LOH discussed in the next chapter.

So the 108-byte VT1.5 superframe spans four SPEs. It is important to realize that the size of the VTG, also 108 bytes, is just a coincidence. A 108-byte VTG is in a single SPE. VT1.5 superframes span four SPEs. Once the four V bytes are subtracted, the VT1.5 has 104 bytes to carry four frames of the DS-1. In four frames the 24 DS-0s inside a DS-1 generate 96 bytes. This leaves 8 bytes for additional overhead.

Once the V5 VTOH byte is located by the V1/V2 pointer field, three other bytes of VTOH can be located. These are the J2, Z6, and Z7 VTOH bytes, according to current Telcordia SONET specifications (other documentation has other names for them). The J2 byte is a way to trace the

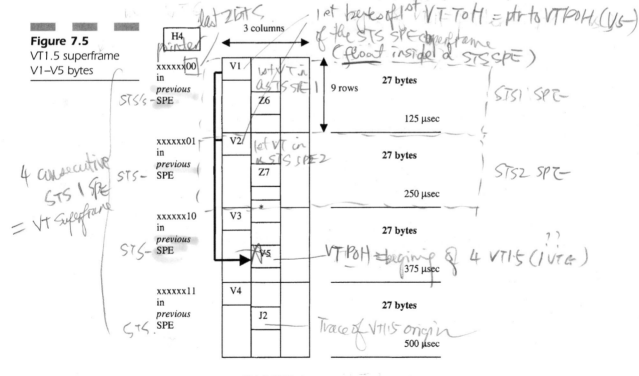

Figure 7.5
VT1.5 superframe
V1–V5 bytes

Total: 108 bytes
(position of V5 arbitrary in floating mode)

origin of the DS-1 inside the VT1.5. The Z bytes are growth bytes for future functions. By definition, once the V5 byte is located, the positions of the J2, Z6, and Z7 bytes are exactly 125 μseconds away. In other words, if the V5 byte is found as the 17th byte of a 27-byte VT1.5 frame within the VT1.5 superframe in an SPE, then the J2 byte is the 17th byte in the following SPE. The Z6 and Z7 byte positions are determined the same way. This is also shown in Figure 7.5.

It is important to realize that this overall VT superframe structure, consisting of four SPEs and the V1 through V5, J2, Z6, and Z7 VTOH bytes, is exactly the same for each of the four VT types. However, only the VT1.5 superframe has a 108-byte structure. Once the 8 VTOH bytes are subtracted, there are only 25 bytes per SPE to carry the 24 DS-0s for each VT1.5 and 1 additional byte that is used mainly for carrying the DS-1 framing bit.

A VT2, which has three E-1s in a VTG, needs four columns, or 36 bytes, for each E-1. So VT2 superframes are 4 x 36 = 144 bytes long. Subtracting the 8 VTOH bytes leaves 136 bytes split among the four—SPE VT superframe sequence, giving 34 bytes per SPE for the 32 DS-0s. This is called the *payload capacity* of the VT superframe. For a VT3, the numbers are 6 columns and 54 bytes for each DS-1C. The VT3 superframe is 216 bytes, with 208 bytes of payload, or 52 bytes per SPE. For a VT6, the numbers are 12 columns and 108 bytes for each DS-2. The VT6 superframe is 432 bytes, with 424 bytes of payload, or 106 bytes per SPE. This comparison in VT superframe size is shown in Table 7.4.

To finish the discussion of VT1.5 superframes, all that is necessary is to map the DS-0 around the 8 VTOH bytes inside the VT superframe. However, there are two ways to accomplish this for VT1.5s. The two ways are *byte-synchronous mapping* and *asynchronous mapping*. Either method may be used and both are equally standardized. It all depends on what needs to

TABLE 7.4

VT payload capacity

SPE Sequence	VT1.5 Payload	VT2 Payload	VT3 Payload	VT6 Payload
1 (00 in previous H4)	25	34	52	106
2 (01 in previous H4)	25	34	52	106
3 (10 in previous H4)	25	34	52	106
4 (11 in previous H4)	25	34	52	106
Total payload capacity, less 8 VTOH bytes:	100	136	208	424

be done to the DS-0s inside the DS-1s riding the VT1.5. If the DS-0 channels need to be found in the SONET NE, probably for cross-connection purposes, then byte-synchronous mapping is indicated. In byte-synchronous mapping, the framing bit is carried in the VT1.5, as are the signaling bits giving status information on each DS-0. In byte-synchronous mapping, each and every DS-0 inside each and every VT1.5 in each and every VTG can be found and extracted from the SONET frame. This is DS-0 payload visibility. Today, asynchronous mapping is more common.

With asynchronous mapping, the channel structure of the DS-1 is neither visible nor preserved. The DS-1 framing bit is not transported through the SONET portion of the DS-1 link, and DS-1s are transported as just 192-bit units which might span DS-1 frame boundaries (or not). When asynchronous mapping is used, SONET NEs can extract and cross-connect DS-1s, but not the individual DS-0 channels inside them. Of course, the extracted DS-1 could always be sent to a separate DS-1 and DS-0 cross-connect for further processing. In fact, this was one reason that asynchronous mappings were invented in the first place: why force SONET NEs to perform a function that legacy equipment still does quite well?

The V5 byte starts off the *VT1.5 SPE.* The VT1.5 SPE is 104 bytes long (108 bytes for the VT1.5 superframe, less the V1-V4 VTOH bytes). The structure of the 104-byte VT1.5 SPE for both byte-synchronous and asynchronous mapping is shown in Figure 7.6.

In Figure 7.6, the VT1.5 itself is represented as a *linear* array 104 bytes long instead of as three columns, but this should cause no confusion. The V1—V4 bytes are shown, but only to emphasize the fact that the VT1.5 unit is now 104 bytes long, not 108 bytes. The J2, Z6, and Z7 bytes are shown in relation to the V5 byte, not the V1—V4 bytes. The notes identify the function of all the overhead bits in both mappings. The asynchronous mapping does not carry the DS-1 framing bit in the overhead. The framing bit is included in the VT1.5 bit stream, of course, but the SONET NE cannot find the frame boundaries directly.

Although byte-synchronous mapping offers DS-0 payload visibility directly to SONET NEs, asynchronous mapping is much more common. This is because cross-connecting at the DS-0 level is usually done by a separate piece of equipment, so cross-connecting is not really all that essential a function of SONET NEs. Also, unchannelized DS-0s, which consist not of 24 channels running at 64 kbps but of a single bit stream running at 1.544 Mbps, cannot be cross-connected anyway. The more unchannelized DS-0s there are, the less sense byte-synchronous mappings make. Byte-synchronous mapping renders the DS-0 channels visible, while asynchronous mapping renders the DS-1 frame visible (but only by examining the VT1.5

cross 4 STS SPE (1st VTOH of each)

VT Superframe

V5 = VT POH byte

Figure 7.6
VT1.5 Byte-synchronous and asynchronous mappings

1st byte of SPE = pointer to V5

V5 = VT POH contains bits #
① BIP-2   3
② REI
③ RDI   8
④ RFI   4

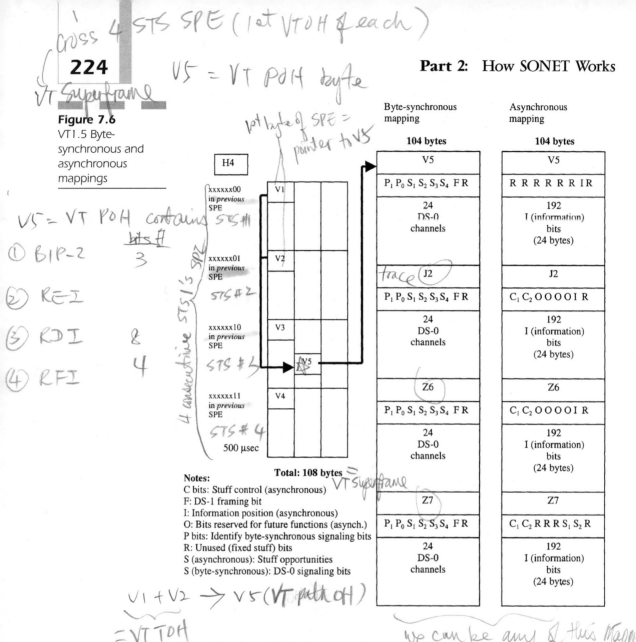

Byte-synchronous mapping	Asynchronous mapping
104 bytes	**104 bytes**
V5	V5
$P_1\ P_0\ S_1\ S_2\ S_3\ S_4$ F R	R R R R R R I R
24 DS-0 channels	192 I (information) bits (24 bytes)
J2 (trace)	J2
$P_1\ P_0\ S_1\ S_2\ S_3\ S_4$ F R	$C_1\ C_2\ O\ O\ O\ O\ I$ R
24 DS-0 channels	192 I (information) bits (24 bytes)
Z6	Z6
$P_1\ P_0\ S_1\ S_2\ S_3\ S_4$ F R	$C_1\ C_2\ O\ O\ O\ O\ I$ R
24 DS-0 channels	192 I (information) bits (24 bytes)
Z7	Z7
$P_1\ P_0\ S_1\ S_2\ S_3\ S_4$ F R	$C_1\ C_2\ R\ R\ R\ S_1\ S_2$ R
24 DS-0 channels	192 I (information) bits (24 bytes)

H4

xxxxxx00 in *previous* SPE   V1
xxxxxx01 in *previous* SPE   V2   STS #2
xxxxxx10 in *previous* SPE   V3   STS #3
xxxxxx11 in *previous* SPE   V4   STS #4

4 consecutive STS1's SPE

V5   STS #1

500 μsec

Total: 108 bytes VT Superframe

Notes:
C bits: Stuff control (asynchronous)
F: DS-1 framing bit
I: Information position (asynchronous)
O: Bits reserved for future functions (asynch.)
P bits: Identify byte-synchronous signaling bits
R: Unused (fixed stuff) bits
S (asynchronous): Stuff opportunities
S (byte-synchronous): DS-0 signaling bits

V1 + V2 → V5 (VT path OH)
= VT TOH

we can be any of this Mapping

bit stream), and this is all that is needed to add and drop DS-1s from the SONET stream. (On a more technical level, byte-synchronous mapping severely limits the ability of SONET clocks to interface properly with DS-1 clocks, due to the requirement to fix the DS-0s into position.)

Some SONET documentation also lists a *bit-synchronous mapping* for VT1.5 and other VT types. Bit-synchronous mapping is just a special case of the asynchronous mapping and uses the same VT1.5 structure as the asynchronous mapping. The only difference is how the stuffing bits (S bits) are used. The mapping type used is determined by the value of the VT Signal Label field in the V5 byte of the VT superframe. The VT Signal Label field is a three-bit field.

A couple of other points should be made about DS-1s and VT1.5. All of the VT1.5 mappings will work with either D4 superframe or ESF DS-1s. All of them will work with either bipolar AMI or the B8ZS T-1 line codes as well. The form of the DS-1 transported inside the VT1.5 does not matter.

So in summary: byte-synchronous mapping lets SONET NEs see the DS-0s, asynchronous mapping lets SONET NEs see a stream of bits, and bit-synchronous mapping lets the SONET NEs see a slightly different stream of bits. Any of the three mappings can be used in VT1.5. It all depends on what the SONET NE needs to do with the DS-1s.

Mappings for Other VT Types

Just as VT1.5 has a distinctive format for the VT superframe, the other VT types also have distinctive VT superframe formats. There are also mappings established for the other three VT types (VT2, VT3, and VT6). It is not necessary here to detail all three to the degree afforded VT1.5, if only because this chapter would then be very, very long. A few words will suffice for VT3 and VT6. More details will be explored for VT2s, but only because E-1 transport is the way in which international circuits must be supported.

VT3s and VT6s have only asynchronous mappings. That is, only the overall DS-1C (VT3) and DS-2 (VT6) frames can be found and delimited by the SONET NE. VT2s, like VT1.5s, allow all three types of mappings: byte-synchronous, asynchronous, and bit-synchronous. The type of mapping employed in the VT is indicated by the value of the 3-bit Signal Label field of the V5 byte. The defined values of the Signal Label field by VT type are shown in Table 7.5.

TABLE 7.5

VT Signal Labels

Value	VT1.5	VT2	VT3	VT6
000	Unequipped	(not used for live traffic)		
001	Equipped for nonspecific payload			
010	Asynchronous DS-1	Asynchronous E-1	Asynchronous DS-1C	Asynchronous DS-2
011	Bit-synchronous DS-1	Bit-synchronous E-1		
100	Byte-synchronous DS-1	Byte-synchronous E-1		

Only the VT2 mappings, intended for E-1 links, running at 2.048 Mbps and carrying 30 DS-0 channels, require some additional comment. First of all, a SONET SPE carrying all VT2s in the seven VTGs transports 21 E-1s (3 VT2s x 7 VTGs) and thus 630 DS-0 channels. The positions of the E-1 columns in the SPE columns is shown in Table 7.6, which might be compared to Table 7.3 for the DS-1 equivalents. Note the absence of SPE columns 30 and 59 in the table. These are always fixed stuff columns and cannot be used in any VT type.

TABLE 7.6

E-1s inside a
VT2 SPE

E-1 No.	VTG No.	VT No. (Letter)	VTG Columns	SPE Columns
1	1	1 (A)	1, 4, 7, 10	2, 23, 45, 67
2	2	1 (A)	1, 4, 7, 10	3, 24, 46, 68
3	3	1 (A)	1, 4, 7, 10	4, 25, 47, 69
4	4	1 (A)	1, 4, 7, 10	5, 26, 48, 70
5	5	1 (A)	1, 4, 7, 10	6, 27, 49, 71
6	6	1 (A)	1, 4, 7, 10	7, 28, 50, 72
7	7	1 (A)	1, 4, 7, 10	8, 29, 51, 73
8	1	2 (B)	2, 5, 8, 11	9, 31, 52, 74
9	2	2 (B)	2, 5, 8, 11	10, 32, 53, 75
10	3	2 (B)	2, 5, 8, 11	11, 33, 54, 76
11	4	2 (B)	2, 5, 8, 11	12, 34, 55, 77
12	5	2 (B)	2, 5, 8, 11	13, 35, 56, 78
13	6	2 (B)	2, 5, 8, 11	14, 36, 57, 79
14	7	2 (B)	2, 5, 8, 11	15, 37, 58, 80
15	1	3 (C)	3, 6, 9, 12	16, 38, 60, 81
16	2	3 (C)	3, 6, 9, 12	17, 39, 61, 82
17	3	3 (C)	3, 6, 9, 12	18, 40, 62, 83
18	4	3 (C)	3, 6, 9, 12	19, 41, 63, 84
19	5	3 (C)	3, 6, 9, 12	20, 42, 64, 85
20	6	3 (C)	3, 6, 9, 12	21, 43, 65, 86
21	7	3 (C)	3, 6, 9, 12	22, 44, 66, 87

The VT payload capacity of the VT2 is not 108 bytes (27 bytes in 4 SPE frames) as are VT1.5s. Because a VTG structured according to VT2 contains only three E-1s instead of four DS-1s, the columns unit is four columns, not three as in VT1.5. So the basic unit for an E-1 is 36 bytes (nine rows, four columns) and the VT2 superframe is 144 bytes long. Naturally, there are the usual V1—V4 bytes, and it is still the position of the V5 bytes that is pointed to by the V1/V2 field. The structure of the VT2 superframe is shown in Figure 7.7, which can be compared to Figure 7.5 for VT1.5. As in Figure 7.5, the position of the V5 byte is determined by the value of the V1/V2 field pointer. The position of the J2, Z6, and Z7 bytes is determined by the position of the V5 byte in the VT2 superframe: each follows 125 µseconds after the other.

In order to complete this brief look at VT2s, all that is needed is to present the byte-synchronous and asynchronous mappings for VT2s. These mappings are shown in Figure 7.8, which can be compared to Figure 7.6 for VT1.5.

Figure 7.8 assumes that the byte-synchronous mapping employ channel-associated signaling (CAS), in which the E-1 frame byte between DS-0

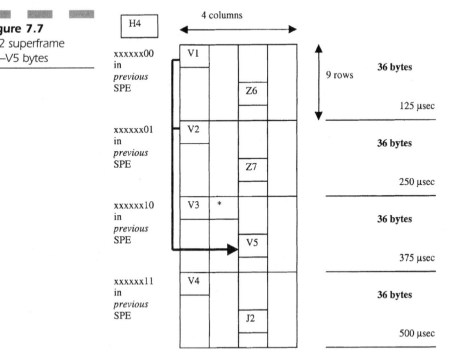

Figure 7.7
VT2 superframe
V1–V5 bytes

Total: 144 bytes
(position of V5 arbitrary in floating mode)

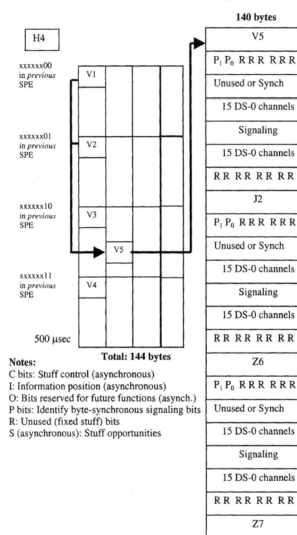

Byte-synchronous mapping

Asynchronous mapping

140 bytes   **140 bytes**

H4

xxxxxx00 in *previous* SPE — V1

xxxxxx01 in *previous* SPE — V2

xxxxxx10 in *previous* SPE — V3 — V5

xxxxxx11 in *previous* SPE — V4

500 μsec

Total: 144 bytes

Byte-synchronous mapping column:
- V5
- $P_1 P_0$ R R R R R R
- Unused or Synch
- 15 DS-0 channels
- Signaling
- 15 DS-0 channels
- R R R R R R R R
- J2
- $P_1 P_0$ R R R R R R
- Unused or Synch
- 15 DS-0 channels
- Signaling
- 15 DS-0 channels
- R R R R R R R R
- Z6
- $P_1 P_0$ R R R R R R
- Unused or Synch
- 15 DS-0 channels
- Signaling
- 15 DS-0 channels
- R R R R R R R R
- Z7
- $P_1 P_0$ R R R R R R
- Unused or Synch
- 15 DS-0 channels
- Signaling
- 15 DS-0 channels
- R R R R R R R R

Asynchronous mapping column:
- V5
- R R R R R R R R
- 256 I (information) bits (32 bytes)
- R R R R R R R R
- J2
- $C_1 C_2$ O O O O R R
- 256 I (information) bits (32 bytes)
- R R R R R R R R
- Z6
- $C_1 C_2$ O O O O R R
- 256 I (information) bits (32 bytes)
- R R R R R R R R
- Z7
- $C_1 C_2$ R R R R R S_1
- S_2 I I I I I I I
- 224 I (information) bits (31 bytes)
- R R R R R R R R

Notes:
C bits: Stuff control (asynchronous)
I: Information position (asynchronous)
O: Bits reserved for future functions (asynch.)
P bits: Identify byte-synchronous signaling bits
R: Unused (fixed stuff) bits
S (asynchronous): Stuff opportunities

channels 15 and 16 is used for signaling the status of each channel. There is also a special case of the asynchronous VT2 mapping for bit-synchronous operation.

Asynchronous DS-3

Compared to the complexities of VT structures, superframes, and mappings, the rest of the SONET payload types explored in this chapter are quite simple. The next SONET SPE content to be investigated in this chapter is *asynchronous DS-3*. In contrast to virtual tributaries, this type of SONET payload occupies all of the SPE. Just like virtual tributaries, asynchronous DS-3 transport is defined only in the basic STS-1 SPE. When the Signal Label POH byte (C2) is set to a value of 04x (04 in hex, or 0000 0100), then the SONET receiver knows that the arriving SPE must be carrying an asynchronous DS-3.

As might be expected, asynchronous DS-3 transport allows the SONET NE no direct access to the DS-1s or DS-0s that might be contained inside the DS-3. This is to be expected, since DS-3 multiplexing is bit-oriented anyway. DS-3s can also be unchannelized as a single bit stream at 45 Mbps. With asynchronous DS-3, any type of DS-3 can be carried inside a SONET STS-1 SPE without regard to channelization issues. Cross-connections can always be done by non-SONET auxiliary equipment, exactly as is done when SONET is not in the equation.

In spite of this apparent DS-1 and DS-0 visibility limitation, the asynchronous DS-3 transport is popular and quite useful. If a service provider has a large national network, for example, a lot of legacy DS-3 and T-3 equipment might still be in place at the regional level, entirely operational, and nowhere near its retirement age. Beyond the cost aspect of DS-3 equipment replacement, there could also be regulatory issues involved with any large-scale equipment replacement scheme. Many regulators limit the amount of physical equipment that can be replaced in any given time period. If this service provider has a large national SONET backbone, then the DS-3s can be transported from region to region over the SONET backbone without concern about the lack of payload visibility. This is really what asynchronous DS-3 transport is for.

When used for asynchronous DS-3 transport, the 87 columns of the SONET SPE are structured according to Figure 7.9. The information (I) bytes are what carry the DS-3 bits themselves. These I bits are spread out row by row into 25-byte units interspersed with control byte columns and reserved columns. It is a little misleading to label these information bits as "bytes" since DS-3 bytes are just bit-multiplexed streams. However, since the basic unit in the SONET SPE is the byte, it makes most sense to represent the I bits as SONET bytes, which is how the columns and rows always appear in SONET. Also, the control bytes in Figure 7.9 (for example, C1—C3) are SONET control bytes, not DS-3 frame bytes. It makes no difference if the I bits are DS-3 framing bits, or C-parity bits, or any other part of the overall DS-3 frame. All DS-3 framing bits are

Figure 7.9 Asynchronous DS-3 mapping

				28 columns					28 columns						28 columns	
	R	R	C1	25 I bytes		R	C2	I	25 I bytes			R	C3	I	25 I bytes
	R	R	C1	25 I bytes	F	R	C2	I	25 I bytes	F		R	C3	I	25 I bytes
	R	R	C1	25 I bytes	i	R	C2	I	25 I bytes	i		R	C3	I	25 I bytes
	R	R	C1	25 I bytes	x	R	C2	I	25 I bytes	x		R	C3	I	25 I bytes
P	R	R	C1	25 I bytes	e	R	C2	I	25 I bytes	e		R	C3	I	25 I bytes
O	R	R	C1	25 I bytes	d	R	C2	I	25 I bytes	d		R	C3	I	25 I bytes
H	R	R	C1	25 I bytes	S	R	C2	I	25 I bytes	S		R	C3	I	25 I bytes
	R	R	C1	25 I bytes	t	R	C2	I	25 I bytes	t		R	C3	I	25 I bytes
	R	R	C1	25 I bytes	u	R	C2	I	25 I bytes	u		R	C3	I	25 I bytes
	R	R	C1	25 I bytes	f	R	C2	I	25 I bytes	f		R	C3	I	25 I bytes
	R	R	C1	25 I bytes	f	R	C2	I	25 I bytes	f		R	C3	I	25 I bytes

87 columns

Overhead byte structure:

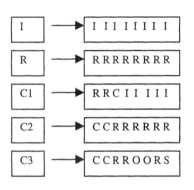

I	→	I I I I I I I I
R	→	R R R R R R R R
C1	→	R R C I I I I I
C2	→	C C R R R R R R
C3	→	C C R R O O R S

Bit meanings:

I: information (payload) bit
R: reserved (fixed stuff) bit
C: stuff control bit
S: stuff opportunity bit
O: overhead communications
   channel

just other bits to SONET. Finally, bit stuffing is necessary, as is usual when the PDH meets the SONET/SDH world.

That is really about all there is to asynchronous DS-3 transport. The main point is that the bits in the I fields are just a meaningless jumble of bits to the SONET NE.

If a service provider actually wanted to carry the 28 DS-1s inside a DS-3 as organized bytes instead of the bit-multiplexed DS-3 stream, it would be necessary to de-multiplex the DS-3 and then load the 28 DS-1s as asynchronous VT1.5s or perhaps even as 673 DS-0s with byte-synchronous VT1.5. But for raw DS-3 transport from here to there, asynchronous DS-3 is the quickest and easiest way to accomplish this task.

There is one other point of interest regarding asynchronous DS-3 transport that will be important when considering ATM cell and POS transport. Although it is not easy to tell by the structure of the SPE in Figure 7.9, the DS-3 bits in the SPE are really organized by *row*, not by *column*. Multiplexing many sources in the virtual tributary world organizes the SPE into columns automatically. But an asynchronous DS-3 is a *single* source of bits, not many multiplexed sources presented to the SONET NE. Any multiplexing of the DS-3 has been done before the DS-3 hits the SONET NE.

Yet Figure 7.9 undeniably seems to organize the SPE into columns. This is only the effect of inserting additional overhead bytes such as C1—C3 systematically through the SPE, which is always sent row by row across the fiber link. This overhead is required for the proper handling of the SPE bit stream in order to recover the DS-3 frame.

However, what if a single source of bits could fill an SPE and have its own framing? Then no column organization of the SPE would be necessary, and indeed would make no sense. And since the single source framed its own information content independently of the SONET SPE, no additional overhead bytes need be inserted into the SPE. This is exactly the philosophy followed by ATM cell transport inside the SONET SPE: organization by rows and no additional columns of overhead.

ATM Cells

The third major type of traffic carried by SONET SPEs is a stream of ATM cells. When the POH Signal Label byte (C2) is set to a value of 13x (13 in hex, or 0001 0011), then the SONET receiver knows that the arriving SPE must be carrying ATM cells.

ATM cells are fixed-length information blocks that are set at 53 bytes in length. ATM cells have a 5-byte header and a 48-byte payload structure, but that is not important to SONET. SONET cannot, will not, must not ever look at an ATM cell header (that is a switch function, and SONET NEs are not switches). ATM is the international standard for *cell relay* technology. Cell relay attempts to speed up networks, not by switching packets (OSI-RM Layer 3 protocol data units, or PDUs), nor by relaying frames (frame relay uses OSI-RM Layer 2 PDUs), but by essentially inventing a new, fixed, small PDU called a *cell.* Cells can carry frames or packets alike through a very fast switch.

ATM performs its magic by heavy use of statistical multiplexing, in which idle connections (ATM is connection oriented) do not occupy "all the bandwidth, all the time," as do traditional time division multiplexing systems. Ownership of cell contents among the active connections is determined by a field in the ATM header. If a transmitter does not have enough live ATM cells to send on a link, then special idle cells are inserted in a process known as *cell rate decoupling.* ATM cells are often positioned as "very small and fixed-length PDUs" compared to frames and packets. It is just as correct, and often more accurate, to position ATM cells as "very big time slots" with variable ownership due to the statistical multiplexing involved.

This is not the place to even attempt an introduction to ATM technology. It is enough to note that ATM cells are fixed in size, and so can be packed directly, head to tail, on a communications link. Once the boundary of one cell tail and head is determined, the rest can easily be found without additional framing, overhead, or control bytes. This process of *cell delineation* is performed by the ATM switch on the receiving side of the link.

In reality then, there is really no need for a fiber or any other type of communications link to carry anything other than a stream of ATM cells. This is sometimes called *raw cell transport,* but it poses a problem for the service provider of the communications link. The ATM gear at each end of the link is presumably customer premises equipment (CPE). How is the link provider to perform OAM&P and detect alarms and failures, if the ATM cells on the link are "invisible" to the service provider? Service providers are not generally allowed to eavesdrop on user traffic, especially in regulated environments.

SONET/SDH offers a perfect solution. The SONET frame offers plenty of OAM&P overhead for the service provider, and the SPE provides plenty of bandwidth for the senders and receivers of ATM cells (SDH is capable of the same feat). The process of putting ATM cells

directly inside a SONET SPE without the need for additional SPE overhead besides the required POH is called *direct cell mapping*. The difference between the use of raw cell transport and direct cell mapping is
the use of a transmission frame, mainly for service provider OAM&P
considerations, in direct cell mapping.

SONET specifications allow for the transport of ATM cells inside an
STS-1 SPE using direct cell mapping. When this mapping is used, the
only alignment required in the SONET SPE is that the bytes of the ATM
cell coincide with the bytes of the SPE. That is, the bit boundaries of the
bytes must be preserved, but this is not a big deal. When used for ATM
cell transport, an STS-1 can carry ATM cells at 48.384 Mbps (51.84 Mbps
minus the required SONET overhead).

The only other important thing to realize is that the SONET SPE is not
an even number of ATM cells long. The 9 rows and 86 information-
bearing columns of an SPE contain 774 bytes. Dividing this number by the
53-byte size of an ATM cell gives ~14.6 cells in each SPE. So a SONET CPE,
assuming an ATM cell starts right after the POH J1 byte that defines the
start of the SPE, will contain 14 full ATM cells and only 32 bytes of the
15th ATM cell. The "tail" 21 bytes of the 15th ATM cell will begin directly
following the POH J1 byte of the next SONET SPE. This next SPE will
have only 11 bytes of the last ATM cell in the SPE (32 − 21 bytes). The position of the ATM cell header keeps shifting from SPE to SPE. So any attempt
to fix the beginning of an ATM cell to a particular place in the SPE is
doomed to failure, since ATM cells are always packed head to tail on a link.
This is one reason that cell delineation (the determination of ATM cell
head-tail boundaries) is a task best left up to the ATM switch. (As an aside,
on links other than SONET/SDH, ATM cell mappings that attempt to fix
the relationship between ATM cell boundaries and transmission framing
require additional overhead known as a *convergence protocol*. The advantage
of direct cell mapping is that no convergence protocol is needed.)

So ATM cells always span the SPEs. This is of no concern to SONET,
since cell delineation is an ATM switch function. SONET links deliver
the bytes in the SPE to an ATM switch. It is up to the ATM switch to
make sense out of the bytes.

All that is left is to look at a SONET SPE carrying ATM cells. A
schematic of ATM cells inside a series of SONET STS-1 SPEs is shown in
Figure 7.10. ATM cell transport is also supported on unchannelized
("concatenated") STS-3c (STM-1) and STS-12c (STM-4) links.

Not too much should be read into Figure 7.10, but it is informative.
Not all 90 columns are shown, of course, and the exact sizes of the ATM
cells are not to scale. Figure 7.10 points out that the full structure of the

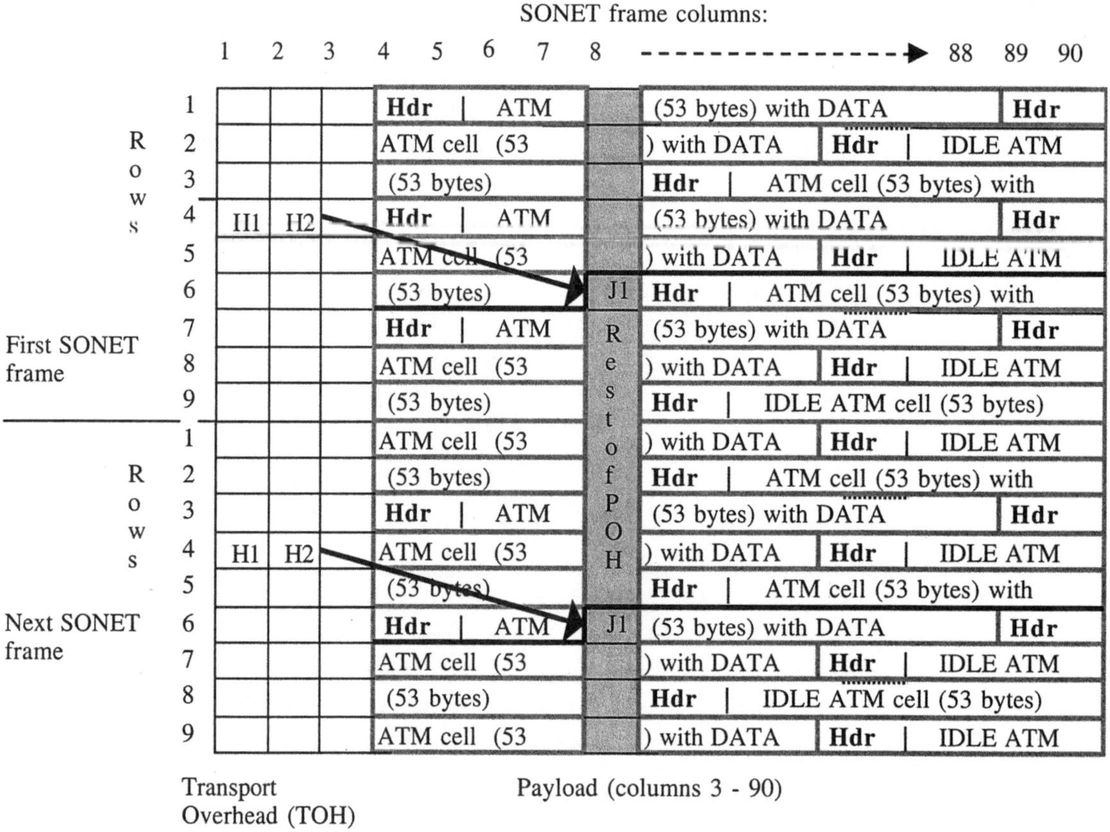

SONET frame columns:

1   2   3   4   5   6   7   8 -------------➤ 88 89 90

			Hdr	ATM	(53 bytes) with DATA	Hdr
ATM cell (53	) with DATA	Hdr	IDLE ATM
(53 bytes)	Hdr	ATM cell (53 bytes) with
III	H2	Hdr	ATM	(53 bytes) with DATA	Hdr
ATM cell (53	) with DATA	Hdr	IDLE ATM
(53 bytes)	J1	Hdr	ATM cell (53 bytes) with
Hdr	ATM	R	(53 bytes) with DATA	Hdr
ATM cell (53	e s	) with DATA	Hdr	IDLE ATM
(53 bytes)	t o	Hdr	IDLE ATM cell (53 bytes)
ATM cell (53	f	) with DATA	Hdr	IDLE ATM
(53 bytes)	P O H	Hdr	ATM cell (53 bytes) with
Hdr	ATM		(53 bytes) with DATA	Hdr
H1	H2	ATM cell (53	) with DATA	Hdr	IDLE ATM
(53 bytes)	Hdr	ATM cell (53 bytes) with
Hdr	ATM	J1	(53 bytes) with DATA	Hdr
ATM cell (53	) with DATA	Hdr	IDLE ATM
(53 bytes)	Hdr	IDLE ATM cell (53 bytes)
ATM cell (53	) with DATA	Hdr	IDLE ATM

First SONET frame — Rows 1–9

Next SONET frame — Rows 1–9

Transport Overhead (TOH)

Payload (columns 3 - 90)

Figure 7.10 ATM cells inside the SPE

SONET frame is preserved, right down to the H1/H2 pointer bytes. It is the structure of the SPE that changes. Figure 7.10 does show the ATM header creep from row to row, and emphasizes the fact that even if an ATM cell header happens to start right after the JI POH byte in a given SPE, this will not be true of the next SPE. Cells also jump over the POH column. The mixture of idle and data (information-bearing) cells is shown as well.

But by far the most important point of Figure 7.10 is the row orientation of the SPE when carrying ATM cells. This is due to the fact that when SONET is used as a link to or from an ATM switch, there is no longer any need for SONET to perform time division multiplexing, which gives the SPE its distinctive column-oriented look. Any multiplexing of the ATM cells is performed by the ATM switch, not by SONET.

The point is that whenever a SONET SPE carries traffic from a single source, it is more efficient to organize the the SPE into rows, not columns. This is especially important when it comes to the last SONET payload type: Packet over SONET.

Packet over SONET (POS)

When Packet over SONET (POS) was originally proposed, some observers remarked that packets had been sent over SONET links since the inception of SONET, so what was the big deal? Take any stream of IP packets, for example, place them inside a DS-0 (64 kbps) or VT1.5 (1.544 Mbps), and away they go. But the key to understanding POS is realizing that POS organizes the SPE into rows, like the ATM mapping, not columns, like virtual tributaries. As long as the packets emerge from one source, like a router port, row-wise SPEs are much more efficient than columnwise SPEs. POS is a way of sending packets between routers at much higher speeds than virtual tributaries allow.

Of course, ATM cells can also carry IP packets. And since ATM cells can fill the SONET SPE, this is another way that IP packets can be sent on a SONET link. However, the issue regarding the use of ATM cells to carry IP packets boiled down to a question of overhead. A comparison of Packet over SONET and ATM with IP inside over SONET overhead is given at the end of this section, after the basics of Packet over SONET have been discussed.

The oddest thing about Packet over SONET (POS) is that, although it is a common enough acronym and term, there is really no such thing as POS. It is proper to talk about "IP-over-SONET" or "PPP over SONET/SDH" or even "HDLC-over-SONET," but technically there is no "packets over SONET." The term POS is just a shorthanded way of saying "allow the SPE to be structured in rows, as it is for ATM cells, but allow for the transport of any OSI-RM Layer 3 packet structure, such as IP packets, inside an appropriate frame." The PPP and HDLC protocols, which *are* supported over SONET, are nothing more than the frame types used to transport the packets. PPP stands for the Internet Point-to-Point Protocol and HDLC stands for OSI-RM High-level Data Link Control protocol. Although intentionally closely related, PPP and HDLC are not exactly the same.

This section therefore makes heavy use of the Open Systems Interconnection Reference Model (OSI-RM), especially the bottom three layers, numbered Layer 1 through Layer 3. The OSI-RM, specified by the International Organization for Standards (ISO), has become the accepted way to refer to data network protocols and architectures. At the bottom of the OSI-RM, Layer 1, or the Physical Layer, defines a serial stream of bits as the basic transport for data. The bit stream has no structure, so Layer 2, the Data Link Layer, creates a structure defined as the *frame*, a variable-length protocol data unit (PDU) from some minimum length up to some maximum length, which varies from implementation to implementation. Both PPP and HDLC define a frame structure and minimum/maximum length. Inside the frames are, by definition, packets. Packets are Layer 3 (Network Layer) PDUs which also have a structure, variable size, and maximum length. Today, the primary example of a packet as Layer 3 PDU is the IP packet without which the Internet and Web cannot be used.

The main difference between frames and packets in the OSI-RM is that frames have headers and trailers, while packets have only headers. The frame trailer is used mostly for error checking, which means that the packet header need not perform error checking at all, or at least not to the extent that the frame trailer provides error checking. Error detection, in OSI-RM talk, is primarily a Layer 2 frame function. Also, frames are defined to flow on simple point-to-point links from NE to NE. Although frames might look exactly the same entering and leaving an IP router, for example, technically they are two different frames. Frames flow "hop by hop" between adjacent systems (those directly connected by serial link).

Packets, on the other hand, flow end to end from source to destination, whether there are 10 or 100 NEs, such as IP routers, in between. This is possible because each packet header has enough information to allow NEs to make routing decisions on a hop-by-hop basis. In IP, for example, this information is the source and destination IP network address. An important part of ISO-RM Layer 3 is the concept of a network address, which exists at no other layer of the model. Generally, OSI-RM NEs (called *intermediate systems*) just extract packets from frames arriving on an input serial link, examine the destination network address, determine the proper output port, and then rewrap the packet inside a frame to queue for sending on the correct output serial link. (If this explanation makes data networking sound very simple, the answer is that data networking *is* very simple! It is the *networks* that are complex.)

So the first and foremost reason that it is not a good idea to attempt to put raw packets on a Physical Layer transport such as SONET is that the frame header and trailer provides error detection functions for itself and also for the packet inside. It would be hard for most packets to provide elaborate error checking for themselves: many packets have no error fields in the header at all, or if they are present, they provide only rudimentary functionality much weaker than the frame error detection (IP falls into this latter category).

There are two other reasons for placing a packet inside a frame before transporting the frame inside a SONET SPE. The first reason concerns the delineation of the packet and the second reason concerns the existence of multiple packet types.

Packets in general rely on the frame structure surrounding them to properly delimit the beginning and ending of the packet itself. Unlike ATM cells, packets and the frames that surround them are almost always of variable length between some minimum and maximum. So frame delineation is an important receiver function (transmitters, presumably, know exactly where a packet and frame begin and end). In HDLC and PPP, frame delineation is provided by means of a special *interframe fill* bit pattern, which is almost universally referred to as an *idle pattern*. When the idle pattern disappears at the receiver, the frame has begun. When the idle pattern reappears, the frame has ended. Everything in between is the frame. Both HDLC and PPP employ as interframe fill 7E in hex, or 0111 1110 in bits. (There are special rules for preventing the 7E pattern and thus *false framing* from occurring within the transmitted frame and packet, but these are not of concern here.)

And not all Layer 3 packet structures are the same. Even IP, the most popular Layer 3 protocol in use today, thanks to the Internet and the Web, employs different packet types at Layer 3. Frequently, the same serial link, employing a single frame type, will have to carry multiple packet formats more or less at the same time. Both PPP and HDLC frame structures provide a method for receivers to determine the precise type of packet inside the arriving frame. This is necessary because the format and processing of each packet type can vary based on the exact type of packet involved. For PPP, the presence of a "normal" data-carrying IP packet inside the PPP frame is indicated by a protocol field in the PPP frame having the value 0021 in hex, or the 16-bit pattern 00000000 00100001.

So the three reasons *not* to place packets directly inside a SONET/SDH payload are:

1. The frame trailer contains more robust error checking.

2. The frame provides delineation for the packet inside.

3. The frame can carry multiple protocols (packet types).

HDLC or PPP?

So far, equal mention has been made of HDLC and PPP. Which one is used when packets are sent across a SONET/SDH link? The issue is important because PPP, like all Internet Protocol Suite (TCP/IP) protocols, is not an official international standard in the sense that ATM or HDLC are official international standards. So Telcordia specifications on SONET, for example, only list "HDLC-over-SONET mapping" as an allowable SPE content. The POH C2 Signal Label assigned to this mapping is 16 in hex or 0001 0110.

But PPP, the serial link protocol most often used with IP packets, is not quite HDLC. HDLC has sometimes been called "the mother of all Data Link Layer protocols," and this is not a bad analogy. HDLC defines a basic frame structure, but many implementations of HDLC vary one or more fields or the header or trailer in size and format. This is not to say that there are no rules; for instance, the *order* of the fields cannot be changed.

PPP was designed as an efficient serial link transport for IP and related packet structures. While PPP simplifies and modifies many of the functions of the entire HDLC protocol, PPP frames were intentionally designed to look as much like a basic HDLC frame as possible. But just to show how different the basic HDLC frame is from a PPP frame, Figure 7.11 shows both.

Keep in mind that many HDLC frames can also have 16-bit (2-byte) address, control, or checksum (also called the *frame check sequence*, or FCS) fields. The PPP frame fixes some HDLC options, such as defining the address filed as 1 byte, and using the all 1s "anycast" address on the PPP link. The PPP control field is also 1 byte and is set to indicate that the content of the frame is an *unnumbered* (i.e. there is no sequence num-

Figure 7.11

Basic HDLC and PPP frame formats

Interframe Fill 0111 1110	Address 1 or 2 bytes	Control 1 or 2 bytes	Data 0 to X bytes	Checksum (FCS) 1 or 2 bytes	Interframe Fill 0111 1110

Basic HDLC frame format

Flag 0111 1110	Address 1111 1111	Control 0000 0011	Protocol ID 1 or 2 bytes	Information Variable/Pad	FCS 2 or 4 bytes	Flag 0111 1110

Basic PPP frame format

bering used on the frame) *information frame*. The most unique aspect of the PPP frame in the presence of the 1- or 2-byte *protocol identifier* field. This is used to indicate the format of the *information* field that follows. This protocol ID field is 0021 in hex when an IP packet is inside the frame. The IP packet itself rides in the information field. Some protocols (not IP) might require some bytes of padding (usually just all 0 or null fields) to end the packet on the proper byte boundary. Finally, the PPP FCS field might be 2 bytes (16 bits) or 4 bytes (32 bits), depending mostly on the speed of the link (the higher the speed, the more likely a 4-byte FCS is used). SONET supports HDLC mapping at the STS-1, STS-3c (STM-1), and STS-12c (STM-4) rates.

The PPP protocol, when used with SONET/SDH, is defined in a series of requests for comment (RFCs) issued by the Internet Engineering Task Force (IETF), which acts as much as a central authority for Internet protocols as the loose structure of the Internet allows.

Three RFCs are especially important for POS:

- RFC1619, PPP over SONET/SDH
- RFC1661, the Point-to-Point Protocol (PPP)
- RFC1662, PPP in HDLC-like framing

It might seem odd that the RFCs defining PPP and its HDLC-like frame come *after* a definition of PPP over SONET. This is because RFC1661 and RFC1662 actually update older documents with numbers lower than RFC1619. (RFCs are never modified. They are always given new numbers and the older number is flagged as "obsolete.") So RFC1661 currently defines PPP procedures, but the PPP frame structure is defined in RFC1662. RFC1619 has the rules for using PPP inside SONET or SDH.

According to RFC1619, there are a few rules that must be followed when a SONET or SDH payload carries PPP frames (which ordinarily would carry IP packets, but not necessarily and not all the time). When SONET is used, STS-1, STS-3, STS-9, STS-12, STS-18, STS-24, STS-36, or STS-48 can be used. The SDH speeds are STM-1 (STS-3c), STM-4 (STS-12c), or STM-16 (STS-48c), and although these are many different speeds, the *basic rate* for PPP over SONET/SDH is actually STM-1 or STS-3c.

The frames inside the SONET SPE are organized into rows, as expected, and cross SPE boundaries. Interframe fill is used between frames. According to RFC1619, the POH C2 Signal Label should use a value of CF in hex (1100 1111) to indicate the presence of PPP frames. The Multiframe Pointer byte (H4) must be set to all zeros as well.

The POS SPE

So either HDLC or PPP can be used to frame packets sent on a SONET link. The difference is in the value of the POH C2 Signal Label: 16 in hex for HDLC and CF in hex for PPP. But as long as both ends of the link understand the frame formats used and process the frames without generating alarms, the Signal Label is not all that important.

Figure 7.12 shows a series of PPP frames inside a SONET SPE. This figure may be compared with Figure 7.10, which shows the ATM equivalent. The major differences are the variable lengths of the frames (packets) and the presence of the interframe fill pattern, or flag pattern of 7Es. Figure 7.12 is not very realistic, however, in the sense that IP packets are routinely 1,500 bytes long and so would easily fill an entire 774-byte

Figure 7.12 PPP frames (IP packets) inside the SPE

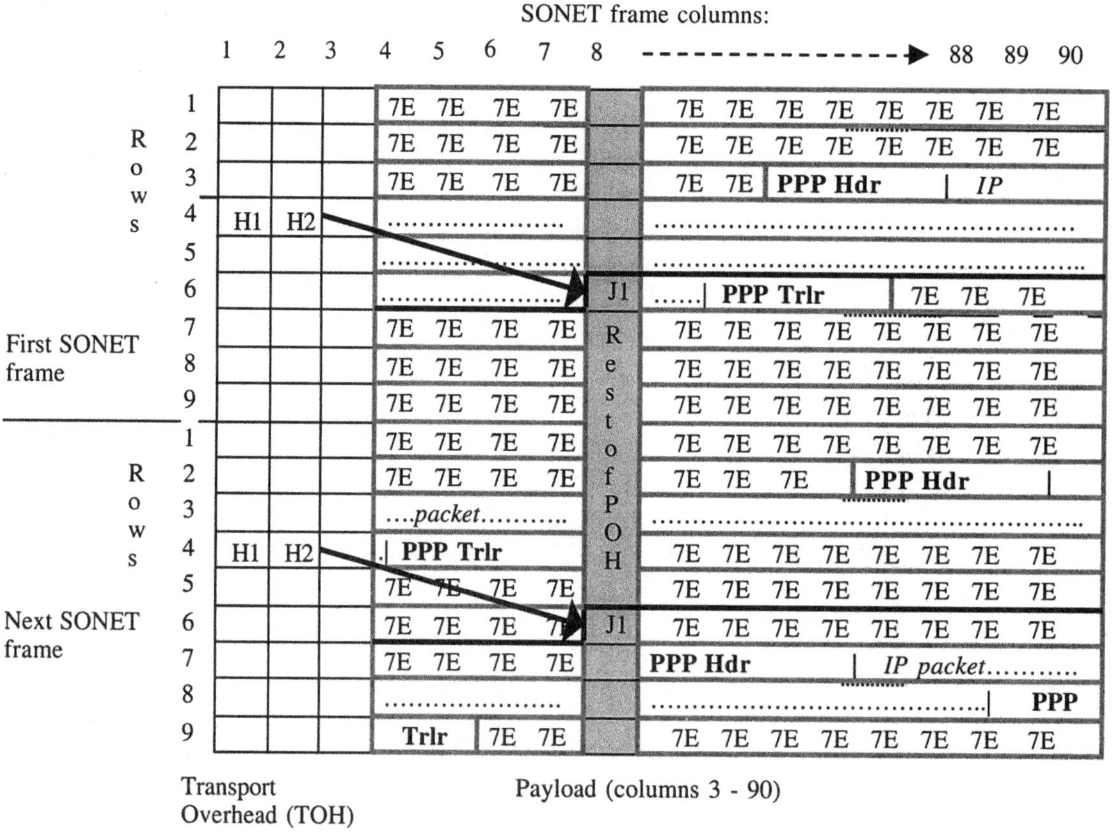

Transport Overhead (TOH)

Payload (columns 3 - 90)

SPE at the STS-1 rate. Even the default maximum IP packet size of 576 bytes is quite large compared to the SPE.

The only matter that should be discussed before closing this chapter is a brief discussion of using ATM cells inside a SONET SPE as opposed to using POS. IP packets can ride SONET inside ATM cells quite nicely. But the overhead needed to place IP packets inside ATM cells was considered much too large by the IETF, which then developed POS to cut down on this overhead burden.

Consider an IP packet to be placed inside an ATM cell and then inside an STS-3c SPE. Assume that the IP packet is 576 bytes long. A special 8-byte field called the *logical link control/sub-network access protocol* (LLC/SNAP) header is first placed on the IP packet. Then a trailer called the *ATM adaptation layer type 5* (AAL5) trailer, which is at least 8 bytes long, and sometimes a lot more, is added to the end of the IP packet. Finally, this whole 592-byte (at least) unit is chopped up into 48-byte cell payloads, each with a 5-byte ATM cell header. The packet will occupy 13 cells (592/48 = 12.333 . . .) and require 13 x 5 = 65 more bytes of overhead, for a total of at least 81 extra bytes to send the original 576-byte IP packet. This translates to only about 80% of the STS-3c line rate being useful for IP packet transfer.

Now compare sending an IP packet on an STS-3c using PPP frames inside the SPE to carry the packets. Only the modest 6 bytes of overhead from the address (1 byte), control (1 byte), protocol identifier (2 bytes), and FCS fields (4 bytes) are needed, compared to 81 when using ATM, to transfer the same size packet. An approximation translating PPP and ATM overhead into line rates is shown in Table 7.7.

Even a quick look at Table 7.7 shows why Internet service providers (ISPs) and other service providers would rather send IP packets with POS

TABLE 7.7

Effective line rates for PPP and ATM STS-3c SPEs

Layer	Line Rate	Less Overhead . . .	Layer	Line Rate	Less Overhead . . .
SONET	155.52 Mbps	90 bytes/frame	SONET	155.52 Mbps	90 bytes/frame
ATM	149.46 Mbps	5 bytes/cell	PPP	149.46 Mbps	8 bytes/PPP frame
AAL	135.36 Mbps	8+ bytes/packet	IP	**147.15 Mbps**	None
LLC/SNAP	126.94 Mbps	8 bytes/packet			
IP	**125.92 Mbps**	None			

than with ATM on SONET. Each line shows the line rate available to each layer, and then subtracts the overhead needed (about 45 cells fit inside an STS-3c SPE). The result is the line rate available to the next layer, and so on. These figures are approximations, but they are representative.

The bottom line is that an ISP "gains" about another 20 Mbps (125.92 Mbps with ATM to 147.15 Mbps) when using POS instead of ATM on the SONET links.

8

SONET Synchronization and Timing

A key feature, if not *the* key feature, of SONET (and SDN) is its synchronous operation. This feature is so important to SONET that it appears first in the name Synchronous Optical Network, even before the fiber optic aspect of SONET. After all, there are many forms of fiber optic network links and communications networks; however, it is the synchronous operation of SONET that makes it truly distinctive.

This chapter explores the synchronous aspects and features of SONET in more detail than has been discussed to this point. This chapter begins with a look at network synchronization in general, and points out that distribution of timing information is not an exclusive feature of SONET, but is indeed necessary in all networks that cross-connect voice. What SONET brings to network synchronization and timing is not so much invention, but rather innovation. SONET is synchronized to a much higher degree than is the older T-carrier network, which is now considered to be part of the plesiochronous digital hierarchy (PDH). SONET, of course, is part of the newer synchronous digital hierarchy.

The chapter also considers exactly how this stringent timing information is distributed in a SONET network, and some of the requirements imposed on SONET equipment to enforce this timing requirement. Processor clocking has improved tremendously over the years, as have the processors themselves; SONET takes full advantage of this fact.

The chapter concludes with the admission that SONET is not perfectly synchronous, and details why. This being the case, SONET's lack of perfect synchronization leads to a reconsideration of the use of SONET overhead pointers, and gives a real-world example of precisely how these pointers are used.

Network Synchronization

The presence of pointers in SONET overhead for payload envelopes and even virtual tributaries was briefly introduced in Chapter 6. Some mention was made of clock phase differences and jitter, but the chapter included no in-depth exploration of SONET synchronization and timing. This chapter is the proper place to investigate this most distinguishing characteristic of SONET. The need for payload pointers in SONET is related to the differences in timing (or, interchangeably, clocking information) in modern digital networks.

It may come as a surprise to many that network synchronization is important in private-line networks, which is what a SONET network is at heart, of course. Many organizations deploy and/or lease private lines with T-carrier and have never considered the issue of network synchronization or the distribution of clocking information (or just "clock" to many) among the many pieces of T-carrier equipment. Yet, the network seems to hum along just fine day in and day out. Other organizations have added a few pieces of T-carrier equipment to an already functioning network and suddenly experienced problems that the organization has never encountered before. Usually after much head-scratching and shoulder-shrugging terms like "timing loops" and "incorrect stratum" begin to make their way into the expert's reports on the situation. However, right down the block, at another organization with just as many T-carrier links, no such problems are encountered—ever. What's going on here?

Clearly, there is more to the situation than meets the eye. What is it about one private-line network that makes timing so critical, while another has no need to worry about timing? It all depends on what the network is trying to accomplish.

A truly thorough understanding of network timing requirements on the part of many telecommunications personnel is about as common as a truly thorough understanding of grounding on the part of many electricians installing telecommunications wiring in a building. Most electricians know that unless a number 6 gauge copper wire is installed from the communications rack to a water pipe or other suitable ground, the building distribution wiring will not work properly. However, a full and insightful explanation of why this should be done is seldom available from the electricians involved. This is not to downplay the expertise of cable installers, which is considerable and more than adequate for the task. Rather, the whole point is that network timing, like proper grounding, seldom becomes an issue until it is critical. (By the way, grounding a communications rack provides a return path, or ground, for normal direct current signals, which will otherwise quickly build up a net charge on the wire and prevent the modest voltages used from pushing signals through the wire.)

Because network synchronization and the distribution of clock generally plays such a large role in private-line networks, and SONET in particular, this chapter will first attempt in simple terms to explain exactly what a network "clock" is doing. Along the way, the configurations in which clock distribution becomes critical will be examined in detail.

Consider two T-carrier, private-line networks. The example is simple, but has all the ingredients needed to understand network timing requirements. The first network is shown in Figure 8.1.

Figure 8.1
A simple T-carrier net-
work

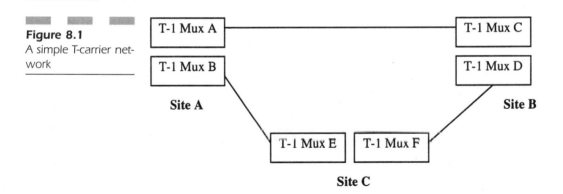

The T-carrier network in Figure 8.1 consists of three T-1 point-to-point links connecting three sites in an organization's network. The customer premises equipment (CPE) at the end of the T-1 link in each case is a T-1 multiplexer (or just "mux") that usually combines the inputs from 24 DS-0 ports and multiplexes them onto the outgoing T-1. Of course, demultiplexing takes the place of the other end of the link. For the sake of simplicity, the 64 Kbps inputs and outputs are not shown in the figure.

As long as a network consists of a series of point-to-point links, there is really no need to "distribute clock" among the network devices in order to "synchronize" the network. Point-to-point links in this configuration can easily use "loop" timing and recover clock from the received data stream. This process needs an explanation.

First, it is important to realize that all digital links, including SONET fiber links, consist of two paths: a transmit and receive path. This is due to the fact that digital signals are not like analog signals. It is easier to send digital signals in a simplex, unidirectional fashion than it is to try to send digital signals full duplex, in both directions, as with analog signals (such as a simple PC, full duplex modem arrangement). In a T-1, these two paths are provided by two twisted pairs of copper wire, one for transmit (typically abbreviated TX) and one pair for receive (RX).

When a T-1 mux sends digital signals from a buffer, which is usually a full frame or more long (193 bits in the case of T-1), timing the output signal is not much of a problem. This means that the sender obviously knows where one bit stops and another begins because the bits are fed serially from the buffer to the transmitter. In a T-1, the 193 bits in a full T-1 frame must be sent 8,000 times a second, or 193 bits in 125 microseconds (1/8,000th of a second). Thus, each bit lasts about 0.65 microseconds. Obviously, the clock in the T-1 mux must be capable of "ticking" faster than 1 million times per second (1 million microseconds = 1 full second) for this to be done properly.

This is not a problem for T-1 equipment, thankfully; The T-1 transmitter merely reads the next bit in the output buffer. When it is a "1" bit, some voltage is placed on the output pair for half of the 0.65 microsecond bit time interval (called a "50% duty cycle"). When the output buffer contains a "0" bit, no voltage is placed on the output pair for the entire 0.65 microsecond bit interval. The point is that the sender always knows when its own bit interval starts and ends.

However, the situation is different on the receiving pair. There is no explicit bit timing and interval information sent along with the data from the transmitter at the other end of the receive pair. It is up to the T-1 receiver electronics to determine when it should check the input pair to see what the voltage is. Presence of voltage indicates a "1" arriving in that bit interval and absence of voltage indicates a "0" arriving in that bit interval. Of course, because even a "1" bit only has voltage on the receive pair for 1/2 of the full 0.65 microsecond bit interval, if the bit interval is not "looked at" by the receiver electronics at the end on the pair at just the right instant, errors will result when "1" bits are mistakenly interpreted as "0" bits. In practice, the process is a little more complex, but not enough to make the discussion invalid.

A hazard does exist: Because no explicit timing information is sent with the data, how is the receiver on the T-1 mux supposed to determine exactly when the bit interval chosen by the sender T-1 mux starts and ends? Each clock in the respective T-1 muxes at each end of the link may be equally accurate, but out of phase. The same may happen when two identical wristwatches show a two-minute time difference, although they are both equally accurate on their own. Clearly, 3 o'clock occurs at different times for each wristwatch wearer. If something vital were to happen at exactly 3 o'clock and would require the presence of both wristwatch wearers, a need would exist to "synchronize watches." Humans can do this by exchanging explicit timing information through direct communication ("I have 5 minutes to three"). But how are T-1 muxes to synchronize bit intervals when there is no contact between them for exchanging explicit timing information? In technical terms, T-1 muxes cannot "recover clock" from the raw, received bit data stream.

Fortunately, there is a simple trick employed by the T-1 muxes in the example network to solve this limitation. The six T-1 muxes can simply employ "loop" timing to provide timing information to the receiver pair. With loop timing, the sending clock "ticks" are used to provide clocking information for the receiver pair electronics. Because both T-1 muxes do the same thing on their receiver pairs, this whole system forms what is called a "phase locked loop" because each sender clock prevents the receiver clock from getting too far out of phase with their counterparts.

In practice, there is a "double-phase locked loop" employed, but the principle is the same. Figure 8.2 shows such loop timing in principle, although the whole process is performed internally and electronically in the T-1 muxes.

Loop timing, locked or not, still has limitations. The clocks can still wander from "ticking" at the proper point in a bit interval, and if this timing jitter were to persist in one direction or another ("faster" or "slower") for long enough, a timing slip would invariably result. In a slip, a bit is either lost or duplicated, because the incoming bit stream is not sampled and interpreted electronically at the correct time. Clocks running slower will invariably lose a bit now and then, and clocks running faster will invariably duplicate a bit sooner or later. This factor shown in Figure 8.3.

In order to prevent persistent jitter faster or slower from inducing slips, T-1 equipment clocks exhibit a characteristic *pull-in range* over which the clocks can adjust their bit intervals and compensate for the effects of jitter. All clocks, therefore, have a characteristic accuracy (which limits slips caused by jitter to a specific number per time interval) and pull-in range (which is how they ensure accuracy).

On a point-to-point T-1 link, the bits originate from a send buffer and finish in a receive buffer. The buffers are filled and emptied by the digital end equipment at each end of the T-1 link. This limits the effects of jitter and their resulting slips. Simple loop timing, therefore, is adequate for point-to-point T-1 link operation. In the example above, the six T-1 mux and three link T-carrier networks operate correctly with simple loop timing employed on each of the three links.

As a result, the organization with this type of T-1 network, with just a series of point-to-point T-1 links, no matter how many, will employ loop timing everywhere and wonder exactly what the debate about network timing is all about. As long as the bits originate at a user end device, and are delivered to another user end device, there is no problem with loop timing, except for occasional slips caused by jitter.

T-1 muxes are not the only devices that may be deployed in a T-1 network. Mention has been made in an earlier chapter of add-drop multi-

Figure 8.2
Loop timing on a T-1 link

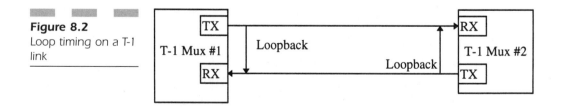

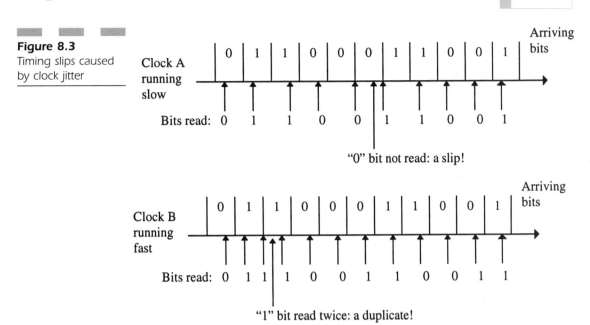

Figure 8.3

Timing slips caused by clock jitter

plexing (ADM) in T-carrier and SONET networks. When applied to T-carrier equipment, adds and drops may be done by equipment normally known as a digital cross-connect system (DCS), but the net result is the same. That is, specific DS-0 channels can be terminated at a site, while others are added to the outgoing data stream, while still other DS-0 channels continue essentially untouched through the DSC or ADM.

Consider the second network shown in Figure 8.4.

Three sites are still connected by three links, as before; however, the end devices are no longer simple multiplexers, but rather more complex DCS devices. These have considerable add-drop capability, and it is possible for any number of DS-0s on any T-1 to be dropped, added, or passed through unchanged at any site, depending on how the DCS is configured by the users. The DS-0s which start out from or end up in a buffer destined for an end user device are not the problem. The problem is with the bits representing the DS-0s that are passed directly from an incoming T-1 frame to an outgoing T-1 frame. These "pass through" DS-0 bits are the reason that other timing arrangements must be made beyond simple loop timing.

Here is why: As soon as one or more of these devices involve the taking of bits off of one link and putting them directly on another link, such as occurs in an add/drop mux (ADM) or digital cross-connect system (DCS), then the relationship between frames starts to become criti-

Figure 8.4
A more complex T-1
network

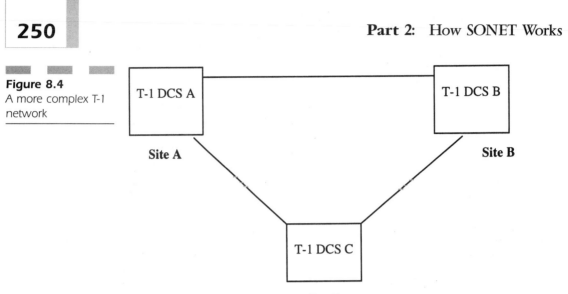

cal, especially in a synchronous multiplexing scheme like SONET. A DS-0 that is dropped and a DS-0 that is added must occupy exactly the proper bit positions in the overall T-1 frame structure. In this case, the receive clock on an input port must agree precisely with the transmit clock on the output port, and not by just a simple loop mechanism.

Therefore, networks that consist of more than just simple point-to-point links, links where bits are terminated in endpoint devices at each end, need to distribute some kind of network timing signal (i.e., distribute clock) among the separate devices on the network.

In order to see this more clearly, extend the wristwatch analogy used earlier. A quick agreement to adjust one wristwatch or the other suffices to allow both humans to engage in some mutual activity at precisely 3 o'clock. Consider this an example of loop timing. But when DCS devices are used, the analogy is more like a conference room without a clock on the wall. When twenty people need a quick break (perhaps they have been reading too much about SONET for one session) and all must return at the same time, they cannot all easily coordinate their wristwatches so they all agree on an time to return. One person's 3:01 may be another's 2:59, and so forth. If everyone had to check with everyone else, it is easy to see that this process could easily consume most of the time allotted for the break. Jitter effects would require the process to be repeated quite often to ensure adequate accuracy for split-second break timing.

However, if there were a master clock in the room (perhaps the instructor's own wristwatch) with which all watches could be synchronized, the process would become much simpler and efficient. This is naturally the whole idea behind *clock distribution* in digital networks. Just to make the analogy more realistic, it could be added that all of the wristwatches have been assigned a certain degree of accuracy, dictating how often they should be synchronized against the *primary reference clock* and how long they could be expected to operate alone without error. It is to be hoped that the instructor's watch would be the most accurate of all, and that some system could be established for ensuring this.

In the example DCS network, each of the DCS devices establishes a special link (normally just another DS-0) to receive timing signals on that keeps their individual clocks within proper operational ranges. This allows each device to accurately take bits directly off of an incoming link and put them on an outgoing link directly, no matter how complex the DCS network. Organizations with many interconnected T-1 links, as in this second example, quickly gain a healthy respect for network timing synchronization and clock distribution techniques.

The need for master clocks and a timing distribution scheme is totally lost on people with experience with IP networks and the Internet. There could be many DS-1s in an ISP's network, but never a thought of timing concerns. There is a simple reason for the IP world's lack of concern about network synchchronization: no one ever worries about "reaching inside" an IP packet and pulling out a particular byte. IP packets are switching, or routed, but always as a whole unit, even when voice is inside the IP packet. In other words, no one cross-connects IP, so there is no need to synchronize IP packet byte streams. But voice bytes inside DS-1s and SONET/SDH links are cross-connected all the time. Equipment routinely requires cross-connecting. This is true whether the SONET/SDH links ever carry voice or not. SONET and SDH require stringent timing mechanisms *just in case* someone wants to use the link for voice.

Where do the DCSs in the private network find the services of a master clock? Organizations that are not service providers (or carriers) with their own ADMs or DCSs can get timing signals from the service provider. Where, then, does the service provider get it? In most modern networks, clock ultimately comes from such standard systems as an atomic clock or LORAN (Long-range Radio Navigation, less common today) or the United States military and civilian versions of the global positioning system (GPS), the master clock for which is in Boulder, Colorado. Use of GPS is more common today. Both LORAN and GPS are navigational aids that fix positions of the earth's surface by radio signal triangulation. LORAN is less common today because it was initially deployed only in coastal areas for ocean navigation, and deployment inland was slow. GPS signals are extremely stable and pulses are very well-shaped. Because there are multiple GPS signals available anywhere on earth (they are satellites), usually at least three at a time, the signals can be cross-checked among each other constantly. As a result, very little wander or jitter occurs in the GPS signal; if you try to position a tank using this signal, jitter potentially can lead the tank to trample your own troops.

Usually in SONET systems, the GPS signal is used to set a primary reference clock (PRC) for the entire network. This primary reference clock is known as the Stratum 1 clock for the whole service provider's network. A Stratum 1 clock is the most accurate in the entire network. When this clock says it's noon, it's noon and that's that. The trick is to let all the other devices in the network know that it's noon. Not long ago, it was common to have only one Stratum 1 clock in the whole network. But with reliable external timing references, such as GPS, it is possible to have many Stratum 1 clocks without running the risk of any of them contradicting the others. AT&T, for instance, has 17 Stratum 1 PRCs, and plans to deploy more in the future.

The network Stratum 1 clock (or clocks) uses regular leased lines (usually of modest speed) to distribute these clock pulses directly to other devices in the network. These are Stratum 2 clocks, and they are directly connected to the Stratum 1 clock. It would be too expensive to hook everything up directly to the Stratum 1 clock, so yet another set of network devices gets clock not from the Stratum 1 clock, but from a Stratum 2 clock. These are, not surprisingly, Stratum 3 clocks. Yet other devices, the Stratum 4 clocks, get their timing from the Stratum 3 clocks.

The whole structure of the network timing hierarchy, which is strictly spelled out in major networking standard documents, is shown in Figure 8.5. This architecture is usually called the *hierarchical source-receiver method* of clock distribution.

In Figure 8.5, networks are shown getting their PRC absolute timing information from LORAN or GPS. There may actually be more than one PRC in a network, used for backup or other purposes. The two networks in the figure may be carrier networks, two private corporate networks, one of each, or even other more exotic combinations.

In the real T-carrier world, most Stratum 2 clocks are in large toll offices for carrier networks or major network nodes in the case of private networks. Stratum 3 clocks are located in smaller end offices, such as a local central office. Finally, the Stratum 4 clocks are usually in CPE devices, such as corporate PBXs or T-carrier muxes. The whole scheme is quite effective, as long as a lower level clock receives timing pulses from a higher level clock. It is possible, when proper care is not exercised, to create situations like "timing loops" where the ultimate source of a clock signal may be a lower level clock. These situations must be avoided at all costs. A simple

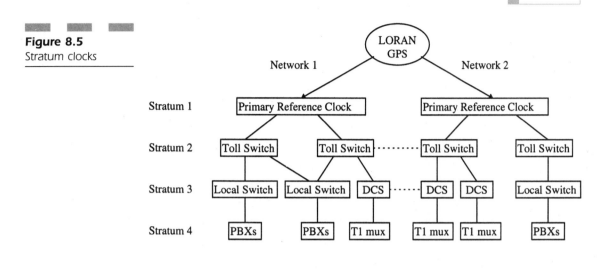

Figure 8.5
Stratum clocks

example of a timing loop occurs when someone is unsure the time on their watch is correct. They ask someone else the time, who asks someone else, who now looks at the *first* person's watch to verify the information! This is not helpful, so the rule is to *never* rely on a lower-level clock for timing signals. This prevents timing loops.

Note that a local switching office may maintain more than one link to a higher stratum clock, which is okay, as long as only one link (the *primary*) is used for timing signals while the other(s) are maintained solely as a backup (the *secondary* or *secondaries*). Note also that the digital network from another service provider would have its own clock distribution scheme. That is also okay. The only time to worry is when a link goes from one service provider to another (i.e., the mid-span meet situation). This is what makes the T-carrier network truly plesiochronous because one network cannot impose its timing on another. Unless, of course, it is a SONET network. Then the more stringent SONET timing makes such a mid-span meet feasible.

What happens when a link to a reference clock is lost? In this case, each stratum clock has guaranteed performance characteristics that have different slip and pull-in parameters, depending on stratum. Naturally, the higher the stratum level, the more accurate the clock must be. The idea is to prevent slips from occurring until the timing links are restored to service. Many service providers maintain physically separate facilities just for clock distribution, but nothing prevents any organization from distributing clock through the T-carrier network itself.

The ITU-T characterizes receiver clock performance based on whether the clock is operating in one of three modes:

- **Ideal operation:**   the short-term behavior of the clock
- **Stressed operation:**   the typically operational mode where timing is received from elsewhere
- **Holdover operation:**   the "rare" cases when all timing references are lost

In ideal operation mode, the receiving clock never loses the input timing reference. This is certianly true in the short term, but is not typical of normal day-to-day operation. But how the clock operates in the short term is a good baseline for overall timing performance. For example, ideal operation displays the short term "noise" of the clock, which will manifest itself as DS-3 level payload pointer adjustments in SONET.

Stressed operation always expects that there will be short interruptions of the timing reference signal, which is certainly true of normal network operations. There can be from 1 to 100 interruptions per day according to the ITU-T. Naturally, during the interruption, the timing reference cannot be used. If the interruption is short, the receiver will restore the timing signal when it reappears. Otherwise, the receiver clock will have to switch reference if the outage is protracted. In either case, there will be some error between the time kept locally and the newly restored or secondary reference. This difference should be less than 1 microsecond. A timing error of even 1 microsecond can cause up to 7 pointer adjustments at the DS-3 level in a short period of time, but SONET can handle this without error. And a 1-microsecond timing difference will not affect the DS-1 level payloads in SONET (or E-1 in SDH). However, timing errors will accumulate and eventually cause a pointer adjustment at lower SONET/SDH levels.

Holdover operation occurs when a SONET/SDH NE loses all timing references, primary or secondary, for an extended period of time. In SONET/SDH, if an NE is on holdover operation or receiving clock from a source in holdover operation, the pointer adjustments come often and at steady intervals.

If all is working as planned, there should be few DS-1 or E-1 pointer adjustments each day. Even in holdover operation, a SONET/SDH NE should have no more than one DS-1 pointer adjustment in 9 seconds, and one per 7 seconds for E-1.

Stratum Clocks

When a link is lost to a higher level stratum clock in the timing hierarchy, what happens depends on at which level of hierarchy the link loss occurs.

Stratum 1 clocks are typically Cesium or Rubidium atomic clocks that derive their signals from the vibrations of atomic nuclei. These clocks, which used to cost $100,000 not too long ago but now go for about $10,000, are accurate to better than .00001 parts per million (i.e., a hundredth of a nanosecond—about the same relationship as one second is to 100 billion seconds, or some 3,000 years). Hence, these atomic clocks may lose one second every 3,000 years. If the link to GPS were lost, there obviously would be little to worry about. Newer atomic clocks have been prototyped at about 100 times more accuracy, or down to the picosecond levels; thus, these may lose one second every 300,000 years.

The Stratum 2 clocks are "only" accurate to .016 parts per million (less than a hundredth of a microsecond). The important thing about Stratum 2 is that if the reference link were lost to the Stratum 1 clock, the Stratum 2 clock would not wander off of the mark far before the link would be restored. Otherwise, the frame sending time would be out of synchronization with the other devices on the network. This wandering would eventually cause a timing slip. The number and interval of such resulting slips are strictly controlled by the stratum network timing standards. Therefore, all Stratum 2 clocks must have less than 255 slips in the first 86 days after a loss of reference. In the case of timing signals, a slip is just a lost or repeated timing pulse. Moreover, the first slip cannot occur within seven days of the reference loss.

A common enhancement is to add GPS capabilities to Stratum 2 or 3E (Stratum 3 "enhanced" clocks, known as "2 GPS" and "3E GPS," which effectively makes them into Stratum 1 clocks as long as the GPS link is functional). Stratum 3E was invented by Telcordia (Bellcore) and is not acknowledged by standards bodies.

Stratum 3 clocks are accurate to 4.6 parts per million (almost 5 microseconds, pretty sloppy in comparison to Stratum 2 clocks). These clocks must have fewer than 255 slips in the first 24 hours after a loss of reference. The first slip cannot occur less than six minutes after the reference loss.

Stratum 4 clocks have no guarantees along these lines. They can slip all over the place and no one can complain. Most CPE devices at end user sites, PBXs or T-1 muzes, are Stratum 4 clocks, as are all PCs and even many wristwatches.

Here is the main point regarding stratum clocks. Even central office switches with plenty of T-carrier links to customer sites and other switching offices usually have only Stratum 3 accuracy. The end-user devices where the T-carrier bits originate and terminate are even worse. Therefore, it is not particularly unusual to have only 192 bits in a buffer instead of

193, or even 194. This is why T-carrier employs "slip buffers" in equipment and must use "bit stuffing" in T-3s. For this reason, SONET uses only Stratum 3 clocks or better to achieve "synchronous" multiplexing.

The Larus company, a common vendor of carrier timing equipment, has extended the stratum hierarchy with its own widely used enhancements to Stratum 3 and 4 clocks. These are known as Larus 3E and 4E, but they are not strictly part of the timing hierarchy.

The major accuracy, pull-in ranges, stability, and time to first slip for all stratum clocks are shown in Table 8.1.

Plesiochronous Clocking

In spite of highly accurate clocks in the T-carrier network, synchronous operation of multiplexers and other pieces of equipment has never been possible, without the need for bit stuffing to make up for these timing differences. There are two reasons for this.

TABLE 8.1

Stratum 1, 2, and 3E characteristics

Stratum	Accuracy	Pull-in Range	Stability	Time to First Slip
1	1×10^{-11}/day	Not applicable	Not applicable	72 days
2	1×10^{-8}/day	Synchronizes to clock with accuracy of +/− 1.6×10^{-8}	1×10^{-10}/day	7 days
3E	1×10^{-6}/day	Synchronizes to clock with accuracy of +/− 4.6×10^{-6}	1×10^{-8}/day	3.5 hours
3	4.6×10^{-4}/day	Synchronizes to clock with accuracy of +/− 4.6×10^{-6}	3.7×10^{-7}/day	6 minutes (255 in 24 hours)
4	32×10^{-6}/day	Synchronizes to clock with accuracy of +/− 32×10^{-6}	Same as accuracy	Not yet specified
4E	32×10^{-6}/day	Synchronizes to clock with accuracy of +/− 32×10^{-6}	Same as accuracy	Not applicable

First, the T-carrier links to be multiplexed may come from two different networks. For example, a DS-4 may need to be created from three input DS-3s. One comes from the service provider's own network, the second comes from another service provider's network, and the third comes from an "asynchronous island," which is another way of saying a private network with its own internal clocking scheme.

Loss of reference also allows the clocks on isolated devices to wander out of phase. They do not necessarily slip, but they may be off just enough to deliver the "wrong" number of bits per frame time. Of course, these devices still need to send and receive data.

Because each of the clocks may not be in phase with each other, and probably will not be, it is necessary to buffer, usually a whole frame, and insert stuffing bits and control bits. The stuffed bits make up for a shortfall in the number of bits in a frame and the control bits enable the receiver to detect the stuffed bits. In the case of a T-1, the slip buffers will repeat an entire frame if the input clock constantly runs slower than the output clock in a DCS. An entire frame will be dropped if the input clock constantly runs faster than the output clock in a DCS. This is just another way of saying that clock distribution and timing accuracy is critical in networks where T-carrier equipment is hooked up back to back, instead of consisting of a small (or even large) number of point-to-point links.

This whole process of adjusting for timing variations translates into much effort and overhead. For example, a DS-4 uses more than 5 Mbps (more than three T-1s worth of bandwidth!) just for overhead to package the plesiochronous DS-3s. The bit-stuffing bandwidth would be much better used delivering users' bits (and adding revenue) to the carrier's network. This bit-stuffing and control process for DS-3 to DS-4 NA is illustrated in Figure 8.6.

SONET Synchronous Multiplexing

More accurate network clocking allows SONET to use more straightforward, byte-interleaved, synchronous multiplexing than the awkward bit-stuffing methods used with the higher levels of the T-carrier digital hierarchy. Thus, three OC-1s can easily be combined into one OC-3 without the use of large buffers, bit stuffing, or much control overhead.

This is not achieved without a price, however. There can be no Stratum 4 clocks in SONET. All clocking in all SONET devices must be Stratum 3 or better. This used to be a major issue and network expense, but the affordability of accurate clocks makes stringent SONET timing that much easier to accomplish.

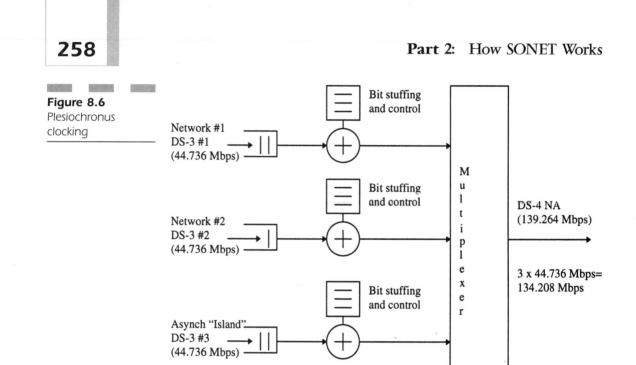

Figure 8.6
Plesiochronus
clocking

The major benefit is shown in Figure 8.7. In the figure, synchronized STS-1 inputs can be directly byte-interleaved onto the synchronized STS-3 output stream. The STS-3 is then converted to optical signals and sent on the SONET fiber. No complex bit-stuffing and control procedures are needed. As a result, SONET speeds are simple multiples of the basic

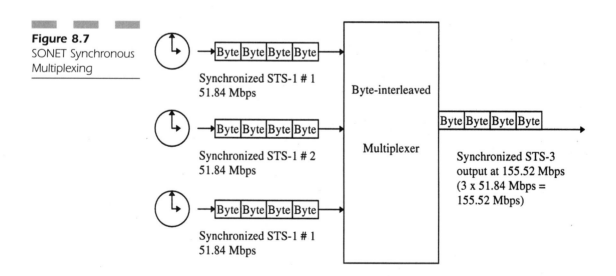

Figure 8.7
SONET Synchronous
Multiplexing

rate of 51.84 Mbps. No additional overhead is ever needed for bit stuffing and control.

Synchronizing a SONET Network

It is important to realize that it is the potential need for DS-1 (and E-1) cross-connecting (pretty much a given in the voice circuit world) that SONET and SDH need to synchronize network clocks. Most telecommunications authorities use a technique called the *hierarchical source-receiver method* to accomplish this.

In the hierarchical source-receiver method, the master clock used in the network is one (or even more) primary reference sources (PSRs). The timing reference is distributed to the rest of the network through a series of receiver clocks. Receiver clocks pass signals along to other receivers in the form of a pyramidal hierarchy.

SONET NEs can use one of five types of timing:

■ **External timing:** There is a direct Stratum 1 clock reference available.

■ **Line timing:** The NE derives its own clock from the scrambled SONET signal input.

■ **Through timing:** The clock in this case is also derived from input, but then the NE sends the signal in a different outbound direction.

■ **Loop timing:** This is the same as line timing, with one major difference. Line timing can be used in an add-drop multiplexer (ADM) used as customer premises equipment (CPE) and there can be other SONET NEs beyond the ADM. Loop timing is used for straight CPE and the SONET ends at this NE. In terms of SONET equipment discussed more fully later, the ADM is configured for Terminal Multiplexer (TM) mode only.

■ **Free running:** The SONET NE has access to an internal clock only.

It would be useful to give an example of how SONET NEs actually use these timing modes.

It is all well and good to describe Stratum clocks and how SONET timing allows for synchronous multiplexing. However, this does not explain exactly how SONET networks distribute clock among the various SONET network NEs.

Figure 8.8 shows how a SONET network is usually synchronized. In the figure, timing for outgoing OC-ns is shown as a dotted arrow. The

Figure 8.8
Synchronizing a
SONET Network

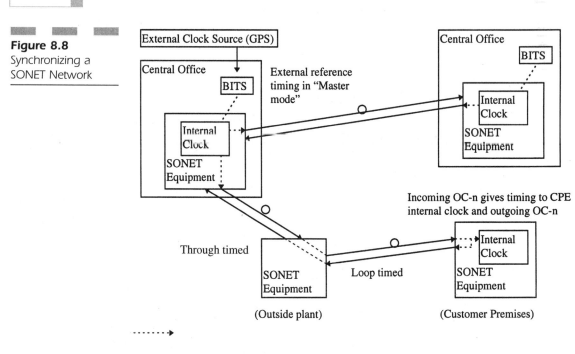

internal clock of a piece of SONET equipment in a switching node (for example, a central office) may derive its timing signal from a building integrated timing supply (BITS). Of course, the BITS system draws its own timing from a higher layer of stratum clocking system. The office may even draw timing directly from the GPS signals. Such building timing systems can be used to provide clock to all of the SONET devices in the switching office.

These BITS supply external reference timing in what is known as "master mode" for each of the internal clocks in the service provider nodes' SONET equipment. This equipment serves as a master for other SONET network nodes, providing timing signals on its outgoing OC-n link at literally any level of the SONET hierarchy. Other switching offices use their own BITS to drive internal clocks and time the outgoing SONET link.

Of course, not all SONET equipment is located in switching nodes where good external timing signals or BITS signals are easily available. For instance, there may be SONET equipment deployed as *outside plant* equipment away from the central office. These other SONET nodes derive their timing from the incoming SONET bit stream in what is called a "loop timed" arrangement, also shown in Figure 8.8. Intermediate SONET equipment, such as ADMs (add/drop muxes), on SONET links use loop timing.

What about the SONET equipment on the customer's premises? This CPE operates in "slave mode" (the term is unfortunate, but universal), which employs "loop timing" for the internal clock. Note that this arrangement times both the incoming and outgoing OC-n signal from the CPE, as well as provides a reference clock for the internal clock in the CPE.

Of course, all clocks in Figure 8.8 should be Stratum 3 or better for SONET because of the lack of adequate performance and accuracy standards for Stratum 4 clocks.

Figure 8.9 shows some details of a BITS operation. For reliability purposes, except in more modest private networks, the timing network has links to both a primary and secondary clocking source. Both should derive from a Stratum 1 source but, of course, other arrangements are possible, including direct GPS synchronization. The secondary link acts as a standby or backup. A selector unit chooses the proper timing signal. If a short period of signal disruption were to occur from both the primary and secondary sources, the building's internal clock could be used. The internal clock must be accurate to + 20 parts per million if used for this purpose.

Within a single building, one of these clock sources is designated as the BITS, which has the highest accuracy. The timing signal generator (TSG) makes the decision. For example, if the primary source were determined to be performing poorly and not meeting Stratum 1 quality, the TSG could tell the selector to try the secondary source. If this were

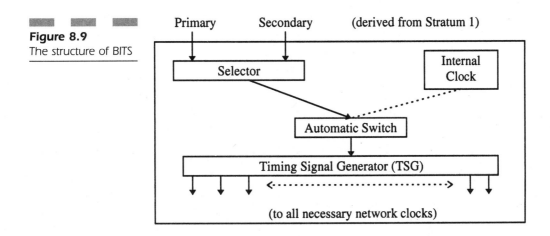

Figure 8.9
The structure of BITS

found to be unacceptable as well, the TSG then could be switched auto-matically to the internal clock as a timing source.

BITS provides all the timing for the digital links in the serving office. These are shown at the bottom of Figure 8.9.

Benefits of SONET Timing

The benefits of requiring Stratum 3 clocking in SONET are numerous; some are listed as follows:

- This is the mechanism that allows SONET to byte multiplex into the gigabit ranges. Without the extremely accurate timing needed for byte multiplexing, SONET would be forced to revert to bit-stuffing and control at very high bit rates.

- Multiplexing equipment is much more compact and simple. The same timing that makes byte multiplexing possible also results in much simpler designs because microprocessors work well with groups of 8 bits (the byte).

- The circuitry in SONET can easily be placed on a simple circuit board, not in a monstrous cabinet; as a result, most SONET equipment is quite modest in size.

- Compared to T-carrier devices, SONET equipment is acquired at a reduced cost and with less complexity. Smaller components draw less power and can be mass-produced much more effectively.

- There is no longer any need for complete demuxing to find an individual DS-0 (which are nothing more than groups of 8-bit bytes) inside a higher bit-rate transport. Everything in SONET is byte multiplexed; therefore, it is easy to find a specific STS-1 inside an STS-n, and just as easy to find a DS-0 inside a given STS-1—when the payload is carrying DS-0s, that is.

Why SONET Needs Pointers

Much of this chapter has emphasized that SONET is tightly synchro-nized and employs accurate clocks to dispense with the T-carrier need for bit-stuffing and control. However, mention has been made that SONET employs pointers in the transport overhead to locate the syn-chronous payload envelope (SPE) within a SONET frame. However, if SONET were truly synchronous, why couldn't the position of the SPE just be "locked" into position and then rely on SONET clocking to keep the SPE there?

The fact is that SONET still may need pointers, even when all of the clocks on the SONET network are derived from exactly the same primary reference source. This is mainly because the speed of light is finite.

The reason for this need of pointers is shown in Figure 8.10. The figure illustrates what happens when a SONET ADM must take payloads from a "West" SONET terminal multiplexer (TM) and send the payload directly to an "East" SONET TM device. This is not an unusual activity for SONET equipment.

First, note that if the fiber link were 40 km long, it would take about 200 microseconds for a frame to reach the SONET ADM. The transmission time is given by the formula $t = d / (c/n)$, where d is the distance in kilometers, c is the speed of light, and n is the fiber's core index of refraction. It is easy to see that if $d = 40$ km (as given) and $n = 1.5$ (typical for SONET fiber), then $t = 40$ km / (300,000 km/sec / 1.5) = about 200 microseconds.

Now, here is the problem: The frame arrives at the ADM after 200 microseconds, but the ADM is sending one frame eastbound every 125 microseconds, as usual, based on the internal clock. Therefore, the ADM may potentially need to wait another 50 microseconds to send the payload on to the east if the payload must always be aligned with the start of a SONET frame. The trouble is that at the OC-48 rate (about 2.488 Gbps), there will be about 20 Mbps arriving at the ADM every microsecond

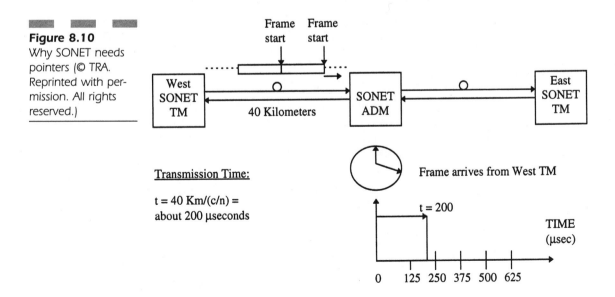

Figure 8.10
Why SONET needs pointers (© TRA. Reprinted with permission. All rights reserved.)

Frame start   Frame start

West SONET TM — 40 Kilometers — SONET ADM — East SONET TM

Transmission Time:

$t = 40$ Km/(c/n) = about 200 μseconds

Frame arrives from West TM

$t = 200$

TIME (μsec)

0   125 250 375 500 625

(2.488 gigabits/125 microseconds)! This is about 1 gigabit in 50 microseconds and would require a buffer of at least 125 megabytes! These buffers also would be required in each piece of SONET equipment along the way on the SONET network.

Nevertheless, some service providers do indeed buffer the payload in this situation, all in the interest of preserving the simple fourth column position of the SPE inside the SONET frame. However, the SONET pointers (i.e., H1, H2, and H3) provide a way to offset the start of the payload eastbound from the start of the SONET frame itself.

Furthermore, the pointers themselves are capable of "positive" and "negative" pointer adjustments (or justifications) due to timing variations in the Stratum 3 SONET network clocks. They do this not 1 bit at a time, but 1 byte at a time. Pointer justification is necessary because the H1/H2 pointers cannot merely point anywhere in the SPE, but must always point to the position of the first byte of the Path Overhead (POH). If this moves due to byte buffer overrun or underrun, so must the H1/H2 pointer to the POH. Thus the use of pointers in SONET reduces the need for buffer space and delays on the SONET links.

SONET Pointer Justification

In spite of the improvements in accuracy of SONET clocking, timing errors do occur on SONET networks and there is still a need for adjusting the positions of the H1- and H2-payload pointers on an active SONET link. Standards control the frequency of these pointer adjustments, known as *pointer justifications*, but they are still necessary periodically to avoid the bit-stuffing pitfalls of the older PDHs, such as T-carrier.

One of the nice things about pointer justification in SONET is that everything is done by the byte, not by the bit. This preserves the ability to do byte multiplexing and easy DS-0 drops and adds while, at the same time, allows high-speed frame alignment without large buffers in each piece of SONET equipment. This section takes a closer look at this process.

Mention has been made about network timing synchronization being a problem in networks where bits make their way directly from an input signal through a buffer to an output signal. The device's internal clock is used to write bits to the buffer and read bits from the buffer. A need for bit stuffing occurs when the buffer is empty at a time when a bit must be read. If the buffer ever fills, bits will be dropped. This process is shown in Figure 8.11.

Figure 8.11
Buffers in network devices

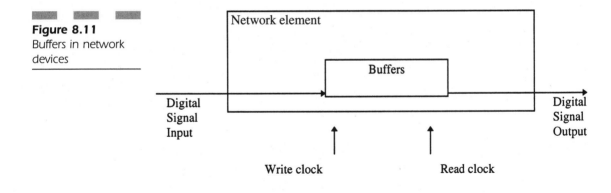

SONET buffers are organized in groups of 8 bits, which are commonly called bytes. When multiplexing occurs, bytes from a variety of buffers must be combined onto a single output signal. However, because of timing variations (jitter), there may not always be exactly 8 bits in the buffer when they are needed for the output signal.

Consider a clock that is accurate to 60 parts per million (60 ppm) on a link operating at a modest bit rate of 1 Mbps (a T-1 sends about 1.5 Mbps). Thus, in a minute, this equipment should send and receive 60,000,000 bits. But at 60 ppm accuracy, the minimum received could be 59,999,940 (60 less) and the maximum received could be 60,000,060 (60 more). This is the limit of timing accuracy. Of course, the goal is to still send 60,000,000 bits per minute, and perhaps the jitter cancels out over time. Nonetheless, when a group of bits or bytes needs to be sent every 8,000th of a second, there is a need to buffer the bits to a greater or lesser extent, depending on the accuracy of the clocks and the speed of the link.

With SONET, there are three situations that must be accounted for in a discussion of pointer justifications. All three involve considering the number of full bytes and "leftover" bits in the buffer. First, when a byte is needed for building the SPE within the frame on the output link of a multiplexer, there may be between 8 bits and 16 bits in the buffer (1 full byte and some bits left over). Next, there may be more than 16 bits, but less than 24 bits in the buffer (2 full bytes and some bits left over). Finally, there may be less than 8 bits in the buffer (less than 1 full byte). This example assumes that no more than 24 bits will ever be in the buffer, perhaps because the buffer is only 24 bits deep, or the timing is accurate even to rule this out.

These possibilities are shown in Figure 8.12, along with the desired result. Normally, it would not be a problem to have a few bits left after a byte is sent. These bits are just the first bits of the next byte, which

Figure 8.12
SONET buffer possi-
bilities

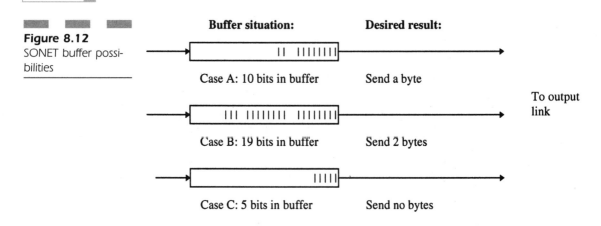

In Figure 8.12:

Buffer situation: Desired result:

Case A: 10 bits in buffer   Send a byte

To output link

Case B: 19 bits in buffer   Send 2 bytes

Case C: 5 bits in buffer   Send no bytes

have arrived a little early. But when there are more than 16 bits to send, it would be nice if a mechanism existed to allow more than 1 byte to be sent during this particular frame time. Naturally, if there were less than 1 byte to send, it would be nice if there were some way to allow no byte to be sent during this frame time. These bits are arriving a little late, and the input clock may need the time to "catch up" with the output clock.

In Figure 8.12, a SONET device is shown taking bytes directly from a input link and placing them on an output link. In the first case, there are 10 bits in the buffer. In the second case, there are 19 bits in the buffer. Finally, the last case shows only 5 bits in the buffer. Jitter errors will only accumulate when there is a mechanism in SONET for allowing not only 1 byte to be output (Case A), but also 2 bytes (Case B), or even no byte (Case C).

A SONET frame is exactly a fixed number of bytes long. An STS-1 frame has 810 bytes, or 9 rows of 90 columns. There are exactly 27 bytes of overhead on the first three columns (9×3) and 783 bytes of payload in the SPE area (9×87). Where could an extra byte be added or omitted when needed?

Fortunately, SONET overhead handles this relatively easily. The H1- and H2-pointer bytes always point to the start of the SPE, which is the position on the first POH byte, by definition. Immediately following the H1- and H2-pointer bytes is the H3-overhead byte. This is the Pointer Action byte, and as may be expected, the pointer action byte plays a role in this process of adjusting the value of the H1 and H2 pointer when more or fewer bytes need to be sent in the SPE.

When more than 783 bytes need to be sent, the H3 pointer action byte location can hold an extra byte. When fewer than 783 bytes need to be sent, the SPE immediately following the H3 byte can be used to hold a special "stuff byte" which is ignored by the receiver. The values of the H1, H2, and H3 determine which is the case. Thus, SONET overhead has a mechanism for sending and receiving SPEs that contain 783 bytes (normal), 784 bytes (when input clocks run fast), or 782 bytes (when input clocks run slow). How often these adjustments are made depends on exactly how frequently send buffers fill and empty and how stable the clocks are in the long run. Naturally, SONET standards cover all of this in considerable detail.

The interplay between the H1, H2, H3 pointer bytes and the input buffers solves the jitter problem in the SONET network. Of course, the trade-off is in processing power at the sender and buffer management techniques. But the prize is the synchronous operation of SONET. Purists may point out that instead of bit-stuffing, SONET merely uses byte-stuffing; however, until all network clocks are equally accurate, some allowance for jitter effects in any network is needed. It is undeniably true that SONET is not "100%" synchronous in the sense that all network clocks guarantee flawless sender and receiver synchronization. This is not the goal of SONET anyway. SONET is a better way of dealing with the reality of phase differences and jitter on the network.

The relationship between the three cases outlined above and these four bytes in the SONET frame (i.e., H1, H2, H3 bytes and following payload byte) are shown in Figure 8.13. One of these will handle any buffer situation as the sender.

In Figure 8.13, the same three cases are addressed. In normal operation, there is always a byte to send in the outbound SPE, and a few bits left over (exactly 8 bits would be a special case). However, when the input clock runs slightly faster than the output clock, there will be more than 16 bits to send. Finally, when the input clock lags behind the output clock, there will be fewer than 8 bits to send (no bits would be a special case in this scenario).

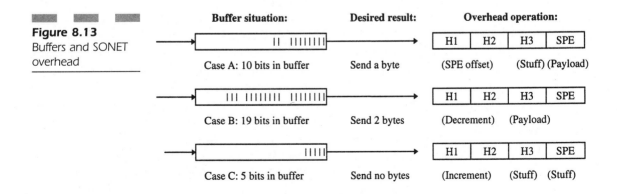

Figure 8.13
Buffers and SONET
overhead

All of these conditions are handled by the interplay of the H1, H2, H3 pointer bytes and the payload byte immediately following. The H1, H2, and H3 bytes are located in the fourth row and first three columns of the SONET frame and, thus, form the initial Line Overhead bytes. The payload byte in question, therefore, is the byte located in Row 4, Column 4 of the SONET frame.

Figure 8.13 shows that, in normal operation, the H1 and H2 bytes taken together form a 10-bit pointer field that indicates the offset of the first byte of the SPE, which is also always the first byte of the POH. In normal operation, the H3 byte is always ignored by the receiver, whatever its value or content, and the SPE byte in Row 4, Column 4 is valid user payload. The SPE is 783 bytes long. However, in the case where two bytes should be sent with the SPE in order to adjust for jitter, the H3 byte now contains user payload as well. The SPE pointed to by the H1 and H2 bytes will now contain 784 bytes. The receiver knows this is the case by the value of the H1 and H2 pointers, which also indicate to the receiver that the position of the SPE in the next SONET frame will move back one position in the SONET frame; thus, the H1 and H2 pointer value must be *decremented* to reflect this change. This is also shown in the figure. Case B is known as a "negative stuff opportunity" or "negative frequency justification" in SONET documentation.

Finally, the figure shows the case where there are less than 8 bits in the buffer. The SPE pointed to by the H1 and H2 bytes will now contain 782 bytes. Not only is the H3 byte ignored as a stuff byte, but so is the SPE position in Row 4, Column 4, which now also contains a byte "stuffed" in by the sender. As before, the receiver knows this is the case by the value of the H1 and H2 pointers. They also now indicate to the receiver that the position of the SPE in the next SONET frame will move forward one position in the SONET frame. So the H1 and H2 pointer value must be *incremented* to reflect this change. This is also shown in the figure. Case C is known as a "positive stuff opportunity" or "positive frequency justification" in SONET documentation.

The fact that SONET contains a mechanism for jitter compensation involving the sending of 1 byte more or less is almost intuitive to grasp. Less intuitive is the need to adjust the value of the H1 and H2 pointers at the same time. After all, why should there be a need to change the pointer value just because 1 byte more or less has been sent? The reason is that the H1 and H2 pointers cannot point just anywhere in the SPE, but must always point to the first POH byte.

To see why there is a need for pointer justification, consider a very truncated SONET frame. Although truncated, this frame has all of the pieces needed to show why the pointer values must change with positive or negative byte stuffing. The special frame is shown in Figure 8.14. It is only three rows by five columns for a total frame size of 15 bytes. The first two columns are overhead, so the SPE is only 9 bytes, while the overhead is 6 bytes. The pointer byte—this frame size needs only a single byte to act as a pointer—is located in Row 2, Column 1. The "stuff" overhead byte occupies the next position.

Just as in a full SONET frame, this special SONET frame contains an SPE. Or rather, the frame holds the beginning of an SPE which, in this case, is only 9 bytes long. In normal operation, this special SONET frame contains the end of one 9-byte SPE and the beginning of the next. The pointer indicates the offset of this first SPE byte. This case is shown in Figure 8.15.

In Figure 8.15, the 9 bytes of SPE occupy the payload area of the SONET frame. The pointer byte contains the value "+3" (the plus indicating that the value is a pointer), which points to the first byte of the SPE (the first POH byte). This +3 offset points to Row 2, Column 4 of the truncated SONET frame, which is correct. The stuff byte is never examined by the receiver.

Now consider another example: There are 2 bytes in the buffer that should be sent in the next outgoing SONET frame. In this case, the stuff byte position is now used to send this "extra" payload byte. This solves the jitter problem and results in a frame that contains not 9, but 10 payload bytes. The next frame then would revert to 9 bytes again. This situation would look like the two frames in Figure 8.16.

In Figure 8.16, the effect on the "extra" byte of payload on the pointer byte is shown. Not only has the stuff byte been used to transport the extra byte from the buffer, but this action has affected the position of the first byte of the SPE where the POH begins. Note that in order to continue to point to the first byte of POH, it is necessary to *decrement* the pointer value (from +3 to +2). It is called a "negative stuff" because 1 less byte of "stuff" is inserted into the SONET frame, which now has 1 more byte of payload.

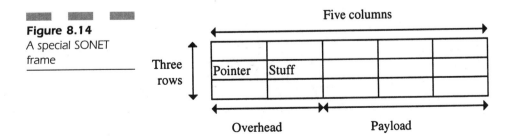

Figure 8.14
A special SONET frame

Figure 8.15
The special SONET
frame in normal
operation

		6	7	8
+3	Stuff	9	1	2
		3	4	5

Overhead   Payload

The "extra" payload byte throws off the start of the SPE in the next SONET frame, making it appear to the receiver that the start of the SPE has moved 1 byte to the left in the SONET frame. The decrementing of the pointer compensates for this problem. All of this is necessary because the pointer must point to the first byte of POH in the SPE, not just anywhere.

The last case is easy to visualize as well. The case where there are only 8 bytes of payload in the buffer and the subsequent effect on the start of the SPE and corresponding pointer value is shown in Figure 8.17.

In Figure 8.17, the effect on the "missing" (i.e., late-arriving) byte of payload on the pointer byte is shown. Not only has the following payload byte been used to transport another stuff byte, but this action has affected the position of the first byte of the SPE where the POH begins. Note that in order to continue to point to the first byte of POH, it is necessary to *increment* the pointer value (from +3 to +4) in the next frame. It is called a "positive stuff" because 1 more byte of "stuff" than usual is inserted into the SONET frame, which now has 1 less byte of payload.

The "missing" payload byte throws off the start of the SPE in the next SONET frame, making it appear to the receiver that the start of the SPE has moved 1 byte to the right in the SONET frame. The incrementing of the pointer compensates for this factor. All of this is necessary because the pointer must point to the first byte of POH in the SPE, not just anywhere.

Figure 8.16
Need for negative
frequency
justification

Frame with "negative" stuff

		6	7	8
(+3)	9	1	2	3
		4	5	6

Frame with pointer decremented

		7	8	9
(+2)	Stuff	1	2	3
		4	5	6

Overhead   Payload

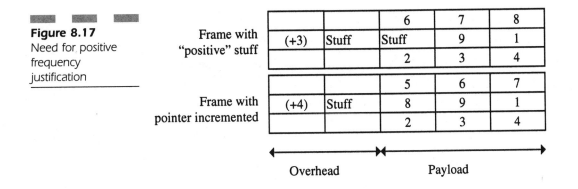

Figure 8.17
Need for positive
frequency
justification

			6	7	8
(+3)	Stuff	Stuff	9	1	
			2	3	4

Frame with "positive" stuff

			5	6	7
(+4)	Stuff		8	9	1
			2	3	4

Frame with pointer incremented

Overhead   Payload

SONET Pointer Values

In a full SONET frame structure, the values of the H1, H2, and H3 pointer bytes tell the receiver when a positive or negative stuff has occurred, and when the pointer values, therefore, need to be incremented or decremented, respectively. This section details the actual values used in SONET operation.

The structure of H1- and H2-pointer bytes is a little complicated, and their use fairly awkward, mostly due to the related needs for both sophistication and yet compactness of operation. The pointer itself is a 10-bit field and, thus, must span the H1 and H2 bytes. Ten bits are needed because 2^{10} is 1,024, which gives more than enough offset value to point anywhere within the 783 byte SONET SPE. Valid values of the H1/H2 pointer are in the range of 0 to 782. Other values indicate error conditions. Offsets do not include the SONET overhead columns, but point to payload areas in the SONET frame exclusively.

The structure of the H1- and H2-pointer fields is shown in Figure 8.18. In the figure, the first 4 bits form the new data flag (NDF) bits. In normal operation, these bits are set to 0110, which means that the 10-bit pointer field is to be interpreted by the receiver as a valid offset to the SPE start. The NDF value of 1001 (which is the exact inverse of 0110) indicates that the previous pointer value is incorrect (due to a positive or negative stuff) and that the correct value is now indicated by the content of the pointer field. All other values are undefined. Any other bit configurations besides 0110 and 1001 are interpreted as one or the other by a simple "3 of 4" rule. Therefore, 1110 is interpreted as 0110, 1101 is interpreted as 1001, and so on.

The next two bits of the H1 byte are reserved and set to 00 during normal SONET operation; this is not strictly required, but is strongly

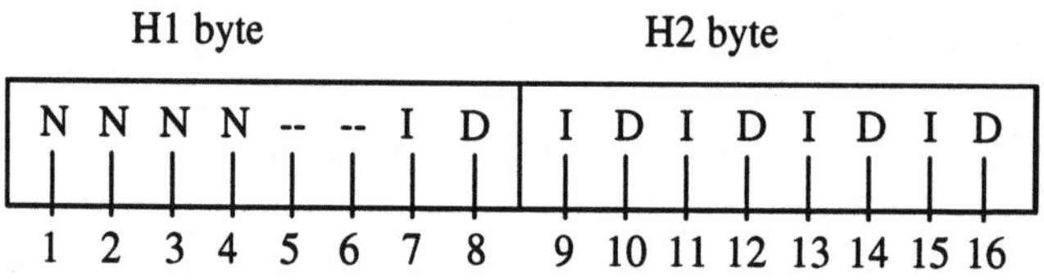

Figure 8.18

The structure of the
H1 and H2 pointer
bytes

H1 byte								H2 byte							
N	N	N	N	--	--	I	D	I	D	I	D	I	D	I	D
1	2	3	4	5	6	7	8	9	10	11	12	13	14	15	16

"recommended." The last 10 bits of the H1 and H2 bytes are the pointer bits, which form a simple binary offset to the start of the SPE. A value of 0 means that the SPE starts in Row 4, Column 4, immediately after the H3 byte. A value of 87 means that the SPE starts in Row 5, Column 4, immediately after the K2 overhead byte, and so on. A value of 89 would be carried by the H1 and H2 bytes as 0110 00 0001011001.

Note that the pointer bits are alternately labeled "I" and "D", which stand for "Increment" and "Decrement," respectively. The compactness of the H1 and H2 bytes is reflected in the use of these bits. When the need for a positive- or negative-frequency justification arises, requiring a new pointer value to be conveyed to the receiver, the I and D bits are used to indicate whether this is to be a positive (I bits used) or negative (D bits used). A positive-pointer adjustment inverts, or flips, the values of the I bits and a negative-pointer adjustment inverts, or flips, the values of the D bits. These changes are easy for electronic components to detect and avoids the need for another field in the H1 and H2 bytes for this sole purpose.

The value of the I or D bits also indicates to the receiver whether there is a positive stuff and, therefore, the payload byte immediately after the H3 byte is to be ignored or whether there is a negative stuff. A negative stuff indicates that the H3 byte itself is now valid payload. Of course, the pointer value of the new SPE offset must be sent as well. This is done by repeating the new pointer value at least three times with the NDF bit pattern (1001). Only then can the pointer change again. Therefore, SONET SPEs can only change position every four frames, and most SPE positions are quite stable because of the accuracy of the network clocks used in SONET.

Conflicts between the values of I bits and D bits are handled by an "8 of 10" rule. This means that if 8 of the bits seem to indicate to a receiver that the I bits are inverted, then that is exactly the assumption that is made. Ambiguous bit configurations will generate "loss of pointer" error conditions when they persist for several frames.

This whole pointer operation with I bits and D bits may sound confusing, but it really is not. Figure 8.19 shows the whole sequence of H1-

and H2-pointer values used when moving the SPE offset from a value of 89 (in decimal) to 90 and back to 89 again. Such movement is not unusual because the long-term stability of SONET clocks typically cancels out persistent movement of the SPE in any direction.

In Figure 8.19, the values of the I and D bits are shown, along with the value of the NDF field. No bit errors are included, but the simple 3 of 4 and 8 of 8 rules would resolve any conflicts. Note that in frames with I or D bits set, there is no valid pointer value. The next frame with the NDF bit pattern of 1001 is to be used for the new value.

Table 8.2 shows the actions performed by transmitters and receivers during positive- and negative-pointer justifications. The table is relatively straightforward but helps to put all of the operations required for frequency justification in perspective.

SONET STS-3c Pointer Justification

Only one more topic needs to be explored in this chapter on network synchronization: It concerns the use of pointers in higher rate payloads, such as STS-3 or STS-12, or any other SONET speed. When the frame structure is "channelized" and the STS-N contains N STS-1s, then the use of pointers is straightforward. Each STS-1 has a pair of H1- and H2-pointer bytes to locate its own SPE. Everything will be multiplexed by columns, of course, making it easy for each STS-1 pointer pair and SPE to be broken out correctly. However, an STS-N with N STS-1s is not the only structure allowed in SONET.

Mention has been made of the concatenated STS-3c frame structure used to transport payloads, such as ATM cells on B-ISDN networks.

TABLE 8.2 Pointer justification operations	Input Running Slow (positive justification)	Transmitter	1) Invert 5 I bits 2) Insert stuff byte after H3 3) Increase pointer value by one
		Receiver	1) Detect inverted I bits 2) Remove stuffed byte after H3
	Input running fast (negative justification)	Transmitter	1) Invert 5 D bits 2) Transmit "extra" payload in H3 byte 3) Decrease pointer value by one
		Receiver	1) Detect inverted D bits 2) Gets and inserts H3 byte into payload

Figure 8.19
Examples of positive and negative frequency justification

Action	DNF value	I D I D I D I D I D	Pointer value	Frame #
Normal operation	0 1 1 0	0 0 0 1 0 1 1 0 0 1	89	X
Positive Stuff	0 1 1 0	1 0 1 1 1 1 0 0 1 1	undefined	X+1
New pointer value	1 0 0 1	0 0 0 1 0 1 1 0 1 0	90	X+2
	1 0 0 1	0 0 0 1 0 1 1 0 1 0	90	X+3
	1 0 0 1	0 0 0 1 0 1 1 0 1 0	90	X+4
Normal operation	0 1 1 0	0 0 0 1 0 1 1 0 1 0	90	Y
Negative Stuff	0 1 1 0	0 1 0 0 0 0 1 1 1 1	undefined	Y+1
New pointer value	1 0 0 1	0 0 0 1 0 1 1 0 0 1	89	Y+2
	1 0 0 1	0 0 0 1 0 1 1 0 0 1	89	Y+3
	1 0 0 1	0 0 0 1 0 1 1 0 0 1	89	Y+4

These frames are carried on OC-3 fiber links (typically called OC-3c) and have 9 rows and 270 columns in the frame structure. There are 27 columns of overhead in an STS-3c and 243 columns of payload, but there is only one SPE in the STS-3c.

Because there is only one SPE, only one pointer and pair of H1 and H2 bytes are needed to locate the SPE. These are in the first and fourth columns of the STS-3c frame, where one would expect to find the first STS-1 in a channelized STS-3. The other H1 and H2 bytes contain a special *concatenation* bit configuration to let the receivers know that this is an STS-3c with only one SPE. This concatenation indicator is an NDF value of 1001 and has a pointer field value of all 1s, which is an otherwise invalid pointer value; therefore, a value of 1001 00 1111111111 would be placed in the second and third pair of H1 and H2 bytes on an STS-3c. Note the use of the NDF 1001 value in this instance.

Now, this seems simple and elegant. There is only one problem, which astute readers may have picked up already. This pointer use for STS-3c cannot possibly work as described. Why not?

Think about the valid pointer values. The H1- and H2-pointer field for an STS-3c is still only 10 bits. This is from 0 to 1023 only. Yet, a full STS-3c frame is $9 \times 270 = 2430$ bytes. Granted, the payload area is only $9 \times 261 = 2349$ bytes, but this is still much larger than 1023. How can the H1 and H2 pointers work with an STS-3c?

The answer is to realize that everything about an STS-3c is three times "bigger" than an STS-1 is every respect. An STS-3c operates three times faster, and the frame is three times larger. It turns out that *the values of the pointers in an STS-3c are also three times bigger than an STS-1.*

This is why the pointers still work in an STS-3c. The single SPE cannot start anywhere in the payload area, as the SPE can in an STS-1. An offset value of "0" still points to the first payload byte after the H3 byte (in this case, Row 4, Column 10); however, an offset value of "1" in the H1- and H2-pointer field points *3 bytes* beyond that. An offset value of "1" points to Row 4, Column 13 in an STS-3c. It is impossible for the pointers to point to Columns 11 or 12 and, therefore, also impossible for the SPE to start in these columns.

Naturally, this means that the bytes are also buffered in "threes" in an STS-3c piece of SONET equipment: This only makes sense. Instead of worrying about 2 bytes accumulating in a buffer as in an STS-1, an STS-3c only performs a negative stuff when more than 3 bytes have accumulated in the buffer. Of course, the 3 bytes are sent using all 3 available H3 bytes, but only one set of H1 and H2 pointers (the first) is ever used. Positive stuffs occur in an analogous fashion, using the three payload locations after the last H3 byte.

Justifications on even high-speed SONET equipment are performed in the same fashion, with N bytes being stuffed for an STS-Nc and the pointer values multiplied by N for normal operation.

This chapter has dealt primarily with SONET timing and pointers. However, many of the concepts introduced apply equally to SDH. Of course, there are differences in how SDH handles the H1, H2, and H3 pointer bytes, due mainly to the fact that the STM-1 frame is like the STS-3 frame in SONET. So in SDH there are essentially always three sets of H1, H2, and H3 pointers to deal with.

A full discussion of these differences is well beyond the scope of this chapter. But it is important to note that in the fourth row of the STM-1 frame, where in SONET the H1, H2, and H3 pointer bytes are found, SDH places something called the *Administrative Unit* (AU). There are two major types of AU in SDH: the AU-3 and the AU-4. The AU-4 pointers are quite similar to the STS-3c pointers described in this chapter, with minor variations. The AU-3 pointers behave in almost the same way as three sets of H1, H2, and H3 pointers would in "channelized" STS-3 frames. However, SDH also defines other structures for the AU-3 pointer, used for SDH Tributary Units (TUs), that must be treated differently than in SONET. The actions of these TU-3, TU-2, and TU-1 pointers mean that vendors of SONET equipment cannot always employ the same chipsets developed to process SONET H1, H2, and H3 pointers to process SDH AU-3 pointers in all cases.

One of the biggest differences between SONET and SDH is that SONET tends to treat the H1, H2, and H3 pointer bytes as just another line overhead function, while SDH treats the same pointer fields as a type of special overhead row called the Administrative Unit. So differences that seem trivial tend to become magnified when moving from SONET and SDH and vice versa.

More detailed information about timing and network synchronization for SONET and SDH can be found on the Web. There are excellent papers in PDF format from Hewlett-Packard (now at the Agilent Web site). They are very helpful real-world guides to understanding the entire issue. The three form a sort of series:

Synchronizing Telecommunications Networks: Basic Concepts
   www.tm.agilent.com/tmo/Notes/English/5963-6867E.html

Synchronizing Telecommunications Networks: Fundamentals
   of Synchronization Planning
   www.tm.agilent.com/tmo/Notes/English/5963-6978E.html

Synchronizing Telecommunications Networks: Synchronizing
   SDH/SONET
   www.tm.agilent.com/tmo/Notes/English/5963-9798E.html

As with all URLs, these references are subject to change.

PART

3

SONET
Network
Equipment

Part Two of this book provided a lot of detailed information about how SONET functions on a byte and bit level. This type of information is essential, naturally. However, SONET networks consist primarily of SONET equipment and components that embody the byte and bit functions as much as an automobile consists of components that embody the functions of a carburetor or radiator. Although it is necessary for mechanics to understand the role of carburetors and radiators in automobiles, most people who encounter a car are unconcerned with the detailed operation of these components. Their main concern is what the entire car can do when all parts are in place and running properly. This part of the book emphasizes the "as a whole" aspect of SONET networks.

This section is intended to be much more "real-world" oriented than earlier portions of this book. Although this book previously described both SONET and non-SONET equipment, the components were examined more or less piecemeal, in isolation from not only each other, but also from the networks they comprise. This section of the book is meant to extend this analysis to the SONET network as a whole.

This section of the book explores the implementation of SONET standards in the various pieces of equipment that make up a network. An attempt is made to classify the diverse SONET components into a small number of categories as a means to better understand the relationship of one type of SONET equipment to another. Admittedly, this scheme is not standard in a sense, but it is instructive nonetheless. Finally, SONET equipment is manufactured by any number of vendors and sold for profit to those who build the SONET networks. This section of the book takes a look at the vendors involved in the manufacture of SONET equipment.

This part of the book is organized into three chapters. The first chapter (Chapter 9) explores how the standards for SONET operation are embodied in the various pieces of equipment needed to build a SONET network. There are different pieces of SONET equipment designed and built to accomplish different networking tasks, and these are all detailed in this chapter. A high-level profile of a typical SONET carrier network is also contained in the chapter, and it concludes with a high-level first-look at SONET rings, their purpose, and operation.

The next chapter (Chapter 10) divides all of the various pieces of SONET equipment introduced into six major categories. The chapter then examines the functioning of each major piece when the device is used in a SONET network. The whole idea is to provide a framework for classifying SONET equipment types according to function.

The final chapter (Chapter 11) deals with SONET equipment providers. The chapter details the vendors that build and supply SONET network components and provides some information about their products. The chapter concludes with a look at several example SONET network configurations that use some of the equipment described in the chapter.

CHAPTER 9

SONET Networks

There is nothing wrong with learning as much as one can about the inner workings of SONET on a byte level and related topics, such as timing distribution; however, SONET networks consist of devices that are real boxes full of hardware. In addition to an intimate working knowledge of SONET bytes and frame structures, it is also desirable to have a intimate working knowledge of what SONET networks look like from an equipment standpoint.

Nothing is more frustrating to network engineers and managers than to have to deal with a person whose knowledge about any technology is limited to "book learning" on an entirely abstract and conceptual level. In the early days of LANs, it was not unusual to come across a supposed expert in Ethernet or Token Ring who had never seen, let alone worked with, an Ethernet hub or Token Ring MAU. They knew all the rules of Ethernet, for instance, from the 64-byte minimal frame size to the strange structure of the IEEE 802.3 version's frame header. But asking them to reset a hub in a telecommunications closet was like asking them to disarm a nuclear weapon. Chances were, nothing good was going to come from the effort.

This chapter is an attempt to address some of these issues. Saying SONET equipment can interface with T-carrier equipment does not say how this can be accomplished in a device. The most detailed study of SONET technology does not begin to scratch the surface as to how all of the nifty features of SONET can be brought to the real world of network hardware and software.

This chapter begins by looking at some of the details of pre-SONET carrier systems like T-carrier. The perspective in this chapter focuses on the equipment used.

Pre-SONET Carrier Systems

Older fiber optic transmission systems were equipment-intense, proprietary, and an operational nightmare. A visit to a central office employing fiber optic trunks revealed a snake pit of coaxial cables leading to and from the fiber optic transmission equipment, a Medusa-like corner with patch panel and physical cross-connections that threatened to entangle an unwary technician, and an utter lack of up-to-date paperwork that was absolutely essential to make sense of it all. Proprietary connectors did not match any other piece of equipment in the building, and it was a truism that any delivered equipment or device would not be able to be hooked up to anything else—in some cases, even from the same vendor—without extensive connector conversion or customized interface components.

Still, deployment of SONET transmission *overlays* in these systems has been cost-effective for two major reasons: An overlay indicates that a particular link or trunk is converted to SONET using as much of the existing physical plant as possible. When the fiber conforms to SONET standards, it can be used. Naturally, as much of the existing input and output multiplexers and cross-connects are retained to minimize expenses. Thus, a SONET overlay replaces non-SONET fiber links and trunks one by one, with no concerted attempt to integrate the SONET equipment into a complete SONET network. The overlaying is common practice in SONET carrier networks.

The two major reasons for the continued use of SONET overlays are easy to describe. First, the capability of SONET multiplexers to accept multiple data rates (i.e., DS-1s and DS-3s) has reduced the equipment necessary for the creation of a high-rate optical link; therefore, the expense of the SONET equipment is easily offset by the reduced amount of equipment needed when conversion to SONET is implemented. The second area of cost reduction is in the operations, administration, maintenance, and provisioning (OAM&P) of the SONET links as compared to the cost associated with the proprietary links. SONET equipment complies with standard operations systems and allows a single point of entry into the network for OS functions.

Figure 9.1 shows pre-SONET carrier systems equipment deployment. The equipment shown is a full complement and is able to *groom* DS-0 channels in a high-speed (i.e., greater than the DS-3 rate of 45 Mbps) fiber optic system. *Grooming* is the process of combining and distributing feeds from many sources operating at various speeds onto one higher speed trunk. Grooming is one giant step up from simple muxing, which is usually point-to-point. (Sometimes the term *hubbing* is used to describe the same process. The *hub* is where grooming takes place.)

Grooming is common practice in carrier systems and is done for reasons of efficiency. Suppose a central office has 10 T-1s coming from one customer location and 7 from another customer at another location. Because the T-1s are coming from two locations, it may seem that two separate carrier systems are needed to send the T-1s onto the trunking network and on to their destination. A single T-3, however, can carry 28 T-1s. It makes much more sense to *groom* the user traffic and install digital cross-connect system (DCS) equipment to combine the total of 17 T-1s onto a single T-3. Therefore, only one T-3 is needed to aggregate the

Figure 9.1
Pre-SONET carrier
systems

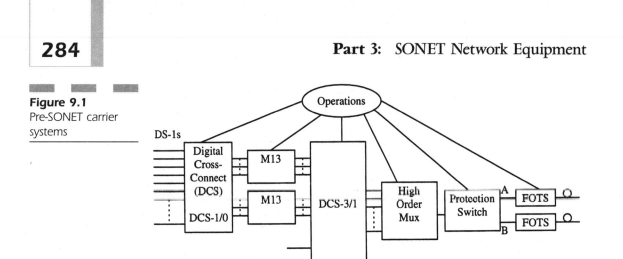

traffic from both locations onto the carrier backbone. Often, when grooming combines traffic streams, it is called *concentration* or *traffic consolidation,* and when grooming is used to split up traffic streams (as may be done at the opposite central office), it is called *distribution* or *traffic segregation.* Whatever the name, the intention is the same: to make more efficient use of otherwise under-utilized high-capacity trunks and links.

In Figure 9.1, there are many pieces of equipment. In order to better understand these devices and the role they play, and not coincidentally how they relate to the equivalent SONET devices, some words of explanation are necessary. In the figure, the equipment and its functions include the following:

- DCS 1/0—DCS-terminating DS-1 trunks and cross-connecting DS-0 channels. The "1/0" designation is used to indicate that the DCS handles both DS-0s at 64 Kbps and the entire DS-1 operating at 1.5 Mbps. Some DS-0s from some customer sites may terminate at other sites served by the same switching office. The DCS-1/0 equipment handles this *turn-around* configuration.

- M13—This is a DS-1 to DS-3 multiplexer. The main task of the M13 device is the multiplexing of up to 28 DS-1 trunks into a single DS-3 trunk. Of course, the same device can demultiplex as well. Not all 28 DS-1s are shown in the figure.

- DCS 3/1—DCS terminating DS-3 trunks and cross-connecting DS-1 trunks. The "3/1" designation is used to indicate that the DCS handles both DS-1s at 1.5 Mbps and the DS-3s operating at 45 Mbps. Note that there may be some DS-3s that do not emerge from an M13. These are *unchannelized* DS-3s that operate as a full 45 Mbps, not as 28 DS-1s or 672 DS-0s. As before, some DS-3s from some customer sites may terminate at other sites served by the same switching office. The DCS-3/1 equipment also handles this turn-around configuration.

- High-Order Mux—Another multiplexer device, but without cross-connect capabilities. The high-order mux combines multiple DS-3 signals into a single, high-speed signal. This is a purely proprietary device, without even a standard output format. Most of these devices combine 9 to 12 DS-3s on a single fiber optic link operating in the 1—2 Gbps range.

- Protection Switch—This device sends the high-speed transmission signal from the high-order mux out on one of two fiber optic transmission media. Naturally, this provides some protection from fiber cable damage and ensuing failure. This protection, in turn, depends on just how far apart the two fiber cables extend, what is known as *diverse routing*. Unfortunately, in the past, diversity was defined as a minimum separation of only 25 feet. This often translated to one fiber down one side of the road, and the other fiber on the other side of the road. Trenching operations across the road frequently severed both cables to the dismay of users and carrier alike. The protection switch is controlled by the operations system or automatically by the condition of the received optical signal. Often the switch is done automatically, but service must be restored manually.

- FOTS—The fiber optic transmission system (FOTS) is the equipment that converts the electrical signal to an optical one and vice versa on the receiver side. It should be noted that the FOTS and the protection switch could be transposed. That is, the output of the high-order mux could transmit directly to a FOTS, which would now feed an optical protection switch. This has the added benefit of requiring only a single FOTS.

- Operations—This is the nerve center for the operation of the entire transmission system, not just the simple link shown in Figure 9.1. All equipment alarms and performance statistics are relayed to this center and all remote control and provisioning is performed from here. Each piece of equipment in the system has a direct link to the operations center. Because each piece of equipment had its own unique brand of network management hardware and software, operations in this environment were complex and awkward.

SONET Carrier Systems

SONET carrier systems can be contrasted with pre-SONET carrier systems in a number of ways; however, the easiest way is to contrast the equipment requirements and layout of the components. A typical SONET carrier system is shown in Figure 9.2.

As can easily be seen in Figure 9.2, SONET carrier systems reduce the equipment complement required for a transmission system to a bare minimum. SONET architecture allows for a reduction in the complexity of the equipment that brings non-SONET digital information streams into the SONET system.

Now there are only the DS-1 and DS-3 DCSs left, and these still are needed to handle getting the T-carrier traffic onto and off of the SONET carrier itself. Of course, the DCS turn-around cross-connection functions for DS-0s, DS-1s, and DS-3s that originate and terminate at the same switching office are still needed as well. However, the bulk of the work done by the many pieces of separate equipment is done by the SONET multiplexer. High-order multiplexing, protection switching, and FOTS functions are all done in this single piece of SONET equipment.

Figure 9.2
SONET carrier
systems

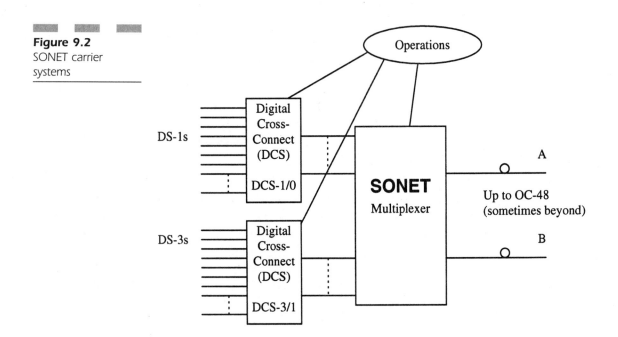

This is another example of the increasing power of electronic components over time.

Another area where SONET reduces complexity is in operations support. SONET standards also specify the OAM&P system for the entire transmission network. All SONET-compliant equipment must follow these standards and implement these functions and capabilities. This allows for a standard operations platform for control of the whole transmission system. Of course, the former links to the existing DCS equipment must be maintained, making the situation less than perfect.

SONET equipment also includes a standard set of craft interfaces for ease of maintenance and repair. Craft interfaces are those connection points where diagnostic equipment can be attached by technicians performing on-site troubleshooting tasks. The terminals or consoles in the operations center where the technicians manage the equipment with software interfaces are also sometimes called a craft interface, but usually the term is reserved for attachments to the equipment itself.

A final advantage of the SONET carrier system is the capability of upgrading the carrier system to higher rates without the wholesale replacement of the transmission equipment. Certain functions of SONET remain the same regardless of the carrier rate. Therefore, the SONET multiplexer box in Figure 9.2 does not need to be replaced to allow the fiber links to operate at OC-48 (about 2.4 Gbps), instead of OC-12 (about 622 Mbps). After all, input ports from the DCSs, protection switching, and network management capabilities are exactly the same regardless of output speed. Typically, a simple processor board upgrade is all that is needed to change the output speed. Most port additions merely involve this simple card upgrade as well, yet another benefit of the spread of microprocessors to the transmission network. This simple upgrade process allows economical upgrades of carrier systems.

SONET in Action

So far, the discussion of SONET networks in this chapter consists of replacing plesiochronous T-carrier components with SONET components: the overlay process. Of course, there is much more to SONET than T-carrier replacement, although this aspect of SONET should never be forgotten. SONET can do more, much more. Figure 9.3 shows the opera-

Figure 9.3
SONET in action

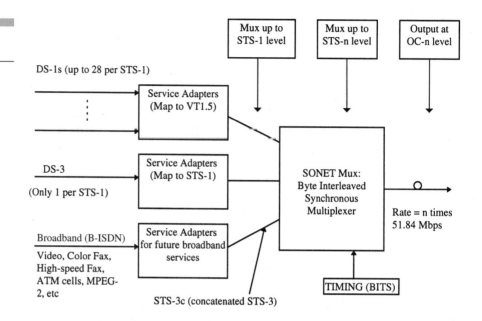

DS-1s (up to 28 per STS-1)

Service Adapters (Map to VT1.5)

DS-3

(Only 1 per STS-1)

Service Adapters (Map to STS-1)

Broadband (B-ISDN)

Video, Color Fax, High-speed Fax, ATM cells, MPEG-2, etc

Service Adapters for future broadband services

STS-3c (concatenated STS-3)

Mux up to STS-1 level

Mux up to STS-n level

Output at OC-n level

SONET Mux: Byte Interleaved Synchronous Multiplexer

Rate = n times 51.84 Mbps

TIMING (BITS)

tion of a SONET link that takes more advantage of SONET's unique capabilities.

There is a lot happening in Figure 9.3, and requires a more detailed examination. On the left side of the figure, the main input possibilities for the SONET link are shown. Note that the familiar DS-1s and DS-3s are still there, but more details are presented regarding their exact operations. The output of an existing DCS-1/0 which is, after all, just another DS-1 (or multiple DS-1s), are aggregated and a group of 28 and sent into a *Service Adapter* which maps the 28 DS-1s into a single VT1.5.

Although the service adapter in Figure 9.3 is shown as a separate physical device, it need not be and often is not. Most frequently, the service adapter is just a type of processor board or card put into a slot in the SONET equipment to handle that particular form of input and output. Vendors have their own names for service adapters, such as port adapters and interface cards. In the case of the VT1.5 service adapter in the figure, the output is an STS-1 frame mapped to VT1.5 and containing up to 28 DS-1s.

Also shown on the left side of the figure is a single DS-3 that may be connected to its own service adapter. Note the difference between adaptation to SONET of 28 DS-1s and a single DS-3. The 28 DS-1s require the use of VT1.5 (mainly for DS-0 payload visibility); however, a DS-3 can be transported directly inside a SONET frame through the process known as asynchronous mapping. This may be desirable when the DS-3 in question is unchannelized itself, or when DS-0 visibility is neither wanted nor needed. Naturally, only a single DS-3 operating at 45 Mbps can be carried by an STS-1 link operating at 51.84 Mbps.

Finally, the figure shows a service adapter to be used for future implementation of broadband services. The breadth and scope of broadband services are astonishing, but most people focus on services like switched digital video (SDV), color faxing, high-speed faxing, transport of ATM cells inside an STS-3c or higher speed frame, MPEG-2 compressed digital video, and a number of other services nearly impossible to provide without SONET. broadband services will have their own service adapter(s), depending on the services provided.

It is important to remember that broadband services transported by SONET links must be handled by SONET links operating at 155.52 Mbps or greater. The speed of 155.52 Mbps is the rate of STS-3. Moreover, for broadband, the STS-3 must be *unchannelized*, or rather *concatentated* in SONET talk, to operate at the full 155.52 Mbps. This is an STS-3c or, more commonly but less technically, an OC-3c. A "plain" STS-3 consists of three STS-1s all operating at 51.84 Mbps. This may be what both the customer and carrier want, but not for the transport of broadband services. This is the exception that the service adapters will multiplex up to the STS-1 level.

Continuing through the figure, the major component is the SONET multiplexer itself. This device does the byte-interleaved multiplexing that distinguishes SONET. The SONET multiplexer also performs the electro/optical (E/O) conversion needed to send the SONET signal over the fiber itself. The BITS timing information is shown in the figure as well. One of the reasons that the service adapters may be separate devices, or even customer premises equipment (CPE), is because in most cases many STS-1s are combined, groomed, and multiplexed onto a single, high-speed SONET link at the switching office. For broadband, the service adapter should be CPE as well.

In any case, the SONET mux combines and grooms the input STS-1s, STS-3s, STS-3cs, or any combination, into a single STS-n frame. The output of the SONET mux at this point will of course be some OC-n. This OC-n will operate at "n" times the basic SONET rate of 51.84 Mbps. It is quite common today for this SONET equipment in a switching office to operate at the OC-48 rate of about 2.4 Gbps. However, higher speed equipment is possible and will become more common in the near future.

Finally, although not shown in the figure, it is good to remember that the output fiber is typically not a single strand of fiber. More often the output from the SONET mux consists of an "A" and "B" fiber pair with SONET protection switching performed on the part of the device or through some operations interface.

SONET as Trunking

Figure 9.3 shows a typical SONET link on a SONET-based network in action. Any type of service, from voice to high-speed data to video, can be accepted by various types of service adapters on the SONET network. However, it is always good to keep in mind that SONET is basically a trunking technology. That is, SONET itself is neither capable of doing any switching based on traffic stream content, nor able to provide any of the neat features of fast packet networks, such as flexible and dynamic bandwidth allocation (also known as "bandwidth on demand"). When a SONET OC-1 pipe is filled with 28 DS-1s, which is the maximum, then the OC-1 is full.

This can be a problem when users are accustomed to seeing SONET services listed in the same categories as frame relay and/or ATM. If frame relay and ATM can give a user bandwidth on demand, why can't SONET? Because frame relay and ATM run in switches as well as across links. SONET can be the link, access or trunk, technology for frame relay or ATM, but SONET can never replace frame relay or ATM switches by itself.

SONET is a trunking technology and a "bit pipe" for traffic, no more and no less. Suppose a SONET multiplexer is used to provision three STS-1s operating at 51.84 Mbps on an STS-3 operating at 155.52 Mbps. The STS-1s are then sold to three customers. Each customer uses only a fraction of the 51.84 Mbps available at any one point in time. A fourth customer's traffic could easily be handled by the STS-3's speed.

However, that's it; there is no fourth STS-1 to sell. No more bandwidth can be provisioned on the STS-3 link, no matter how underutilized it may appear. The only solution is to add a frame relay, ATM, or some other kind of switch to the ends of the SONET link. Only then will features like flexible bandwidth allocation for "bursty" data applications help the service provider to support more than three users. Of course, this is not a trivial task. Figure 9.4 shows the situation with pure SONET.

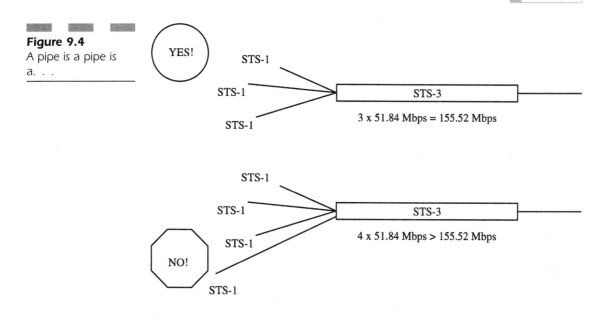

Figure 9.4
A pipe is a pipe is
a. . .

Figure 9.4 shows that up to three STS-1s can be provisioned on an STS-3. The math involved is simple but absolute. A fourth STS-1 cannot be added or sold.

SONET Links in Summary

These last few sections have been full of alternative arrangements and complexities. This may be a good place to briefly summarize the operation of SONET equipment at the end of a SONET link.

All types of service adapters can map the incoming signal from one service type or another into the SPE of a SONET link. New services and signal types (e.g., digital video) can be transported by adding new service adapters at the edge of the SONET network.

All inputs are eventually converted into the basic format of a synchronous STS-n signal. Lower speed inputs, such as individual DS-1s, are first multiplexed onto VT 1.5s. Several synchronous STS-1s can be multiplexed together in either a single-stage or two-stage process to form an electrical STS-n signal. Higher speed signals, such as those from newer broadband services and cells from ATM networks, can be carried on an STS-3c.

STS multiplexing is performed by the byte-interleaved synchronous multiplexer. The bytes are interleaved in such a way as to provide "lower speed payload visibility" on the SONET-based network. No additional signal processing occurs, except for a direct conversion from the electrical STS format to the optical OC format for fiber transmission.

SONET and ATM

Much has been written in books and articles about the precise relationship between SONET and ATM. Some of this relationship has been mentioned earlier in this book. Because one of the last major points made in this book is that SONET is a trunking technology, and not a switching technology like ATM or even frame relay, this is the place to discuss the relationship between SONET and ATM equipment more fully. There may be a false impression implied by this switching/transport split. The impression may be that there is no relationship between ATM and SONET. In fact, the ATM/SONET relationship is deep and precise. This section describes the ATM/SONET relationship in more detail, and from an equipment point of view.

Simply, SONET is intended to be the transport network for an ATM network. This network, with both ATM switches and SONET links, provides broadband services to users. If the broadband services comply with a series of services defined by the ITU and the switches are ATM and the trunks are SONET, then the whole network complies with the international B-ISDN standard. All service providers today are aiming toward providing full B-ISDN services in the future because they figure that B-ISDN will ultimately be very popular with potential customers. Naturally, these service providers want to build a network that does not have to be re-engineered for B-ISDN at some future date. The price of replacing all the switches at the end of SONET links with ATM, or replacing all links between ATM switches with SONET links will be much too high, even in the remote future. It is better to build the B-ISDN combination of ATM switches and SONET links once and be done with it.

For now, the issue is a little bit more complex. The situation is a little different, mainly due to the lagging of B-ISDN services behind the SONET and ATM technologies and equipment available. A full set of defined B-ISDN services does not exist, except at a very high level in some ITU documents. Users and service providers today cobble together broadband networks and services based on ATM with implementation agreements from sources like the ATM Forum. This is not necessarily bad, but it does point out that SONET and ATM have become "decoupled" from B-ISDN, and even from each other. That is, it is not unusual to see ATM "services" provided over transport networks other than SONET, or see SONET offered as just another high-speed leased line "service," without ATM involved.

The reason for combining ATM switching and SONET links into B-ISDN service networks is simple: Broadband networks are characterized by low and stable network delays to support applications that require interactive operations (e.g., voice and videoconferencing) and high bandwidths on access lines and trunks to support applications that are bandwidth-intensive (e.g., client-server networked graphics applications or videoconferencing). Note that some applications (e.g., videoconferencing) share the twin requirements of low and stable delays as well as high bandwidths.

Network delays have two causes. First, the bandwidth may not be sufficient for bits to travel quickly enough through the network from source to destination. Such applications are *bandwidth bound* and will not function properly without plentiful bandwidth. Obviously, this is SONET's function in a B-ISDN network: to supply the bandwidth needed for otherwise bandwidth-bound broadband applications. SONET's scaleability certainly helps. When OC-3 operating at about 155 Mbps is not enough bandwidth, then perhaps OC-12 operating at about 622 Mbps will work.

Some applications are more affected by the delay through the network from a source to a destination. If the network delay were too high, or too variable and unstable, the application would not work. Voice is always a good example. The needs of voice in terms of bandwidth are quite modest by today's standards (it was not always so). The 64 Kbps needed for voice is easily provided by a number of means. Applications, which are more sensitive to low and stable network delays than bandwidth, are known as *delay-bound* applications.

Some delay can be eliminated with increased bandwidth, but most delay-bound applications, such as voice, will not benefit from more bandwidth. Voice at 128 Kbps will not operate better than voice at 64 Kbps when the delay is not low and stable enough to begin with. Thus, low and stable network delays are affected by more than bandwidth, of course. One of the two components that contributes to network delay is the propagation delay. No matter what the bandwidth is, the signal will take longer to reach a destination that is farther away than one that is close by. This is simple physics. When international voice traffic began to be routed over geosynchronous communications satellites, many argued that the delays were much too high for voice applications. In spite of concerns, nothing could be done about long uplink and downlink satellite delays.

The other component of network delay is nodal processing delay. Whether the network nodes are routers, central office switches, or something else, it takes a finite amount of time for bits to enter an input port on a network device and emerge from an output port. Here is where ATM fits in. A full discussion of ATM is far beyond the scope of this chapter and even this book. However, because ATM and SONET are intimately related, a brief overview of what ATM switches add to SONET links is not out of order.

ATM is the international standard for cell relay technology. In cell relay, data units are chopped into small pieces (the cells) to facilitate network node processing. The end result is more effective switching and multiplexing. ATM provides the low and stable nodal processing delay and, thus, the overall low and stable network delay needed for the whole spectrum of broadband services.

This relationship between ATM switches and SONET links is illustrated in Figure 9.5.

Figure 9.5
Why and how
SONET and ATM are
related

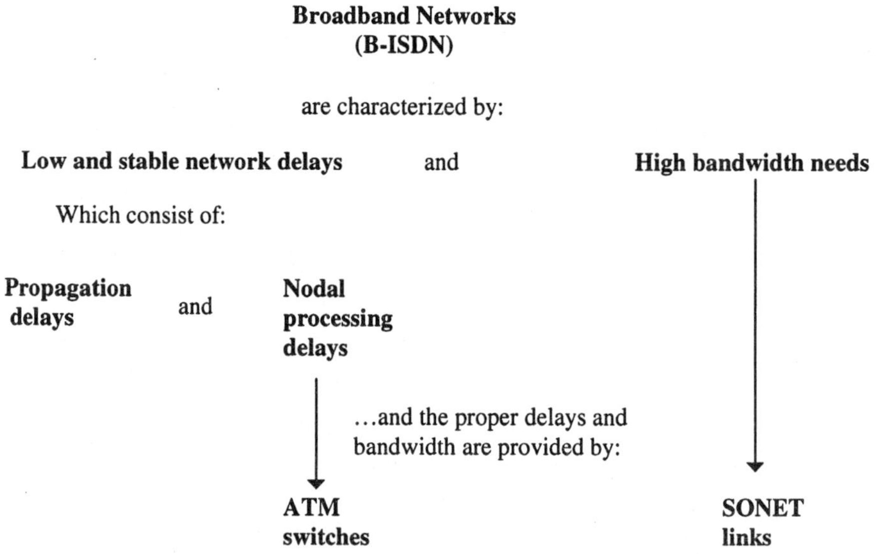

Broadband Networks
(B-ISDN)

are characterized by:

Low and stable network delays   and   **High bandwidth needs**

Which consist of:

**Propagation   **Nodal
delays**   and   processing
delays**

...and the proper delays and
bandwidth are provided by:

**ATM   SONET
switches   links**

SONET Components

So far, most of the discussion of SONET equipment has revolved around single links, and (if the truth be told) only the sending side of these links. Not that receiving is any more complex. Almost any figure in the text could have a mirror installed in the right-hand margin to reflect the fiber link and all to the left of it. All SONET multiplexers also demultiplex, and service adapters not only adapt from T-carrier to VT1.5s and the like, but also adapt VTs back to T-carriers.

However, SONET networks consist of more that single links. In fact, SONET networks can be quite complex. This chapter introduces the overall components of a typical SONET network, and the next chapter details their individual operation. Figure 9.6 shows the major components of a multiple-link SONET network, such as may be deployed by a major carrier in a metropolitan area today. SONET documentation refers to these components as *Network Elements (NEs)*.

Since there are more than single links involved in the network, clock distribution and network timing become all the more critical when DS-n cross-connecting is needed. Note that some of the SONET compo-

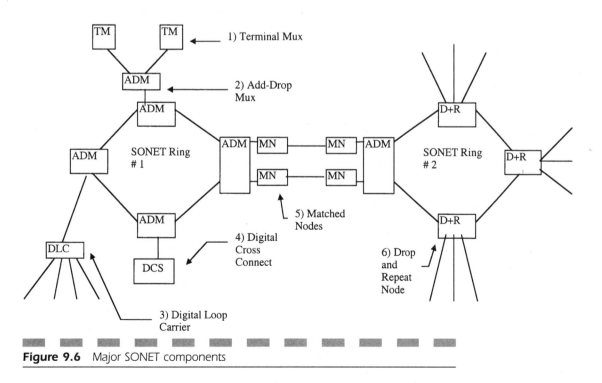

Figure 9.6 Major SONET components

nents form rings in the diagram. SONET rings have become a distinguishing feature of SONET, much as touch-tones have become a distinguishing feature of telephones. Both are simply a much better way of doing things. A full discussion of SONET rings deserves a chapter of its own. But this is the place to introduce them.

As shown in Figure 9.6, there are six main network elements that can be used in SONET networks. These are labeled in the figure as follows:

1. *Terminal multiplexer (TM).* An "end point" device on the SONET network. This device gathers bytes to be sent on the SONET network link and delivers bytes on the other end of the network. TM is intended to be the most common type of SONET CPE, but ironically is still rare at the customer site. In most current SONET configurations, TM is still found in the serving office.

2. *Add/drop multiplexer (ADM).* Really just a "full-featured" TM. However, it is more accurate to refer to the TM as an ADM operating in what is known as "terminal mode." This device usually connects to several TMs and aggregates or splits (grooms) SONET traffic at various speeds.

3. *Digital loop carrier (DLC).* This SONET device is used to link serving offices with serving offices with ordinary analog copper-twisted-pair-local loops in order to support large numbers of residential users in what is known as a carrier serving area (CSA). The devices are also commonly referred to as remote fiber terminals (RFTs) by many vendors.

4. *Digital cross-connect (DCS).* This SONET device can add or drop individual SONET channels (or their components in a VT environment) at a given location. It is basically an even more sophisticated version of the SONET ADM.

5. *Matched nodes (MN).* These SONET devices interconnect SONET rings. They provide an alternate path for the SONET signals in case of equipment failure. This feature is commonly known as signal protection, but it has a variety of other names.

6. *Drop and repeat nodes (D+R).* These devices are capable of "splitting" the SONET signals and sending copies of bytes onto two or more output links. The devices will be used to connect DLC devices for residential video (or even voice) services.

Figure 9.6 shows a more complete network deployment that exploits the features of SONET. It should be noted that there are no "official" designations for the various pieces of SONET equipment shown in the figure. Although the figure generally follows terminology of such SONET equipment vendors as Nortel Networks, it is not uncommon for vendors to have their own names for the devices shown in the figure. However, the emphasis here is on function, not so much on the names themselves. The equipment found in this network is separated into central office terminal equipment and outside plant equipment.

The central office equipment consists of a SONET-compatible switch (i.e., digital cross-connect switch) and digital loop carrier equipment. This equipment forms the network terminating point for SONET transport services.

The outside plant equipment consists of a fiber ring topology for survivability, and access multiplexers to this ring. The multiplexers are functionally separated into two categories, TMs and ADMs. TMs are functionally similar to channel banks in the T-1 networks. They provide conversion from the non-SONET transmission media to the SONET format. Add/drop multiplexers are an integral component of the SONET architecture; they allow access to the SONET transmission network without fully demultiplexing the SONET signal. Non-SONET signals may be added to or taken from the SONET transmission signal via this equipment. The differences between SONET TMs and ADMs is often only a difference in network location and in function, rather than a fundamental or intrinsic difference. Figure 9.7 shows the main difference between a SONET ADM and a SONET TM, which is really just an ADM operating in "terminal mode."

In Figure 9.7, the major difference between a SONET ADM and TM is that the SONET ADM has two network connections that must be equal. Channels, or even other OCs (as long as they are less than the network OC), can be dropped or added as desired. These channels or OCs can lead to SONET TMs, or course. On the other hand, a SONET TM has only one network connection. The "terminal mode" just multiplexes various bit rates onto and off of this single network link. The whole point is to allow (relatively) "low-speed" user access to the SONET network. The user access may be at DS-3 or even OC-3 rates, as long as the network link can handle the speed.

Physically, the SONET equipment described is of quite modest size. The solid-state electronics that characterize SONET result in components that are typically rack-mounted in standard communications cabinets. The power requirements are correspondingly modest as well.

Figure 9.6 can be misleading in one sense. The figure implies that the six major SONET NE functions might all be found in all large SONET

Figure 9.7
SONET ADM and TM

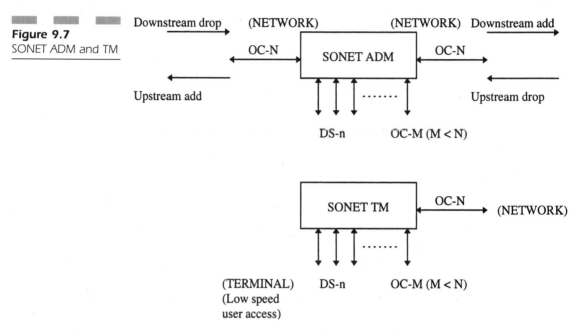

networks and with equal likelihood. In fact, the truth is just the opposite. The SONET NE packages offered by most equipment vendors are ADMs, first and foremost. Many of the other NE functions are added (or subtracted) from the basic ADM package to yield the required NE for the application. So an ADM with non-SONET interfaces and a single SONET interface becomes a SONET TM package. DCS functions can be added to the ADM in the same way, but this is seldom done. Most DCSs remain separate-gear-attached to the ADM. If the ADMs have multiple links used to connect their rings, this makes that pair of ADMs into Matched Nodes. In other words, the MN function is just another feature of the ADM.

The ADM/TM/MN package is normally what a SONET equipment vendor emphasizes, and that might be it. If a SONET DCS NE is offered, it is usually not an ADM feature, but separate equipment altogether. The D+R and DLC functions can be addressed in the same way. The D+R NE is a special ADM. The DLC is a special TM optimized for residential voice applications.

In summary, SONET equipment vendors typically offer a basic ADM package in larger or smaller sizes. Small ADMs make great TMs and DLCs with the correct line cards and associated functions. Larger ADMs make terrific D+R nodes and the MN function is usually just another interface card in the ADM unit to interconnect rings. So the same basic product line can be ADM, TM, D+R, MN, and DLC device all in one. It just depends on the precise hardware and software configuration. The odd NE out is the DCS. Cross-connecting has always been a complex function to configure and control. In SONET, it is often best to leave this task for more specialized equipment.

Oddly, this means that many a SONET ADM is busily extracting DS-1 and DS-0s from SONET frames just so the DS-1s and DS-0s can be sent over to the legacy DCS across the room for cross-connecting. The rearranged traffic then comes back to the ADM for repackaging into outgoing SONET frames.

SONET Transmuxing

It has not escaped many SONET equipment designers that many of the SONET network elements or components share many common functions and characteristics. The close relationship between SONET ADMs and TMs has already been noted. One of the main driving forces behind the creation of the SONET standard was that T-carrier networks required many different pieces of equipment to accomplish what were relatively modest goals.

The more ambitious goals of SONET make such device efficiency more paramount. The situation is most critical due to the fact that users neither often have—nor do they often want—SONET-customer-premises equipment. Carriers have almost exclusively SONET-based networks today, but customers still order, purchase, and configure the same DS-1s and DS-3s that they are more familiar with. As a result, a carrier's switching office is still full of a tangle of T-carrier cross-connect and multiplexer devices which, in turn, feed the SONET equipment.

This T-carrier-to-SONET interface can be handled in one of two ways. The carrier can take all of the traffic through a cross-connect and back to the central office to make the connection. This results in the clutter mentioned above. Alternatively, the carrier can interface closer to the customer with racks and shelves of low-speed multiplexer equipment to essentially back-haul everything. This requires back-haul bandwidth and/or involves paying for co-location space and equipment. However, there is even a third possibility: SONET transmultiplexing.

SONET transmuxing allows for the direct linking of T-carrier to SONET networks. It takes advantage of two SONET characteristics. First, SONET transmuxing makes the most of the fact that SONET devices share many functions in common. Second, it takes advantage of the fact that most SONET functions are performed in solid-state electronics chipsets. Transmuxing moves the T-carrier to SONET (and back) interface directly to the silicon.

SONET transmuxing is both more and less than simple SONET multiplexing. Instead of the complex building up and breaking down of other multiplexing schemes, SONET transmuxing devices simply strip off or add on the SONET overhead. Transmuxing devices function at the DS-3 or DS-1 level. The DS-3s may be handled as 28 DS-1s or as an unchannelized bit stream.

Transmux devices may actually become quite common at the local carrier to interexchange carrier interface. Here, many DS-1s and DS-3s from the local carrier can be handed off to an interexchange carrier's SONET network without the need for other equipment. The SONET transmux is basically a box with T-carrier interfaces on one side and SONET on the other. All M13 multiplexing functions, DS-1 framing, and even cross-connects, can be done by the firmware in the transmux.

Transmux device will become more common and popular in the future. One possible hurdle is the conventional wisdom that some network devices and components should be separate to avoid the most common single-point-of-failure problems. For example, there is no reason that LAN servers and routers need to be separate devices. There is enough horsepower in modern computers to do both. However, most users are reluctant to put themselves in a situation where the loss of router means the loss of a server, and vice versa.

Nevertheless, the reliability of solid-state electronics is far greater than that of other devices based on the more simple boards and wires of the past. Transmuxing is a promising new possibility for SONET (and T-carrier) networks.

SONET Rings

A distinguishing characteristic of SONET links is their capability to be deployed in a ring topology and configuration. A full discussion of SONET rings and their various forms deserves a chapter all its own, but this section introduces the concept.

Fiber optic links, like T-1 twisted pair and most other digital transmission links, are inherently unidirectional and not full duplex. All SONET fiber links consist of a transmit fiber and a receive fiber, as T-1 links consist of a transmit-and-receive twisted pair. SONET ADMs have a minimum of four fiber interfaces: one for *upstream* and *downstream* transmission in each direction. What happens to these fiber links at the "ends" of a series of SONET ADMs? An interesting configuration results when the loose "ends" are looped around to form a closed loop, or ring, of SONET ADMs connected by the fiber links. This is shown in Figure 9.8.

In Figure 9.8, the simple SONET ring consists of several ADMs linked by their upstream and downstream fibers. The ring may span only a few miles, or stretch to literally thousands of miles, depending on its purpose. Note that there are still SONET TMs, with users attached to feed the ring with traffic. Naturally, the aggregate traffic from the TMs cannot exceed the capacity of the SONET ring.

What is the advantage of deploying SONET in a ring configuration? Quite simply, rings provide greater reliability by furnishing two separate paths for digital signals between the ADMs. Even this simple SONET ring provides what is known as path protection switching (PPS), or just *path switching*, between the SONET ADM nodes.

Typically, the two fibers are deployed in the same cable sheath and are laid in the same conduit. Consider what would happen if the fibers between SONET ADM #2 and ADM #3 in the figure were severed by some environmental disaster (e.g., flood, tornado) or even a construction "incident" (known in the industry as "backhoe fade"). Signals can still travel between ADM #2 and ADM #3 through ADM #1. There is still a fiber path between ADM #2 and ADM #3.

SONET rings allow for repairs to be made without disrupting customer service. This is a valuable capability that justifies the added expense of the extra fiber needed to "close the ring" of SONET ADMs. Today, it is rare to see a SONET deployment of any size without rings.

The essence of path switching is to provide an alternate path for signals; however, in this simple ring, spare capacity must be available on the

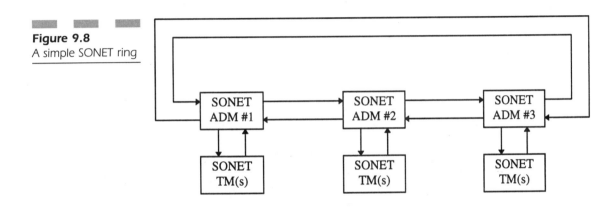

Figure 9.8
A simple SONET ring

links between ADM #3 and ADM #1 to allow for this capability. This only makes sense. There are more elaborate ring configurations that will be explored later. For now, it is enough to point out that the ring illustrated in Figure 9.8 is a 2-fiber, unidirectional, path-switched ring. SONET rings are often referred to as "self-healing" rings, although this may imply that the rings somehow fix themselves and restore failed links to service on their own. They do not.

Typically, a SONET ring consists of several ADM devices connected together in a ring topology. Users, however, get access into and out of SONET using TM devices. This is not a problem because ADMs can easily interface with one or more TM components at many points around the ring. ADMs take the form of a simple "card cage" with a power supply (often with backup) and several slots for interface cards. In their most basic form, ADMs have at least two cards for the input and output fiber links of the ring itself. Adding TMs to a SONET ring of ADMs is as easy as adding cards to card cage slots. Of course, when a device's slots are full, no more TMs (or other devices) can be added to the configuration.

It is also possible to extend the reach of a SONET ring by strategically placing another ADM not in the path of the ring, but off of an ADM on the ring itself. This allows for the connection of more TMs, of course. At the same time, however, this process cuts down on the amount of protection a customer receives from the SONET ring. It makes no difference to a customer if the carrier has a SONET ring or not when the problem is the single fiber link from the TM or ADM to the ring itself.

What would the typical components of a service provider's SONET ring look like when used to provide more reliable and efficient DS-1s and DS-3s? Where would the typical components be located? Figure 9.9 shows such a layout, with the dual fibers rendered as single lines to reduce the complexity of the figure.

In Figure 9.9, the carrier has installed a SONET ring by linking the ADMs in several office locations (i.e., central offices and, in this case, a wire center). A wire center is a more centralized switching office for trunk connectivity to interexchange carriers and other long-haul network elements. The most common speeds used for this ring are in the OC-12 and OC-48 ranges, although higher speeds will become more common.

TMs that handle the customer interfaces to the SONET ring may be located in the carrier's space in a large office building, or even on the premises of a particularly large customer. Note that the customer's TMs may feed an ADM which, in turn, is connected to the ring. These ADMs

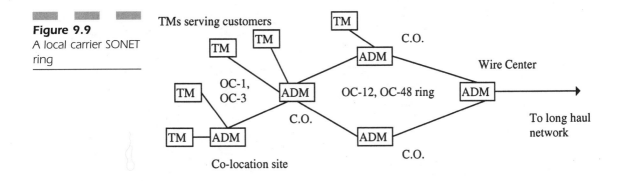

Figure 9.9
A local carrier SONET ring

are most often located in a co-location space servicing a large area, but may also be located in the carrier's space in an office building. Frequently, one building in an office campus complex holds the ADM for the whole complex, while individual buildings hold TMs. Of course, the campus ADMs may form a ring themselves.

TMs typically feed the ring at OC-1 or OC-3 rates. There is not "bandwidth on demand" in action here, and the sum of the OC-1s and OC-3s cannot exceed the capacity of the SONET ring. Thus, when 24 OC-1s and 8 OC-3s feed an OC-48 ring, the ring then is completely full because simple math shows that (24 × 51.84 Mbps) + (8 × 155.52 Mbps) = (1.24416 Gbps) + (1.24416 Gbps) = 2.48832 Gbps (or 2488.32 Mbps), which is the capacity of an OC-48 fiber link. It makes no difference whether the links are in a ring or not. As an aside, even at full capacity, the SONET ring may still provide some failure protection. This subject is discussed in more detail later in Chapter 14.

Rings in Action

What would a working SONET ring look like and be used for? SONET is not only for metropolitan areas. SONET rings exist in rural areas as well. To balance the perception of SONET rings in heavily populated city areas, this section describes a SONET ring in a more sparsely populated area, yet a ring that is still easily cost-justified.

A typical rural SONET ring may service about 300-square miles in the Midwestern United States. It would link five telephone exchanges and service perhaps 2,000 business lines and 6,000 residential subscriber lines. Operational costs are a big factor in rural areas. SONET rings can help to control these costs and at the same time maintain the high-service levels customers need and have come to expect.

The SONET ring is cost-justified as a way of simultaneously upgrading the copper feeder infrastructure while, at the same time, providing a basis for new high-speed services and controlling maintenance costs. Businesses are rapidly entering more rural areas around the country, and if the incumbent carrier cannot provide the services these new businesses need, they will look elsewhere.

The SONET ring itself may only span about ten miles. In addition to the protection features, the maintenance costs for SONET are about 75% less than the cost of copper cable upkeep. The TMs feeding the ring are placed around the 300-square mile service area. They provide both voice and data services at a variety of speeds. This keeps the cost of new circuits down and provisions service very quickly.

For example, a business ordering 100 new access lines does not need new cable laid from the serving office. The carrier links the new lines to the nearest TM, where software provisions the circuits. The carrier can provision the links in a day or so, compared to two weeks previously. The TM may combine some (or even all) of the features of a SONET DLC (sometimes called an RFT). In that case, any dialtone service, voice, or data, can be provided as well as leased lines. Sophisticated SONET equipment is just basically an *overlay network* superimposed over the existing copper network.

Network management takes advantage of the standard SONET set of features for this purpose. Outages are pinpointed more accurately, and service due to ring-link outages can be restored while most customers are unaware that there has been a problem. Naturally, the degree of protection depends on the details of the overall SONET ring topology; however, any SONET ring gives much better protection than does the most sophisticated T-carrier point-to-point links.

The same business may order 100 lines initially, expect 300 lines in five years, and 500 lines in seven years. The SONET infrastructure can support whatever growth is needed over time. The success of one business often encourages others in the same area. More customers mean more revenue for the service provider.

Eventually, the ring can grow to include more offices and an enlarged customer service area. A ten-mile ring could grow to 100 miles. TMs linked to rings by unprotected, single-fiber links could be "upgraded" to full ADM operation with new fibers to make the TM a node on the SONET ring itself. All of this can be managed with SONET-specific software and network management capabilities.

This chapter has opened up some new areas for further exploration. The example SONET ring just described should be examined even further. The operation of each piece of SONET equipment should be explored in more detail. Finally, the makers of various pieces of SONET equipment should be investigated to try to determine how each of them seek to distinguish themselves in an area where, after all, everybody is selling the same thing.

CHAPTER **10**
SONET
Equipment

This section of the book began with an overview of SONET networks. Some of the various types of SONET equipment were identified and their overall function on a SONET network was outlined for each. There were six overall classes of equipment defined, but the categories are anything but hard and fast. Indeed, any new technology goes through an early phase of specialized equipment that can only perform one task. This phase is followed by another that is characterized by rapid equipment evolution and convergence of form and function as an electronics evolution overtakes the standard. Consider early LAN bridges and routers that could only handle Ethernet or Token Ring, but not both.

Thus, even as the SONET standard evolves, there has been something of an evolution of equipment. This chapter explores the current state of the SONET equipment marketplace. It details the function of the various pieces of SONET equipment needed to build a SONET network. Finally, this chapter attempts to project the evolutionary trend of SONET devices.

First, a slightly different perspective of SONET devices is in order.

SONET Networks

SONET networks can be roughly divided into four main parts. These same four parts are not unique to SONET networks; they have been around for more than 100 years in one form or another. However, the application of SONET to the structure is the topic of discussion.

The four portions are distribution (or local access), the feeder portion, the local backbone, and the long haul backbone network. These sections vary in equipment capabilities as well as in scope. The overall structure of the four portions is shown in Figure 10.1. From user to user, the parts may be reflected in the sense that there must be user access on each end of the network, and the communications between local users close enough together may not involve all four portions of the SONET network, but the general picture is clear.

The local distribution system takes signals to and from homes and businesses. The link to the home and business location—frequently known as "the last mile," although this can stretch to nearly five miles in many areas—is not necessarily fiber. Most often, even in many business situations, the link is still coaxial cable or twisted-pair copper wire. With few exceptions, the link to the home is twisted-pair copper. SONET equipment interfaces with these traditional distribution systems at this point.

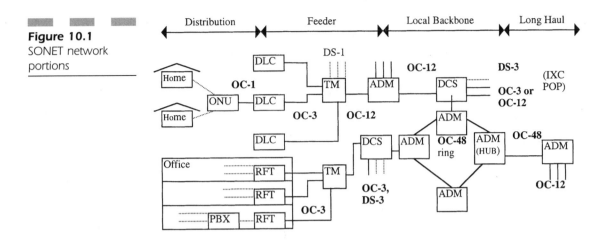

Figure 10.1
SONET network portions

The next portion of the SONET network is the feeder system. Feeder links may stretch for miles as well, and generally cover whatever portion of the distribution system that is not serviced by older coaxial of twisted-pair access links. As its name implies, the feeder system aggregates and distributes the SONET traffic from and to the distribution (or access) system. The feeder system equipment is characterized by modest SONET speeds and environmentally hardened casings and enclosures. This last aspect is due to the fact that, in most cases, SONET feeder equipment is deployed in the field in *outside plant* locations. As such, it is subjected to extremes of weather conditions and the rigors of all other environmental hazards. Nevertheless, few SONET rings are yet deployed at this level of the network except in major metropolitan areas.

The third portion of the SONET network is the local backbone. Here is where SONET rings are first typically deployed by local exchange carriers (LEC). The ring may be a mixture of outside plant and office space (*inside plant*) equipment locations. The ring protects the SONET network from single point-of-failure concerns. This feature, however, is at least somewhat offset by the common single-link feeder systems that are the rule today. Local rings may span only ten miles, or stretch to about 100 miles across. Generally, because they are found mostly in metropolitan areas, local rings of larger sizes are few and far between.

Finally, the local backbone (or backbones because several may be linked at this stage) connects to a long-haul backbone. This last portion also may be furnished by the local service provider, but it is more likely to belong to one of the interexchange carriers (IXCs) in the United States like MCI or Sprint or one of the dozens of others. The long-haul backbone may consist of more rings or single links, depending on the

SONET or other capabilities of the IXC. These rings and links may span the country, or even be international in scope.

The Distribution Network

Taking a more detailed look at each segment in Figure 10.1 is instructive. The distribution, or access, network consists of two main types of users. These are business and residential (home) users. This split has endured since the early days of telephony, when the telephone was marketed as a business tool and sold only grudgingly to home users, who were used to writing letters instead. (One of the earliest residential telephone users was Mark Twain in Hartford in 1878. Twain was also said to have invented the complaint letter, since his protests about the service were so strident and frequent.)

In any case, these homes are serviced by special Optical Network Units (ONUs) which convert SONET fiber signals to ordinary twisted-pair copper-wire signals. The ONUs most often connect to the rest of the SONET network at low fiber speeds, and may not even support SONET. When ONUs do support SONET speeds, this is most often just an OC-1 at 51.84 Mbps. Of course, other speeds are possible, depending on the number and density of the homes served. ONUs are interesting enough devices in their own right, but their relationship to SONET is only peripheral; therefore, they will not be discussed further.

Naturally, if fiber were to extend right to the home in the so-called, "fiber to the home" (FTTH) or even, "fiber to the curb" (FTTC) configuration, the ONUs may be mounted on the side of the house (FTTH) or on the nearest utility pole (FTTC). Neither of these configurations is common yet and will be ignored for the remainder of the discussion.

The ONUs interface with the first piece of SONET gear in Figure 10.1. This is the SONET digital loop carrier (DLC) device. DLCs are also called remote fiber terminals (RFTs) by many vendors and service providers, but the figure distinguishes between them. In the figure, DLCs are positioned as residential units supporting ONUs, while RFTs are depicted as business SONET CPE devices. Neither designation is right or wrong; they merely reflect a possible system for telling them apart easily in this context.

DLCs interface with many ONUs, with the exact number varying among vendors and models. The link to the ONU may be non-SONET fiber, SONET fiber, or (most commonly) some form of T-carrier, such as T-1 operating at 1.5 Mbps. The evolution of this DLC (and RFT) is interesting as well.

The use of DLCs and RFTs in the local loop is the culmination of years of evolution in wire technology. Previously, miles of twisted-pair copper wires bundled together snaked their way across the landscape on poles and formed both the distribution and feeder portion of the networks. One pair of copper wires was needed for voice service and dedicated for this purpose all the way back to the local serving office (i.e., central office: CO).

In the 1960s, analog loop-carrier systems evolved to combine many voice signals on a few pairs of wires. These systems provided a way to gain back some of the pairs used for each and every user. Because of this, they came to be known as *pair gain* systems. By the 1970s, digital technology was used for the same purpose for a variety of reasons, not the least of which were improved quality and reduced costs. In 1979, AT&T, then the "national telephone company" in the United States, introduced the most successful DLC equipment, the Subscriber Loop Carrier 96 (SLC 96) device.

Network growth throughout the 1960s and 1970s led to a need to limit the amount of new copper placed in and above the ground. This has been the highest historical cost in the telephone system. The SLC 96 from AT&T (now from an AT&T component called Lucent Technologies) uses four pairs for transmission to the CO and four pairs from the CO, with a spare pair in either direction for control and backup.

The SLC 96 system supports up to 96 voice lines on the ten pairs of cable. In the case of a new housing development an SLC 96, or equivalent, is placed at a central point. From that point, individual copper pairs will be connected to the customers' homes and businesses. It allows the network to support more customers with minimal deployment of new cable plant. This approach is ideal when providing telephony services. As they look to add new, higher bandwidth services, service providers will have to upgrade the local loop. After many years of being viewed as having limited significance, the local loop is an area of great concern in upgrading the network to support a service network.

Thus, an SLC 96 uses four T-1 lines for 96 voice channels, and the fifth T-1 provides protection switching. Without SLC 96, 96 homes would need 96 pairs of wires. Unfortunately, several design features made SLC 96 unsuitable as a long-term solution. Compromises in design were made to allow all types of copper pairs to interface on the user side, and more compromises made to interface with all types of service providers' switches. Throughout the 1980s, the quest was on to provide a better DLC solution than SLC 96.

In the 1980s, Bellcore defined a new, all-digital, T-1-based, loop-carrier system called *integrated digital loop carrier* (IDLC; Bellcore IDLC-TR303) just from digital and computerized serving offices. In the late 1980s, a migration strategy to SONET to deploy fiber in the loop (FITL; an umbrella term for FTTH and FTTC and others as well) was defined by Bellcore as a supplement to IDLC-TR303.

Today, almost all DLCs or RFTs are SONET-compatible. As time goes on, more and more SONET fiber will extend nearer and nearer to the end user. This makes a lot of sense. SONET offers a consistent and standard interface for vendors between the access network to homes and the feeder network to the serving office. SONET will eventually allow for the replacement of copper cable with fiber, which provides enormous advantages. This will not be inexpensive, but it will definitely be worth the effort when the SONET fiber provides the basis for new broadband services.

Figure 10.1 also shows a business arrangement using SONET equipment. In this case, the device is shown as an RFT, but there are few differences between DLCs and RFTs. In the figure, the RFT at the business location can interface with a variety of other devices, from PCs with analog modems for data services, to the organization's PBX for voice services.

Naturally, digital interfaces can be provided for user data equipment as well. The RFT can often directly interface with the organization's building wiring when it meets current specifications. The RFT provides a digital path to the serving office at full SONET capacity. The RFT cabinet can provide a business or organization with interfaces operating at 64 Kbps (DS-0) to 1.5 Mbps (DS-1) to 45 Mbps (DS-3) speeds. Naturally, for DS-3 speeds, a SONET OC-12 instead of an OC-3 may be desirable.

The use of both DLC and RFT equipment will increase rapidly as SONET moves into the local loop and access portions of the network.

The Feeder Network

The feeder portion of a SONET network is the area where subscriber-line multiplexing technologies, like SLC, 96 are employed. In SONET networks, the key component of this portion of the network is the terminal multiplexer (TM). The TM is essentially a SONET add-drop multiplexer (ADM) running in *terminal mode*. Previously in this book, the TM has been positioned more as a CPE device or piece of equipment that is SONET on one side and something else on the other.

However, the emergence of DLC and RFT equipment with SONET interfaces has changed this picture somewhat. In fact, it is probably just as accurate to say that newer DLCs and RFTs with SONET interfaces are now a kind of SONET TM, but most people seldom talk this way. It is possible that a SONET TM may be located in a large office building, especially a building of 20 or 30 stories in a major metropolitan area. In many cases, however, there are different vendors and needs for SONET TM equipment and RFT equipment, enough to make the retention of the distinction a wise idea. As time goes on, and prices decline, the DLC and RFT device will merge rather quickly into one SONET TM device.

In Figure 10.1, the SONET TM can feed either a DLC for residential distribution or a RFT for business access. The biggest difference is usually in the speed. Even heavily populated residential areas can be fed from a few OC-3s, each providing more than 2,000 voice channels. Businesses, on the other hand, can easily require 2,000 phone lines in a modest-sized office building. Moreover, the needs of businesses for speeds beyond 64 Kbps and 1.5 Mbps will consume more bandwidth more rapidly. Thus, TM equipment deployed here would support up to OC-12 rates; each has the equivalent of some 8,000 regular telephone lines.

In a SONET network, the main job of the TM is to aggregate and groom the traffic passing through it. In large business applications, it is even possible to have a SONET TM as a true user CPE, although this application of the TM has been rare to date.

The TMs at the end of the feeder network for residential services are connected to ADMs, which are pretty much the same type of equipment as the TMs, with minor differences in operation and capabilities. The ADMs further adjust the traffic because residential areas are characterized by many customers spread out in a wide area. Although the links to and from these residential ADMs are typically OC-12s, this is not a problem. The density of residential users makes this feasible and the speed of the link can always be increased if necessary.

Note that the business customers' TM usually links directly to an ADM on a ring. The amount of traffic to and from a business location makes it desirable to directly feed a SONET ring at the local backbone level. Typically, these links can operate at OC-12 speeds, although speeds of up to OC-48 are becoming more common in some cases.

Most links to and from the TM and ADM are linear links and not rings. There may be multiple links with diverse routing, but neither offers the same type of protection from failure that SONET rings offer. Ironically, most service providers will play up the presence of SONET rings on the local backbone and never mention that the distribution network, and most likely the feeder network as well, are not deployed in ring configurations. Naturally, this is where an outage affecting customer service is the most likely to occur and where environmental risks are the greatest.

The Local Backbone

The local backbone is where SONET rings are most often deployed. In Figure 10.1, there are only four SONET devices in the ring, but this is only a representative ring and should not be taken to imply any degree of precision. There may be five, ten, or up to 16 ADM nodes on the SONET ring. The configuration depends on the size of the ring and the density of users in a given area. Typically, each node on the ring will be located in a serving office. This may be a telephone company local switch (CO, quite commonly), a cable TV company's headend, or almost any commercial office space. SONET equipment requires little special air handling or backup commercial power supplies, although both are a good idea.

There are two main types of SONET equipment employed on the local backbone ring. These are the ADM and the SONET digital cross-connect system (DCS). The differences between ADMs and DCSs, in terms of functions and features, are significant. The ADM can add and/or drop signals at a variety of speeds, including both SONET and T-carrier speeds. That is, the traffic that the ADM adds and drops is SONET traffic in SONET standard formats, such as STS-1, STS-3c, as well as T-carrier traffic in DS-1, DS-3, or even other formats.

A SONET DCS is sometimes called a "full-featured" ADM. The DCS can interface with other speeds and types of equipment, and also the same forms of T-carrier traffic, such as DS-1, DS-3, and so on as an ADM. A DCS can also have SONET interfaces; thus, a DCS can connect to an ADM or TM as well, as shown in Figure 10.1. Generally, a SONET DCS has more management features and interconnection options between the input and output ports than an ADM. That is, instead of a passive dropping of DS-1 as in an ADM, a DCS can interconnect the DS-1 in the same device, a feature sometimes called "hairpinning" (as in "hairpin turn" on a road).

The main differences between SONET TM, ADM, or DCS, are in the relationship between the interfaces supported. ADMs usually do not "hairpin" a channel in an outside plant location. Their main task is to get the traffic to the serving office, where the DCS can decide where the channel should go. Of course, a DCS on a local backbone can still easily add and drop a DS-3 to a customer site or somewhere else. This direct DS-3-to-SONET-local-backbone arrangement is often used when a customer needs (and can afford) SONET service in a location where deploying TMs and RFTs does not make sense.

Consider a customer with many sites located in a downtown area, but one site (a factory perhaps) located some distance away. Obviously, the customer will want to have SONET-based services everywhere. The question is how to reach the factory site. The other sites may be served by RFT-TM-DCS arrangements to reach the local backbone. The factory may need to use an existing DS-3 to reach the ring. Without a direct DS-3 interface to the DCS on the ring, tying the factory in to the rest of the network would be awkward and difficult.

A DCS is also typically used to carry traffic over SONET links to other service providers, such as IXCs or even local competitors. Naturally, other service providers may have their own SONET rings, but the link to their rings are still commonly not rings themselves. The speeds are usually at DS-3 (672 voice channels), OC-3 (about 2,000 voice channels), and even OC-12 (about 8,000 voice channels). The channelization does not need to be 64 Kbps units.

Most SONET rings operate at OC-48 speeds, with higher speeds in the near future. Some service providers are starting to fill their SONET rings with traffic, and some have looked at adding a second ring in the same geographical area. This is fine and even offers another measure of protection beyond that provided by a single ring; however, OC-96 and even OC-192 will be operational soon.

One other type of ADM is employed on the local backbone ring. This is the *hub* ADM, most often simply called a SONET hub, although it is not really a distinct piece of equipment. A hub also interfaces between other ADMs and DCSs. Typically, all of the links into and out of a hub operate at a very high SONET speed (usually OC-48), which helps to distinguish the hub device from other ADMs and DCSs.

The hub is used to interface to another SONET ADM device off of the local backbone. The main purpose is to extend the SONET services into remote areas, usually further afield than those reached with feeder network arrangements. Alternatively, and as shown in Figure 10.1, the hub may feed an IXC's long-haul SONET links.

Admittedly, there is a fine line between ADM, DCS, and SONET hub. Some vendors have different products and names for all three. Others have only one product and rely on interface and configuration differences to tell them apart in the field. Still others deny any difference exists. In the future, it seems certain that differences will become even more blurred. In a LAN environment, few distinguish between a bridge and a router today, but they began as completely different devices for completely different purposes.

The Long-Haul Backbone

The last piece of the SONET network is the long-haul backbone. Usually, this portion of the SONET network is controlled by the IXC, and may be accessed not only at SONET speeds (typically OC-3 or OC-12), but even at relatively modest DS-3 speeds. When T-carrier is employed, as in the case of DS-3, the nice feature is that the IXC is not required to have any SONET equipment available, although all of the T-carrier limitations are now in force. Long-haul backbones have become a very active area of SONET deployment, where speeds of OC-48 (2.4 Gbps) and even OC-192 (10 Gbps) are common.

The long-haul network is not always ringed, but it may be. More IXCs are building their own SONET rings. As local competition on a facilities basis (local lines owned by the IXCs) becomes economically feasible, the IXCs may begin to build their own local backbone rings, although some have criticized this as a needless duplication of effort.

All four portions of the SONET network—distribution, feeder, local backbone, and long-haul backbone—function together to bring the benefits of SONET to greater and greater numbers of customers. The only thing missing from the figure would be a mirror image of the local, feeder, and distribution network somewhere across the country to complete the picture.

SONET Components

The more that the world of SONET TMs, ADMs, and DCSs is explored, the more that they come to be viewed as the same basic SONET device. What distinguishes them are the number and type of interfaces they support. Any input ports can be linked to any output ports with a little configuration effort.

What do these devices actually look like? Surprisingly, the devices are quite small and compact, yet another example of the triumph of electronics miniaturization. Those used to seeing a room full of cables, cabinets, and cross-connect panels for T-3 services are often astonished at the tiny size of SONET gear. A SONET ADM can easily be overlooked in a large equipment room, and SONET shipments have actually been misplaced in the shipping room. No one believed it could be so small.

The basic look of a basic SONET device is shown in Figure 10.2. Although a lot of SONET equipment comes in its own cabinets, the basic component is the *card shelf*. These are just standard rack-mounted units that are installed in cabinet enclosures or a standard communications rack. Most are about one foot high, but smaller units are also sold. The key component of these units is the control module or main card—or whatever name the vendor chooses to give this component. This board contains the heart of the device, and usually controls whatever SONET or other type of processing that is possible with the unit. It has its own power supply and interfaces with the backplane of the unit to allow communication between all other cards. This main board may be an integral part of the card cage, or a separate board itself. Separate boards can be changed without disturbing the rest of the device, which can be a plus.

None of the foregoing should be taken to mean that SONET equipment is not often found as 7-foot-high racks. It certainly is common to find standard telecommunications racks filled with shelves of SONET gear. But each shelf of the rack is capable of performing the tasks previously reserved for a whole room full of equipment. That is the point here.

The rest of the card shelf can be filled with interface cards to accomplish the intended task for the device. Most of the card slots are typically filled with various input and output ports with standard SONET connectors on them. Amazingly, standard SONET connectors are simply small plugs that are not much different in appearance than stereo jacks. The types designated SC or ST are most common, although there are others as well. The more cards slots that are occupied, the greater the capacity of the device. When all of the slots are full, upgrades may be made by adding a second shelf if the vendor allows this. Otherwise, a complete box swap may need to be made.

Figure 10.2
A typical SONET
device

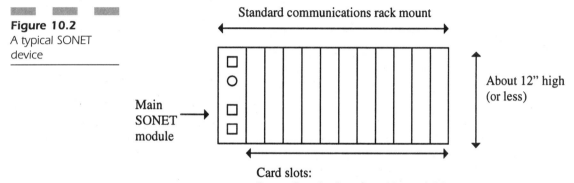

Standard communications rack mount

Main
SONET
module

About 12" high
(or less)

Card slots:
Power Supply, Interface, Network Management, etc.

Input and output ports are not the only type of card that can fill the card slots. Other types can be used to add to the capabilities provided by the main board (often at a considerable price). One of the most common additions is a backup power supply for the unit. Such redundant power supplies can keep the unit functioning for extended periods of AC power outages, sometimes for hours. The modest energy demands of electronics means that the elaborate batteries and diesel generators of the recent past can effectively be replaced by battery packs not dissimilar to those used in laptops.

In many cases, even the main electronics board can be made redundant with a card. A special network management module may be available, but most units will include at least the standard SONET network management functions and interfaces on the main controller board. Some SONET network elements come with a *hot-swappable* feature. This means that cards in the device can be removed for repair, or others installed, without shutting down or powering off the device. Needless to say, this is a very nice feature for repair-conscious service providers with a need to maximize availability.

Although specific vendors will add their own favorite features and functions to this basic package, the general appearance of SONET equipment will remain fairly consistent. The card cage and shelf approach is nearly universal.

SONET Network Element Functioning

Most SONET equipment looks fairly similar, whether the device is an ADM, DCS, or some form of hub. However, differences do occur in what the SONET Network Elements do internally or even externally. This section takes a closer look at what this statement implies.

All of the components of a SONET network are shown in Figure 10.3, with added details highlighting which arrangement is more suitable for business and residential customers; however, nothing precludes the use of either arrangement in a given area.

Figure 10.3 shows a fairly complete network deployment that exploits the features of SONET. Such an arrangement would typically be employed by a local service provider, such as the local telephone carrier of even a cable TV company. Of course, this SONET network would not be

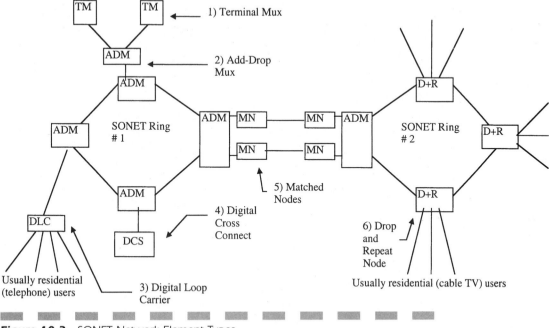

Figure 10.3 SONET Network Element Types

limited to these entities; the list is merely representative. Power companies have a lot of SONET in use internally, and there is little to stop them in the current regulatory environment from extending the SONET network to their customers.

The equipment found in this network is separated into CO terminal equipment and "outside plant" equipment. Admittedly, these are telephone company terms, but they are widely understood and used in the industry and not necessarily limited to a telephone networking environment.

The CO equipment consists of a SONET-compatible switch (i.e., DCS) and DLC equipment. This equipment forms the network terminating point for all SONET transport services. Note that the DCS is not capable of "switching" based on the bit configuration of the user's traffic. A traditional voice switch (for telephony services), router (for Intranet traffic or Internet access), or some other form of more exotic switch (such as ATM) must be used for this purpose. SONET is still very much a trunking—not a switching—technology.

The outside plant equipment consists of a fiber ring topology for survivability, and access multiplexers to this ring. These multiplexers are functionally separated into two categories, TM and ADM. However, SONET standards specify that the TM is just an ADM operating in *terminal mode.* Nevertheless, in many cases, vendors manufacture and market two distinct products for these two network elements.

The TMs are functionally similar to channel banks in the T-1 networks. They provide conversion from some form of non-SONET transmission media and format, such as T-carrier, to the SONET format. The ADMs are the nuts-and-bolts component of the SONET architecture. The ADMs allow access to the SONET transmission signals, even down to the DS-0 level, without fully demultiplexing the SONET signal. Non-SONET signals may be added to or taken from the SONET transmission signal with ADM equipment.

As shown in Figure 10.3, and mentioned previously, there are six main network elements that can be found in SONET. These are:

1. Terminal Multipexer (TM)

2. Add-drop Multiplexer (ADM)

3. Digital Loop Carrier (DLC)

4. Digital Cross-connect System (DCS)

5. Matched nodes (MN)

6. Drop-and-repeat nodes (D+R)

It is important to remember that the names of some of the SONET network elements are covered by standards, but some are not. On the list, the names TM, ADM, and DCS are well defined and universally used. However, names like DLC, MN, and D+R have different names depending on the manufacturer. The names used here were chosen for their descriptive powers and should not appear to be an endorsement of any particular vendor or vendor's terminology. But whatever the name, the most important aspect of the device is where it is employed in the network and what function it must perform.

The access node into a SONET network is most often a multiplexer of which there are two basic varieties. The simplest, the TM, acts as a concentrator of multiple DS-1 or DS-3 signals onto a SONET electrical STS-1/3 or optical OC-N backbone. It is analogous to the asynchronous M13 multiplexer that forms the entry point to the DS-3 digital signal hierarchy. The TM is usually a less redundant configuration of the SONET network element known as the ADM.

The next few sections take a closer look at the internal functioning of the SONET network elements.

Terminal Multiplexers

One major SONET network component is the TM. TMs are used to terminate several DS-1 signals and package them together as STS payloads. For example, one type may terminate 28 DS-1 signals, thus eliminating the need for today's M13 multiplexers. Yet another variety may terminate 84 DS-1 signals to generate an OC-3 signal.

Figure 10.4 shows the basic internal structure of a TM that takes 28 DS-1s, each with 24 DS-0s, and produces an STS-1 frame for transport on an OC-1 fiber link. Note that multiple groups of DS-1s and multiple OC-1s may be produced in the same device, depending on the number of card slots available for use.

Figure 10.4 shows that a SONET TM consists of a number of T-carrier ports that are multiplexing using normal time division multiplexing into SONET payloads. The payload is given the complete set of SONET overhead bytes and out it goes. The same happens in reverse on the receiving side. Note that each OC interface must have two fibers, one for input and one for output, as is common in digital systems.

The individual fibers are not shown in the figure. The TM is a point-to-point device; thus, even though there are actually two output paths, the two paths must terminate on the same SONET device. Two physically separate fiber cables would be used in between devices, with each cable containing a receive and transmit fiber pair.

Although not strictly limited to point-to-point SONET links, it is unusual to find a SONET TM on a ring. This would essentially make it a DCS. Although no harm is done to the network, confusing TMs with DCSs (or even ADMs) is a misuse of terminology and generally not acceptable.

In Figure 10.4, the "DSX-1" should not be confused with a "DCS." The digital system cross-connect for DS-1 (DSX-1) is not really a piece of elec-

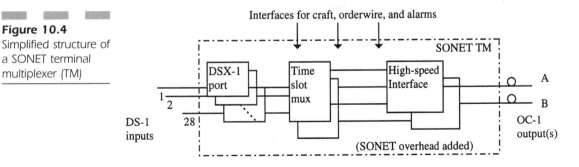

Figure 10.4
Simplified structure of a SONET terminal multiplexer (TM)

tronic gear by itself, as is a DCS. A DSX interface is more likely to be a physical patch panel arrangement attached to a rack or even on plywood, nailed to a wall.

- A DSX-1 is usually a series of terminated RJ-45 jacks that can be patched with jumpers to provide access to the SONET TM (i.e., the "DSX 1 Port" on the visual).

- A DSX-3 is usually a series of terminated coaxial cable jacks (less frequently, mini-SCSI connectors) that perform the same function for DS-3s.

The key is that DSX is an *interface*, not a piece of equipment. It is possible, but not necessary, to have a real DCS feeding a SONET TM DSX port. This can be confusing and many people use the terms DSX/DCS interchangeably for this very reason.

The large amount of information that is transported by SONET devices requires a high degree of redundancy in order to eliminate any single point of failure with regard to the fiber path, internal circuits, and power supplies. There is too much information flowing too fast on the SONET links to permit a single point of failure. The concern for reliability is reflected in the architecture of the TM.

The controller in the main board plays a key role. The complexity of the SONET protocol favors the manufacture of a software-driven product whose configuration may be determined by software downloaded from a central network site. This type of design simplifies the addition of new feature sets at the modest expense of advanced microprocessors and large amounts of memory. Operating instructions stored in the device's memory provide the necessary intelligence for provisioning, monitoring, status reporting, and sounding alarms.

The DSX-1 ports (transceivers) provide access to the lower-speed channel signals. For example, up to 672 DS-0 signals are received in the form of 28 DS-1s by an STS-1/OC-1 multiplexer transceiver, which processes them in a form acceptable by the pair of redundant time slot multiplexers, or interchangers. There are usually four additional DS-1 channels: two are used for maintenance and the other two for the European E-3 frame.

The Time Slot Interchanger (TSI) in the TM handles the low-speed signals to a redundant pair of high-speed interfaces, A and B, which normally operate at the SONET STS-1/3 electrical or OC-N optical rates.

The high-speed interface controls the SONET output links. One of the two signals, either A or B, is transmitted while the other is held for back-up purposes. Sometimes both A and B signals could be transmitted simultaneously and it could be up to the receiver which to use. The optical signal is simply "bridged" at the electrical level in order to transmit the identical signal from the TM. The unit monitors both the in-service and protection optical signal independently.

The TM has other specialized interfaces as well. Craft, orderwire, and alarm units (called the "COA" interface) selects whether the A or B SONET link will be active, provides for craft (technician) interfaces, analog input and output (if used), and forwards the alarms to remote network operations centers.

As an aside, the clocking for TM is derived from either an internal source or a received optical signal. Figure 10.4 illustrates that a SONET TM can take existing DS-1s and DS-3s, whether employing DSX interfaces or not, and convert them to and from SONET.

SONET TMs, as well as ADMs, transport existing DS-n signals without alteration. The low-speed DS-n ports may feature bit- or byte-synchronous operation (locked VT) or asynchronous (floating VT) traffic. Bit-mode provides transparent transport of DS-1 traffic and is useful for transport of unframed DS-1 or framed DS-1 with ZBTSI coding. Byte-mode transports D4-framed DS-1s with superframe or ESF formats. When configured as a CPE, the TM becomes more specialized. There will be SONET multiplexer "gateways," to use the term loosely, for Ethernet and Token-Ring LANs, ISDN primary rate access, National Television System Committee/Phase Alternate Line TV, HDTV, Switched Multimegabit Data Service (SMDS) for metropolitan area network access, and ATM cells.

Add/Drop Multiplexers

The most crucial SONET network element is the SONET ADM. This is because the ADM is not only an essential piece of the SONET network, but it also can be viewed as a kind of generic SONET device that can be easily modified to function as a TM or ring interconnection device or something else as the deployment scheme requires.

The ADM is a device unique to SONET with no real counterpart in the T-carrier world. The SONET ADM replaces the back-to-back M13s and associated manual DSX patch panels often used for DS-1 cross-connections. The ADM is a fully synchronous, byte-oriented multiplexer capable of adding and dropping DS-n signals onto one of two SONET links. DS-n payloads within the OC-N signal also may be added or dropped within the ADM. A simplified schematic of the ADM is shown in Figure 10.5.

Figure 10.5 shows less internal detail than does the previous TM figure. The emphasis here is on overall function, but all of the internal components still exist. SONET ADMs have two sets of OC interfaces instead of one, as in the TM. ADM features include DS-n electrical low-speed interface, OC-N high-speed optical interface, DS-n/OC-N multiplexing, DS-n add/drop, and the usual COA management interfaces.

The input and output STS-1/3 interfaces may be used to interconnect within an equipment frame or building site, but are unable to drive signals any significant distance over copper wire. For this, an electrical-to-optical conversion must be added to the ADM. Then the customary SONET physical distance rules apply. For example, a 1,310 nm fiber optical signal may span 26 miles before requiring regeneration.

The ADM interfaces with existing T-carrier equipment and facilities by way of multiple DS-n electrical interfaces. The incoming payloads that are not terminated by the ADM pass through to the OC-N interface on the other side of the ADM. Each DS-n interface may drop information from an incoming OC-N and add information to the outgoing OC-N. Therefore, an ADM may add or drop DS-1 signals within an STS-1 or STS-3. In addition, some ADMs are capable or performing DS-0 "groom and fill," which aggregates and distributes normal voice channels. This goes beyond the specifications of the SONET ADM, as described in the Bellcore technical references, which only require ADM capabilities at the DS-1 level.

SONET ADMs can be placed in series along a SONET path, so that non-SONET signals can be added to, or dropped from, the STS payload. ADMs operate bidirectionally and component terminations can be accomplished from either direction. Typical implementations of ADM employ protection-switching capabilities, providing an active fiber and a standby fiber (A and B), as do the TM devices.

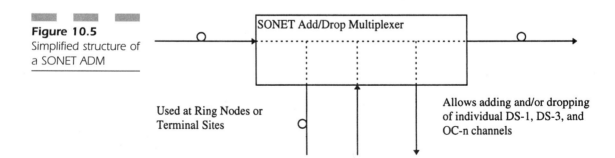

Figure 10.5
Simplified structure of a SONET ADM

SONET Add/Drop Multiplexer

Used at Ring Nodes or
Terminal Sites

Allows adding and/or dropping
of individual DS-1, DS-3, and
OC-n channels

ADMs can also be equipped with time-slot interchangers to allow digital cross-connect functions between channels, carried as part of the STS-1 payload. When optioned in the add/drop mode, ADMs eliminate the need for back-to-back multiplexers—the common solution in T-carrier networks—which allow component signals to be broken. Of course, theADM then effectively becomes a SONET DCS.

The SONET DCS can be an add-in card to the overall SONET equipment rack. Alternatively, the DCS can be a separate physical device altogether.

Because they interface with both the optical network and the conventional electrical network, ADMs need a low-speed electrical interface and a high-speed optical interface. Additionally, they require remote- and local-operation interfaces, data communications channel for Section Overhead functions, local nonvolatile memory for backup purposes, and distribution frame and facility maintenance capabilities.

SONET Digital Loop Carrier

The SONET DLC, like any loop carrier, is designed to minimize the assignment of dedicated plant to the loop environment. The IDLC combines the features of intelligent remote digital terminals (RDT) and a feature of the digital switch called integrated digital terminals (IDT); the two are connected by a digital transmission facility. The primary goal of this architecture is to allow RDTs from one manufacturer to efficiently and transparently interface with switches (i.e., IDT) from another manufacturer, thus providing a generic interface of sorts.

For most of the digital revolution, the local loop has remained primarily analog. This will have to change with the introduction of fiber in the local loop and the deployment of SONET DLC equipment. The current local telephone environment is ideal for SONET DLC deployment.

Within a Local Access and Transport Area (LATA) are numerous *central offices*, or *wire centers*, which serve specific communities of interest. Wire centers handle trunk-to-trunk traffic, such as for IXC access. A wire center may be a 50,000-line CO or even what is known as a community dial office (CDO) serving only a few hundred subscribers. Service may also be provided by remote terminals (RTs) which give commercial tenants of a distant office park access to the telephone network.

The SONET DLC may be considered a concentrator of low-speed services before they are brought into the local CO for distribution. In remote areas, SONET DLC is actually a system of multiplexers and switches designed to perform concentration from the remote terminals to the CDO and, from there, to the CO.

Typically a multiple-port OC-3 device, the SONET DLC provides direct access to its constituent DS-0 signals. Available DLCs support up to 2,016 DS-0s and are most economical where the general demand falls between 200 to 2,000 lines. The DLC supports both voice and data traffic and can provision, test, and maintain POTS, coin phones, multiparty lines, digital data service (DDS), and ISDN basic and primary rates.

The SONET DLC is really a short-term transition device meant to increase the capacity of the existing network until fiber is deployed almost everywhere. Initially, the DLC acts as a helpful addition to the CO switch by terminating an OC-N data stream and applying its embedded DS-0s to the switch. As the switch is upgraded or replaced by a SONET-capable switch or SONET DCS, the DLC may be moved elsewhere in the carrier serving area (CSA).

A SONET TM may be deployed at the customer premises, but the SONET DLC is intended for service in the CO or a controlled environment vault that belongs to the service provider. Bellcore describes the generic IDLC, which consists of intelligent Remote Digital Tenninals (RDTs) and digital switch elements called IDTs, which are connected by a digital line. The IDLCs are designed to more efficiently integrate DLC systems with existing digital switches.

To facilitate the introduction of SONET, additional feature groups have been included that provide needed functions. Feature Set B provides such features as remote test unit (RTU) integration, remote switch unit (RSU) integration, and integrated network access (INA).

Feature Set C, called the IDLC/SONET Interface, generates savings in the areas of facility terminations, information transport, and operations expense. These savings come from a variety of sources, most notably reductions in DSX terminations, intra-office cabling, and signaling terminations (e.g., framing circuits, timing signals).

SONET Digital Cross-Connect Systems

The SONET DCS is a key component in any SONET network that performs cross-connecting. Sometimes called a "full-featured" ADM, the DCS does it all.

One major difference between a DCS and an ADM is that the DCS can be used to interconnect a larger number of STS-n links. The DCS can be used for grooming (i.e., consolidation and segregation) STS-1s or broadband traffic management. For example, a DCS may be used to segregate the high-bandwidth from the low-bandwidth traffic arriving from a SONET ring and to send them separately to a high-bandwidth switch (e.g., ATM switch for videoconferencing) or a low-bandwidth switch (e.g., voice switch). This is the synchronous equivalent of DS-3 DCS and supports hubbed network architectures.

In the serving office (and eventually perhaps even on the customer premises) SONET digital cross-connects will replace much if not all of the current back-to-back multiplexer configurations. This will reduce the number of intermediate distribution frames and manual jumper panels between these frames. Cross-connection is a fundamental capability of SONET standard network element descriptions. SONET DCSs will be used at the serving office to terminate an OC-N and separate switched and non-switched DS-n traffic. Then any switched DS-0s can be brought into a standard voice switch. Even better, by taking advantage of the SONET remote configuration features and capabilities, selected signal streams can be cross-connected even at intermediate sites as needed, although most often DCSs will be deployed on centralized office-based rings.

The SONET DCS can become a large interoffice *hub.* This type of DCS is capable of managing large amounts of traffic transmitted at the DS-1, DS-3 or OC-N rate, and primarily at OC-3 through OC-12. DS-1 cross-connections will be performed by cross-connecting floating Virtual Tributaries (VTs). Higher-level cross-connects will use electronically alterable memory maps to groom the various VT signals into OC-N signals.

Figure 10.6 shows, in some detail, the structure of a SONET DCS. The detail is warranted in this case due to the position of the SONET DCS as a "superset" of the ADM and TM devices. The DCS still physically looks like a shelf or card cage device, but the figure shows a more functional organization. The DCS must still contain a SONET timing module, and will have other interfaces used for craft, operations (technically, Operations Support Systems: OSS), and alarms. There will usually be some backup functions included for memory, power supply, or other redundancy.

Figure 10.6 also shows the common controller module that oversees the general functioning of all input and output operations, as well as the internal operation of the DCS itself. A variety of interfaces are supported, from DS-1 through DS-3 and up to some large OC-N level. To be accurate, the DCS actually "terminates" the DS-3 or OC-N and is not

Figure 10.6
SONET DCS
architecture

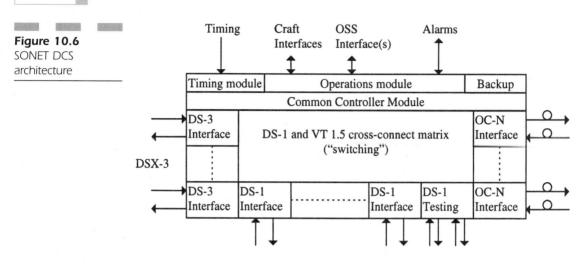

merely an *interface*; however, the term *interface* is more common among equipment vendors. In practice, the availability of SONET "level N" service depends on the availability of SONET DCS equipment from a vendor to support it.

The heart of the SONET DCS is the cross-connect matrix. This is how signals from an input port are "switched" (or more accurately a "shuffle") to an output port. Some models can cross-connect all the way down to the DS-0 level, but this is not common. All ports can be connected through the matrix.

If the input and output signals are at the same level or not, whether SONET or T-carrier, the cross-connection is simple due to the synchronous nature of SONET. Grooming and hubbing are automatic and transparent.

There are two distinct types of SONET DCS. They are called the broadband DCS and the wideband DCS. They differ in the types and speeds of signals that they cross-connect.

Broadband DCS

The SONET broadband DCS (B-DCS) is similar in function to a DS-3 cross-connect in the T-carrier hierarchy. The B-DCS has interfaces for full-duplex SONET signals, as well as DS-3 clear-channel interfaces. The basic function of the SONET B-DCS is to make two-way cross connections at the DS-3, STS-1, and even the concatenated STS-N (STS-Nc) levels.

The B-DCS, thus, provides a type of "switching" at STS level and DS-3 level. However, this "switching" is not the same as one would encounter in a voice or frame relay switch. Rather, the "switching" is similar to that achieved by DS-3 cross-connect. In other words, no input-to-output port correspondence is established by the content of the information stream. The correspondence is strictly established by configuration software and remains in effect until changed.

The general architecture of a B-DCS is shown in Figure 10.7. The B-DCS provides *transparent* cross-connections between DS-3 interfaces (actually, the DSX-3) and between DS-3s and OC-Ns terminating at the DCS. In this context, the term *transparent* (or clear channel) means that any DS-3 at the normal DS-3 rate can be cross-connected, whether it is carrier synchronously or asynchronously on the SONET network. Thus, the B-DCS does not need to frame align on the incoming DS-3 unless enhanced-performance monitoring is required.

The B-DCS is also fully capable of performing STS-1 and STS-Nc cross connections as well. As a consequence, the B-DCS must be able to frame align with the incoming OC-N. It must also identify and gain access to the desired constituent SPE. Again, this *terminates* the OC-N because the SONET transport overhead is terminated and processed. The B-DCS is also capable of generating new outgoing OC-N with valid transport overhead. In general, this makes the DCS a natural focal point for gathering network information and performance statistics.

Wideband DCS

The main distinction between a B-DCS and wideband-DCS (W-DCS) in SONET is that the W-DCS goes "deeper" into the digital hierarchy, all the

Figure 10.7
Simplified structure of a SONET B-DCS

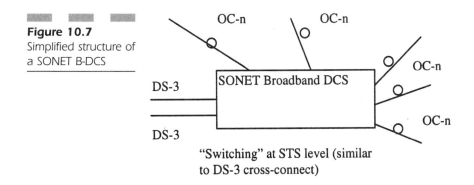

way down to the VT level. In this respect, the W-DCS resembles the DCS-3/1 cross-connect in T-carrier. Thus, a W-DCS terminates SONET and DS-3, while cross-connecting DS-1 signals. The W-DCS DS-1 and DS-3 interfaces meet the DSX-1 and DSX-3 specifications.

The general capabilities of a SONET W-DCS are shown in Figure 10.8. This is a simple, but accurate, picture. Note the presence of the DS-1 speeds into and out of the device. The ability to cross-connect at the DS-1 level (actually, the VT 1.5 level) makes the W-DCS a very useful SONET device.

In Figure 10.8, it is easy to see that the major difference between a broadband and a wideband SONET DCS is that the W-DCS operates at the VT 1.5 level. This device, therefore, is similar to a DS-3/1 DCS because it accepts DS-1s and DS-3s, and is equipped with optical interfaces to accept optical carrier signals. The W-DCS is suitable for DS-1-level grooming applications at hub locations, such as the serving office. One major advantage of deploying W-DCSs is that less demultiplexing and multiplexing are required because only the affected VTs are accessed and switched.

The W-DCS network element cross-connects floating VT-1.5 signals between OC-N terminations and transparent (clear-channel) DS-1 signals. It also provides transparent DS-1 cross-connections between DS-1 interfaces and DS-3/OC-N terminations. The term "transparent," when applied to cross-connection, means that the W-DCS is capable of handling any type signal at the normal DS-1 rate. This is important because DS-1s come in a variety of framing formats and digital coding methods. The W-DCS should be able to handle them all, regardless of the nature and format of the signal.

This transparency is accomplished by framing the incoming OC-N to identify and access the target VTs. This, in turn, requires that the section, line, and path overhead of the OC-N be regenerated from scratch using

Figure 10.8

Simplified structure of a SONET B-DCS

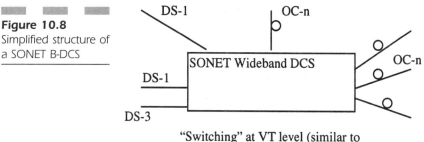

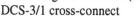
"Switching" at VT level (similar to DCS-3/1 cross-connect

new STS-1 SPEs and new section, line, and path overhead. This is why both the broadband and wideband SONET DCS network elements are said to "terminate" the signals.

SONET Matched Nodes

SONET networks today are usually deployed in a ring architecture, but how do the rings talk to each other? That is, how can one set of users on one ring communicate with another set of users on another ring? One of the most common ways of accomplishing this is with a set of SONET matched nodes. It should be noted that this device is usually some form of SONET ADM deployed in pairs for this task. Some vendors and service providers consistently refer to this arrangement by this name and, therefore, requires a few words of explanation.

In a SONET matched-node configuration, whether the devices used are sold as "matched nodes" or not, SONET traffic is duplicated and sent over two links between two rings. One of the signals is selected and continues around the rings. If the connection were lost between the active modes, the backup link could be used until repairs were made. Figure 10.9 shows this use of matched nodes.

Rings are not even necessary for matched-node operation in SONET. Even on point-to-point links, as is common from a customer site TM to a ring ADM, matched nodes can be used. Matched nodes can be configured with diverse routing to the serving office, separate entrance facilities at the customer site itself, backup and redundant power supplies, and even totally redundant common control electronics to make SONET links almost impervious to failure, except in the most catastrophic circumstances.

Figure 10.9
SONET matched
nodes in action

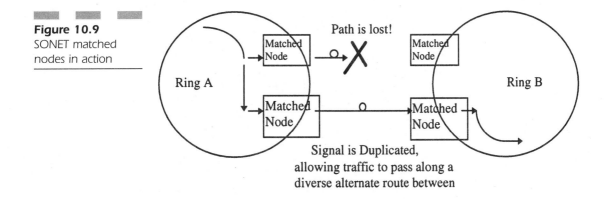

SONET Drop and Repeat

Like matched nodes, SONET drop-and-repeat nodes are not really separately defined SONET network elements. Both matched and drop-and-repeat nodes add a feature to duplicate traffic received on a input port and send the same bits out of two (or even more) output ports. They are merely SONET devices deployed in a certain configuration to accomplish a specific task.

When SONET is deployed in a ring architecture, drop-and-repeat SONET equipment provides alternate routing for traffic passing through a higher level ring from lower level feeder rings. For instance, several OC-12 rings with traffic from individual customer sites may be linked with an OC-48 ring configured as a regional ring. This use of SONET drop-and-repeat devices is shown in Figure 10.10.

At each node in Figure 10.10, a drop-and-repeat SONET device "drops" the signal off of the main ring onto the feeder ring, yet keeps a "copy" of the signal going around the main ring. This is obviously one giant step beyond the simple "matched-node" configuration.

In multiple-node distribution applications, such as cable TV, drop-and-repeat is essential. One transport channel can efficiently carry traffic between multiple distribution nodes. For example, when transporting cable TV video signals, each programming channel can be dropped (i.e., delivered) onto the feeder ring at the node and yet repeated for continued delivery around the ring. Not all of the bandwidth (i.e., program

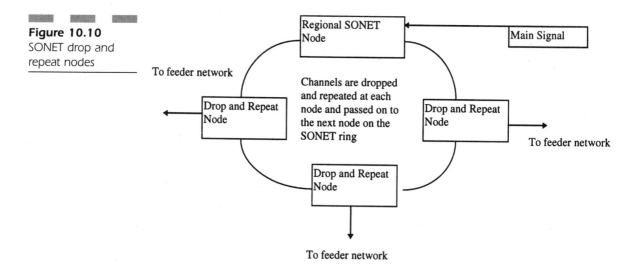

Figure 10.10
SONET drop and repeat nodes

channels) need be delivered at all of the nodes. Channels not dropped at a node are passed through without physical intervention to the other nodes.

The SONET drop-and-repeat feature is sometimes called *drop and continue.* It depends on the service provider and equipment vendor. In each case, operation is identical.

11

SONET
Equipment
Providers

The last chapter in this section of the book takes the major categories of SONET network elements established in the previous chapters and examines just how equipment vendors have implemented them. It is one thing to investigate standards and bit configurations and frame structures, but it is quite another to take these specifications and build them into products that people will buy to implement SONET networks. SONET equipment must embody the SONET standards. This chapter looks at how and to what extent vendors have done this.

All of the SONET product offerings from the major SONET vendors are explored in this chapter. Every attempt is made to compare products on an apples-to-apples basis. However, some vendors are more open regarding information about their products than others. Many make a lot of detailed information, including pictures and technical specifications, available at a product Web site accessible over the Internet. Many are more guarded about exact product functioning in many key areas; therefore, a complete feature-by-feature comparison is not possible.

Another area of concern that is not addressed in this chapter is interoperability. It may sound strange that products based on exactly the same set of international standards would be anything but interoperable, but this is not always the case. This is an especially sensitive issue in SONET, one of the major goals of which was to provide the interoperability that the T-carrier hierarchy lacked. However, not all standards are rigid.

The term, *interoperability* encompasses more than simple bit communication. The term is more inclusive. How are issues like network management, troubleshooting, and even performance monitoring handled when two or more vendors' products are included in a SONET ring? These are the basic concerns of interoperability, beyond the respect for standards. Unfortunately, these compatibility issues cannot be explored in this chapter for the simple reason that not many unbiased and wide-ranging studies have been made on SONET product interoperability issues.

This chapter makes an effort to be as up-to-date as possible. In all cases, Web sites for further research to determine the latest information from a vendor are given.

The Big Six

When it comes to SONET vendors, there are really six major vendors. Combined they account for more than 90% of all SONET equipment sold to date. The six vendors include (in alphabetical order): Alcatel Network Systems, Fujitsu Network Transmission Systems, Lucent Technologies (formerly AT&T Network Systems), NEC Transmission, Nortel Networks, and Tellabs. This is not meant to slight any of the other vendors with SONET products on the market. For example, DSC Communications and Siemens make fine SONET products as well. The prominence given to the selected vendors is just an acknowledgment of the current market mix and an attempt to narrow the focus of the chapter.

Here is a little overall information on each of the vendors. Those concerned with corporate histories and present financial status are referred to each vendor's Web site. Strangely, all of the vendors once highlighted these products at their Web sites. It is surely a sign of the times that now all of these vendors prominently feature DWDM and other optical networking products. SONET equipment is now sometimes buried in the back pages of the Web site. This is a sign of both changing times as well as the maturity of SONET equipment.

Alcatel Network Systems (*www.ans.alcatel.com*) became a major player with the acquisition of the transmission product arm of Rockwell in 1991. Rockwell was a respected supplier of asynchronous transmission gears and products, but was very conservative when it came to new technology. Alcatel inherited the conservative outlook of Rockwell and brought out SONET products on a leisurely schedule, supporting OC-3 initially and OC-12 later. Alcatel now has a SONET product line that includes OC-3, OC-12, and OC-48 support. Alcatel is known for its fiber ring products as well as digital cross-connects for SONET networks.

Fujitsu Network Transmission Systems (*www.fnc.fujitsu.com*) was the first company to market Phase 2 SONET-compliant products. Phase 2 included such welcome features as the data communications channel protocol (but not yet a full message set), network management systems, and quite a few other things. Fujitsu has been a pioneer in SONET ring products, as well as the user-friendly adjunct software packages to make it easier to add features and functions to SONET networks and products.

Lucent Technologies (*www.lucent-optical.com*), formerly AT&T Network Systems, was formed after a reorganization of AT&T into more realistic lines of business. Billed as the beneficiary of the technologies developed by Bell Labs, the idea that the best SONET products should come from the organization that invented the laser in the first place is a powerful marketing tool. Lucent/AT&T was late to enter the SONET ring market, but its latest ring products are second to none. Whatever the name, AT&T is still AT&T.

NEC Public Networking Group (*www.necpng.com*) is another company that has excelled in SONET ring products. Its products specialize in providing high capacity in a small package. NEC has generally introduced products at higher and higher OC speeds as quickly as the speed becomes economically and technically feasible. Its products are quite versatile, with special emphasis on reliability.

Nortel Networks (*www.nortelnetworks.com*), formerly Northern Telecom, arose from the former Bell System in Canada after 1984. Nortel Networks is notable for its wide range of SONET products and has also emphasized the network management services that are needed to configure and run SONET networks. Nortel Networks has also been quick to support higher and higher OC speeds quickly. The products are tightly integrated and easily upgraded or reconfigured as network requirements change.

Tellabs (www.tellabs.com) was founded in 1975 and might seem out of place rubbing shoulders with giants like Alcatel and Fujitsu and Nortel Networks. Nevertheless, Tellabs has been successful in the SONET field by finding its niche and exploiting it quite well. Tellabs has evolved from being a maker of analog-based telecommunications products to become a truly global equipment provider, mostly in the field of digital crossconnects. Lately, Tellabs has begun to branch out into the main lines of SONET gear and has even begun to offer optical networking gear and integrated ATM switching. But the main strength of Tellabs remains its TITAN line of digital cross-connects.

Several of the companies mentioned here should be singled out in this book for reasons other than their SONET products. Alcatel provided otherwise hard-to-get product information promptly and gladly. And Nortel Networks' numerous white papers and product information provided a welcome basis and framework for this book as a whole.

It should also be noted that out of the companies on the SONET list, Alcatel and Lucent are the two which make complete product lines that can be integrated seamlessly with their SONET products. In other words, from cross-connect to long-haul ADM, both Alcatel and Lucent can provide equipment that does it all.

Product Overview

Each of the major vendors introduced makes a number of SONET products. And each one of these SONET products comes in several sizes and has various features. So before detailing the key aspects of each individual product, it might be a good idea to briefly take a look at some of the products from each of the vendors. A representative sampler of these products will then be detailed.

Table 11.1 is certainly not meant to imply that the vendor in question makes only these SONET products and no others. Such a claim to be exhaustive would be quickly out of date anyway. Again, the intention is

TABLE 11.1

A sampling of SONET vendors' products

Vendor	Product	SONET Device*	OC-n Levels Supported
Alcatel	1631 SX	W-DCS	OC-1, OC-3, OC-12
	1633 SX	B-DCS	OC-3
	1630 CSX	W-DCS	OC-1, OC-3
	1603/12 SM	ADM, TM	OC-3, OC-12
	1648 SM	ADM, TM	OC-48
	1603 SMX	ADM	OC-3, OC-12, OC-48
	1603 SE	ADM, TM, DCS	OC-3
Fujitsu	FLASH-192	ADM, B-DCS	OC-12, OC-48, OC-192
	FLASH-600 ADX	ADM & ATM	OC-3, OC-12, OC-48
	FLM-2400	ADM, TM	OC-48
	FLM-6	TM, DCS	OC-3
	FLM-150	ADM, TM, DCS	OC-3, OC-12
	FLM-600	ADM, TM	OC-12
Lucent	DDM-2000	ADM,TM	OC-1, OC-3, OC-12
	FT-2000	ADM, TM	OC-48
	SLC-2000	DLC	OC-3
NEC	IMT-150	ADM	OC-3
	ITS-2400A	ADM, TM	OC-3, OC-12, OC-48
	ITS-600	ADM, TM	OC-3, OC-12
Nortel Networks	S/DMS Access Node	DLC	OC-3, OC-12
	S/DMS Transport Node	ADM, TM, D+R	OC-3, OC-12, OC-48, OC-192
Tellabs	TITAN 4500GS	TM, DCS	OC-3, OC-12
	TITAN 5500	DCS	OC-3, OC-12
	TITAN 6500	DCS & ATM	OC-3, OC-12, OC-48

* Main product use. Does not imply that product cannot be used in other SONET network situations.

to profile a sample of the vendor's SONET products as representative of the line as a whole. All of the information in the table comes directly from the individual vendors' Web sites. The intent is mainly to give an overall flavor of the range and depth of each vendor's SONET offerings. Some of the packages are small enough to sit on a desktop. Others are full 7-foot-high equipment racks. This table is only a starting point for exploring the world of SONET equipment.

This table also excludes DWDM and other optical networking equipment. Many vendors have begun to package SONET ADMs and DWDM equipment together to provide multiple SONET links on one fiber (e.g., Fujitsu and Lucent). This list is restricted to pure SONET equipment without optical multiplexing considerations.

Even a quick look at Table 11.1 shows that all varieties of SONET network elements are well represented, especially ADMs. It is worth mentioning that many of the vendors' ADM products can be configured to perform other SONET equipment functions, such as Drop and Repeat and Matched Nodes. In many cases, the matched node ring interconnectivity can be handled by a digital cross-connect or by even more imaginative uses of SONET devices. Note that two of the products listed, the Fujitsu FLASH-600 ADX and the Tellabs TITAN 6500 Multiservice Transport Switch, combine SONET functions with ATM cell switching.

The Product Sampler

Before exploring some of these products in more detail, a few words of caution. First, when researching or comparing SONET products (or any other networking products, for that matter), there is no substitute for up-to-date information. As current as books or articles may wish to be, vendors are constantly introducing new products, updating old ones, and prototyping the next generation of SONET equipment on an ongoing basis in labs and even in the field. It would not be surprising, therefore, that specifics might differ on some products. In all possible cases, the vendor's Web site should be consulted for the latest information. This section is intended as a general guide, not as a substitute for vendor product information.

Second, since the information presented is of a general nature and is, after all, dependent on the openness of the vendors themselves, there is a variation in the amount of detailed information available. Although every effort has been made to present information in a balanced manner, in some cases this has not been possible. Just because a product profile does not specifically mention a feature like remote memory backup and restoral, this does not necessarily mean that this feature is unavailable.

Usually, the differences in product coverage are in matters of detail, so the overall value of this section should not be diminished. Not every vendor, for instance, will list all ring configurations or component requirements for their products. But this would not add substantial value to a general description of the products, at least from a general perspective.

No effort has been made to evaluate the products in terms of absolute or even relative value or utility. Only two products from each vendor are profiled. No pricing information is included. This is not a marketing pitch. This is not the intention of this chapter.

Alcatel Network Systems

Alcatel Network Systems is based in Richardson, Texas. This is the headquarters to Alcatel Telecom's North American organization. A leading supplier of telecommunications products and services, Alcatel provides basic network infrastructure to all major long distance carriers, to virtually every local telephone company in North America, and to private networks, government agencies, and telecommunications systems throughout Latin America and the Pacific Rim.

Alcatel has been around for over 100 years. Alcatel Network Systems is part of Alcatel Alsthom, of Paris, France. In North America, the roots go back to Rockwell International's Commercial Telecommunications Group, the Collins Radio Company, and ITT.

Alcatel 1633 SX Broadband Digital Cross-Connect

The Alcatel 1633 SX B-DCS provides good performance and is the largest SONET-based Broadband Digital Cross-Connect System (B-DCS) available. The 1633 SX is modular and has a growth path while in service that allows the system to start small and expand as needed up to the current capacity of 4,096 DS-3/STS-1 ports. Future planned growth will increase the size of the 1633 SX to over 8,000 ports.

The switching matrix is nonblocking and transparent. This ensures full connectivity and broadcasting, and is designed to ease the transition from asynchronous non-SONET network interfaces to SONET. All of the SONET functionality has been designed into the switch matrix, switching all signals at the STS-1 bit rate. The available input and output interfaces include DS-3 and STS-1 electrical port cards, as well as OC-3 and OC-12 optical ports. A single 1633 SX can support any mixture of these interface types and provide SONET gateway connectivity between asynchronous DS-3 facilities and SONET facilities.

The 1633 SX allows networks to survive catastrophic failures, such as fiber cable cuts. Several advanced features make it easy to reroute critical traffic with minimal impact on end users. Since the 1633 SX is such a valuable survivability and restoration tool, it has also been designed for robust operation. In fact, no single-point failure within the cross-connect is capable of dropping even a single DS-3 circuit. All critical subsystems, as well as the optical drop interfaces, are fully protected. It can be coupled with the more conservative protection schemes for the electrical DS-3 and STS-1 interfaces.

Key features of the 1633 SX include 4,096 equivalent STS-1 capacity (future growth to over 8,000 ports); STS-1, DS-3 and OC-12 ports; and full broadcast capability for video applications. The 1633 SX also supports concatenated connections (STS-3c, STS-12c, STM-1, and STM-4), optical interrack cables, and ATM specific enhancements.

Alcatel 1648 SM

The 1648 SM is a flexible OC-48 SONET device. There are new interfaces for payload selection and grooming at the OC-3 or OC-3c and OC-12 or OC-12c levels. The interfaces are software configurable for a mixture of SONET and European SDH payloads, a nice feature for international links and networks. The mixture of electrical and optical tributaries in the same tributary shelf is virtually unrestricted, along with unlimited grooming capabilities of the working traffic within the tributaries.

Features added in the latest release of the 1648 SM (Release 5.0) include full ADM grooming, enhanced performance monitoring, synchronization and timing features, flexible protection switching options, bidirectional section and line orderwire support, and mixed tributary cabling options. The emphasis on the whole package is flexibility.

The drop and insert capabilities of the 1648 SM are what one would expect. The ADM grooming feature means that any STS-1 from a working OC-48 link can be assigned to any working STS-1 time slot regardless of the tributary type. For example, a single working OC-12 containing an STS-1 can be assigned to an OC-48 and the same can be done in the other direction. There is an interface that can handle either DS-3s or STS-1s as well. An OC-12 on the 1648 SM can be a mixture of STS-1s, STS-3s or STS-3c's, or even STM-1s.

Performance monitoring allows for the retrieval of all BER values. Path and line far end performance are standard features for the OC-3 and OC-12 interfaces. Threshold crossing alerts can be generated for BERs of up to 10^{-11}. Synchronization capabilities now include arrangements for line timing, selection and switching prioritization for incoming network clock sources, and DS-1 framing. Line timing derives clock directly from an incoming OC-n signal and times all outgoing signals from this. Switching prioritization allows for the device to choose between two timing sources with equal quality based on some configured priority scheme. DS-1 framing is used for T-1 tributaries, of course.

Protection switching can be provided by allowing for two full-duplex channels, one active and the other standby, where both channels transmit simultaneously. There are two conditions that can cause a switch to occur. First, a signal failure (usually a fiber break) causes the receive terminal to detect a loss of signal and switch to the standby channel. Second, a signal degradation (typically higher than threshold BER calculations) switch involves the receiver constantly monitoring the duplex channels and selecting the one with the best signal. A switch will occur if the standby channel has a higher quality than the active channel. The BER threshold is service provider selectable over the range from 10^{-5} to 10^{-9}.

Fujitsu Network Transmission Systems

Fujitsu has been around since 1935 in Japan as the Communications Division of Fuji Electric Company. Starting in 1949 Fujitsu was listed on the Tokyo stock exchange and in 1953 they began to make radio communications equipment. In 1954 the company manufactured the first commercial computer made in Japan (the FACOM 100). The combination of communications and computers proved to be a fruitful one, and Fujitsu grew rapidly throughout the 1960s. Overseas offices opened in New York (1967) and California (1968), which grew into Fujitsu America.

The company has a laboratory that researches supercomputing and artificial intelligence applications and semiconductor technology, as well as their communications products. Fujitsu is now a huge multinational organization specializing in communications systems, computers, and other electronic devices. In the field of communications, Fujitsu makes digital switches, satellite communications equipment, and, of course, all manner of SONET fiber-based systems.

Fujitsu FLASH-192

The FLASH-192 is a member of the whole Fujitsu family of SONET products. All are designed to service the high-capacity transport applications of interoffice networks and interexchange carrier networks. The FLASH-192's speed of 10 gigabits per second (10 Gb/s) provides the greatest SONET transmission bandwidth available today.

The emphasis here is on high capacity and economies of scale. A single FLASH-192 system offers significant equipment and fiber optic cable savings for transporting 192 STS-1 circuits when compared with multiple lower-rate transmission systems operating in parallel.

FLEXIBLE CONFIGURATIONS USING SAME BASIC MIX-AND-MATCH SHELVES The FLASH-192 unit consists of basic shelves which can be mixed and matched, along with appropriate plug-in units, to support all of the following SONET network configurations:

- *OC-192 Transport:* The shelf contains high-speed optical line units and system control equipment and integrated OC-48 optical tributary interfaces.

- *OC-48/12 Transport:* The shelf provides a combination of OC-48 and OC-12 optical tributary interfaces.

- *FLM 2400 ADM (as tributary shelf):* The shelf provides OC-12, OC-3, EC-1, and DS-3 tributary interfaces.

- *OC-192 Routing Complex:* This provides fully flexible cross-connect matrix, unrestricted time slot assignment, and STS-1 cross-connect granularity with hairpinning. (Hairpinning is the cross-connecting of tributaries so they go right back out of the SONET device.)

The FLASH-192 has a number of more technical advanced features. For example, the unit supports in-service upgrades of the embedded OC-48 multiplexers while the device is operational. The FLASH 192 supports the Fujitsu FLM 2400 ADM (point-to-point or ring) and other point-to-point OC-48 equipment.

The device has the same basic shelf support in all configurations—Terminal Mode or Drop and Repeat, or as a ring ADM. There are several survivable SONET ring architectures supported, and the FLASH-192 allows full access and routing of individual signals within the entire STS-192 bandwidth, including hairpinning between tributaries running at STS-1, STS-3c, STS-12c, and STS-48c speeds. A full range of network management and integrated operations capabilities are supported, including software downloads of new features and enhancements, and remote memory backup and restoral.

The FLASH-192 has an onboard Stratum 3 timing source and features synchronization status messaging as well as DS-1 BITS primary and secondary clock output and input. The unit is extremely compact, with OC-48 tributaries in one-half of a 7-foot bay, and with OC-12 tributaries in a single 7-foot bay. The total transmission capacity is 129,024 DS-0 equivalent channels. The tributary interfaces run at OC-48 or OC-48c (from the OC-192 transport shelf); OC-48 or OC-48c and OC-12 or OC-12c (from the OC-48/12 transport shelf); and OC-12 or OC-12c, OC-3 or OC-3c, STS-1 and DS3 (from the FLM 2400 ADM).

Fujitsu FLM 2400 ADM

The FLM 2400 ADM is Fujitsu's SONET OC-48 multiplexer. The device is flexible enough to construct a multitude of network configurations, especially when used with the FLASH-192. The FLM 2400 can be used to build point-to-point networks, linear chains, several types of rings, and even interconnected rings. The FLM 2400 ADM can also be used to construct a high-capacity SONET backbone that interworks seamlessly with the FLM 150 ADM, the FLM 600 ADM, or other Fujitsu products to create transport networks that are very versatile.

These networks are easily managed because Fujitsu supports complete OAM&P access from a single location across the entire SONET product line. In order to accommodate all the new broadband services, the FLM 2400 ADM transports concatenated STS-3c and STS-12c payloads.

The FLM 2400 supports several ring configurations. Each configuration has advantages, depending on traffic distribution. With Fujitsu, both are supported from the same shelf, so you can upgrade in-service from one type of ring to another—a very nice capability. Fujitsu's OC-48 rings also give you the flexibility to expand in-service to OC-192 in some configurations.

Other features of the FLM 2400 include software downloads to allow easy feature upgrades using industry-standard protocols. Also, remote memory backup and restoration can be used for quick provisioning of similar nodes or disaster recovery. In some ring configurations, an automatic squelch table update can be used to prevent misconnected traffic during ring failure scenarios.

The FLM 2400 has several possible configurations. The unit can run in Terminal Mode, as a hub, as a linear add/drop, as several ring configurations, and as a matched node. In-service upgrades can change the FLM 2400 from TM to ADM, from linear ADM to ring, from OC-48 terminal to OC-192 terminal, and from OC-48 ring to OC-192 ring.

Interfaces supported include OC-48, DS-3, EC-1, OC-3 or OC-3c, and OC-12 or 12c. In terms of capacity, the unit supports 48 STS-1s in various flexible combinations, time slot assignment, 48 STS-1s, 16 STS-3s or STS-3c's, 4 STS-12s or STS-12c's, or combinations of the preceding.

Lucent Technologies

Lucent Technologies builds on the developments of Bell Laboratories. They have locations and offices or distributors in more than 90 countries and territories around the world, and Bell Laboratories facilities in 17 countries. Their headquarters is in Murray Hill, New Jersey.

Lucent Technologies was created in 1996 as part of AT&T's decision to split into three separate companies. Lucent Technologies combines the systems and technology units that were formerly a part of AT&T with the research and development capabilities of Bell Labs.

Lucent started out as a Fortune 40 company already. They provide both the hardware and software for worldwide telecommunications networks. Lucent also builds local networks, business telephone systems, and consumer telephones that access global networks. They also make the microchips and related components needed to run a host of products and systems, from digital cellular phones and answering machines to advanced communications networks.

Lucent DDM-2000 OC-3/OC-12/OC-48 ADM

The DDM-2000 OC-3 and OC-12 ADM product is Lucent Technologies' main entry in the SONET race. The DDM-2000 also supports OC-48, but this section will focus on the OC-3 and OC-12 models. The OC-3, OC-12, and OC-48 models share a lot of characteristics, and the DDM-2000 for OC-12 is basically an enhanced version of the OC-3 product in terms of interfaces. For example, the maximum number of DS-3s supported in the OC-3 model is 3, while the number supported in the OC-12 model is 12, as one would expect.

In terms of tributary support, the DDM-2000 with OC-3 interfaces supports DS-1, DS-3, STS-1, or OC-3c tributaries, while the OC-12 supports DS-3, STS-1, and OC-3c. Note that the DDM-2000 for OC-12 does not support a direct DS-1 interface. There is also the DDM-2000 Fiber-Reach product that is an OC-1 SONET multiplexer.

Both product models include continuous DS-3 performance monitoring and offer full-line protection switching. This extends to full one-to-one protection in point-to-point configurations. There are downloadable software updates available for each model, and the equipment itself is fully protected by a variety of redundancies.

The DDM-2000 product can be deployed in point-to-point or ring configurations. In a point-to-point deployment, one DDM-2000 would typically reside in a central office, while the other end was housed in a remote terminal location, perhaps attached to a SONET Digital Loop Carrier (DLC, which Lucent also makes, of course). The DDM-2000 can also be used in a hubbing and add/drop configuration, which take full advantage of the ADM capabilities of the product. In a hubbing arrangement, a DDM-2000 at a remote location can extend an OC-3 or OC-12 to other remote locations, all feeding DLCs for hubbing and grooming purposes. But the real power of the DDM-2000 product naturally comes into play when SONET ring configurations are deployed.

The products are surprisingly lightweight and compact, especially the DDM-2000 with OC-3 interfaces. In some cases, the full shelf is not even needed and may be covered with a faceplate to keep out dust and other contaminants.

Lucent SLC-2000 Access System DLC

Lucent makes a number of SONET products. Just in the interest of variety, the second product profiled in this section will be a Digital Loop Carrier (DLC) SONET offering. The product is the SLC-2000 (SLC stands for Subscriber Line Carrier and the 2000 is Lucent's common designation for a lot of their SONET products). A SONET DLC device interfaces residential subscriber lines (local loops) onto a SONET carrier system. Obviously, the DLC is a key piece of equipment for SONET deployment, yet one that gets little attention nor generates much excitement.

The SLC-2000 can handle 768 subscriber ports, more than enough for a good-sized neighborhood, even when second lines are considered. Most DLC products come in twos, one for the central office (known as the COT, or Central Office Terminal), and one for the remote site where the local loops are serviced (known as the RT, or Remote Terminal). In the SLC-2000, the identical plug-in boards are used for both COT and RT—a nice feature. The RT can be handled with remote inventory and diagnostic software—another real plus.

The SLC-2000 can support up to 28 DS-1 lines, all of which can be powered ("wet T-1") and protected by backup links. This is attractive for business districts and applications where DS-1s are common. The SLC-2000 has an integrated OC-3 interface and has a compatibility mode that complies with the Bellcore specifications for DLC SONET products (TR-008 and TR-303).

NEC Transmission

NEC, a Japanese company, known as Nippon Electric Corporation, first established a U.S. presence more than 30 years ago when it opened a sales office in New York in 1963. Since that time, NEC has broadened its operations in the United States by expanding into manufacturing, research, and software development operations, by employing 7,000 people and by establishing extensive marketing, sales, and service networks nationwide with revenues exceeding $5.9 billion.

NEC is a huge, sprawling operation, with a web of U.S. subsidiaries and diverse operations. NEC's SONET products are part of its first U.S. subsidiary, NEC America. This organization oversees a vast range of telecommunications operations from its Melville, New York, headquarters, including manufacturing in Hillsboro, Oregon.

NEC develops, manufactures, and markets a complete line of advanced communications products and software for public and private networks, as well as SONET products. NEC America also oversees software and hardware development operations for various communications technologies in Dallas, Texas; Herndon, Virginia; San Jose, California; and Hillsboro, Oregon.

NEC ITS-2400A ADM

The NEC ITS-2400A is a truly formidable product. The ITS stands for Intelligent Transport System and the 2400 refers to the SONET OC-48 speed (about 2,400 Mbps). The maximum number of DS-3s that can be serviced is 96, and the tributaries can run at DS-3, STS-1, OC-3, and OC-12. However, as is common for most OC-12 ADM devices, there is no direct DS-1 interface. There is a field repeater for the ITS-2400A available as well. All DS-3s are continuously monitored for performance and line protection if one-to-one is desired. The ITS-2400A features downloadable software updates, and the device can be used in SONET ring configurations.

The ITS-2400A can be deployed in five configurations, as a terminal multiplexer, an ADM, and a drop and repeat device, and in two slightly different (two-fiber and four-fiber) ring configurations. The compact design accommodates mixed low-speed (relatively) interfaces in the same shelf and converts DS-3s, STS-1s, OC-3, OC-3c, OC-12, and OC-12c signals into OC-48 signals without loss of shelf capacity.

Network management can be done right down to the port level. Both an X.25 and an Ethernet (IEEE 802.3) interface are provided for this purpose. The unit features a prewiring and preprovisioning option to cut down on deployment time.

The ITS-2400A can add, drop, or distribute and groom up to 192 DS-3s or their equivalents. Up to 30 of the devices can be strung out in the drop and repeat mode. In a ring configuration, up to 16 nodes maximum can be linked together to form the ring. There are various line and equipment redundancies available for protection purposes.

NEC IMT-150 ADM

The second NEC product profiled is the NEC IMT-150 ADM. The IMT stands for Intelligent Multiplexer Transport and the 150 refers to the OC-3 rate (about 150 Mbps). This product has a lot in common with many other SONET OC-3 ADMs. That is, the device supports three DS-3 interfaces and tributaries of DS-1, DS-3, and STS-1. The DS-3s are continuously monitored for performance and line protection if one-to-one is desired. The IMT-150 features downloadable software updates. Naturally, the device can be used in SONET ring configurations.

The IMT-150 can be used wherever OC-3 transmission is needed: interoffice, for loop transport, or even on the customer premises. In this terminal mode, the device acts as a gateway between the world of DS-1 and the world of SONET. The IMT-150 can also serve as a central office terminal for NEC's DLC SONET product, the ISC-303.

All modes, point-to-point, add/drop, and rings are served from a common shelf and can mix low-speed (sub-OC-3) interfaces in the same shelf. The unit is quite compact and has a front access option to make maintenance much easier. Network management can be done right down at the port level, and both an X.25 and an Ethernet (IEEE 802.3) interface are provided for this purpose.

The IMT-150 can add or drop the equivalent of 48 DS-1s in various combinations. Both line and equipment protection are available in redundant configurations.

Nortel Networks

Nortel Networks (Northern Telecom) marked its 100th anniversary in 1995. The company has evolved from being a small Canadian telephone equipment supplier to become a global player in networks. Nortel Networks really began in 1880 when the Bell Telephone Company of Canada was formed, based on Alexander Graham Bell's Canadian patents. The Manufacturing Branch especially prospered and in 1895 was incorporated as a separate company, called Northern Electric and Manufacturing Company Limited.

For the next 50 years, Northern Electric essentially manufactured equipment based on designs and processes licensed from AT&T's Western Electric. It made products primarily for the use of the Bell Telephone Company of Canada, although it did make some consumer electronics products. The Western Electric association endured until 1956, when Western Electric had to terminate its patent and licensing relationship with Northern Electric. Northern Electric created its own research and development facilities known as Bell Northern Research, based in Ottawa, Ontario, and in 1959 established Northern Electric Research and Development Laboratories in Ottawa, Ontario.

Since 1984, Nortel Networks has grown into a respected global supplier of products, with large markets in Canada, the United States, Japan, and all around the world.

Nortel Networks S/DMS TransportNode OC-192

Nortel Networks poses a special challenge when it comes to SONET products. Nortel Networks has been so influential in the whole SONET industry in terms of white papers, configuration information, and just the amount of product information that many of the ideas in this book and much of the terminology itself owes a great deal to the Nortel Networks influence on SONET thought.

In any case, the Nortel Networks S/DMS TransportNode OC-192 product offers up to 20 Gbps of optical tributary access (two OC-192s) in a single frame, along with very sophisticated management. The S/DMS system is a single SONET network element that can groom, segregate, and transport literally millions of phone calls at once. The S/DMS can be configured in protected linear or ring configurations and offers centralized control through a graphical network management system.

The S/DMS TransportNode OC-192 system is a member of a family of products that runs at other SONET speeds as well: OC-3, OC-12, and OC-48. The S/DMS OC-192 interworks fully with its OC-3, OC-12, and OC-48 counterparts, as well as with other S/DMS products. It also provides centralized, end-to-end network management.

The S/DMS terminates up to 20 Gbps of tributary bandwidth. These tributaries can be fed to the transport side or looped back into other tributaries. The S/DMS can be used to provide service with full-survivability form self-healing, interconnected rings using the Matched Nodes feature of the S/DMS. There are linear deployments as well, all with OC-48, OC-12, or OC-3 tributaries with bandwidth management down to the STS-1 level.

The S/DMS TransportNode OC-192 system is a fully modular system in a single 7-foot rack. There are shelves and plug-in circuit packs that make it easy to deploy varying amounts of capacity, allowing for easy expansion. A typical OC-192 version of the S/DMS would include the following elements:

- *Control shelf:* This unit supports the breakers and power units for all battery connections and a number of controller components. The controller units support such functions as timing interfaces, SONET overhead processing, the Orderwire and DCC channels, the hard drives (and even tape drives) for software downloads and backups, various remote access ports, and Ethernet ports.

- *Local craft access panel:* This provides the user interface for local craft personnel. There is an RS-232 port for a VT-100 asynchronous terminal, a handset and headset jack for orderwire communications, and alarm cutoff switches.

- *Main terminal shelf:* This is the heart of the SONET transport system. The shelf is used in all configurations to support OC-192, OC-48, OC-12, and OC-3 interfaces. The top half of the terminal shelf contains the interfaces for the OC-3, OC-12, and OC-48 tributaries. The bottom half contains the OC-192 line interface.

- *Extension shelf (optional):* The optional extension shelf can house additional OC-12 and OC-3 tributaries. It can also support optional OC-192 line interfaces if desired.

- *Environmental control panel:* This houses the fan units which control the temperature of the main terminal shelf. The control shelf above is cooled by convection.

The S/DMS can be deployed in a wide range of configurations. In linear configurations, the S/DMS can support 10 Gbps on the active OC-192 and another 10 Gbps of traffic on the protection link. In rings, the same 20 Gbps of capacity can be deployed and protected by the rings.

Nortel Networks S/DMS AccessNode DLC

Nortel Networks has not neglected the lower end of the SONET product market. The S/DMS product line includes a couple of AccessNode products intended to extend enhanced services to customers in smaller areas with lower customer densities. At the heart of these offerings is the S/DMS AccessNode FST (for Full Services Terminal). The S/DMS FST is a remote SONET DLC (Digital Loop Carrier) device for linking customers to a central switching office with OC-3 or OC-12 links. The FST links by means of DS-1s or OC-1s to a S/DMS Remote Fiber Terminal (RFT). It is the RFT that links to the central office with the OC-3 or OC-12. This section describes the FST device, intended to be deployed downstream of the RFT itself.

The FST can handle residential customers with analog telephone service, business customers with PBXs, or anyone with ISDN lines. Up to 96 lines can be supported on one FTS, and a single RFT can handle up to 7 FSTs, for a total of up to 672 lines for telephony services. Provisioning is done in conformance with Bellcore SONET DLC standards (TR-008 and TR-303).

The FST houses a single copper distribution shelf which can support two 48-line "drawers" for a total of 96 lines. The unit also can contain 2 optional OC-1 multiplexers for the transport services, or up to 6 DS-1s in other configurations.

The feeder DS-1s that lead from the FST to the RFT device can also be delivered on OC-1 fiber, a nice flexibility feature. The FST is a nice fit for entry-level, in-building business applications. The device is a small, compact, wall-mounted package. There is an optional battery power supply backup, typically mounted below the FST or in an equipment room or closet on the customer premises. The FST can also be mounted in a standard rack or frame.

Figure 11.1 shows the S/DMS AccessNode FST in a typical configuration.

It is perhaps fitting that this survey of SONET products, which began with and has tended to emphasize OC-192 ring products, includes a small SONET access device. It is worth remembering that OC-192 and gigabit links are fine, but the success of SONET is also dependent on backward-compatibility and just how easy it is for everyone to take advantage of SONET benefits.

Tellabs

Tellabs is headquartered in Lisle, Illinois, and began in 1975. In the first year, the company ended up with a total of 20 employees and sales well under a half-million dollars. Since then, Tellabs has carved out a position in the digital cross-connection field that has resisted all assaults from larger and older companies with telecommunications roots going back more than 100 years. Originally specializing in analog products, Tellabs is now a $2 billion global company. Tellabs recently began to explore more of the SONET/SDH market than just digital cross-connects, but this remains the core business.

Tellabs acquired two new companies towrd the end of 1999. Alcatel's Co Communications businesses in Europe became Tellabs Denmark, and in the United States Tellabs acquired NetCore Systems.

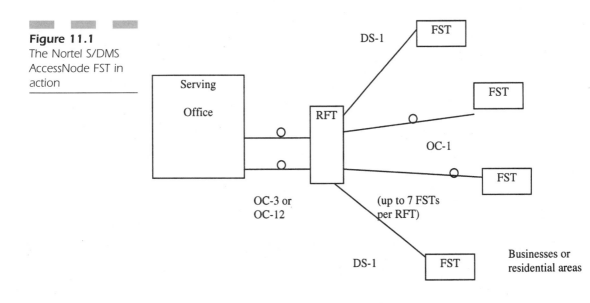

Figure 11.1
The Nortel S/DMS
AccessNode FST in
action

Tellabs TITAN 5500 DCS

The TITAN 5500 family of digital cross-connects is intended for all types of telecommunications networks, including local telephone service, long-distance networks, wireless networks, private networks, the emerging networks in corporate America and networks around the world. The TITAN 5500 family has many applications in a wide variety of network types and has very good migration capabilities. Telecommunications competition makes it imperative for service providers to upgrade networks to cut costs, improve quality and guard against equipment obsolescence. TITAN systems satisfy the demand for new wideband and broadband services. TITAN systems allow remote and centralized performance monitoring and control.

The TITAN 5500 system replaces multiple pieces of colocated equipment such as digital patch panels (DSX), M13 add/drop multiplexers (ADM), fiber-optic terminals (FOTs), and the manual wiring techniques previously used to cross-connect and manage DS-3, DS-1, and OC-n transmission facilities. A TITAN 5500 system can grow from a small TITAN 5500S system with 32 ports to 1,024 DS-3/STS-1 equivalent ports and beyond in increments of 32 DS-3/STS-1 equivalents. The growth path is accomplished without manual reentry of database information and without rendering current equipment investments obsolete. With a single architecture that can grow far beyond 1,024 ports, large hub locations no longer have to colocate, or tandem smaller DCSs that have reached maximum capacities.

The maximum configuration of the TITAN 5500 is quite impressive. It can cross-connect a maximum of 2,048 DS-3s, with 57,344 DS-1s representing 1,376,256 DS-0s. However, a separate TITAN device, the TITAN 532L, is required to cross-connect at the DS-0 level.

Tellabs TITAN 4500GS

It is fitting to end this product sampler with the TITAN 4500 Global Services Delivery System because this product demonstrates a growing trend in SONET product offerings. SONET, after all, is just the North American version of SDH. So SDH is the true global market, not SONET. So many vendors of SONET equipment also offer a more or less complete line of SDH. Tellabs has emphasized the growing problem of SONET to/from SDH networking and addressed the issue well with this product.

The TITAN 4500GS is intended for both SONET and SDH networks. The TITAN 4500GS system is a good fit for the new global service providers who own and operate the entire network: two distinct national networks and the international connection between them. The TITAN 4500GS systems sit at the service delivery endpoints, so these global network operators can provision and manage services on an end-to-end basis across the SONET and SDH domains. Nothing else is needed. Of course, the TITAN 4500GS system can also be used by service providers having only a national network.

The nice thing about the TITAN 4500GS is that it does not support only VT1.5- and STS-1-level cross-connections. It additionally supports VT2s to handle E-1 traffic and also handles E-3 traffic within the SONET payload. So when the TITAN 4500GS is configured as an SDH network element, it can support and carry DS-1 traffic in its most obvious form within the SDH payload, unlike some older SDH access equipment, which handled this task awkwardly. Not only that, but support for VT1.5, VT2, and STS-1 payloads within a single SONET interface is simultaneous. All in all, the TITAN 4500GS can be used to simplify the payload transformation between SDH and SONET. This does not imply that other products from other vendors cannot do much the same things, of course.

Sample Product Configurations

It is one thing to outline product features and interfaces. But it is quite another to see just how all of the vendors' offerings might fit together to solve a network problems with SONET and how typical SONET deployments might look on a smaller or even a larger scale. By small scale, a fairly restricted area is meant, as opposed to multicity or even nationwide SONET network deployments. Large scale includes rings or links that might span the country.

Each of the vendors' products is shown in a typical SONET deployment. It should be noted that this does not imply that the product is necessarily limited to the configurations shown, especially when it comes to SONET rings. An example showing a small SONET network does not rule out the use of the product for large SONET networks, naturally. The aim is to be informative with respect to the various SONET network elements, rather than be exceedingly precise about all of a product's features and possible scenarios.

Alcatel Network Systems

Alcatel makes a full range of SONET-based products, including wideband and broadband DCS products and SONET ring devices that operate at a range of speeds. Figure 11.2 shows how several of these products can be used to implement SONET ring configurations.

In Figure 11.2, the Alcatel 1603 Enterprise Node (a form of SONET Terminal Multiplexer) concentrates and interfaces with customers' T-carrier, OC-3, Ethernet, Token Ring, ISDN, frame relay, SMDS, or even ATM voice and data equipment. This traffic is carried back to the serving office by an OC-3 or OC-12 ring with 1603 SM (for OC-3) or 1612 SM (for OC-12) SONET ADMs. Once at the serving office, the signals can be cross-connected with the 1631 SX wideband crossconnect network ele-

Figure 11.2

Example SONET network using Alcatel products

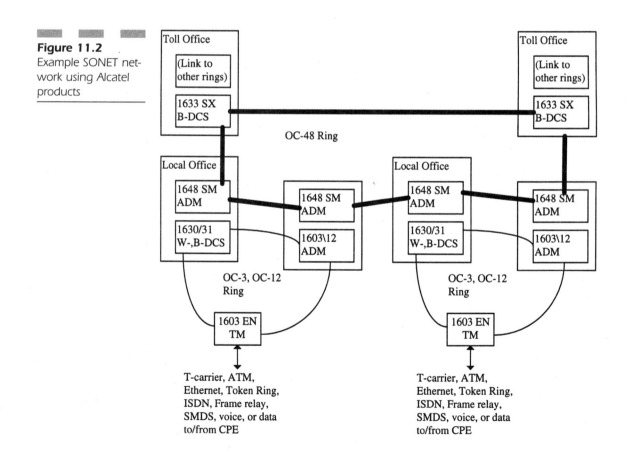

ment. Other Alcatel cross-connect products can be used here as well. Traffic can be switched—especially voice traffic—or transferred to a higher-speed SONET ring for longer hauls. Typically, 1648 SM would be used for this purpose, running at the OC-48 rate. Further signal processing can be done at this level with the 1633 SX broadband DCS.

Although not shown in Figure 11.2, it is possible to feed an OC-192 ring for covering even larger areas. Also, the Alcatel product line includes a network management system for all of the components.

Fujitsu Network Transmission Systems

Fujitsu makes a very respected line of SONET TM and ADM products for implementing all types of SONET rings. The line culminates with the FLASH-192 product, which is both powerful and versatile. Figure 11.3 shows how the FLASH-192 product can be used to implement a very high speed SONET ring. Naturally, there are more Fujitsu SONET products in the whole family, but the FLASH-192 can incorporate the functionality of many of them as optional modules.

In Figure 11.3, the FLASH-192 network elements, variously equipped, both handle the SONET ring links running at OC-192 and at the same time cross-connect down to the OC-1 (STS-1) level. Network management is a key part of the product as well.

Figure 11.3
The Fujitsu FLASH-192

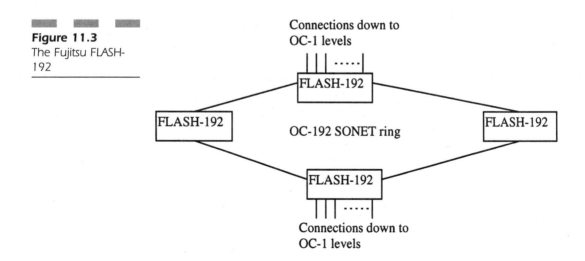

Lucent Technologies

Lucent (AT&T) makes a full line of SONET-based products. In the interests of variety, a rather simple arrangement of Lucent equipment is shown in Figure 11.4. Of course, the DDM-2000 can also be used in a wide range of ring configurations.

In Figure 11.4, the Lucent DDM-2000 is shown in a typical metropolitan area service configuration. In this point-to-point configuration, the OC-3 ADM is located at a central office location. The DDM-2000 has DS-1 links to a digital cross-connect—in this case, a DACS IV-2000—and a digital central office switch. Another OC-3 ADM is located at a large customer location, perhaps an office building. Two sets of fibers are employed, a working pair and a protection pair. In the building, the ADM connects to a remote terminal linked to telephones, again with a DS-1 interface. To service other floors, fiber cables can be installed to other remote terminal units through another DS-1 interface. The OC-3 links can deliver other than voice services, of course. Video is only one such possible application.

NEC Transmission

NEC makes a range of SONET products well-suited for ring configurations. Figure 11.5 shows a modest but complete SONET ring application using NEC products.

In the figure, the ITS-600s operating in terminal mode (the ITS-600 can also be configured as an ADM) gather traffic from user devices at customer locations. These can be linked by OC-3 or OC-12 links to the

Figure 11.4
The Lucent DDM-2000

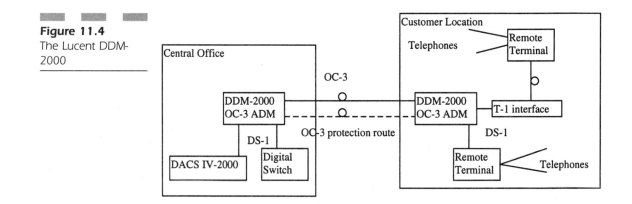

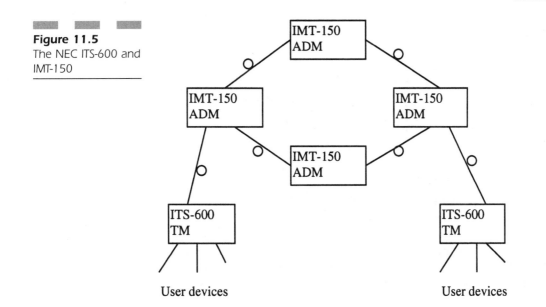

Figure 11.5
The NEC ITS-600 and IMT-150

ITM-150 ADMs at the serving offices. The ITM-150s can be linked together to form SONET rings running at OC-12, OC-48, or even OC-192 rates.

Nortel Networks

Nortel Networks' prominent position when it comes to SONET need not be commented on further. Figure 11.6 shows a SONET ring implemented on the Nortel Networks S/DMS products.

In the figure, the S/DMS products, which run at a range of SONET speeds, are configured in an OC-12 or OC-48 SONET ring. Each can deliver up to 48 DS-3s or the equivalent (on an OC-48 ring) to cross-connect or digital switches. Other configurations are also possible, naturally, and the traffic capacity can be handed off to other rings or other SONET terminal multiplexer arrangements.

Tellabs

Tellabs has staked out its claim in two major SONET categories: easily upgradeable digital cross-connects and multinational transport networks with SONET on one side and SDH on the other. Tellabs has also begun

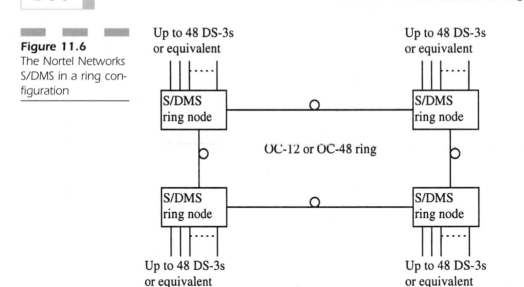

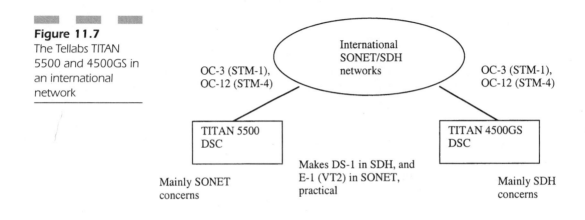

to explore the marketplace for new offerings in the SONET ADM arena and ATM switching category. This section will emphasize Tellabs' traditional role, however.

Figure 11.7 shows how the TITAN 5500 DCS and the TITAN 4500GS can be used to create a multinational SONET/SDH environment. While there are other Tellabs products that could be used in this figure as well, the figure is intended to showcase the role of the TITAN 5500 and 4500GS.

SDH and DWDM Equipment Providers

This book is about SONET and this chapter has been about SONET equipment providers. However, the recent trend in the optical products industry has been to emphasize SDH and DWDM products, often at the expense of the older, more mature SONET product line. This is only natural: vendors are always keeping an eye on the future. The global Internet and Web have made it clear that the future belongs to international networks and higher-bandwidth solutions. This translates to SDH and DWDM. It is often difficult to navigate the SONET vendors' Web sites and isolate the SONET equipment details, since many of the newer offerings are intimately tied to SDH and DWDM.

A full survey of SDH and DWDM equipment to the depth provided on SONET equipment would not only take many pages, but be far beyond the scope of this book. Nevertheless, it seems only fitting to close this chapter with a look at the SDH and DWDM equipment providers.

First of all, it has already been pointed out that many of the vendors listed as SONET equipment providers also make a lot of gear for SDH and DWDM. But the top players in the SONET field are not necessarily the same in SDH and especially in DWDM. New markets always attract smaller players who seek to aggressively carve out their own territory within the newer technology niche.

The following is a list of the top equipment providers in the SDH and DWDM arenas, in alphabetical order. Again, the list is not exhaustive, but is representative of the market leaders. The Web site for each vendor is given. Interested readers are urged to contact the companies directly for the latest product information.

SDH Equipment Providers

Alcatel (www.ans.alcatel.com)

Fujitsu (www.fnc.fujitsu.com)

Hitachi (www.hitachi.com)

Lucent (www.lucent-optical.com)

Marconi (www.marconi.com)

NEC (www.necpng.com)

Nortel Networks (www.nortelnetworks.com)

Siemens (www.siemens.com)

The SDH market appears to be a three-way race between Alcatel, Siemens, and Lucent. These companies combined had about 65% of the SDH market at the end of 1999.

DWDM Equipment Providers

Alcatel (www.ans.alcatel.com)

Ciena (www.ciena.com)

Lucent (www.lucent-optical.com)

NEC (www.necpng.com)

Nortel Networks (www.nortelnetworks.com)

Pirelli (pirelli2000.pirelli.com)

The DWDM market also appears to be a three-way race, this time between Lucent, Nortel Networks, and Ciena (in 1998, an intended merger between Ciena and Tellabs came to nought). These three companies combined had about 80% of the DWDM market at the end of 1999.

PART

4

SONET
Advantages

P revious parts of this book have mentioned the distinct advantages SONET has over other forms of transmission networking. However, the discussions to this point have been scattered throughout earlier sections on the SONET technology and compared SONET almost detail by detail with alternatives such as T-carrier. This section is an attempt to be more systematic about the advantages that SONET offers both customers and service providers, through a thorough examination of SONET as more than just a way to overcome the limitations of the existing digital hierarchy. This section is on a higher level than the bits and bytes of the technology itself.

In fact, this section goes beyond a simple recitation of SONET benefits for customers and service providers. Many of the benefits of SONET are in the area of more efficient and comprehensive network management. Service providers are concerned with much more than simple network management, naturally. Networks must be provided, configured, and run on an ongoing basis as well as managed. In the United States, service providers refer to the daily tasks involved in running a network of transmission links as Operations. The record-keeping tasks that involve things like billing the customer are known as Administration. The repair and preventive maintenance done to keep the system up and running are referred to as Maintenance. Finally, the tasks required to configure and setup the network initially are known as Provisioning. Usually, these tasks are gathered under the umbrella acronym of OAM&P (i.e., Operations, Administration, Maintenance, and Provisioning).

The OAM&P tasks are often called OAM (for Operations, Administration, and Maintenance, or just for Operations and Maintenance) in various standards documents. This does not imply that OAM involves no provisioning, or administration, just that the term OAM&P is more explicit.

The investigation into SONET OAM&P leads into a fuller discussion of SONET rings than given before. More details on the actual equipment and operations used in the forming and running of SONET rings are presented.

This part of the book is divided into three chapters. Chapter 12 looks at the overall advantages that SONET brings to customers and carriers alike. The chapter points out that although SONET was essentially invented to overcome the limitations of existing digital hierarchies, especially in the United States, SONET has gone far beyond this simple goal and possesses many other advantages beyond the obvious technological ones.

Chapter 13 explores one of these major SONET advantages in detail: the distinctive OAM&P procedures and protocols built in to the SONET overhead. The amount of overhead defined in SONET has been a perceived liability in the past, but newer concerns for reliability have made SONET OAM&P a crucial part of modern networks. This chapter also explores the handling of alarm conditions in the SONET environment. Alarms signal the presence of error conditions and, thus, are an important part of SONET OAM&P.

Chapter 14 in this part of the book details methods for addressing this increased concern for reliability in networks. This chapter examines one of the key features of SONET: the deployment of formerly point-to-point transmission systems in ring architectures. Not all SONET rings share a common architecture or method of protection switching. This chapter explores the architectures of 2-fiber and 4-fiber ring configurations and compares the characteristics of each.

12
Customer and Carrier Advantages

Some technologies mainly benefit the users and customers that employ it. The technology may require a great deal of expense on the part of the service provider to employ, with marginal financial benefit. In some cases, there may be only the intangible benefits of increased user satisfaction, but that may hopefully lead to increased revenues in the long run.

One is tempted to put ISDN in this category. Upgrading analog copper local loops and central office switches to support ISDN proved to be enormously expensive, and yet customers were unwillingly to pay much more for ISDN services which, in most cases, still amounted to making telephone calls. The renewed interest in ISDN for Internet access has not changed the basic economic perception of ISDN on the part of carriers. It may still be seen as an expensive upgrade that does not substantially increase revenues in proportion to its expense. As a result, ISDN is still unavailable in many areas.

Other technologies mainly benefit the service providers and carriers that deploy it. The benefit of the new technology may be invisible to users and customers. If users and customers were to benefit at all, it would be indirect and typically in the form of vague assurances of "improved service" without more specifics.

One is tempted to put things like T-carrier in this latter category. The first deployments of T-carrier increased the trunk capacity of service providers and carriers, but had little to no direct impact on or benefit to users. Carriers would often imply that the new digital trunks made phone calls sound better, but with at least two noisy analog local loops on both ends, these claims lacked the conviction they might otherwise have had if the local loops had been digitized as well. This change would have to wait for ISDN.

Happily, SONET falls into neither of these categories. SONET benefits customers and carriers alike. This has virtually assured that SONET be both widely and quickly deployed and popular with customers as well. This has proved to be the case in all areas where SONET is available.

Naturally, the benefits of SONET fall into a number of categories. Some involve cost reductions in deploying network equipment to support a given mix of services, others involve the possibilities of enhanced service providers, while still others involve a direct reduction of customer network costs. Various features of SONET contribute to each of these benefits.

Key Benefits of SONET

The key benefits of SONET for customers and service providers are given in tabular form in Table 12.1. The table reflects a lot of information compiled by Nortel Networks, but differs in many details from similar tables. The table is organized into five main features of SONET, and the benefits provided in each of three categories are given a "YES" entry when the feature provides that particular benefit.

In Table 12.1, the key benefits are categorized into five broad areas or sets of features. Each of these major features produces a benefit, either in terms of reduced capital expenditures for service providers building SONET networks, reduced costs (service pricing) to customers using the SONET network, or enhanced revenue streams for service providers. This new source of revenue usually comes from the new data services SONET can support, but not always. Sometimes a feature benefits service providers and users in all three categories.

TABLE 12.1

Key Benefits of SONET

Feature	Reduced Capital Expense for Network Builders	Customer Cost Benefit	Service Provider Revenue Benefit
1. In Multi-point Configurations:	YES		
A) Grooming			
B) Reduced back-to-back muxing	YES		
C) Reduced DSX panels and cabling	YES		
2. Enhanced OAM&P:			
A) OAM&P integration		YES	
B) Enhanced monitoring capabilities		YES	YES
3. Integrated operations and monitoring		YES	
4. New service offerings			YES
5. Optical interface ("Mid-span Meet")	YES	YES	YES

The five main features listed in the table include the following:

1. *Multipoint Configurations.* SONET, unlike many T-carrier networks with strictly point-to-point links, is frequently deployed in *multipoint configurations.* This means that several sources of SONET bits can be combined or distributed without terminating the digital stream to recover and process the constituent signals, such as the basic 64 Kbps channel structure. This combination of multiplexing and cross-connecting in the same device is known as *hubbing.* Many SONET ADM devices are deployed at hub locations to perform this essential SONET task. The SONET device that combines these functions is called the *hub* and the whole process is generally known as *grooming.* At the hub, traffic is combined or distributed to "spurs." Grooming is not unique to SONET equipment. Much the same was and is done with T-carrier devices; however, SONET grooming requires much less equipment, reduces the need for linking multiplexers and digital cross-connects back-to-back as in T-carrier, and reduces the need for cabling between T-carrier terminations and patch panels (i.e., DSX panels). All three aspects of SONET multipoint configurations help to reduce the capital expenditures needed to build and deploy SONET network links. Grooming can concentrate traffic and service more customers with fewer links than would otherwise be needed without grooming. The reduced need for back-to-back multiplexers indicates that one SONET device can perform the functions of several T-carrier pieces of equipment, saving even more capital. The reduced need for cabling and DSX cross-connection panels not only saves money with regard to this equipment, but also reduces the number of personnel needed to run the network and makes more efficient use of those who have responsibility for the day-to-day operation of the network.

2. *Enhanced OAM&P.* SONET not only greatly enhances the OAM&P capabilities available in digital systems like T-carrier, it also tightly integrates them into all SONET network elements. The integrated and enhanced OAM&P overhead and procedures built-in to all SONET equipment helps in two ways. First, the integration means that service providers can offer SONET at lower cost to customers. Second, the enhanced performance monitoring in terms of errors can not only lower customer costs, but also can offer a means for service providers to earn more revenue by making more efficient use of SONET network resources. This is possible because the

integrated SONET OAM&P can help service more customers with the same physical resources.

3. *Concurrent Operations and Monitoring.* Operations is the process of configuring and running a network on a daily basis. Monitoring is needed to make sure that everything is running correctly. Ideally, monitoring should result in a network that detects and corrects problems even before users are aware of them. Without close monitoring, a network operations center can resemble a hospital emergency room, with stressed out network doctors scurrying around to treat the most serious cases, while many painful but less life-threatening situations go untreated. Because the OAM&P capabilities in SONET are enhanced and integrated, service providers can consolidate their operations centers and minimize the equipment needed to support a community of users. Although not necessarily a true cost reduction in terms of capital, this consolidation will help to lower SONET users' costs.

4. *New Service Offerings.* A SONET network's huge amounts of bandwidth in the gigabit ranges, can support new services that were unheard of in the T-carrier world. Business services will more than likely lead the way. Businesses needing bandwidth to support video applications, or even modest speed LAN interconnections at 4, 10, or 16 Mbps are often at a loss for affordable solutions to their networking problems. With SONET, however, routine support of video, 100 Mbps LAN interconnection, color faxing, and the like, will become commonplace. Customers will gladly pay for reliable support for these needed services. Many of these new services will be tied to asynchronous transfer mode (ATM) networks and B-ISDN deployments. Therefore, the customer is not directly buying SONET, but an ATM or B-ISDN service delivered on a SONET network. However, the continued reliance of customers on private network solutions will enable a service provider to immediately increase revenues as a direct result of SONET.

5. *Optical Interface ("Mid-Span Meet").* Finally, the last benefit is one that is frequently considered the biggest and most important, but is really less of an issue as time goes on. In the near future, it will seem as unusual to think that fast-fiber systems were once mostly proprietary as to contemplate the fact that routers from different vendors would once not work together at all. But the mid-span meet is still important. The fact that SONET is a standard which

defines a standard optical interface means two things. First, it is possible to deploy SONET in a multivendor environment with the transmitter from one vendor and the receiver from another. Second, it is even possible to have two different service providers operating the two ends of a SONET link. The responsibility is divided in "half" at the middle of the link, hence the term "mid-span meet." In the current environment of deregulation in both the long-distance and local areas, this service provider's standard optical interface is an important one. The existence of a SONET standard optical interface provides benefits in all three areas defined in Table 12.1. There is less outlay of capital because equipment procurement can easily and confidently be put out to a competitive bid. Customers can do the same with customer premises equipment. Finally, service providers can potentially earn more revenue because competition keeps equipment prices down for everyone.

This list of SONET benefits could easily be extended in both dimensions. The ones chosen are those deemed to have the greatest impact. The remainder of this chapter is a more detailed look at each of the categories.

SONET Multipoint Grooming

The practice of *grooming* refers to the consolidating (i.e., combining) or segregating (i.e., distributing) of traffic to make the most efficient use of the existing facilities. With grooming, traffic to and from customer sites is carried on lower speed SONET links, such as OC-3, and can easily be serviced with a higher speed SONET link. In some cases, the link may even operate at the same speed as the links to the customer site.

Although grooming is not unique to SONET, an argument could be made that *efficient* grooming is unique to SONET. Grooming was certainly done in T-carrier networks. In a simple form, T-1s carried on two T-3s from several customer sites could be "groomed" onto a single T-3, assuming the total number of active T-1s was not greater than 28. Without grooming in this situation, two T-3s would be required all the way to a cross-connect location or other processing site. However, grooming was often inefficient and difficult to do in T-carrier, given the limited range of speeds available and amount of T-carrier equipment needed to groom.

SONET, with its simple and step-like series of defined speeds, always offers numerous opportunities to groom as signals make their way through the service provider's network. Figure 12.1 shows the general process of grooming in a SONET multipoint link configuration.

In Figure 12.1, network traffic that originates at Node A must travel to Node D, but so must the traffic from Node B. The easiest way to accomplish this is to establish two direct point-to-point links from Node A to Node D and also from Node B to Node D. This was typically how the situation was handled on T-carrier networks. With SONET, however, instead of connecting two links from each site (Node A and Node B) to Node D directly, it is easy to establish an intermediate hub site (Node C in the figure) to handle both traffic streams.

Before SONET, T-carrier could also accomplish this same thing with extensive multiplexing equipment at Node C. This usually meant at least one pair of T-carrier multiplexers at the appropriate speeds, manual DSX patch panels, and a digital cross-connect system (DCS). This substantial equipment requirement made it much more difficult to establish a site with adequate floor space, power supplies, and so on for T-carrier hub sites. With T-1s and T-3s to deal with, it was not unusual to backhaul a large amount of traffic to already established sites for hubbing and grooming purposes. The term, *backhaul* indicates that a service provider must take links from a local user to a remote site to provide a service or accomplish what is desired. For example, with T-carrier links, much of the traffic from Node D to Node A may actually have been routed to

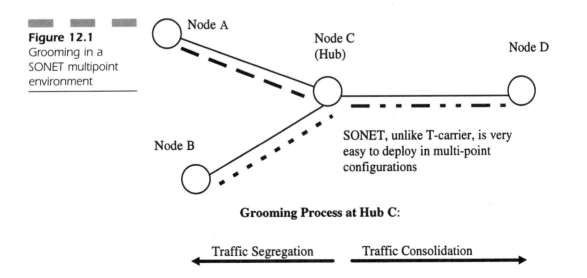

Figure 12.1
Grooming in a
SONET multipoint
environment

Node A

Node C
(Hub)

Node D

Node B

SONET, unlike T-carrier, is very
easy to deploy in multi-point
configurations

Grooming Process at Hub C:

← Traffic Segregation   Traffic Consolidation →

Node B, where the required grooming equipment had been located previously for other purposes. Node C—if it existed in the T-carrier network and played any role—was just another link in the chain in spite of its central geographical location. No one wanted two hubbing sites and their associated expenses.

When it comes to grooming or hubbing, SONET can combine or distribute traffic at the STS-1 or VT 1.5 level, right down to the individual DS-0, with ease. SONET grooming is not limited to mere manipulation of anonymous bit streams. Grooming in SONET can involve *service* combination or distribution as well. For example, video-conferencing links, data, or even voice channels can simply and effectively be handled to or from a number of sites with a SONET hub.

Thus, with SONET, grooming is much more efficient. In fact, SONET gear is so compact that in the future there may be no need for a "central office" or other location for switching or grooming! A *central* office was needed to provide a controlled environment for sensitive and bulky switching equipment. Why switch every 1,000 lines with one switch when a central location with one switch could handle 10,000 lines?

SONET microprocessor-based equipment is robust and inexpensive, especially in comparison to a central office. Why not move as much of the handling of signals as close to the customer as possible? Although this may make a lot of sense, there has been little incentive—financial or otherwise—for service providers, especially local exchange carriers, to decentralize the central office. However, this may change with local service deregulation.

Reduced Back-to-Back Multiplexing

The integration of SONET equipment into compact, shelf-mounted, microprocessor-driven units has led to a revolution in the way that signals are combined onto fiber links. No need exists any longer for an array of multiplexers to build up high-speed bit streams to funnel onto fiber links. SONET marks the culmination in a long evolution of multiplexing equipment. With SONET, the use of such integrated devices reduces the need for back-to-back multiplexing equipment needed in T-carrier to build up fiber optic speeds to access individual DS-1 or DS-0 streams. This reduces the equipment cost dramatically.

The evolution to SONET has been a long time coming. Multiplexing environments that combine signals onto fiber links before SONET are usually known as *fiber terminals*. The evolution of pre-SONET fiber terminals to SONET is shown in Figure 12.2.

In Figure 12.2, fiber optic terminal evolution is shown in three stages, culminating with SONET. Initially, multistage multiplexing devices, each with its own footprint and power supply, were needed to combine the standard number of 28 DS-1s to the DS-3 level. Then, several DS-3 electrical signals, the exact number varying from vendor to vendor, were combined and converted into a proprietary optical format.

The next stage in the evolutionary process was to combine the multiplexing into a single device with a smaller footprint and reduced power requirements. However, the output was still a proprietary fiber optical signal format. Moreover, great care had to be exercised when routing the links in order to avoid multiplexing and demultiplexing the fiber signal too many times in too many places for cross-connection purposes because the total electronics package needed, and their associated capital costs, had to be installed wherever a DS-1 signal was processed. The net result was a fair amount of backhauling to the adequately provisioned sites.

Finally, on the far right portion of the figure, the SONET arrangement is shown. With SONET, not only is the equipment footprint and

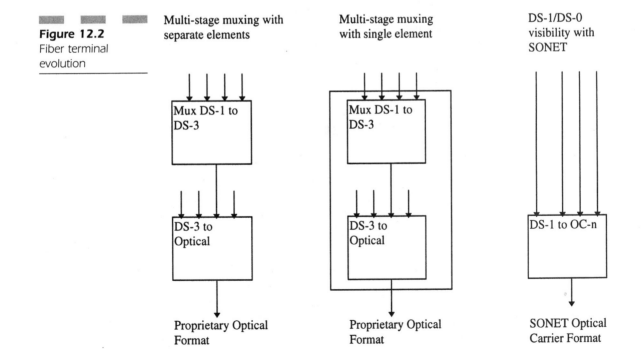

Figure 12.2
Fiber terminal
evolution

Multi-stage muxing with separate elements

Mux DS-1 to DS-3

DS-3 to Optical

Proprietary Optical Format

Multi-stage muxing with single element

Mux DS-1 to DS-3

DS-3 to Optical

Proprietary Optical Format

DS-1/DS-0 visibility with SONET

DS-1 to OC-n

SONET Optical Carrier Format

power needed quite small, but the whole process is done at once with dedicated microprocessors and produces a standard SONET optical signal, not a proprietary one. This prevents vendor "lock-in" and means that SONET signals can be groomed practically anywhere, not just at large switching nodes, typically central offices. SONET's synchronization features also mean that only the required STS or VT signals (with the needed DS-1s or DS-0s) need to be demultiplexed, not the entire bit stream, as in T-carrier.

It is this development of the SONET fiber terminal that leads directly to the reduced need for back-to-back multiplexing in SONET. A roomful of equipment has become a small unit that can be deployed in many more types of locations.

T-carrier systems are characterized and even dominated by the presence of back-to-back terminal and multiplexing equipment because the "asynchronous" or plesiochronous T-carrier hierarchy is inefficient and less cost-effective for other than simple point-to-point configurations.

The contrast is stark. Figure 12.3 compares how DS-1s are accessed in a fiber T-carrier environment and a SONET environment.

The upper part of Figure 12.3 represents a T-carrier fiber environment and the lower part represents how SONET accomplishes the same task. In T-carrier, a fiber optic termination system (FOTS) is needed to convert the optical signal to an electrical one. The FOTS is naturally a separate piece of equipment. The recovered DS-3 signal (known as a "DSX-3" signal at this point because it flows with simple electrical cod-

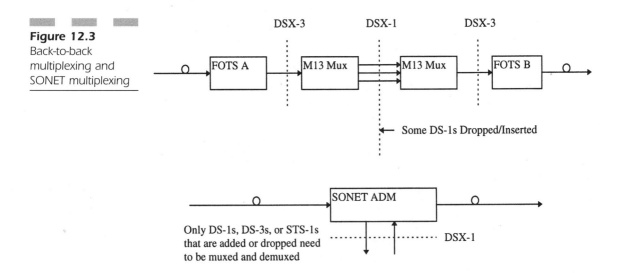

Figure 12.3
Back-to-back
multiplexing and
SONET multiplexing

ing on short cables) must next be demultiplexed to access the 28 DS-1s inside. These follow a signal format known as DSX-1, similar to the DSX-3 format in the sense that the coding is simple and distance quite limited.

Once recovered, the required DS-1s are dropped and/or inserted. These DSX-1s are typically taken to a DCS or some other switching device, which is shown in Figure 12.3. The DCS is usually a separate device. The recovered and processed signals must be recombined into a single DSX-3 and reconverted to an optical signal at the outgoing FOTS.

Contrast this with SONET. With SONET, all of the above is accomplished in one, simple, cost-effective package. A SONET add/drop multiplexer (ADM) can convert the optical signal, access the bit stream, drop and insert the required DS-1s, DS-3s, or even STS-1s, and recreate the optical signal. Nothing could be simpler.

In fairness to T-carrier, it should be noted that the SONET ADM does not actually cross-connect the required signals. In both cases, a DCS or some other switching device must be used when cross-connecting is needed. SONET would require the use of a wideband or broadband DCS (W-DCS or B-DCS). The DCS could still be a card on a SONET shelf, of course. In neither the T-carrier nor the SONET portion of the figure is the DCS device shown.

Reduction of Cabling and DSX Panels

In the FOTS-dominated world of pre-SONET T-carrier multiplexing and demultiplexing, a huge amount of equipment is needed to transport a signal from one end of a network to another. Even in extremely simple cases where an electronic DCS or switch is not needed, the expense can be considerable because when a number of DS-3s simply need to be combined or distributed to or from several sites (the process known as grooming), an array of spaghetti-like coaxial cables "nailed up" on DSX-3 cross-connect panels is needed.

The panels are expensive, in terms of the panels, bays, cables, and labor needed to install them. Congested cable racks not only take up a lot of floor space, they all pose a mild hazard to employees through job-related injuries (e.g., cable trips). Obviously, there is a financial incentive to dispense with the need for this labor-intensive environment when possible.

Fortunately, SONET provides the answer. Figure 12.4 contrasts the use of DSX panels for grooming with the use of a SONET hub for the same purpose. The hub is just another form of SONET ADM; therefore, no special SONET network element needs to be deployed for this purpose.

Physical cross-connect panels, such as those used in T-carrier grooming, are undesirable for a number of reasons unrelated to expense. First, the cables are exposed to environmental damage due to dust, moisture, and other hazards. Second, considerable harm can come to them through accidents (e.g., careless cleaning personnel) or even sabotage (e.g., disgruntled employees). No technical expertise is needed to cause damage by ripping cables off of a piece of plywood, whether with a misguided broom or well-placed screwdriver. All in all, DSX panels are bulky, expensive, and difficult to maintain. DSX panels are an inelegant solution to a common problem.

Moreover, the most common high level of the T-carrier hierarchy was DS-3. Grooming usually consisted of trying to groom underutilized DS-3s—in terms of the number of "live" customer DS-1s they carried—with traffic from more highly utilized DS-3s. In other words, even though these fiber links may operate at 405 Mbps or 565 Mbps, the difference was in terms of the number of DS-3s (groups of 28 DS-1s) they carried, not in the speed of the DS-3s themselves. This did not make much sense, but it *had* to work this way. There was really no place to go above DS-3!

SONET hubs can combine the ADM capability with the DCS capability, all in the same box, which offers a cost-effective method of doing

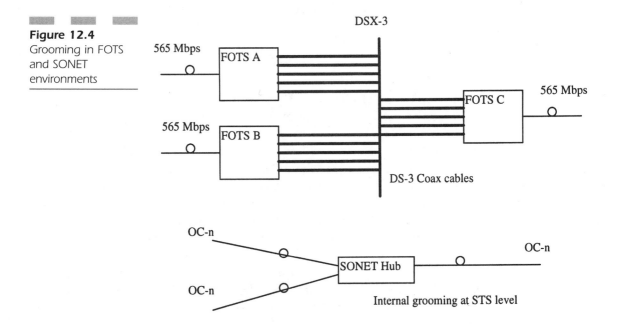

Figure 12.4
Grooming in FOTS and SONET environments

grooming *internally*, within the hub itself. This reduces the need for back-to-back equipment, except when needed to interface with existing T-carrier equipment. Of course, a SONET network element used for this purpose can still be a pure ADM, with a separate (usually preinstalled for T-carrier) DCS used to cross-connect signals at the hub site.

The availability of various OC levels also makes the whole grooming process more efficient. Instead of trying to squeeze a number of DS-1s from various DS-3s into an outgoing DS-3 or group of DS-3s, the signals from various OC-3 (for instance) can be groomed onto a single OC-12 or even OC-48, depending on the amount of the traffic at the hub or the number of sites served.

Enhanced OAM&P

OAM&P (sometimes just OAM especially to the ITU) is the art of controlling and managing an extended physical network consisting of many links in diverse geographical locations. OAM&P procedures were added haphazardly in the T-carrier hierarchy, due to a lack of any way to integrate such procedures with the link itself. The end result was a system of OAM&P that had modest capabilities and many limitations.

SONET integrates and enhances OAM&P, mostly through the inclusion of dedicated overhead bytes reserved for this purpose. This section compares and contrasts OAM&P in T-carrier with OAM&P procedures in SONET.

OAM&P in T-Carrier Networks

In a T-carrier network, OAM&P was and still is an adventure. Separate operations systems are required to provide a centralized, single-ended maintenance group. This means that craft technicians need to be trained and well-versed in using a variety of management packages and techniques for providing OAM&P functions for FOTS equipment, DCS devices, and so on. The craft technicians must have access to a number of terminals because no single package can monitor the network as a unified whole. If that terminal were in use, a problem would have to wait.

Each vendor of a specific device (e.g., FOTS or DCS) makes use of different management software. For example, any encryption devices used for secure communications links (heavily used in military and large corporate environments) must use that vendor's proprietary management software.

Even worse, a separate data link network is used to connect the network operations center terminal to every manageable network element. Some service providers, making a virtue of necessity, extolled the "diversity" that the separate OAM&P network supposedly provided. But this was an illusion. Typically, the *links* are used for internal OAM&P, but the physical path is shared along with the traffic-bearing links. When one is lost, so is the other anyway.

In T-carrier, both the bandwidth and types of information available to the network operations center are extremely limited. The extended superframe format (ESF) for DS-1 and C-bit parity for DS-3 help the situation, but not much. A problem continues over not having enough overhead in the DS-1 and DS-3 frame structures for OAM&P purposes.

This limitation is not merely an annoyance. It has real impact on users and service providers alike. Because of this limitation, even a simple OAM&P function, such as performing a bit error rate (BER) test, is awkward. Suppose a customer has complained about end system retransmissions. A possible cause is an elevated BER on the link. The customer may even suspect that a contract or tariff is being violated with regard to these errors. Performing the BER test requires taking the link out of service for the duration of the test; frequently, the BER test shows that the link is fine. Usually, the problem turns out to be at one of the endpoints, making it the customer's responsibility. The link, however, still had to be taken out of service and tested intrusively.

OAM&P in SONET

With SONET, the art of controlling and managing an extended physical network consisting of many links in diverse geographical locations for OAM&P is much simpler than in T-carrier networks. SONET makes all of the OAM&P procedures an integral part of the SONET standard. In other words, a SONET network cannot be built without these OAM&P pieces as well. With SONET, network management operations systems are not added on, but are built-in. The SONET standard uses much more bandwidth, and has much more bandwidth available for OAM&P than does T-carrier and the information available is very sophisticated.

SONET OAM&P makes use of the same physical network as the SONET links facilities themselves. At the network operations center, a local area network is included in the SONET OAM&P standard to allow multiple terminals (actually, just PCs with OAM&P software) to have access to the same information instead of needing access to a specific terminal, as in T-carrier. None of the OAM&P methods are proprietary in SONET.

This means that OAM&P in SONET networks is at once easier to implement and yet much more robust than the OAM&P procedures in T-carrier networks. One connection can reach all manageable network elements. SONET OAM&P also includes a means of remote provisioning and configuration of SONET network elements. This means that SONET networks can easily be centrally maintained, reducing travel expenses for maintenance personnel. Additionally, the substantial amounts of information available in the SONET overhead allows for much quicker troubleshooting and detection of failures before the network is degraded to unacceptable levels.

For instance, SONET uses nonintrusive BER testing, rather than the intrusive type of BER testing needed in T-carrier. There are three distinct bit-interleaved parity (BIP) check fields in SONET. BIP checks exist for the SONET payload, as well as the line and section levels. The three BIP fields allow for a simple and quick table lookup and conversion from BIP error counts to BER levels. Because all SONET equipment must record these BIP errors, the BER for the last hour, day, or even week can be easily determined from a central location without disturbing the users' service or the link.

Consider the customer situation outlined above. In this scenario, a customer has complained about end-system retransmissions. As before, one possible cause is an elevated BER on the link. However, instead of needing to take the link out of service to perform a BER test, the technician can employ the craft interface for OAM&P to gather up-to-date BIP counts from the devices in question. The OAM&P software can then perform a simple table lookup and present the BER test result in both directions to the technician in seconds. This process can be made to be as easy as possible for the technician. Network maps, point-and-click interfaces, and graphs can make the task much more efficient.

Integrated Operations and Monitoring

Not only are the OAM&P procedures in SONET a great improvement and enhancement over the methods of performing the same activities in

T-carrier networks, but SONET integrates operations and monitoring capabilities. This can be a tremendous advantage when the twin concerns of operational tasks and performance monitoring are considered.

The general needs of OAM&P procedures in T-carrier are illustrated in Figure 12.5. The figure shows a relatively simple arrangement of FOTS and DCS devices, with a multiplexer thrown in for variety.

Figure 12.5 shows the complexity of OAM&P in a T-carrier network. In T-carrier, separate craft technician interfaces (i.e., terminals or PCs) are needed to access the various types and pieces of equipment in the network. This can be a problem when, for example, two technicians are trying to troubleshoot two separate problems involving FOTS devices on the same parts of the network. Usually there is only one PC capable of gathering FOTS OAM&P information from FOTS devices with one particular vendor. When one technician is using it, the other must wait, and so must the customer.

Even worse, with T-carrier OAM&P procedures, the overhead OAM&P communications are separated between the various systems. In other words, no FOTS information can pass through a DCS to reach the operations system in the network management center. This requires a separate data link to each of the devices or portions of the network.

The overall effect is that OAM&P in T-carrier networks is disjointed and complicated. Craft technicians typically scurry around to locate the correct terminal or section of the room, wasting time and effort.

Contrast this with the SONET-integrated OAM&P shown in Figure 12.6. The SONET philosophy of OAM&P is to integrate network management functions within the SONET network.

In SONET, separate terminals are not required to control various network elements or physical portions of the network. Because the concept

Figure 12.5
OAM&P
requirements in
T-carrier

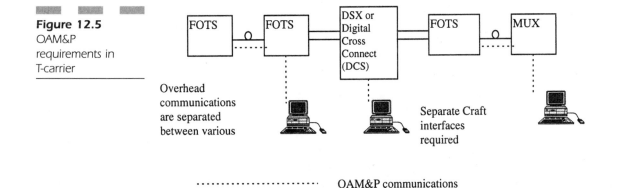

OAM&P communications

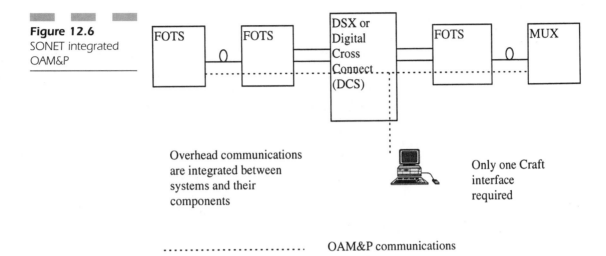

Figure 12.6
SONET integrated
OAM&P

of the LAN is included in the operations system at the network operations center, there is no need to run around trying to find the proper terminal (which may be in use to diagnose another problem) or section of the room before settling down and trying to troubleshoot the problem. This often occurs in T-carrier networks.

Moreover, because the standard SONET overhead is passed along from device to device on the SONET network, regardless of type, the craft technician can just as easily manage a FOTS device as a DCS from one network link to almost any SONET network device.

This makes OAM&P in SONET not only more efficient, but more cost-effective. Craft technicians only need to be trained in SONET, not in a diverse array of devices with different OAM&P techniques and procedures.

New Service Offerings

The types of things that users seek to do with their networks today have been dealt with earlier in this book and need not be repeated in detail here. Briefly, the kinds of services that users are beginning to demand in quantity all share at least one major characteristic: They require more bandwidth than ever before.

One of the most important benefits of SONET is the ability to position the network for carrying new revenue-generating services unheard

of before. Nothing but SONET will be able to handle the new bandwidth-hungry services and applications that users are beginning to
deploy in a networked environment. SONET, with its modular, service-
independent architecture and enormous bandwidth, provides vast capabilities in terms of services flexibility. High-speed, packet-switched services, such as ATM, LAN interconnection, and video applications, are
supported by SONET. Figure 12.7 shows how SONET will enable service
providers to support these new services, while supporting higher numbers of older applications and services as well. Note that the growing
trend to do voice, video, or data over IP is of no concern to SONET.
SONET carries IP packets or ATM cells with ease.

   Figure 12.7 shows a variety of devices and services that may be used
by applications and users in many organizations today. Users need public or private carrier services to support these applications. In the figure,
a SONET OC-3c or OC-12c transport can be used to handle the traffic
generated by an ATM multiplexer or switch. Organizations still need
voice service support from the PBX, and SONET can support these traditional services with ease. But SONET can also help to support newer
services like video, represented by the camera, videoconferencing, and
high-speed LAN interconnection.

Figure 12.7
SONET service
support

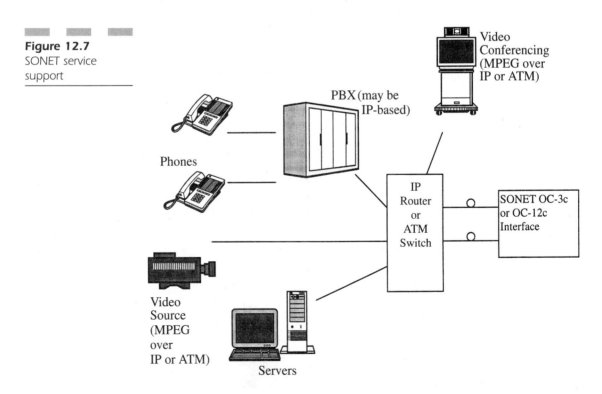

Users in large business and more specialized areas have begun to explore the potential of applications and services that involve the use of interactive multimedia. These combine the traditional use of networks to distribute data with the ability to add video or audio to the application. Advanced graphics and other niche applications, such as computer-aided design and manufacturing, also play a role. None of this rules out higher-speed pure data uses, such as bulk file transfer or remote server backup.

Many of these *broadband* services may use ATM, a fast packet-switching technique using short, fixed-length packets called cells. When used to transport ATM cells, SONET fills the payload with a cell stream that can be generated and switched as necessary. Because of the bandwidth capacity it offers, SONET is a logical transport network for ATM. In fact, the ITU specifies SONET as the transport for ATM cells in a B-ISDN network. Even if IP renders ATM totally obsolete, SONET can still be used as a transport for IP packets.

In principle, ATM is quite similar to other packet-switching techniques—but ATM switches cells, not packets. However, the details of ATM operation is somewhat different. Each ATM cell consists of 53 octets, or bytes. Of these, 48 octets make up the User Information (or payload, but different than the SONET payload) field and five octets make up the header. The cell header identifies the "virtual path" to be used in switching (or "routing") a cell through the network. The virtual path defines the connections through which the cell is switched to reach its destination.

An ATM-based network is bandwidth-transparent, which allows handling of a dynamically variable mixture of services at different bandwidths, as shown in Figure 12.7. ATM also easily accommodates traffic of variable speeds. An example of a service that identifies the benefits of a variable rate interface is that of a video code-based service. The video signals can be coded into digital signals and packetized within ATM cells. The number of cells generated per second may vary, reflecting the variable speed. Because ATM is essentially packet-switching with fixed-length packets, support for this variable bandwidth requirement is easy.

During periods of low activity within the video camera's field of view, the rate of packet transfer decreases. Hence, an average rate of transfer may occur below the maximum capacity of the physical transport. During peaks in activity, the rate of packetization exceeds the normal maximum. This *flexible bandwidth allocation* (sometimes called "bandwidth on demand") is a feature that distinguishes ATM networks from the fixed, channelized bandwidth that SONET links provide, as large as it is.

SONET and ATM, therefore, can work hand-in-hand to make more effective use of SONET bandwidth to support more customers than when SONET is used to provide point-to-point private lines. This is because a customer having an OC-3 link (for instance) operating at 155 Mbps may use it to support three DS-3s operating at 45 Mbps each. However, each of these DS-3s must terminate at some device or multiplexer on the customer premises. The multiplexer can aggregate lower-speed signals onto the DS-3, or a single device can use the full bandwidth of a DS-3 by itself. Once all of the high- and low-speed channels are assigned, the SONET link is full.

ATM is different in the sense that all of the devices' combined can share the entire OC-3 link's 155 Mbps dynamically. The low-speed devices' combined traffic generates a lesser number of cells, and the single device formerly linked to a full 45 Mbps DS-3 generates a greater number of cells, but all of the cells are multiplexed onto the OC-3c (for concatenated) based on the devices' *instantaneous bandwidth demand*. SONET helps to meet peak bandwidth demands while supporting many average demands for service at the same time. It should be noted that many of these ATM characteristics are shared by IP networks as well. Both IP and ATM can form an upper switching and multiplexing layer for a SONET link.

High-speed LAN interconnection is another area where service providers have used SONET to support new services. LANs form a special challenge for private-line buyers. LANs are characterized by high speeds, usually at least 10 Mbps and upwards of 100 Mbps. However, LANs are also characterized by extremely restricted network distances, usually much less than 1,000 feet or so. The challenge is to link LANs with network links that are long enough and yet have enough bandwidth to enable LAN users to perform common tasks efficiently across these interconnected LANs. Leased private lines have no distance limitations, but readily available and easily affordable private lines typically top out at 1.5 Mbps, the DS-1 speed. This is a mere fraction of the native LAN speed, and can act as a significant bottleneck when interconnecting LANs.

In some cases, a customer with particularly deep pockets may purchase a SONET OC-1 operating at 51.84 Mbps, or even an OC-3c operating at 155 Mbps. SONET speeds help to alleviate the LAN-to-private-line bottleneck. Even when a customer cannot afford the full 155 Mbps bandwidth represented by an OC-3c, the customer may be able to buy a DS-3 on SONET at a price lower than otherwise possible if the DS-3 were not delivered on SONET.

SONET will help to make DS-3s available in areas where no private lines at speeds greater than DS-1 have ever been available before.

Optical Interconnect: "Mid-Span Meet"

Because of different and proprietary optical formats among vendors' T-carrier products, it is not possible to optically connect one vendor's fiber terminal device to another. For example, one manufacturer may use a line rate of 570 Mbps, while another uses a line rate of 565 Mbps. Both may package the same number of DS-3s in the aggregate signal, but the amount of overhead bytes and their exact format do not match. This precludes any effort to interconnect them.

The two benefits provided by SONET optical mid-span meets is shown in Figure 12.8.

As can be seen in Figure 12.8, a major advantage and benefit of SONET is that it allows for a mid-span meet for multivendor compatibility. Today's SONET standards contain definitions for fiber-to-fiber interfaces at the physical (photonic) level. These low-level aspects of the SONET standards determine the optical line rate, wavelength, power levels, pulse shapes, and coding for bits on the fiber link. Above the physical transmission level, the current SONET standards also fully define the frame structure, overhead, and payload mappings. Other aspects of the standards continue to be developed to define the messages in the overhead channels to provide increased OAM&P functionality.

SONET allows optical interconnection between network providers regardless of who makes the equipment. The network provider can purchase one vendor's equipment and conveniently interface with other vendors' SONET equipment at either the different carriers' locations or customers' sites. Users may now obtain the OC-n equipment of their choice and "meet" with their network provider of choice at that OC-n level.

Figure 12.8 shows how one service provider can easily decide to deploy one vendor's equipment and still interface optically with others. For instance, the two other service providers use other vendors' SONET equipment. However, a mid-span meet is still possible. In the figure, a customer using a direct SONET interface (for ATM perhaps) also can have another vendor for the customer premises equipment. The need for a mid-span meet was one of the original goals for developing the SONET standard.

It is fair to say that the benefits of SONET far outweigh the expense of deploying SONET links. This chapter presented a more or less systematic look at all of the benefits. However, it is still worth remember-

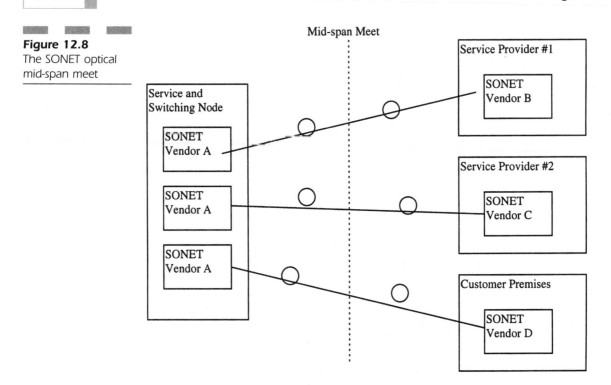

Figure 12.8
The SONET optical
mid-span meet

ing that these are only the major benefits and the ones upon which most people would readily agree. As time goes on, what were once perceived as incidental aspects of SONET may emerge as major benefits in their own right. For instance, T-carrier, once strictly perceived as a way to deliver digitized voice for internal carrier use, turned out to be virtually indispensable for sales to customer interested in building private networks. It took the upheaval of divestiture in 1984 to give carriers the incentive to tariff and offer T-carrier to the general public. The success of T-carrier in this added role came as somewhat of a surprise to some who still saw DS-1 as only the vehicle to deliver DS-0 links at 64 Kbps to scattered users. Perhaps SONET will emerge as ideal for situations yet undreamed.

CHAPTER 13

SONET OAM&P

One of the most powerful features of SONET is the presence of built-in standards for operations, administration, maintenance, and provisioning (OAM&P). Also know as OAM in an international context, SONET OAM&P involves all aspects of the day-to-day operations and fault detection in all parts of the SONET network. When things go wrong, it is up to the SONET OAM&P procedures to detect the problems and put them right. Even when things do not go wrong, operations is involved with configuration issues, performance statistics, and so forth. It is tempting simply to equate OAM&P with "network management," but OAM&P involves more than the traditional activities associated with a network management center. SONET OAM&P is not only network management, but also customer support, trouble tracking, performance evaluation, configuration management, technical support, and billing.

The difficulties of performing OAM&P in T-carrier networks have been dealt with earlier and need not be repeated here. Briefly, T-carrier networks were characterized by inadequate bandwidth for OAM&P needs, lack of OAM&P standards, and even many key operational areas where OAM&P procedures were lacking at worst, or haphazard at best. This apparent oversight in T-carrier was not due to a lack of interest or attention. At its inception, T-carrier pushed the envelope of equipment performance to such an extreme that simply making the network actually work was enough of an accomplishment at the time. It was only after more exacting OAM&P needs became apparent that T-carrier added what functions it could. Limiting factors included the absence of adequate overhead bits to carry OAM&P information, alarms, and commands; and the inability of equipment to handle the processing needs required by extensive performance statistics and monitoring information.

SONET labors under no such restrictions. SONET overhead is more than adequate to handle the most sophisticated OAM&P indicators and messages. SONET equipment has internal processors that are powerful enough to deal with all internal, functional monitoring. This chapter is a systematic approach to SONET OAM&P, including all major aspects of this key SONET feature.

SONET Overhead Revisited

It is not necessary to repeat a discussion of the full structure of the SONET overhead bytes. It is enough to point out that some of the

SONET overhead bytes at the path, section, and line levels are used for OAM&P procedures. Several bytes are constantly monitored by receiving SONET equipment for fault detection and others are used to try to correct and compensate for these errors.

Figure 13.1 shows the SONET overhead bytes at the line, section, and path SONET levels. The bytes directly relating the SONET OAM&P are as follows:

- *The A1/A2 framing bytes:* The bytes are monitored by receiving equipment to ensure frame alignment. Loss of these expected framing bit patterns for 3 milliseconds results in a loss of frame (LOF) error condition or alarm.

- *The D1, D2, and D3 bytes:* These bytes comprise the SONET section Data Communication Channel (DCC) and are used to convey network management messages between a SONET Operation System (OS) and the particular section SONET device being managed.

- *The H1/H2 pointer bytes:* These bytes are monitored by receiving equipment to locate the Synchronous Payload Envelope (SPE). The values must be valid. If invalid pointer values were received for eight, nine, or ten consecutive frames (all three values are allowed), this results in a Loss of Pointer (LOP) error condition or alarm.

- *The K1/K2 automatic protection switching (APS) bytes:* These bytes are monitored by receiving equipment in order to determine when the SONET signal should be switched to an alternate physical path. Specific bits within these bytes are also monitored to detect inconsistencies and outright errors within the APS bytes themselves.

- *The D4 through D12 bytes:* These bytes comprise the SONET line DCC and are used to convey network management messages between a SONET OSS and the particular SONET line device being managed.

- *The S1 synchronization byte:* This byte can be used to determine the source and stratum level of the SONET timing clock. As such, it can be used for trouble-shooting and reliability purposes.

- *The M0/M1 byte:* This byte is used to convey bit error rate information to the sending device. At the STS-1 level, this is the M0 byte, while at the other STS levels, this is the M1 byte.

- *The C2 signal path byte:* The signal path byte is monitored by receiving equipment at the end of a SONET path (typically the Terminal Multiplexer) and is used mainly to determine if the path is used, or

Figure 13.1
The SONET
overhead Bytes

A1 Framing	A2 Framing	J0/Z0 (STS-ID) Trace/Growth
B1/undefined BIP-8	E1/undefined Orderwire	F1/undefined User
D1/undefined Data Com	D2/undefined Data Com	D3/undefined Data Com
H1 Pointer	H2 Pointer	H3 Pointer Action
B2/undefined BIP-8	K1/undefined APS	K2/undefined APS
D4/undefined Data Com	D5/undefined Data Com	D6/undefined Data Com
D7/undefined Data Com	D8/undefined Data Com	D9/undefined Data Com
D10/undefined Data Com	D11/undefined Data Com	D12/undefined Data Com
S1/Z1 Sync Status/ Growth	M0 or M1/Z2 REI-L/ Growth	E2/undefined Orderwire

Section and Line Overhead

J1 Trace
B3 BIP-8
C2 Signal Label
G1 Path Status
F2 User Channel
H4 Indicator
Z3 Growth
Z4 Growth
Z5 Tandem Connection

Path Overhead

equipped (has a sender and receiver) in the first place. If equipped, the signal path byte gives the format of the content of the SONET SPE (e.g., ATM cells, virtual tributaries).

■ *The G2 path status byte:* The path status byte is monitored by receiving equipment at the end of a SONET path (typically the Terminal Multiplexer) and is used to convey bit error rate information to the sending device at the other end of the SONET path (usually the sending Terminal Multiplexer).

There are additional monitoring features and alarm conditions built into the Virtual Tributary (VT) levels of SONET. However, because these add little understanding to SONET OAM&P functions, the VT level OAM&P functions will not be detailed in this chapter.

Most of this chapter makes repeated reference to the SONET section, line, and path overhead bytes and the conditions they represent. It is important to realize that it is one thing for a SONET network element (NE) device to realize that the framing has been lost from the upstream device. It is quite another thing to make this information known to some network management center or to automatically take action until human intervention is possible.

The very speed of SONET links make network management especially challenging. Even if some corrective action could be taken in a second, which would be much faster than humanly possible, some 10 Gigabits of information, most of it (presumably user information) would be lost at the OC-192 rate.

What Could Go Wrong?

What could happen on a SONET network that would affect users? SONET standards define several major failure conditions and their associated alarm indicators. The alarm indicators are used to inform SONET network operations that the failure exists. The major failure types and indicators include the following:

1. *Loss of Signal (LOS):* When a SONET receiver detects an all-zeros pattern for 19 ±3 microseconds or longer, this constitutes an LOS failure. LOS also occurs when the power of the recieved signal drops below a preset threshhold. It indicates that the upstream transmitter has failed. Due to the nature of SONET finer links, however, it is important to realize that the reverse path to the

upstream device may still be operational. This condition is cleared when two consecutive valid frames are received.

2. *Loss of Frame (OF):* The absence of a valid framing pattern for 3 milliseconds (called Severly Errored Framing) leads to an LOF failure condition. This is cleared when two consecutive valid A1/A2 framing patterns are received.

3. *Loss of Pointer (OP):* The absence of valid H1/H2 pointer bytes in eight, nine, or ten consecutive frames leads to an LOP failure condition, or when there are "excessive" New Data Frame (NDF) values. This is cleared when three consecutive valid pointers are received.

4. *Alarm Indication Signal (AIS):* This condition can occur in response to one of the failures above. It is the indication to other SONET devices that a failure condition exists. AIS is sent *downstream* from the NE device that has detected the failure.

5. *Far End Receive Failure (FERF):* This is another type of failure indication condition. FERF is sent *upstream* from the NE device that has detected the failure because the failure may not have affected this return link.

6. *Remote Alarm Indicator (RAI):* This is an indication of an error condition which is sent from one end of the SONET path to the other. Thus, the RAI is an "end-to-end" failure indication.

Basically, each of the first three items on the list is a major failure condition. The last three items on the list are the indicators of the error condition sent and received by the SONET NEs. Naturally, as the failure affects more than a single fiber link or piece of equipment, these indicators may be going off into the great bit bucket, but they must be generated anyway.

There are other failure conditions defined in the SONET standards. These include a simple *equipment failure,* which covers a variety of situations and is more or less implementation dependent, because one vendor's equipment may differ radically from another's (e.g., Backup power supply on? Cooling fan failure?). Equipment failures fall into five categories: Service Affecting (SA), Non-service-affecting (NSA), Critical (CR), Major (MA), and Minor (MN). There is also a failure state for Loss of Synchronization, defined as a loss of timing source for periods of 100 to 1,000 seconds, and the failure of the protection switching and DCC. These last two have still not been fully defined in SONET standards.

The relationship between the major failure states detected and the alarm indicators raised and sent in consequence is an important one.

Figure 13.2 shows the relationship between the failure states and alarm indicators in one a simple SONET point-to-point link. For the sake of completeness, VTs are shown although not discussed in detail.

In Figure 13.2, several types of SONET equipment are shown to illustrate the differences in SONET section, line, and paths. Two types of paths are shown in the figure, the STS path from SONET TM to TM, and the VT path from one digital loop carrier to another. Failure detection is illustrated by a solid circle, while the error indication procedures are represented by open circles.

Note than a failure condition is detected by one piece of equipment or another and results in one or more error indications sent both upstream and downstream on the network. In each case, the actual SONET mechanism used to convey the error indication is shown. For

Figure 13.2

Failure states and
alarm events

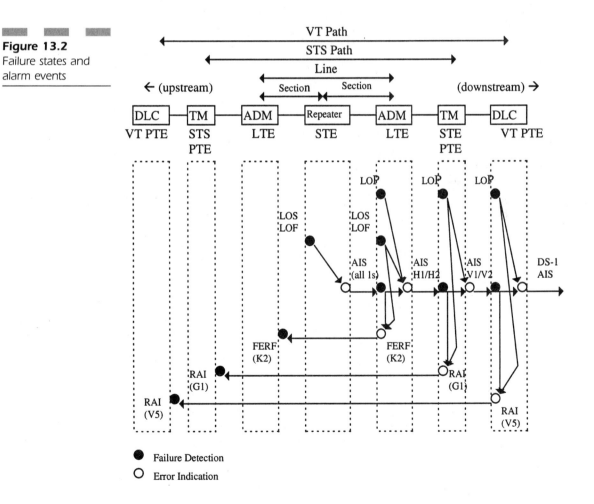

instance, the FERF is conveyed by the value of some K2 overhead byte bits, RAI is in the G1 path status bits, and so on. Sometimes multiple failure events will result in the same indicator. For example, a SONET ADM detecting a LOS, LOP, or LOF failure may respond with an AIS in the H1/H2 bytes to a downstream device.

Generally, the detection of LOS, LOF, or LOP at the section-terminating equipment (STE, a SONET regenerator in the figure) or line-terminating equipment (LTE, a SONET ADM in the figure) will cause the generation of alarms on that NE's output port to the downstream NE, or in the right portion of the figure. An LOS detected at the repeater would generate an AIS consisting of all 1's to the LTE (ADM). The ADM would, in turn, generate an AIS to the path terminating equipment (PTE). At the same time, the ADM would indicate a FERF to the upstream LTE (ADM) using the K2 byte. The other PTE devices would indicate RAI in the upstream direction as well.

A major goal in the SONET failure and indication system is to avoid "cascading" error conditions wherein a single failure can generate an ever-increasing circle of error indications. The rules for detecting failures and raising indicators in SONET prevents this from happening.

OAM&P Signals and Layers

SONET OAM&P functions include all aspects of trouble detection, repair, and service restoration. In order to support these functions, SONET is designed with a number of alarm surveillance operations designed to detect a problem or a potential problem. However, before the specific surveillance operations are explored, the terms, *state*, *indication*, and *condition* should be defined further.

In SONET, the term *state* describes any occurrence in the network that must be detected. A SONET NE enters a specific state whenever this occurrence is detected and leaves the state when the occurrence is no longer detected. The detection of an occurrence may lead to an alarm condition being signaled by the NE device, which is called an *indication*. An indication represents the presence of a certain *condition* in the NE. Indications are not always reported, but are rather made available for later access by an OS. Others indications can be reported immediately, either as an alarm or a nonalarm indication.

As mentioned previously, the purpose of the AIS is to alert *downstream* equipment of a problem that an *upstream* NE has detected. Different types of AISs are reported by the various SONET layers. Figure 13.3 shows the relationships between the SONET layers and various AIS indications.

Figure 13.3 shows the interplay between the two aspects of OAM&P, which means that both higher layers and peer layers in downstream devices must be informed of the failure condition. On the left portion of the figure, vertical arrows and their associated OAM&P indications show how higher layers are informed. The purpose of these AIS conditions is to inform the downstream SONET NE peer process of a failure. In the figure, the failure is detected at the section level (not atypical) and the AIS flow is upstream to downstream. The vertical arrows indicate that the following events are taking place:

1. The upstream SONET STE (e.g., a regenerator) detects an error condition and the event triggers a line AIS. This informs a downstream LTE (e.g., an ADM) of the failure.

2. The upstream STE now informs the downstream PTE (e.g., a TM) of the failure by generating an STS path AIS and an STS "yellow alarm" using the path G1 byte.

3. The upstream STS PTE detects the line AIS or path AIS. Once this happens, the upstream STS PTE will inform the downstream STS PTE of the failure using either a VT path AIS (in the V5 byte), a DS-3 AIS, or even a DS-0 AIS, depending on the specific payload in the STS SPE.

4. If the payload in transporting DS-n signals, then the SONET NE informs a downstream NE of the failure, or a termination of the DS-n path using the proper DS-n AIS.

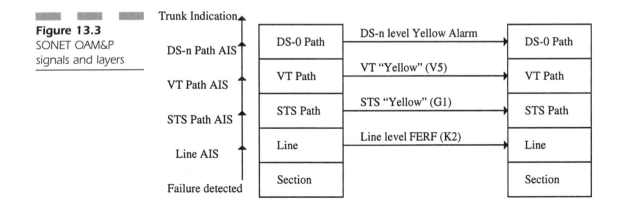

Figure 13.3
SONET OAM&P
signals and layers

The figure also shows some "yellow" signals between some layers. Those familiar with T-carrier will recognize the term "yellow alarm" instantly. In the SONET world, a "yellow alarm" loosely applies to the error indications sent to the OS in response to a "red alarm," which is the initial failure detected. In SONET, the FERF is the SONET line layer maintenance signal, and SONET "yellow signals" are STS and VT path layer signals. As in T-carrier, yellow signals can be used for *trunk conditioning* (recovery procedures) or by a downstream terminal to inform an upstream terminal to initiate trunk conditioning on the failed circuit. These yellow signals are used for troubleshooting and trouble isolation.

The position of the horizontal arrows in Figure 13.3 illustrates the following events. These events now flow not upstream to downstream, as the vertical flow shows, but rather downstream to upstream:

1. A downstream LTE informs an upstream LTE about a failure along the downstream SONET line level using the line FERF (in the K2 byte).

2. The downstream PTE informs an upstream PTE that a downstream failure indication has been declared along the STS path using the STS path "yellow alarm" (in the G1 byte).

3. Because virtual tributaries are in use in this example, the downstream VT PTE informs the upstream VT PTE that a failure indication has been detected along the downstream VT path using the VT path yellow (in the V5 byte).

4. Any required DS-n T-carrier yellow alarm signals are generated due to the failure, or for any DS-n paths that are terminated using the proper T-carrier DS-n yellow alarm. A detailed description of T-carrier yellow alarms is not necessary here because the emphasis is on the SONET portion of the link.

The series of steps above can be summarized in a simple illustration. The important events occur between SONET NEs, typically ADMs. Figure 13.4 shows a simple example of how SONET NEs react to a failure. This example is a brief summary and is not intended to show all possible situations; thus, the figure shows only a few actions from among the wide range of possible actions. The exact types of AIS, FERF, and yellow signals sent between the NEs will vary widely, depending upon the nature of the failure.

Two other common OAM&P terms appear in Figure 13.4 and should be explained. They are *red alarm* and *performance monitoring parameters*. In SONET, a red alarm is generated when an NE detects a

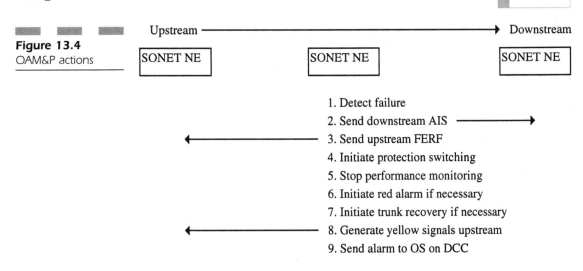

Figure 13.4
OAM&P actions

failure state that persists for 2.5 seconds or longer, or when the NE is experiencing continuous but intermittent failures. Performance monitoring statistics include parameters, such as the number of bit-interleaved parity (BIP) check errors. The collection and recording of these performance monitoring parameters are suspended during the handling of a failure state. This only makes sense: The SONET NE has more pressing things to do and there is no meaningful performance to report anyway.

Users and OAM&P Operations

In a perfect world, SONET OAM&P surveillance and monitoring would be so good that users would never have to worry about outages or detecting SONET path level failures. In the real world, however, users still want to feel that they have some control over the network, especially because SONET, at heart, is basically just another kind of private line.

Figure 13.5 shows an example of several OAM&P information *flows* in relationship to user equipment. The term *flow* just means that a sequence of OAM&P indicators is *flowing* between the SONET network elements.

In Figure 13.5, the user device terminates the SONET section, line, and path. The user, thus, has access to the following SONET overhead bytes and the OAM&P information they contain:

Figure 13.5
OAM&P user
information

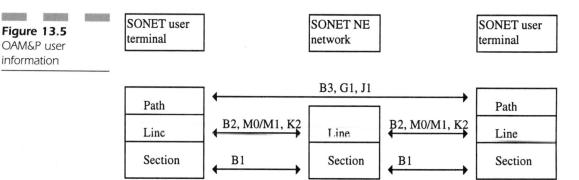

J1: Verifies a continued connection between the two user devices at the ends of the SONET network (user must have access to the SONET TM).

G1: Indicates an STS yellow signal when byte content includes a 1 value in bit 5.

B3: The BIP-8 calculation results on the previous STS SPE using even parity.

B2: The BIP-8 calculation results on the previous line overhead and STS-1 envelope using even parity.

M0/M1: Informs the originator of the error counts from the B2 calculations using a count in the M0 octet of an STS-1 in an OC-1 or the M1 octet of third STS-1 in an OC-n.

K2: Indicates a line FERF and that the line is entering a LOS or LOF state. The FERF is indicated by the value 100 in bits six, seven, and eight of the K2 byte.

B1: A BIP-8 calculation results on all bits of the previous STS-N frame using even parity.

So OAM&P in SONET involves a SONET NE monitoring and processing the arriving overhead bytes at the section, line, path, and even the virtual tributary levels. Based on the overhead processing at the receiving NEs, certain *events* can occur which raise alarm conditions in the detecting NE. These events are covered generally by the blanket alarm indication signal (AIS) state in the detecting NE. The AIS state is used to convey events *between* SONET layers in the same NE device. Sometimes, events are relayed downstream to other NEs and upstream to the sending NE (assuming the upstream link is still capable of functioning) in other overhead bytes. In the case of section overhead, the upstream device could be a regenerator, but no events are reported to regenerators. The events typically report some error condition, but the severity of the error is not always catastrophic, as has already been pointed out.

The end result is a kind of cascading series of events generating AIS conditions all the way through the SONET layers. Now, some events can take place at the path overhead (payload) level, for instance. The SONET frames are formed and arriving just fine, but the payloads inside are hopelessly confused and error-ridden. This event will generate alarms only at the STS path and virtual tributary levels, if the payload is used for VTs. The complete structure of all SONET events and the overhead bytes monitored to detect them (except for a catastrophic and complete loss of signal) is shown in Figure 13.6. The whole figure is quite complex, so a few words of explanation are in order.

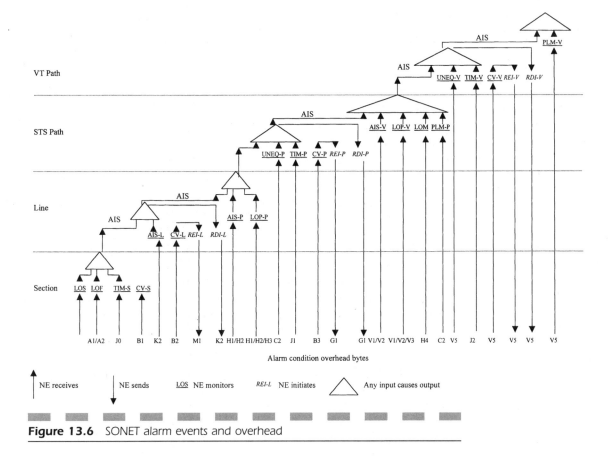

Figure 13.6 SONET alarm events and overhead

At the bottom of Figure 13.6 are the overhead bytes that are processed to detect the event. The event detected by each set of overhead bytes is shown at the end of an arrow. An upward arrow represents events that are detected either by processing the bits in the overhead byte(s) or by monitoring the bits within the byte for a certain pattern (which means the event is actually occurring at the *upstream* device). The six downward arrows represent events that are reported to other SONET NEs by bit patterns within the overhead bytes indicated. The triangle symbol represents a logical process by which any event entering any triangle symbol from below generates the actions indicated at the top of the triangle. The whole thing is hard to talk about in the abstract, but easier to understand by examining some concrete examples.

For example, suppose a SONET NE detects coding violations (BIP errors) at the line level based on the processing of the B2 bytes. This CV-L event will result in a remote error indication at the line level (REI-L) being reported to the sending NE in the M1 byte. The details of how the overhead bytes are used to detect and report events are shown in Table 13.1.

One further, more complex, example will suffice to show how SONET events and alarms are related.

Consider a SONET NE that suddenly finds that the J1 Path Trace byte, which contains the network address of the SONET NE that packaged the payload, does not match the address expected in the link configuration table. This change is consistent—not the result of random bit errors—perhaps because a new PTE has been turned up and is not configured properly. The receiving Path Termination Equipment (PTE) will realize that this trace identifier mismatch at the path level (TIM-P) has occurred and this realization triggers two actions at the path level.

First, this remote defect indication at the path level (RDI-P) is conveyed to the PTE across the SONET link(s) that packaged the SONET payload in the G1 Path Status overhead byte. The RDI-P is conveyed in bits 5-6-7 of the G1 byte. Appendix B contains the further information that the bit pattern of 110 in these bit positions is interpreted as a *remote connectivity defect* to the other PTE (the UNEQ-P event is conveyed in the same fashion). Now the sending SONET NE knows that the receiving PTE does not like the network address in the J1 byte or that the path is *unequipped* (the receiving PTE is not configured to do anything with the payload contents).

Second, the TIM-P event triggers an AIS to the virtual tributary level above (as, obviously, do a lot of other things). This AIS triggers a remote defect indication at the virtual tributary level (RDI-V) since, obviously, all

TABLE 13.1

SONET events and meanings

Overhead Byte	Event	Meaning	Comment
N/A	LOS	Loss of signal	All 0s or below receive threshold
A1/A2	LOF	Loss of framing	Framing pattern lost
J0	TIM-S	Regenerator section trace identifier mismatch	The J0 network address of the regenerator is not what was expected
B1	CV-S	Coding violation—section	BIP errors detected in B1 byte
K2	AIS-L	Alarm indication signal—line	Bits 6-7-8 of K2 byte set to 111
B2	CV-L	Coding violation—line	BIP errors detected in B2 byte
M1	REI-L	Remote error indication—line	Count of B2 BIP errors
K2	RDI-L	Remote defect indication—line	Bits 6-7-8 of K2 byte set to 110
H1/H2	AIS-P	Alarm indication signal—path	Excessive New Data Flags (NDF)
H1/H2/H3	LOP-P	Loss of pointer—path	Invalid payload pointer values
C2	UNEQ-P	Unequipped STS at path level	All 0 in C2 byte: no user information
J1	TIM-P	Trace identifier mismatch at path level	The J1 network address at the path level is not what was expected
B3	CV-P	Coding violation—path	BIP errors detected in B3 byte
G1	REI-P	Remote error indication—path	Bits 1—4 of G1 report B3 error count
G1	RDI-P	Remote defect indication—path	Bits 5-6-7 of G1 code path events*
V1/V2	AIS-V	Alarm indication signal—VT	Excessive VT pointer adjustments
V1/V2/V3	LOP-V	Loss of pointer—VT	Invalid VT pointer values
H4	LOM	Loss of multiframe—VT	0 thru 3 payload sequence in H4 is lost
C2	PLM-P	Path label mismatch—path	Payload content not what expected
V5	UNEQ-V	Unequipped VT at VT level	Bits 5-6-7 of V5 set to 000
J2	TIM-V	Trace identifier mismatch at VT level	The J2 network address at the VT level is not what was expected
V5	CV-V	Coding violation—VT	BIP errors detected in bits 1-2 of V5 byte
V5	REI-V	Remote error indication—VT	Bit 3 of V5 reports any BIP error
V5	RDI-V	Remote defect indication—VT	Bit 4 of V5 reports VT event
V5	AIS	Alarm indication signal—VT	Alarm reported to affected tributary

*See Appendix B for details.

of the assumed VTs in the payload are affected by the J1 mismatch (or unequipped status of the receiver). Table 13.1 shows that this RDI-V is conveyed to the sender by setting bit 4 of the V5 byte in each VT. Naturally, an AIS condition is also raised to whatever device is attached to the VT at the receiver.

By using Figure 13.6, Table 13.1, and the additional information in Appendix B, almost any SONET event and alarm can be traced from cause to reaction, and from layer to layer. This is true of even the complex messaging protocol used in the automatic protection switching (APS) K2 byte.

Note that the event triggered by the J1 path level event resulted in no events at the line and section levels. This is as it should be, since the line and section SONET NEs are happily shuttling perfectly valid (to them) SONET frames end to end along the path. The hierarchy of events is enforced from bottom to top. So a section-level event triggers AIS at all SONET levels above, since lines, paths, and virtual tributaries should all be informed of a lower-level failure or alarm condition. So a section-level loss of framing (LOF) triggers RDIs at the section, path, and VT levels. However, early SONET networks were plagued by "alarm storms" where one low-level-event triggered an avalanche of alarms that swamped network operations centers and obscured the low-level root cause. Improved network management software and systems have better isolated the lower-level events that result in higher-level alarm conditions.

Timing Problems

Next to outright equipment failures, perhaps the most vexing thing about SONET networks is timing problems. This not only affects the synchronous nature of SONET, but makes it impossible to cross-connect tributaries or to retrieve the information inside the SPE.

When timing problems occur, a SONET network administrator or manager can look for two possible sources of the problem. First, the NE may be at fault, or second, the timing network feed may be faulty. Fortunately, tests can be run with a variety of test sets to determine where pointer adjustments are being made on the SONET network. With a simple strategy of working back toward the upstream NEs, the tests can reveal the point where pointer adjustments are not being made. That is, that particular NE will have proper pointer adjustments on the input side, but not on the output side.

Once the trouble area has been pinpointed, the next step would be to determine whether the clock recovery board in the NE is faulty, or whether the BITS timing has somehow failed, isolating the NE.

SONET and T-Carrier Network Management

Surveillance is the practice of monitoring a network for various alarm (or "alert") conditions that may arise. Alarms typically are used by NEs to notify the OS of some critical problem or condition in the device, or on the links between them. Just as in SONET, LOS and AIS are examples of alarm conditions on T-carrier networks.

T-carrier faced all of the same problems when it came to network management as SONET did. But SONET greatly improves on T-carrier's surveillance capabilities. This section explains how and why.

Obviously, it is not enough just to have one network element aware of a problem. Networks consist of many devices that are not always under the control of a single service provider. For example, a T-carrier network may have had a DS-3 link between an interexchange carrier and a local access provider, each with their own fiber optic terminal system (FOTS). Presumably, the mid-span meet is not a problem because the same vendor's equipment is used on both ends of the link between them. When a problem occurs on the interexchange carrier's network, it would be nice if the local access provider had a way of determining this as quickly as possible, because users will typically call the local access provider once the end-to-end bit stream is lost due to the failure. Figure 13.7 shows how this process would work in a T-carrier situation.

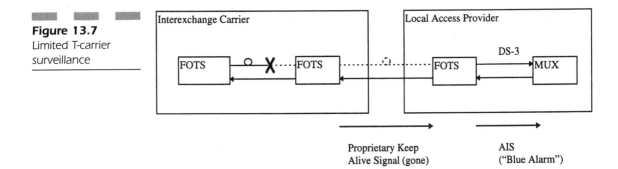

Figure 13.7
Limited T-carrier
surveillance

In Figure 13.7, if a FOTS link were lost within the interexchange carrier's network, then the symptom would be the loss of the incoming bit stream to the local access provider. The link is still there physically, of course, but there is nothing on the fiber that makes sense. Each vendor will have a proprietary "keepalive" signal between these devices to let the "downstream" device know that the link is still there, but the information bits are now gone due to the upstream failure.

The loss of the data stream triggers the generation of an AIS signal (usually called "Blue Alarm" in T-carrier) to other downstream devices. This suppresses a flurry of other downstream alarms at lower levels. This blue alarm process allows the local access provider to quickly determine that the problem is not a local one.

There is no standard mechanism for informing the *upstream* devices in the interexchange carrier network that the problem even exists, when these devices must ultimately deal with the problem. Of course, the local access provider at the other end of the link will be fielding calls from local end-users and not even have a clue as to what is going on.

Now consider how SONET improves the situation. T-carrier networks are limited in the types of alarms they support (Yellow, Red, and Blue) and the way that alarm conditions are propagated throughout the network to all *upstream* and *downstream* devices affected. This limitation is necessary for two reasons: first, to prevent the original failure from propagating a storm of other alarms from downstream devices through the network; second, to alert these other devices to the problem in case recovery steps need to be taken (e.g., protection switching).

SONET is much more robust when it comes to network surveillance than is T-carrier. This is illustrated in Figure 13.8. In the figure, a SONET hub at an interexchange carrier site has a failed OC link from a SONET terminal mux (TM A) to the SONET hub—the SONET hub combines add/drop multiplexing (ADM) and digital cross-connect system (DCS) capabilities. Naturally, the SONET hub at the local access provider's site will want to notify the downstream devices about the failure.

With the full SONET overhead to employ, an STS Path AIS condition is raised and passed along. The SONET TM B device will respond with a downstream DS-3 AIS. But SONET does more. On the upstream path, which is not affected by the failure, an "STS Path Yellow Alarm" (technically the Remote Alarm Indicator: RAI) condition is raised and passed along to the SONET TM at A. In addition, the SONET hub will raise its own "STS Line FERF" to let the SONET TM at A know that the problem exists.

The key here is that all sites are aware of the problem upstream and downstream and all devices are alerted in exactly the same way: STS

Figure 13.8
Improved SONET
Surveillance

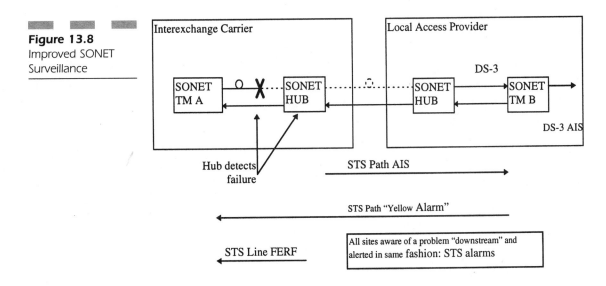

Figure 13.8
Improved SONET
Surveillance

alarms in the SONET overhead. Note that the alarms flow between different service providers as well.

SONET Network Management

SONET networks today are in a period of transition when it comes to network management issues. Both the ISO and ITU-T have been working on the development of several Open Systems Interconnection (OSI) network management standards for a number of years. These standards revolve around the Common Management Information Service Element (CMISE), which defines the network management applications, and the Common Management Information Protocol (CMIP), which is the protocol used for CMISE applications.

Unfortunately, when SONET deployment began, neither CMISE nor CMIP was in any condition to be used for production equipment and networks. Early SONET vendors were forced to adapt the standard Transaction Language 1 (TL1) interface from Bellcore for SONET use.

Not that vendors of early SONET equipment were reluctant to implement TL1 network management for their products. Quite the opposite, as many saw CMISE/CMIP as a needlessly complex solution to the situation of a technician sitting at a network management terminal sending messages to, and receiving alarms from, a SONET NE. And this was entirely the point of the whole TL1-versus-something-better debate: if there is a human being at one end, then TL1 is fine. But once SONET (and SDH) networks grow to the point where the whole network of NEs cannot be seen and managed from a single terminal with a human operator, the quest for some other form of network management becomes critical.

What is needed to perform OAM&P in modern SONET/SDH networks will be considered a little later in this chapter. For some, TL1 is still the answer, at least for the foreseeable future. For others, TL1 may remain, but only as part of an evolutionary transition to something better (what that something better might be is still debated as well). For now, it is time to take a closer look at just what the attractions and limitations of TL1 are.

TL1 is still the most widely used telecommunications management protocol, although TL1 has been assailed on all sides in recent years. A lot of the advantages of TL1 in relation to its rivals mirror the advantages of the IETF's Internet Protocol (IP) compared to the more full-blooded international standard protocols proposed in the 1980s. TL1 is not an international standard, but is vendor neutral, simple, extensible for new features, and easy to implement. TL1 can be used to manage almost any vendor's product, so service providers do not always have to buy proprietary implementations for each vendor's gear.

However, TL1's biggest drawback is that it was originally intended as a "man-machine language" with a human sitting at a terminal at the Operations Support System (OSS) location and the managed NEs at the end of the communications links. It was typical to have a separate link to each managed device. Today the need is for OSS components to be computer programs themselves, that gather information on and perform routine diagnostics automatically, and then both store the results in a historical database and post the results to a Web site for authorized personnel to see.

Network management languages and the protocols they used first appeared in the 1970s, once networks with intelligent (i.e., computerized) NEs began to appear in force. These NEs not only could flash a red light on their outer cover when disturbed, but could actually send data messages to a central OSS. Technicians did not have to be in the same room as the equipment at all times. The OSS center could be almost anywhere. This was a much more efficient arrangement, and the human operator could even compose messages, such as configuration commands or diagnostic commands to the device, at the terminal.

By the early 1980s, the ITU had looked at this whole arena in telecommunications and established the Z.300 series for user interfaces to the human-and-computer OSSs. Like many ITU umbrella specifications, the Z.300 series was a mass of choices and options, and did not directly address interoperability issues. So two vendors of a similarly acting and configured M13 device (for example) could be totally compliant with Z.300 and yet still require service providers with both devices to buy, use, and maintain two totally separate-looking and -acting network management hardware and software.

Bellcore (now Telcordia) came along in early 1984 to bring research and development skills to the newly emerged local pieces of the former Bell System—the RBOCs. They decided right away to specify a standard language to control transmission network devices (SONET/SDH was still a few years off)—mostly T-carrier. Most of TL1 was essentially done in January 1985, but pieces were added throughout 1985 and 1986. Firmly based on Z.300, TL1 embraced both language and messages to and from all telephone company devices except the circuit switches themselves (interoperability and standardization was much less of a problem in the highly conservative central office marketplace).

The TL1 specifications had no interest in the *lower*-layer protocols used to shuttle the TL1 message to and from NEs. After all, there was usually a direct link from the OSS to the NE. Packets and frames were of little importance except as packages for the TL1 messages and higher-layer protocols. But in 1988, Bellcore specified CMIP as the protocol to be used for TL1 messages. No one really worried, since there were TL1 implementations that used CMIP, X.25, or even IP as the protocol package for the TL1 messages.

There was nothing wrong with using TL1 for SONET. After all, TL1 worked. However, there were and are several limitations that TL1 suffers from that are beginning to become a problem in networks.

First, TL1 defines a *human-machine interface*. This means that TL1 expects to have a human being sitting at a terminal in an operations center somewhere typing commands and looking at alarms.

It is not easy today to realize how awkward the type-and-read interface is. SONET NEs used to ship with about 3 feet of documentation. Most of it went into a bookshelf or cabinet and stayed there. But one document was never far out of reach: the TL1 command book for that particular device. Figure 13.9 shows what a typical SONET alarm looks like on an OSS console.

This message reports a loss of pointer (LOP) condition at an ADM on a SONET OC-12 ring. Of course, a command to the SONET NE in distress had to be composed, formatted, and typed properly in just as arcane a format. There were some point-and-click front ends built for

Figure 13.9
A typical TL1 alarm
message

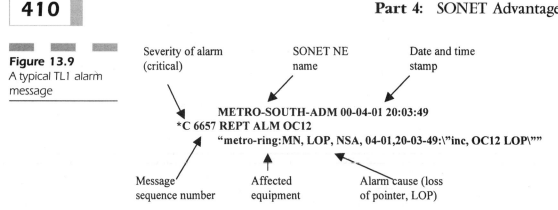

TL1-based OSSs, but all these did was interpret and formulate the TL1 on behalf of the human operator. Besides limiting the range of possible TL1 messages to what was programmed, these graphical front ends did not enhance TL1 in any way.

In the simpler world of T-carrier, this was not a particular problem; however, in the SONET/SDH world, things happen faster. Not only are line rates higher, but protection switching should happen in milliseconds. SONET/SDH would be better served by *machine-machine interfaces*. This would allow a management process (or program) running in a SONET/SDH NE to communicate with a management process (or program) running on a computer in the network management center. Then decisions could be automated to a large extent and the whole task made more efficient. CMISE includes such interfaces.

The second problem is that TL1 is a Bellcore and T-carrier specification. As international links become more common, a transition from TL1 to some form of CMISE would be welcome. Indeed, this is happening now. CMISE/CMIP use is also the foundation for the ITU's standard for Telecommunication Management Networks, or TMNs. TMNs will be discussed more fully a little later. However, this chapter still explores some aspects of TL1 due to its position in SONET networks today.

As if TL1 and CMISE were not enough, there is a third contender for network management protocols. A major effort into network management standards has been through the activities of the Internet Engineering Task Force (IETF). The current importance of the Internet need not be commented on here. The emphasis is on network management and the Internet.

The initial Internet was organized and built through the ARPANET research project that originated in the United States. In 1971, the Defense Advanced Research Project Agency (DARPA) continued the work of the earlier organization. DARPA's work in the early 1970s led to the development of the Transmission Control Protocol/Internet Protocol, known as TCP/IP. Early network management techniques on the Internet were proprietary and quite spotty. In the last ten years or so, the Internet has assumed the lead in setting standards for network management for all devices, not just Internet routers and other components. This is the Simple Network Management Protocol (SNMP).

SONET vendors and service providers are faced with the difficult decision of sticking with TL1, or migrating to something else. But what? SNMP or CMISE? Both are open standards, but only CMISE is currently an official ISO and ITU-T international standard. Yet SNMP is much more widespread, especially as the Internet has grown around the world. For now, most vendors seem to be backing TMNs for the long haul.

Of course, the key goal of all network management standards is to develop an integrated set of procedures and standards that will work equally well across different vendors and products. Both CMIP and SNMP offer enhanced network management services beyond the simple alarms and indications of the basic SONET OAM&P. Both CMIP and SNMP can run inside the SONET DCC bytes to actually try to manage failed equipment. For example, DCC can be used to run a remote diagnostic on the failed device and report the results to the network manager.

Both SNMP and CMISE use one other component that is essential for all standard network management systems. This is called the management information base (MIB). In spite of its name, the MIB is a database of network management information that resides in the managed device. Technically, the MIB is just a description of the database fields and contents. When implemented in a piece of equipment, the MIB becomes the *managed object*, but most just use the term, MIB. Therefore, MIB defines the content and structure of a database that is common among similar network devices in order to provide information about the managed network elements.

With either SNMP or CMISE, the managing process, or management workstation, is responsible for directing the actions of the managed system or agent. The agent software would typically reside in a SONET NE. The managed system contains a remote manager as well as the managed objects defined by the MIB. This remote manager is called the *agent*. The agent is software that is responsible for receiving the network management messages from the managing process and making sure that proper access control measures are taken regarding the managed objects (so that no one can hack their way into the device). The agent controls the logging, or recording, of network management information as well. Through the use of a process known as *trapping* in SNMP, or an *event-forwarding discriminator* in CMIP, the agent makes decisions about whether messages are to be returned to the managing process.

Although there may be a one-to-one relationship between a managing process and a managed system, in the sense that the managing process always manages all managed systems, the managing process itself can also be an agent (managed) process. No restriction applies to the roles that these two entities play. In fact, the roles can even be exchanged. Most commonly, a device with modest processing capabilities and a small agent will be managed by another process which, in turn, is an agent managed by the network management system. This is called the *proxy agent* arrangement because the device is not directly managed by the network management system.

OSI and SNMP Network Management Layers

Although they share many similarities, such as agents and MIBs, OSI and SNMP network management standards are quite different in implementation. The keyword in SNMP is *simple*. Ironically, the simplicity in SNMP that led to its widespread acceptance eventually proved to be a liability. There was little in SNM in the way of security, authenticity, or even reliability. These have been added to form an SNMP version 2 (SNMPv2), but this effort has been marred by political infighting among Internet factions and so has languished. The net effect has been to turn SONET vendors towards OSI and CMIP.

OSI Network Management

If SNMP seems overly simple for complex network management tasks, then OSI seems overly complex for even simple network management tasks. The OSI network management model is consistent with the overall OSI application layer architecture, layer 7 of the OSI reference model.

In addition to defining a series of MIBs, or managed objects, OSI also provides several standards for the use of these objects in widely used management areas. The OSI management standards help in defining standard ways to solve common network management problems. There are several system management functions (SMF) that are designed to support the more generalized system management functional areas (SMFAs) and include fault, configuration, accounting, performance, and security management. Each of the SMFs, along with the standard objects, can contribute to one or more of the functional areas.

Several standards address the set of SMFs, ranging from low-level object management to more involved application areas, including security, accounting, and performance. They provide a useful framework for the use and application of the base MIB object classes, which are defined in the X.721 standard. The standardized SMFs include the following:

- Object management (X.730), which includes basic object manipulation services. These primarily consist of the CMIS functions, and do not contain extensive object models.

- State management (X.731), which defines standard mechanisms for management of object states. Services include state monitoring, state change notifications, and state change commands.

- Relationship management (X.732), which defines ways in which managed objects relate to one another.

- Alarm reporting (X.733), which defines the mechanisms for alarm reporting. This includes the mechanisms for alerting objects that an alarm has been detected along with the characteristics of the alarm that can be used to better understand the nature of the condition being alarmed.

- Event management (X.734), which provides a set of mechanisms to control the distribution of event reports within the management system.

- Log control (X.735), which provides mechanisms to manage the recording of management information.

- Security alarm reporting (X.736), which defines services for managing event-forwarding discriminators that are responsible for managing the flow of security-critical alarm notifications.

- Confidence and diagnostic testing (X.737), which defines services and objects that can be used in managing the execution and reporting of system tests.

- Summarization function (X.738), which provides a framework for the definition, generation, and scheduling of system information summary reports.

■ Workload monitoring function (X.739), which manages and controls workload monitoring services.

There are additional ISO standards that have been developed to address management functions including testing and metrics. Additional SMFs may be included in the already significant list of standards as the technology matures.

The object management SMF defined in X.730 identifies the services that a management entity should provide in the handling of system objects. These services have been mapped into two sets. The first are "pass through" services that are directly mapped from the CMISE services. The second are "direct services" that provide additional services beyond the basic CMISE services. The "pass-through" services map directly to the MIB object creation, deletion, action, set, get, and event report primitives in CMISE. The "direct services" consist of additional reporting services, and are specified to assist managers in tracking the changes that can occur to objects within a managed system.

Obviously, OSI network management is quite comprehensive, although somewhat complex, especially when compared to SNMP. The attraction of CMISE for SONET is the ready implementation of an international standard that applies to SDH as well as SONET.

Of course, these standards all rely on the underlying OSI application layer standards. These are shown in Figure 13.10.

In Figure 13.10, the systems management application service element (SMASE) creates and uses the protocol data units (PDUs) transferred between the management processes of the two devices: manager and

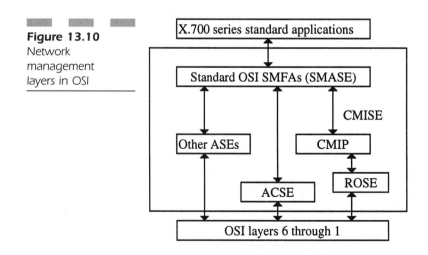

Figure 13.10
Network management layers in OSI

managed. These data units are called management application data units (MAPDUs). The SMASE messages have to do with the five OSI SMFAs of fault, performance, configuration, accounting, and security operations. The CMIP protocol supports these messages through the use of a common set of procedures (e.g., Get, Set, Create).

The SMASE component may use the communications services of various application services elements (ASEs) or the CMISE. The use of CMISE implies the use of CMIP with either ROSE or ACSE, which are just other elements of the full OSI application layer standards.

The purpose of the ACSE is to set up an association, or session, between the managing process and the agent. Once the session has been established, ROSE is used by the SMASE and CMIP to invoke operations and exchange information between the managing process and the agent.

An example of the OSI SONET message flow for OAM&P operations is shown in Figure 13.11. Here a managed system, usually the agent software in a SONET NE, is prompted or is configured to periodically send some performance statistics to the OS at the management center. These statistics are application independent; thererfore, it does not matter what the SONET link is transporting. There are reports on traffic conditions at the NE, error reports, and so on. Next, these statistics are passed to CMIP, which creates a header to indicate just what type of network management message is being conveyed to the peer CMIP process at the OS site. This header also informs the companion CMIP how to process the traffic: in other words, what type of message CMIP is to generate as a result of the message. In the example in the figure, it is an Event-Report type message.

Next, the performance information and the CMIP header is passed to the ROSE process, which creates an Invoke header and appends this header to the data unit originated from the upper layers, in this case the

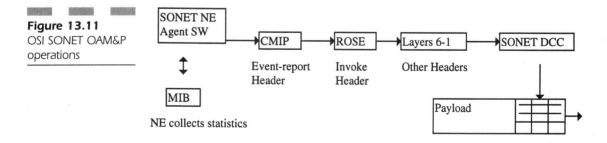

Figure 13.11
OSI SONET OAM&P operations

performance data itself and the CMIP header. This header is used by the peer ROSE process at the OS site in order to determine its actions. This could be whether it is to report back to the SONET NE about the success or failure of the operation at the OS, or other possible actions.

Next, this information is passed down through the lower OSI layers of the SONET equipment. When the SONET layer receives the information it places these bytes into the DCC bytes of the SONET header. It could also use the payload of the SONET envelope, but the DCC method is preferred and more common in order to keep the payload area exclusively available for user traffic. Finally this information is sent to the OS management process in a series of SONET frames, since each DCC, line or section, only accommodates a few bytes per frames.

At this OS computer, the entire encapsulation process is reversed. The various layers of headers and data units are passed up through the OSI layers to reach the performance and management process itself, which processes the data in accordance with some standard SMFA application-specific requirement.

SNMP Network Management

Compared to OSI, the layering for SNMP atop the Internet suite is much simpler than the OSI suite. The SNMP protocol forms the foundation for the Internet network management architecture. However, the network management applications (the Internet equivalent of the OSI SMASE) are not defined in the Internet documentation and specifications. In SNMP, these applications consist of vendor-specific network management modules, such as fault management, log control, security, and audit trails. It is somewhat ironic that SNMP is everywhere, but has no standard network management applications, while OSI network management applications are standard and full, but OSI implementations on which to run them are difficult to find. The utter simplicity of the SNMP architecture is shown in Figure 13.12.

As shown in Figure 13.12, SNMP is used by the custom-written (usually by the SONET equipment vendor) network management application and uses the User Datagram Protocol (UDP). UDP is connectionless, and the wisdom of using a connectionless protocol for network management has been repeatedly questioned; much can be learned simply from whether a connection to a network device can be made or not. SNMPv3 (SNMPv2 suffered from many problems) allows the use of TCP, which is

Figure 13.12
SNMP Architecture

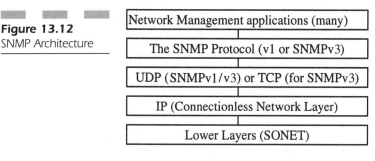

| Network Management applications (many) |
| The SNMP Protocol (v1 or SNMPv3) |
| UDP (SNMPv1/v3) or TCP (for SNMPv3) |
| IP (Connectionless Network Layer) |
| Lower Layers (SONET) |

connection-oriented. In most cases, with SNMP, UDP is used and uses IP packets to form its messages. IP uses the lower layers of the Internet protocol suite, which correspond to the data link layer and the physical layer of OSI.

When used for network management, SNMP's function (along with some other software) is to keep track of the ongoing SNMP operations (get put) between the agent and manager process. UDP directs the messages to and from the network management ports, and IP packet addresses are used to identify and route the traffic being exchanged between the two devices.

The figure merely shows some generic "network management applications" running on top of SNMP. None of these is defined by Internet standards, but nothing would stop an enterprising vendor from rewriting all of the SMFA standards for SNMP. To date, however, most vendors have chosen to gear their network management products closely to their specific products, making them difficult to use in a multivendor environment. This will slowly change as OSI network management becomes more common in SONET devices.

The interactions of the SNMP layers with respect to SONET are almost identical, if simpler, than the OSI sequence. In this case, SONET maps the SNMP management traffic instead of OSI into the DCC fields or the payload area for transmittal to the management process or OS site.

Standard Objects and SONET

It should come as no surprise that standard technologies like SONET demand standard network management techniques like SNMP and OSI. The major benefit of any standardized network management package is the ability to identify, in an unambiguous manner, any SONET network element. Actually, this is true of *any* kind of network element. To do so, the network management standard must include an associated naming and identification registration program.

The concept of a registration program simplifies internetworking between different network components that need to know about the location and identities of other network components, like network management software applications. It is a concept similar to the telephone system in which there is a standard and worldwide agreement on a hierarchical numbering hierarchical scheme. After all, a telephone number is just a network address of a device on the voice network.

Both SNMP and OSI use a standard object "tree" for designating devices and the MIB objects they contain. ISO maintains this international registration and naming hierarchy. Vendors can apply for a "branch" in this standard naming and numbering scheme associated with this registration. Once registered, a customer with any standard network management package, SNMP or OSI, can access a SONET network management database MIB. The objects (fields) in this database are easily obtained by using the registered names (the object identifiers) that are registered and known by the network management software vendor and, thus, the customer as well.

OAM&P, the ITU, and SONET

There is more to SONET OAM&P than failure conditions and alarm indications. There is even more to SONET network management than SNMP or OSI protocols. What should be done with the information gathered by these means? What exactly is the OS doing in a SONET network? This section looks at the overall architecture for SONET OAM&P at a higher level.

The overall architecture for OAM&P communications is defined by the International Telecommunication Union (ITU), an agency of the United Nations based in Geneva, Switzerland. This architecture, known as the Telecommunication Management Network (TMN) architecture, is shown in Figure 13.13. TMN provides a framework for an OS at a network operations center to communicate with the physical network and a way for alarms to be sent to the OS. An OS is a sophisticated collection of software applications that manages the entire network.

An OS can gather information from the network along with alarm, status, and performance statistics. The OS can perform remote testing and network configuration tasks. An OS may also maintain information

■■ ■■ ■■■
Figure 13.13
The TMN
architecture

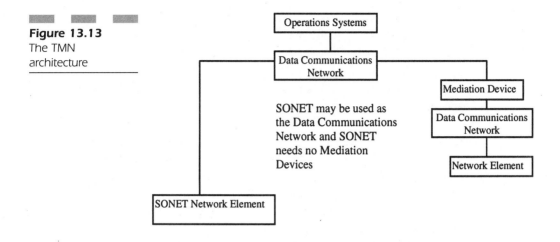

of facilities available for provisioning and may be used to provision resources by sending remote commands to enable or disable links.

As shown in Figure 13.13, the OS communicates with the managed NEs either directly over a data communications network or through a *mediation device.* A mediation device can support multiple functions, such as consolidation of data links for devices in a specific geographical portion of the network (east coast/west coast), convert between proprietary OAM&P techniques for different devices, and handle a collection of performance statistics into a database.

The advantage of SONET is that *no separate data communications network is needed,* because all SONET OAM&P information flows on the network itself in the SONET overhead bytes. Because SONET is entirely standard, *no mediation devices need be used,* unless the service provider wishes to still provide for separation of OAM&P functions (east coast/west coast).

Telecommunications Management Network (TMN)

TMN is a huge, sprawling topic all on its own. Only those pieces of TMN of direct concern for SONET/SDH are discussed here. This is not intended as a TMN tutorial, many of which are available both in print and on the Internet/Web. It is enough to note that the most important component of TMN for managing SONET/SDH NEs is known as the

Q3 (often, q3) interface. The Q3 interface is used for communications between a TMN-compliant operation system (OS) and various other components of TMN. Only devices equipped with a Q3 interface can be directly managed by the TMN OS. The Q3 interface specifies that TMN messages using CMISE are sent between NEs using CMIP. In other words, there is no place for TL1 in Q3.

So how can legacy SONET/SDH NEs that still understand only TL1 be managed? There are two possible ways. First, the SONET/SDH NE can employ a Q-adapter (QA) which understands Q3 on the OS side but still understands TL1 internally. A QA therefore could be placed in each and every TL1-managed SONET NE to make it compliant with TMN. This is usually a combination hardware/software upgrade (sometimes firmware upgrade) that many SONET/SDH vendors have made available for their NEs.

Second, the Q3 interface can terminate at a TMN mediation device (MD). Not only can the MD employ TL1 to directly manage non-TMN-compliant NEs with a Q-adapter, the MD can also be used to gather regional information about a cluster of SONET/SDH NEs (perhaps a SONET/SDH ring). So the MD can act as a kind of proxy for the TMN OS. The MDs support Q3 on the OS side and can use various Qx interfaces to communicate with QAs or NEs directly. The important thing is that any non-Q3 interface be shielded from the view of the OS itself. The use of these interfaces is shown in Figure 13.14.

Figure 13.14
The use of the TMN
Q3 and Qx interfaces

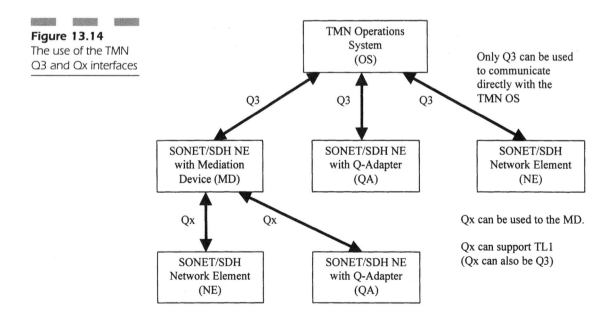

Some SONET/SDH vendors began experimenting with MDs as well as QAs for their NE devices. However, the MD became a real issue because the presence of the MD was a massive database and associated processing burden without offering any real incentives. MDs complicated and congested the network. Many vendors therefore emphasized the direct QA approach for TMN compliance. QAs exist today for all four of the common variations employed to convey TL1 messages back and forth to SONET NEs. These are shown in Figure 13.15.

In spite of the complexity of QAs in and of themselves, the emergence of TMN at least has offered the potential for service providers to buy SONET/SDH NEs from different vendors using different transport protocols for TL1 messages and to manage them all through one OS.

It is important to realize what the SONET/SDH DCCs really do for network management. Because of the presence of a full OSI stack in each NE, the SONET/SDH network becomes more or less a router network for network management purposes. The routing tables in each SONET/SDH NE are all updated by a routing protocol, so the SONET/SDH network (e.g., a SONET ring with attached SONET TMs) maps its own topology automatically. All the OSS operator needs to do is inject a message into one SONET/SDH NE directly, and the DCCs take care of routing the message to the proper NE. The same goes for the reply. Figure 13.16 shows how a SONET/SDH ring could be managed today.

Figure 13.15
TMN Q-Adapters for
TL1

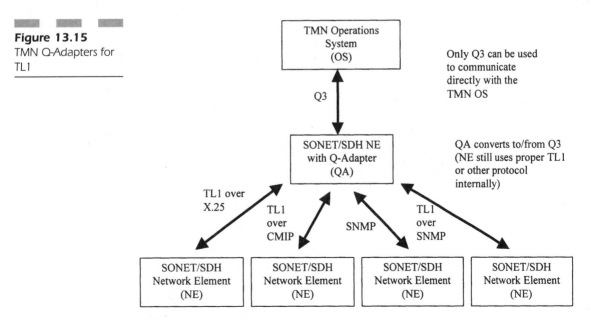

Figure 13.16
Managing
SONET/SDH NEs

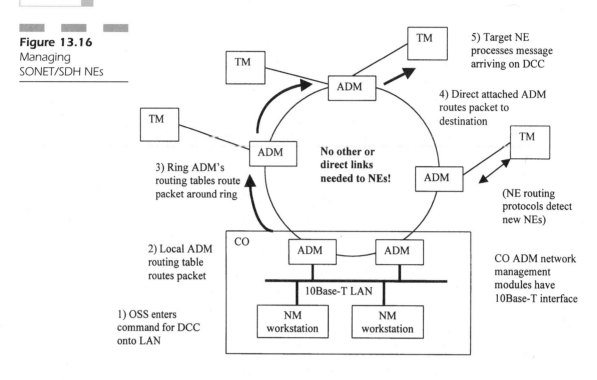

The figure shows:
- 5) Target NE processes message arriving on DCC
- 4) Direct attached ADM routes packet to destination
- No other or direct links needed to NEs!
- 3) Ring ADM's routing tables route packet around ring
- (NE routing protocols detect new NEs)
- 2) Local ADM routing table routes packet
- CO ADM network management modules have 10Base-T interface
- 10Base-T LAN
- 1) OSS enters command for DCC onto LAN
- CO, ADM, TM, NM workstation

Figure 13.16 shows a simple ring with two ADMs located in a central office (CO) and three other ADMs on the ring itself. The remote ADMs have TMs attached to gather and groom customer traffic. Each CO-based ADM has a network management module with a common 10Base-T Ethernet interface. Two network management workstations are attached to the LAN. This arrangement provides redundancy for both workstations and ADMs.

Now, suppose an operator needs to reconfigure the VTs on the TM in the upper right of Figure 13.16. No other links (or direct links) are needed with SONET. The operator simply points and clicks on the icon of the TM on the screen, and the proper TMN or TL1 message is packaged up (perhaps "put yourself into configuration mode") and sent across the LAN to the ADM that is active for network management (there is no need to duplicate the message). This ADM has a complete routing table to shuttle such messages around the ring. The message is inside a packet, just like everything else today. The message inside the packet is routed onto the ring in the DCC (because the TM is a line-level device and command, even though the complete operation will affect the VT paths, so the line DCC is used). Intermediate ADMs have their own DCC routing tables, so the packet is passed through to the ADM directly attached to the target TM. Once its own packet address is recognized by the TM, the packet is parsed and the message retrieved. Any reply is routed through the collection of NEs in the same way. Routing protocols running between the NEs take care of automatically detecting and mapping the topology of newly installed TMs and ADMs.

Nothing would stop the LAN shown in Figure 13.16 from being extended to other rings and locations in a variety of ways. So a remote OSS could easily gain access to the SONET/SDH ring as well. But local management is often provided by a standard and common 10Base-T Ethernet LAN interface built into the network management module of at least one of the SONET/SDH NEs. Both network management and access to the ring should be redundant.

Security is not much of a concern, at least not to the extent it is on other packet networks like the Internet, because users have no access to the SONET/SDH overhead bytes at all. Remote access ports on the management LAN *are* vulnerable, however, and steps must be taken to reduce the threat of unauthorized access by this means.

SONET Management Information: 1/7/24

All SONET equipment, whether repeater or Terminal Mux (TM) or Add-Drop Mux (ADM), must comply with certain Bellcore (Telcordia) network management standards for SONET. All equipment must gather network management information and performance statistics for the previous hour, previous seven hours, and previous twenty-four hours (known as "1/7/24"). This information is stored in the equipment itself, and gathered by some centralized network management center on a periodic or as-needed basis. What follows is basic management information. The amount of information that SONET NEs must gather has been extended. Interested readers are referred to relevant Telcordia and ANSI specs for current details.

What are SONET network managers able to monitor in this fashion? The information goes above and beyond alarm conditions, of course, which are usually the triggers for some network management activity that gathers performance statistics.

At the Photonic level, network managers can check the laser bias current relative to some initial, tuned installation value. Lasers tend to "age" and need periodic preventive maintenance, usually every six months or so, for continued acceptable performance. The received optical power also must be monitored to detect a sudden drop, which could indicate either tampering or damage to the fiber.

At the Section level, there are three things to monitor. First, the section BIP is checked for Coding Violations (CVs), which are the "bipolar violations" of SONET. Up to 8 CVs may be recorded per BIP, of course. Next, Out-Of-Framing (OOF) seconds are recorded. These are seconds during which framing is lost. Naturally, the framing will normally only be lost for a fraction of a second. Finally, the number of Errored Seconds (ESs) is tracked, which is defined as seconds with more than one CV.

Most of the monitoring action takes place at the Line level. In addition to the CVs and ESs, defined this time on the Line BIP, the Line level records for Severely Errored Seconds (SESs), which are defined as "x" number of CVs in a second (where "x" is greater than two, but not universally defined). In addition, the Line level watched the H1/H2 pointer justifications, which cannot exceed one in every four frames. All of the preceding are also stored in SONET equipment for the previous six days, showing their importance.

The Line level also records Unavailable Seconds (UASs) during which the SONET equipment is simply out of action. Degraded minutes used to be monitored, which are minutes containing one or more SESs (degraded minutes are no longer tracked). Finally, the time during which the SONET link or ring is Protect Switched is recorded.

At the Path level, SONET equipment keeps many of the same statistics as the Line level, with a few differences. CVs, ESs, SESs, and UASs are based on Path overhead BIPs, of course. Instead of watching H1/H2 pointer justification, the Path level keeps track of VT pointer justifications, when the VTs are allowed to float as the SPEs are allowed to (usually VTs are locked). Also, the Path equipment does not record Protection Switched duration.

All of this information is summarized in Table 13.2. Many SONET/ NEs today are capable of capturing and storing much more information. Table 13.2 illustrates only the basics.

All of the SONET equipment network management information, from Coding Violations to Protection Switch duration, is only useful when made available to network managers in the network management center. The problem is to allow network managers to query the SONET equipment in some standard fashion, regardless of vendor or service provider, in order to gather the information either periodically or as needed. The results of the queries may be used to diagnose problems, verify tariff compliance, or just to maintain a historical database.

TABLE 13.2

SONET
management
information

Information Type	Photonic Level	Section Level	Line Level	Path Level
Laser bias	YES			
Optical Power Received	YES			
Coding Violations		YES	YES	YES
Out of Frame (OF) Seconds		YES		
Errored Seconds (ES)		YES	YES	YES
Severely Errored Secs (SES)			YES	YES
Pointer Justifications			H1/H2	VT
Unavailable Seconds (UAS)			YES	YES
Protection Switch Duration			YES	

Fortunately, the Section and Line DCCs have been set up by ANSI and the ITU-T for exactly this purpose. The SONET Section DCC (D1-D3 bytes) operates at 192 Kbps and the SONET Line DCC (D4-D12 bytes) operates at 576 Kbps. Both DCCs are supposed to follow the full seven-layer OSI-RM at all layers. Many of these standards exist in both OSI and ITU-T "flavors," often referencing exactly the same text with different titles; of course, unique pieces do exist.

For instance, at the Data Link Layer, the DCCs employ ITU Q920/Q921 Link Access Protocol for the D channel (LAP-D). The default option is to be Acknowledged Information Transfer Service (AITS), which will perform error recovery of each SONET link; however, implementers are required to offer Unacknowledged ITS (UITS) as an option. This will help to speed things up on the DCC.

At the Network Layer, the DCCs employ ISO 8473 Connectionless Network Layer Protocol (CLNP), which is ISO's "version" of IP. DCC information is independently routed network by network. The Transport layer uses ISO 8073/8473, which describes a connection-oriented protocol similar to TCP. In fact, it describes a method of supporting ISO's Transport Protocol 4 (TP4) over CLNP, which forms ISO's version of TCP/IP, or TP4/CLNP.

There is considerable DCC documentation overlap at the higher layers. At the Session layer, the DCCs use ISO 8073/8473 and/or X.215/X.225, which provides sessions to recover from transport failures and access validation for security. At the Presentation layer, the DCCs use ISO 8327 and/or X.216/X.226 and X.209, which describes a common Abstract Syntax Notation language (ASN.1) for internal data representation differences on the part of SONET equipment vendors, as well as compressed ASCII coding and encryption.

The Application layer uses the usual ACSE/ROSE/CMISE/SMASE network management structure. ISO 8650 and/or X.217/X.227 defines the ACSE, ISO 9072 and/or X.219/X.229 defines the ROSE, and ISO 9595/9596-1 defines the CMISE for SONET. These pieces only essentially define an API for network management, not the network management application itself.

TMN brings order to this appaent chaos. Many vendors run TL1 in the F1 "user channel," which is totally in line with TMN.

TMN Issues

In a perfect world, TLI would evolve to TMN and that would be that. Vendors could migrate from using TL1 messages inside X.25 or SNMP or whatever by using Q-Adapters and eventually native TMN messages and protocols. However, just as TMN is reaching a certain level of maturity and TL1 is entrenched enough to have considerable momentum on its side, the whole arena of telecommunications network element management is being challenged by changes in the way OSSs are deployed and used. The new challenges arise from considerations of the use of Java, the Common Object Request Broker Architecture (CORBA), the Microsoft distributed component object model (DCOM), and the Web itself. Certainly there is nothing in TMN that addresses any of these newer concerns. Some have gone so far as to pronounce TMN dead on arrival, with the use of the CORBA Internet bus model as the most logical successor.

This section is intended to at least introduce some of the latest ideas in the OSS field. It seems premature to write off TMN entirely at this point in time. But the whole point is that it looks more and more like TMN alone cannot accomplish all of the goals the service providers are seeking to achieve with their networks today. The root of the issue is not TMN (or even TL1) itself. The messaging protocol used in all OSSs simply has to be understood by each end of the conversation. The issue revolves around the infrastructure used at the lower layers of the OSI model to shuttle these messages through the distributed communications network linking all the NEs together.

Distributed computing and networking today means the Internet protocol suite (TCP/IP). CORBA adds distributed object computing support to TCP/IP. A full discussion of exactly what "objects" are or are not would fill many pages and is not important here. What is important is that objects can be defined once for some purpose (e.g., network management) and then used over and over again in other forms to provide information in billing, provisioning, order tracking, and related systems.

As an example of the attraction of objects today, consider the case of simple error tracking for a VT provisioned on a SONET ring. The bit error rate (BER) the customer is promised is a key part of the tariff or contract. But how can a customer possibly know what the BER on a SONET VT is? One way is to take the alarms or errors recorded by the NEs involved (which go to the OSS, of course) and post them in mutated form to a Web site. Customers could access the Web site in a secure fashion and yet see only the VTs that belong to them. All of the appearances of the BER-VT association, from network management workstation to Web site to customer Web browser, are just different manifestations of the same object. CORBA uses TCP/IP to connect all these instances of the BER-VT object. No conversions are necessary to relate all of these views of what is, at heart, the exact same information. CORBA allows the Web site software application to communicate with the OSS software regardless of vendor or whether they support TMN. (Purists might cringe at this extremely simple example of objects, but the underlying reasons for object use are the important thing here.)

It is somewhat ironic that CORBA is mostly used with the UNIX (and now LINUX) operating system and the C++ programming language, while many clients and servers on the Internet today are Windows-based devices. Microsoft's DCOM supports DOS and Windows, as well as Visual Basic. The common link between the two environments is Java, which both DCOM and CORBA buses can use, although Microsoft also employs ActiveX in place of Java in many places, and many Windows applications are written in C++. These are just generalizations based on most common usage. The lure of CORBA and to some extent DCOM to enhance or at least complement TMN is shown in Figure 13.17.

In Figure 13.17, all traditional network element management functions are still performed with either TMN and/or TL1. But in addition, CORBA is used to allow direct Java or C++ access to this information. An application could be written using CORBA, for example, to allow a field technician with a personal digital assistant (PDA) or browser-enabled cell phone to access the information stored at the network management workstations connected to a SONET ring. For back-office

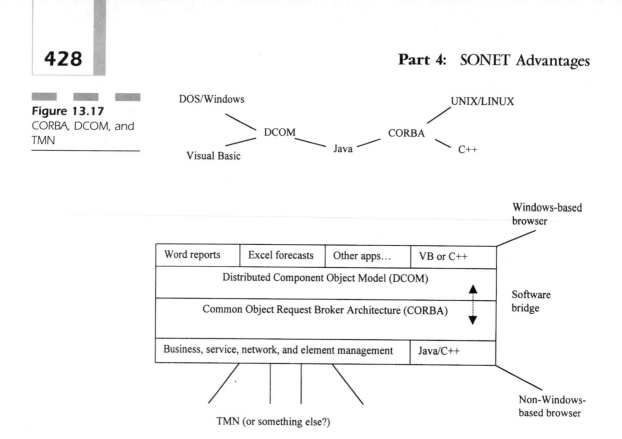

Figure 13.17
CORBA, DCOM, and TMN

operations, a simple DCOM-to-CORBA bridge could be written to allow direct access to this information from any Windows browser. For reports, modeling, and so on, all of the TMN or TL1 information could be directly piped as objects into spreadsheets or reports. Those with a bleak view of the TMN future see little point in duplicating CORBA/DCOM capabilities into a TMN-compliant distributed network. Better to extend the CORBA/DCOM structure into the TMN/TLI world of SONET/SDH NE management with TCP/IP, they argue.

Now, *object servers* can be added to TMN and TL1 as well, to take information from NEs and massage it for Web sites and servers. But CORBA is complex enough and powerful enough to consider extending CORBA to the OSS rather than interfacing the OSS to CORBA.

After all, most service providers today use CORBA (and more) on their own IP networks for e-commerce and e-business. They have all the hardware, software, security, procedures, and so on, for this environment already in place and running. Why attempt to recreate the whole thing just for TMN purposes? It is even possible to redefine TMN to ride on top of CORBA rather than CMISE/CMIP. It all seems to boil down to a question of what is more useful and valuable to the organization as a whole: CORBA or TMN?

What would happen to TMN in a CORBA/Web world? TMN is really three things. First, TMN is a layered model with systems for viewing the network as a series of business, service, network, and element management pieces. This is all part of the TMN distributed approach. Second, TMN establishes a system of common names and relationships for all these items, from NEs to trouble tickets. This piece is mainly for vendor independence. Finally, TMN establishes a distributed network for connecting these consistently named and addressed components. Some of the TMN standards rely on OSI (CMISE/CMIP) for this last piece.

At least replace OSI with CORBA in the third component of TMN, say the critics. Why reproduce (for instance) the security already in place for CORBA with unproven, untested, and redundant OSI security? The next step would be to replace TMN and OSI naming and addressing with IP/Internet naming and addressing. Why not give everything an IP address and be done with it? And once these two pieces of TMN fall to the IP empire, why not just do everything with some secure version of SNMP?

And so the debate rages. There are those that argue that as long as TMN clings to its first mission on business and element and other types of management, TMN is not dead, just changed. Others see any crack in the TMN architecture as a whole as a sign that the whole ediface will come crashing down. Still others see TMN as providing a certain level of security because if it is based on something other than IP, no one will bother to hack into it! (Obscurity never stopped hackers from getting into telephone company management systems like the LMOS, the loop maintenance operating system.)

The whole issue of the proper relationship between TMN and CORBA and the whole arena of getting SONET/SDH NE information to and from other parts of the enterprise is just starting to crystallize. Stay tuned.

Transaction Language 1 (TL1)

This chapter closes with a more detailed look at the TL1 language. Until full OSI implementations of CMIP, the X.700 standards, and TMN become common, TL1 will occupy an important place in SONET network management. This section provides a fairly detailed overview of TL1, based on the Bellcore documentation (SR-NWT-002723).

TL1 is a generic Bellcore network management human-machine interface for managing network devices. Because the decision was made to support a transition in SONET to OSI management based on CMISE quite early in SONET deployment, only a limited set of TL1 messages was defined for SONET NEs. This message set allows service providers to establish services, to monitor their SONET networks for failures, and to perform common and needed administrative tasks associated with maintaining and monitoring NEs in all simple SONET architectures. These include point-to-point, daisy-chained, tree, and path-switched ring architectures.

TL1 for SONET has the following two objectives:

1. Provide Bellcore's own view of how equipment vendors and suppliers can map SONET requirements to specific TL1 messages and support implementation of these specific requirements.

2. Identify the distinctly SONET requirements for which TL1 messages do not exist. In some cases, the existing TL1 message set was extended to support the requirement. In others cases, new messages were needed.

In order to perform their tasks, network OSs generate, accept, and process subsets of the existing generic TL1 message set. Equipment vendors and suppliers use something known as the Operations Systems Modification for Intelligent Network Elements (OSMINE) process in order to get support for non-generic TL1 messages in Bellcore compliant OSs.

Why bother with TL1 at all? Bellcore hoped that if suppliers supported TL1, the SONET NEs used by a typical service provider would be more consistent in security, provisioning, and maintenance with regard to OS/NE interface implementations. This level of standardization would result in reduced operational expenses for deployment of these products in a SONET network. If everyone were to follow these OS/NE message mappings, it also may reduce the expense of determining conformance to the proposed Bellcore requirements in product testing and analysis.

The focus of TL1 messages for SONET is on SONET NEs. This is hardly surprising. However, although they identify the core set of TL1 messages applicable to SONET NEs, Bellcore did not mean that all NEs must support the entire TL1 message set. An NE's TL1 message set is dependent on the interfaces and the features that are supported. A good example is the conditional requirement for exercising protection switching. If the feature were not provided, then the associated messages should not be needed. All additional features in an NE, or features for which messages have not been provided, may need other TL1 messages beyond the basic set.

SONET ADMs are often configured in several types of self-healing rings to provide network survivability. These NEs require special equipment functionality and the exchange of additional information across the OS-NE interface. The associated TL1 messages for rings are discussed in Chapter 14.

There are also two specialized applications for SONET technology for which TL1 messages have been defined. These are the Fiber In The Loop (FITL) and Integrated Digital Loop Carrier (IDLC). For the sake of completeness, the TL1 messages required by FITL and applicable for IDLC systems are mentioned in this section.

TL1 MESSAGES All of the TL1 messages were developed to provide a generic human-machine interface language between NEs (where the machine is) and OSs (where the human is) and meet a series of generic network operations requirements. All of these TL1 messages fall into two general categories: *command-response,* and *autonomous.*

In the command-response type of messages, the OS initiates commands (usually by simply typing them in or clicking on some icon) for the NE. The SONET NE must provide a response message. Conversely, an autonomous message is initiated by the NE to inform an OS about its current status. In this case, the OS does not respond to an autonomous message itself.

The TL1 language has its own semantics and syntax, but it is not necessary here to detail the full language. Instead, the emphasis here is on the use of TL1 for SONET networks.

The specific TL1 message set for OS/NE communications falls into two general areas: memory administration and network maintenance. Both are defined in four Bellcore TRs. These are Security Administration Messages (TR-TSY-000835), Memory Administration Messages (TR-NWT-000199), Surveillance Messages (TR-NWT-000833), and Access and Testing Messages (TR-NWT-000834). Moreover, some messages applicable to SONET can be found in two Bellcore TAs: Memory Administration Messages (TA-NWT-000199), and Surveillance Messages (TA-NWT-000833).

In TL1, security is technically part of Memory Administration because it deals with databases resident in an NE's memory. These security messages are found in separate Bellcore documents because the security database is independent of the data necessary for the functional operation of the NE. That is, security is a separate issue from TL1, as important as it is.

In spite of efforts to cleanly divide TL1 areas, there are instances of overlapping functions between memory administration and network maintenance, and between network maintenance and testing. For instance, setting performance monitoring parameter thresholds (e.g., for errors) has aspects of both memory administration and surveillance functions. TL1 messages, therefore, are provided in both areas.

However, there are significant differences in the messages defined for these two areas. For example, the "EDIT" TL1 command in memory administration can change all parameters for a specific entity. That is, all parameters (e.g., monitored threshold values, the state of service) are changed with this single EDIT command. In contrast, the network maintenance messages have specific commands for more individualized functions. The SET-THRESHOLD command, therefore, sets performance monitoring threshold values, the RESTORE command places an entity into service, and so on.

SONET AND TL1 MESSAGES   A good place to begin a discussion of TL1 messages is with security. All TL1 messages involve security considerations. Security in TL1 involves, among other things, logins, passwords, and security options to control access to NEs and their management databases. The TL1 security requirements focus on both the users of the NE and the security administrator for the NE. A login ID is the "name" by which an NE recognizes a valid user. The claimed identity of the user can be verified by several methods in TL1, including passwords. Security options may restrict a user from executing specific TL1 commands, or may limit the user to using only certain fields and parameters within a specific command. Security may also restrict the operations channels and resources available to users. Security features also include tools to provide audit trails, administer the security database, and perform consistency and reliability checks. TL1 security requirements apply to all SONET NEs in all configurations. TL1 security messages can be originated by one of three sources: the user, the administrator, or the NE itself.

In TL1, *users* are defined as those people (or systems) who have a need to access the NE for performing OAM&P functions on the NE itself or the overall network. A SONET NE should recognize the following TL1 user commands:

- ACTIVATE-USER. This permits a user to initiate a session with an NE. The command supports the important requirement for user identification.

- EDIT-PID. This allows users to modify their personal identifiers.
- CANCEL-USER. This permits the user to terminate the session. This is part of the session, or system access, control.

Other related user functions, not explicitly required for SONET NEs, allow users to examine their permissions and the security permission levels associated with the current channel used for TL1 messages.

In order for a SONET NE to recognize the user's IDs, the NE must have a database. Thus, the requirement that SONET NEs administer user IDs also requires the NE to support the security messages designed for the security administrator. This only makes sense.

TL1 Messages and Memory Administration

In TL1, the messages that fall into the class known as *memory administration* are those that control and access the data in the databases of a SONET NE. These databases contain configuration information critical to the services provided by the NE, such as cross-connection information. These messages also contain information involved in maintaining a certain level of quality for the services. Thus, memory administration deals with the addition, deletion, editing, and retrieval of information resident in the NE database. In addition to the manipulation of this information, memory administration also deals with memory backup, restoration, and system administration.

The most common command verbs used in memory administration are ENTER, EDIT, DELETE, and RETRIEVE. These are abbreviated in TL1 as ENT, ED, DLT, and RTRV, respectively. In the following discussion, the abbreviated verb form is used. There also are references to *data dictionaries*. The TL1 data dictionaries contain information about the SONET NE that memory administration commands can access. In this way they resemble MIBs, but not completely.

A TLI message consists of a verb and one or two modifiers: the identity of the SONET NE for whom the message is intended, the entity to which the message is addressed, and the data to be transmitted. The modifier often identifies the data dictionary, which contains the list of permitted keywords and values. SONET data dictionaries in TL1 include:

OC-n. Optical Carrier at line rate n (where n = 1, 3, 9, 12, 18, 24, 36, or 48). This specifies keywords for the SONET Section and Line levels at an optical interface.

EC-m. Electrical Carrier at line rate n (1 or 3). This specifies keywords for the SONET Section and Line levels at electrical interfaces.

STS-p. Synchronous Transport Signal, where p is the rate of the STS synchronous payload envelope (p = 1, 3C, or 12C). This specifies keywords for the SONET STS path level.

VT-x. VT, where x is the rate of the VT synchronous payload envelope (x = 1, 2, 3, 6, or 6C). This specifies keywords for the SONET VT path.

ULSDCC. The Upper Layer Section Data Communication Channel specifies upper layer protocol stack parameters for the Section DCC.

LLSDCC. The Lower Layer Section Data Communication Channel specifies lower layer protocol stack parameters for the Line DCC.

There also are TL1 messages for SONET that do not require data dictionaries. These messages include the modifiers TADRMAP and OSACMAP and are for the verbs ENT, ED, DLT, and RTRV. All of these may be used by what are known as *gateway NEs* to route TLI messages, while OSs and NEs are in transition from TL1 to OSI/CMISE network management.

Provisioning SONET NEs with TL1

TL1 memory administration messages support *provisionable parameters.* These are associated with the SONET DCC protocols, linear automatic protection switching, and information associated with performance monitoring and alarm reporting that involve provisioning.

Much OAM&P activity concerns the establishment of SONET *termination points.* These determine just where a SONET link begins and ends. With TL1, SONET termination points may be created by use of the ENT-OCn and ENT-ECm commands for an optical facility or electrical facility, respectively. As an example, the ENT-OCn command may be used to specify that the DCC is or is not supported on this link, that the OC-n is or is not a protection line, and whether it is the working line in a protection group, and so on. Another key function of these commands is to specify whether the interface is drop (customer or tributary) or line (trunk) side. These commands could also indicate the type of board in the NE, if the equipment can contain multiple cards, each with different SONET OC-n characteristics.

The TL1 messages ED-OCn, RTRV-OCn, and DLT-OCn allow a user to modify, retrieve, and delete information about the identified OC-N, respectively. A similar set of messages exist for the ECm view.

Payload Mappings and Cross-Connections

When it comes to terminations, SONET NEs (especially TMs, naturally) can support a variety of types of facility terminations, both SONET and non-SONET. Non-SONET signals, or tributaries, are mapped into a SONET VT payload for transport through a SONET network. SONET DCSs and ADMs will also permit the service provider to define the SONET time slot in which the signal is transported.

Both the payload mapping and cross-connect functions are accomplished with a single TL1 command. This is ENT-CRS-rr (i.e., enter-cross-connect-rate). When the cross-connect requires payload mapping, the rr modifier identifies this non-SONET rate. Each of the data dictionaries for non-SONET tributary rates (DS 1, DS 1C, DS2, DS3) contains the keyword MAP and a list of permitted values. The value assigned to the MAP keyword defines just how the signal is to be mapped and gives the content of the appropriate path level signal label in the SONET payload. Of course, messages to cross-connect non-terminated VT-SPEs or STS-SPEs through a SONET NE, such as an ADM, do not require mapping information. Mapping information is required only at path terminations.

Both SONET ADM and DCS NEs have several optional features for which supporting TL1 messages have not been provided. For example, an optional feature is one-way broadcast services. This feature is not supported in the limited TL1 message set, and will probably have to wait for OSI/CMISE.

PROVISIONING AND SECTION DCC With TL1, an individual OC-n (or EC-m) can be provisioned to indicate support of the DCC. The Section DCC uses a full OSI, seven-layer protocol stack. Each layer of the stack contains user-obtainable parameters. The protocol stack has been divided in TL1 into two data dictionary views. These are the Lower Layer Section Data Communications Channel (LLSDCC) and the Upper Layer Section Data Communications Channel (ULSDCC). Only layers one and two are covered in the LLSDCC view, while the remaining layers (three through seven) are in the ULSDCC. The ENT-LLSDCC, ENT-ULSDCC, ED-LLSDCC, and ED-ULSDCC commands are used to create and modify parameters at each of the layers, including the address at each layer. The state of the DCC can be controlled by means of the EDIT command and the primary state (PST) keyword. The RTRV-ULS-DCC and RTRV-LLSDCC can retrieve the current value of any keyword or the entire data dictionary.

Until such time as both the OS and SONET NE support CMISE interfaces, the NEs may communicate via TL1 with the OS. In order to do so, special *gateway NEs* must maintain additional data to direct automatic messages to the intended OS and to route OS messages to the correct NE beyond the gateway. The TL1 commands ENT-OSACMAP, DLT-OSACMAP, ED-OSACMAP, and RTRV-OSACMAP enable the network service provider to maintain a table at gateway NEs that map an OSI application "context identifier" (similar in function to a network address) to an OS's X.25 network address.

During this transition period from TL1 to OSI/CMISE, a gateway NE also needs information to correctly route messages from the OS to the NE. The command modifier TADRMAP may be used with ENT, ED, DEL, and RTRV to maintain a table that maps an NE's TL1 target identifier (TID) to its network address (NSAP, the OSI equivalent).

TL1 and System Backup

The use of TL1 allows the OS to provide for *memory administration* of a SONET NE. This is like a system backup function for the NE's memory, which contains configuration information, of course. The NE must notify the OS of any automatic or local changes to this configuration. Some documents call these "hidden updates."

The REPORT-DATA_BASE_CHANGE message, which is abbreviated as REPT-DBCH, provides this function. What kinds of changes would be reported? Assume that a maintenance person has made some changes to the local database. These changes may include changing the service state of a termination point (using the REMOVE and RESTORE TL1 messages), changing the threshold of a performance parameter (SET-TH), or changing a cross-connect (ENT-CRS, DLT-CRS, ED-CRS).

TL1 Network Maintenance and Alarms

With TL1, the service provider defines which SONET maintenance signals (e.g., AISs, FERF) are to be reported to the OS. When reported, the service provider must decide whether they are reported as alarms or events. This is done in the TL1 network maintenance environment by way of the SET-ATTRIBUTE command. The SET-ATTRIBUTE command is also used to define how other trouble conditions (e.g., failures) are to be reported. The choices are critical, major, minor, and whether or not they are service affecting.

In some cases, the network provider may want an NE to report environmental conditions at the NE's location to the OS. The SET-ATTRIB-UTEENVIRONMENT command is used for this purpose.

All SONET NEs are required to detect and report automatically incoming signal failures, which only makes sense. These are serious problems like loss of signal and loss of frame, and other troubles. Thus, the SONET NE must support the autonomous REPORT-ALARM, REPORT-EVENT, REPORT-ALARM-ENVIRONMENT, and REPORT-EVENT-ENVIRONMENT messages. The first two of these messages also report trouble in the incoming SONET APS protocol channel (bytes K1, K2 in the line overhead).

In the TL1 message set, one area of SONET requirements that is not accounted for is the ability for the user to specify the amount of time a defect should be present (or absent) before the NE reports the failure (or the clearing). These are covered by SONET standards.

In addition to all of the autonomous messages, the NE must allow the network service provider to retrieve current alarms and other information about the state of the NE. The RETRIEVE-ALARM and RETRIEVE-CONDITION messages meet these needs. Other commands are used to retrieve the attributes of the various trouble conditions and to suspend and resume the generation of autonomous messages from the NE. If OC-12 SONET link has just been installed and has no live service on it, but a bad connection is causing trouble reports every ten minutes, the network managers may want to suppress these messages until the final work is completed and live traffic is present. This would be done with the INHIBIT-MESSAGE command. The condition is enabled again with the ALLOW-MESSAGE command. Also, the SET-ATTRIBUTE and RETRIEVE-ATTRIBUTE TL1 messages tell an NE to set or retrieve the notification codes associated with the specified event. The event is a TL1.

Monitoring Performance

The point of monitoring the overhead in a SONET signal is to provide early detection of trouble and to help isolate the cause of customer-reported complaints. In order for this to be effective, the service provider must have the ability to set threshold values for the various performance parameters and the NE must be capable of reporting threshold-crossing messages automatically when the thresholds are crossed. The SET-THRESHOLD and REPORT-EVENT messages provide these two functions. The monitored types (designated as in the TL1 messages) supported by these commands were expanded for SONET.

Just to show how important TL1 monitoring can be, the following list contains the TL1 for each defined SONET performance parameter and the name of the parameter. The symbols (R), (CR), and (O) mean that the monitoring in this area is required, conditionally required, or form a performance objective, respectively.

LBCN	Laser Bias Current Normalized (R)
LPTN	Optical Power Transmitted Normalized (CR)
LPRN	Optical Power Received Normalized (CR)
CVS	Coding Violation count - Section (CR)
CVL	Coding Violation count - Line (R)
CVP	Coding Violation count - Path (R)
ESS	Errored Second count- Section (CR)
ESL	Errored Second count - Line (R)
ESP	Errored Second count - Path (R)
SESS	Severely Errored Second count - Section (CR)
SESL	Severely Errored Second count - Line (R)
SESP	Severely Errored Second count - Path (R)
SEFS	Severely Errored Framing Seconds (R)
PJC	Pointer Justification Count (O)
PSC	Protection Switching Count (R)
PSD	Protection Switching Duration (R in revertive systems)
UASPUn	Available Seconds - Path (R)

Naturally, pointer justification counts are unique to SONET and the associated performance monitoring is identified as an objective. Note that there is no special differentiation between STS line and VT path pointer justifications. This has been a source of some concern.

All SONET equipment must recognize the following TL1 performance monitoring messages:

INITIALIZE-REGISTER

SET-PERFORMANCE MONITORING MODE

RETRIEVE-PERFORMANCE MONITORING SCHEDULE

RETRIEVE-PERFORMANCE MONITORING

RETRIEVE-PERFORMANCE MONITORING MODE

RETRIEVE THRESHOLD

SCHEDULE-PERFORMANCE MONITORING REPORT

ALLOW-PERFORMANCE MONITORING REPORT

INHIBIT-PERFORMANCE MONITORING REPORT

Protection switching is initiated by an excessive BER or a degraded signal in addition to failures. The level of bit errors that determines an excessive BER and a degraded signal BER level can be set by the user, as may be expected. The SET-THRESHOLD command accomplishes this by specifying the monitored types as BERL-LT for excessive BER and BERL-HT for degraded signal.

All SONET NEs are required to support performance monitoring. Performance monitoring is not turned on or off, but autonomous messages can be suppressed with the INHIBIT-MESSAGE command.

TL1 and Testing Process

TL1 documentation defines several functions to help with trouble isolation on carrier networks, but these were not included in the TL1 subset for SONET. In particular, TL1 surveillance messages have not been defined for the following:

- Activation of a corrupted BIP.
- Retrieval of optical power transmit and optical power received values.
- Examination of path trace byte at nonpath-terminating NEs.
- Retrieval of the content of the STS path or VT path signal label.

However, things are not as bleak as they may seem. The existing TL1 DIAGNOSE and DIAGNOSE-DETAIL commands are generic and may be used, in some cases, to meet the needs of SONET networks in this area. The ability to initiate either a facility or terminal loopback is provided in the OPERATE-LOOPBACK command. The REMOVE and RESTORE messages are used to change the service state of the line being looped back.

This final section on TL1 may seem overly long; however, the intention was to give the reader an appreciation of how TL1 operates and at
the same give a feel for what TMN systems in a SONET network must
be capable of doing. Whether TL1 of OSI or SNMP, all TMN systems
can be quite complex.

14

SONET Rings

Perhaps the most distinctive feature of SONET networks, in general, and the most obvious way that SONET differs from T-carrier is in the area of rings. The ability of SONET to be deployed in ring architectures, rather than as strictly point-to-point or multi-point architectures, has become the defining feature of SONET to date. This chapter explores all aspects of SONET rings, from the very basics to the more sophisticated timing methods that must be employed with them.

SONET rings can be a few miles long and span a few city blocks or stretch to literally thousands of miles and span continents. Rings of international proportions are not uncommon, and small rings in downtown metropolitan areas are no longer newsworthy. SONET rings are built by local exchange carriers (LECs), competitive access providers (CAPs) or competitive local exchange carriers (CLECs), and interexchange carriers (IECs). There are even a few totally private SONET rings intended for the exclusive use of a single organization. All share the distinctive protection characteristics of all SONET rings.

SONET rings, however, come in many shapes and sizes, and all have differing characteristics that make them suitable for one application or another. This chapter deals with all such characteristics.

This chapter begins by outlining the incentive for building SONET rings in the first place. After all, protection switching was not unknown in T-carrier networks. However, although the linear protection afforded by T-carrier was adequate for less exacting applications and services in terms of reliability and connectivity, the newer applications and services delivered by SONET networks place new demands on protection and reliability schemes.

All of the variations involved in SONET ring architectures are detailed, and a summary comparing the major architectures is given. Finally, the chapter concludes with several sticky issues that come to the forefront in circular telecommunications networks. The two most critical issues concern SONET rings delays and timing. Both are given full treatment at the end of this chapter.

SONET Rings and Fiber Failures

SONET links are sometimes deployed in a straight point-to-point fashion. Of course, SONET has distinct multi-point features that make grooming and hubbing much easier for service providers. In spite of SONET's multipoint configuration capabilities, a SONET network may still consist of a number of point-to-point trunks and links. The prob-

TABLE 14.1

Causes of fiber
system outages

Failure Cause	Percentage
Fiber cable dig-ups (backhoe fade)	51%
Fiber non-dig-ups (aerial/electronics)	24%
Other causes or equipment	15%
Digital cross-connects	7%
Synchronization timing	2%
Internal power components	1%

lem with single, point-to-point links is that the entire path of the link is vulnerable to failures.

Table 14.1 and Figure 14.1 show the causes of fiber optic facilities failures from July 1, 1992, through June 30, 1995, a three-year period. The statistics come from the Alliance for Telecommunications Industry Solutions. This information might seem somewhat dated and in need of an update. However, the point of the table and figure is exactly that before 1995 or thereabouts, service outages from fiber failures were common. The large number of SONET rings deployed since 1995 would not reflect the same thing.

Obviously, if some way could be found to allow service to continue uninterrupted when the fiber cable fails, this would be a big benefit to users and customer providers alike. The elimination of 75% of all fiber system failures would be a huge plus for SONET technology. Tariffs and contracts reliability terms would be respected, Public Utility Commissions would field fewer complaints, and the risk of the service provider being fined for inadequate service would be minimized.

Figure 14.1

Causes of fiber
system outages

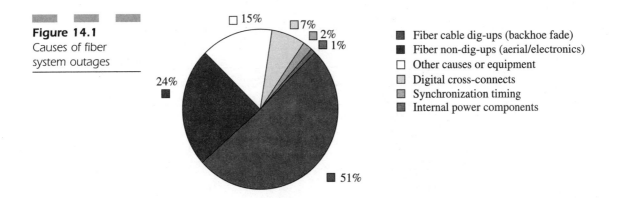

The only question is over the form that the protection should take. T-carrier was not immune to cable cuts and the like. Protection switching began with T-carrier. Perhaps a quick look at protection in non-SONET architectures may help to better appreciate the contribution SONET rings make to overall network reliability.

Nonring Automatic Protection Switching

It has already been noted that Automatic Protection Switching (APS) did not begin with SONET. Many T-carrier links were protected from cable failures by APS systems as well. Typically, these were not employed with the T-carrier links in a ring architecture, and just sought to backup one or more links with spares in a rather simple fashion.

The first form of protection switching used in T-carrier was known as *1:7 switching*. This is usually pronounced as "1 by 7" or "1 to 7" protection switching. There is nothing magical about the number "7" here. The second number can theoretically be any number, and the general form is 1:n protection switching. The term just means that "n" active links are protected by 1 standby link in case of a failure. The reason that n = 7 is so common had to do with the way that T-1 was initially deployed in the United States.

It was common practice in T-carrier networks to install the transmitting and receiving equipment in standard 7-foot high communications bays with 21-inch-wide shelves. Eight transmitting and receiving modules or boards could fit on an individual shelf. Usually, seven of these were used as active or *working* links. The eighth formed the protection or standby link in case any of the others failed, giving the now familiar 1:7 APS pattern. This scheme is illustrated in Figure 14.2.

Figure 14.2
1:7 protection
switching

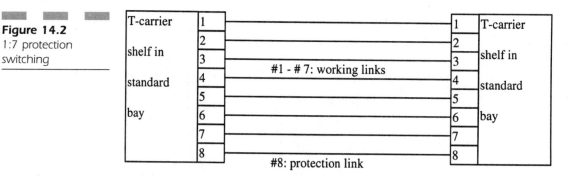

#1 - #7: working links

#8: protection link

There is nothing wrong with 1:7 protection switching. It is still common and respected. In some cases, two full shelves could be protected this way, giving a maximum "n" of 14. In fact, many SONET products allow the 1:7 form of APS. Most commonly, the protection switching from a working link to the protection link is quick and automatic. No human intervention is required. When used for telephony applications, a brief disruption in the conversation is all that was encountered. Even when used for newer applications, such as live football and other sports broadcasts over DS-3 facilities, a quick freezing of the picture that quickly clears is the only symptom of this APS procedure. Typically, the *restoral* of service from a protection link to a formerly working link once repairs have been made requires human intervention, but not the initial APS engagement.

APS permits the network to react to many conditions that threaten service quality. Not only outright failures can trigger APS action, but also errors and overall poor signal quality in general. If an interface board were to fail, the net effect still would be a failed link. A network manager could also initiate APS on a link and switch it from a working facility to a standby facility for routine maintenance, bit error rate testing (perhaps due to user complaints), and so forth. Most commonly, APS operations occur as a result of lost connections or poor signal quality on connections in eminent risk of failure.

Not that 1:7 protection switching is perfect—far from it. Consider what would happen when more than one link of the working group is lost at the same time. Why should this happen? Frequently, several links will follow the same physical path (and sometimes use the same feeder bundle or cable sheath) for a considerable portion of the entire link, and perhaps even the entire distance. This meant that when one link in a shelf was lost, generally several were lost at that same time.

Obviously, it is impossible to protect many T-carrier working links with one single protection link. Only one working signal can be diverted to the protection link at a time. A big step forward came as the cost of operating T-carrier facilities came down and the cost of the end interface electronics also dropped in price. These cost reductions eventually resulted in the deployment of 1:1 protection-switching methods in high-risk areas: usually major downtown metropolitan areas where constant street work was common. In 1:1 protection switching, every working link has its own individual backup link.

In general practice, the 1:1 protection operation involves the sending of the same signals over both links. That is, a single set of information is duplicated and sent over both the working and protection paths.

Both signals are constantly monitored for signal quality by the receiver and the higher quality signal is chosen for live reception. The active link is reverted, or switched, to the standby link when the connection is lost or signal quality degrades beyond a predetermined threshold. Naturally, a network management center is notified when the automatic protection switch takes place. The return to the active link is usually done manually.

To make 1:1 protection switching function ideally, it would be nice to have the working and protection links taking different paths from source to destination; however, this is not always possible. Separate cable sheaths may be involved, but may employ the same cable conduit. Even if the conduit were different, the feeder route may be shared by several conduits, due to right-of-way or congestion considerations. This common *banking* of conduits from different service providers on the same feeder route is a common practice in many parts of the United States, and not merely in urban areas.

Figure 14.3 shows some of the options available to service providers and often used in T-carrier networks. In most cases, extra cost to the customer is involved in one or more of these route diversity situations.

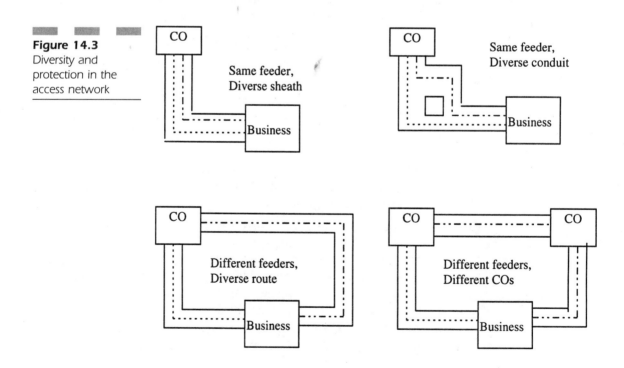

Figure 14.3
Diversity and protection in the access network

In Figure 14.3, several types of 1:1 protection-switching options and possibilities are illustrated. In its simplest form, this type of 1:1 protection switching offers physical diversity because a single business location is connected to a central office for T-carrier services with the same feeder, but employs links in two different cable sheaths. Obviously, if road work or a backhoe were to disrupt services on the working link, the protection link would almost invariably be lost as well.

The next step is to use the fact that feeders are typically laid out in a grid pattern and route the same feeder (in the sense that the feeder links the same source and destination) in two separate conduits for as much of the route as feasible. Note two things about such 1:1 protection switching. First, *route diversity* is usually defined as having twenty-five feet of separation between the conduits; therefore, usually one goes down one side of the road and the other goes down the other side of the road. Second, the two building's *entrance facilities*, the places where cable actually enters and leaves the premises, are still weak links in this arrangement. Many construction trenches stretch all the way across the street, and many building projects (e.g., new steps, sidewalks, ramps) can affect entrance facilities. Thus, this type of protection is better, but still lacking in many respects.

Next, it is possible to try to separate everything; that is, separate feeders, conduits, entrance facilities, and even separate central offices. Feeders routes in densely populated areas are good candidates for this scheme. This arrangement offers both customer and carrier the best 1:1 protection switching available. This solution is neither cheap, nor even possible in many circumstances. Usually separate entrance facilities are too expensive and require too much effort (including architectural and structural waivers and easements) to make total separation possible. Of course, the diversely routed cable must still be laid to the serving central office, as shown in Figure 14.3.

All of these 1:1 protection variations are quite common in T-carrier. All are lacking in one way or another. When SONET was standardized, however, the SONET architects noticed something quite interesting about this last "loop shaped" protection arrangement. What if the architecture were extended to include not just one customer site, but many? What if two or more serving offices were routinely included in the network design? What if backbones were created instead of access networks by linking serving offices of varying kinds, such as central offices, wire centers, and toll offices?

To make a long story short, this was indeed what was allowed and actually encouraged with SONET. Thus, SONET rings were born as a means of standardizing and extending 1:1 protection switching to its logical conclusion.

SONET Ring Basics

This section discusses SONET ring basic configurations and automatic protection switching. The discussion begins with a definition of a SONET ring, explores various types of ring architecture, and examines the definitions of *unidirectional* rings and *bidirectional* SONET rings. It then examines the differences between a *2-fiber* and a *4-fiber* SONET ring, *line* (or *ring*) switching versus *span* switching, and the characteristics of these forms of SONET ring APS. Finally, application examples of Bidirectional Line Switched Ring (BLSR) and Unidirectional Path Switched Ring (UPSR), and the definitions and functions of the SONET K1 and K2 overhead bytes for ring APS applications are discussed in some detail.

A *SONET ring* is defined as a collection of more than two SONET network elements (or just *nodes*) forming a *closed loop*. That is, the SONET NEs are linked together with the last forming a closed loop by being connected to the first. Each NE is thus connected to two *adjacent* nodes by means of at least one set of fibers operating in a duplex (i.e., one transmit, one receive) fashion.

A SONET ring provides many benefits to the service providers who deploy them and the customers that are served by them. They can provide redundant bandwidth for protection, redundant network equipment and electronics, and in most cases, both. A SONET ring is commonly known as a *self-healing* ring, but the term is little misleading because it implies that the rings somehow repair themselves. However, because the rings' services can be automatically restored following a failure or degradation in the network signals, the term has stuck for better of worse.

SONET rings are intended to bring *loop diversity* to networks. Although ring architectures are distinctive to SONET, recall that loop architectures and the protection they provide have been around since the T-carrier days. The preferred way of meeting SONET ring loop diversity is by deploying the fiber links with three major guiding principles. These are by employing a separate fiber sheath, separate conduits, and taking different physical routes from source to destination. This provides what is known as *route diversity* in SONET rings. Loop diversity can also be achieved by employing a separate fiber sheath, separate conduits, and taking the same physical route from source to destination, but this is not the preferred method. This is known as *structural* diversity, and is not quite as good as true route diversity, but may be the only form of SONET ring diversity feasible in many metropolitan areas.

It should be noted that both forms of loop diversity, route or structural, often must still use the same entrance facilities in either a central office (the cable entrance facility: CEF) or a business location (also known as the entrance facility). Route diversity is defined as separate routes from the first diverse point achievable at the source to the last achievable diverse point at the destination. Usually, this is within a few hundred feet of the endpoints, but not universally. In any case, route diversity is much better than no diversity at all.

It has already been mentioned that there are several SONET ring architectures from which to choose. In fact, three main features characterize all SONET rings, each with two alternatives. These three basic distinguishing attributes are listed in Table 14.2.

Naturally, when there are three attributes, each with two choices, the total number of combinations is eight ($2 \times 2 \times 2$). Thus, there are eight different SONET ring configurations that differ in at least one of the major attributes. Usually, the ring architecture designation merely strings together the three major attributes to name the SONET ring type. Therefore, one hears of "2-fiber, unidirectional, line switched rings" and "4-fiber, bidirectional, path switched rings" with regularity.

All of this is a mouthful, so naturally various abbreviations are used for the ring types. Unfortunately, the abbreviations do not mention the

TABLE 14.2

Types of SONET rings

Attribute	Choices
Number of fibers per link	2-fiber
	4-fiber
Direction of the signal	Unidirectional
	Bidirectional
Level of protection switching	Line switching
	Path switching

fiber numbers, information about which must be provided separately, but it still helps. The abbreviations include:

■ Unidirectional Line Switched Ring (ULSR)

■ Bidirectional Line Switched Ring (BLSR)

■ Unidirectional Path Switched Ring (UPSR)

■ Bidirectional Path Switched Ring (BPSR)

Any of these rings can be either a 2-fiber or a 4-fiber ring. Thus, there are *2-fiber UPSRs* and *4-fiber BPSRs* and so on. SONET (and SDH) standards provide for all eight types of rings for network element node interconnection.

With eight architectures from which to choose, things may seem confusing for service providers ready to build SONET rings and customers ready to buy the services delivered on them. In actual practice, however, with very few exceptions, only three of these eight types of rings have been built on a large scale outside the field trial stage, including:

■ 2-fiber, unidirectional, path switched (2-fiber UPSR)

■ 2-fiber, bidirectional, line switched (2-fiber BLSR)

■ 4-fiber, bidirectional, line switched (4-fiber BLSR)

Other types of rings are possible, of course, but most documentation addresses these three architectures. The ANSI specification on ring protection switching, for example, emphasizes 2- and 4-fiber BLSRs.

Most local exchange carriers have tended to favor 2-fiber rings of the unidirectional sort with either line or path switching. Interexchange carriers and competitive carriers, on the other hand, have favored 4-fiber BPSR for reasons that will become clear later.

Unidirectional Versus Bidirectional

Although SONET rings typically mention the number of fibers first, it may be wiser to start with the directionality of the rings while exploring the major ring attributes. Because the terms *unidirectional* and *bidirectional* are exclusively SONET ring terms and not exactly obvious in meaning, a few words of explanation may be helpful. These definitions come from the official ANSI documentation. In a unidirectional SONET ring, normal routing of the working traffic is done so that both directions of a connection between two endpoints travel around the ring in the same direction. Therefore, each connection uses bandwidth capacity along the entire circumference of the ring. For example, the SONET ring network in Figure 14.4 is a 2-fiber, unidirectional ring.

In Figure 14.4, each fiber link or span has been labeled for clarity. When a user attached to the NE1 SONET node connects to a user attached to the NE2 node, then Span # 1 in the figure is used for that traffic. In order for the user at NE2 to talk back to the user at the NE1 node, Spans # 2, # 3 and # 4 are used, not Span # 5, as may seem more logical. Note that both transmission paths (from NE1 to NE2, and from NE2 back to NE1) are clockwise. That is, for two-way communication, only one direction, or a *unidirectional*, transmission path has been established. By the way, all working traffic in SONET unidirectional rings travel clockwise by convention.

Note that Spans # 5, # 6, # 7, and # 8 are not used for working traffic. What then are they there for? These are the protection spans or links. They are used when one of the working (clockwise) spans fails. Unidirectional rings are quite common when the geographical area to be spanned is not too large, for example a metropolitan area.

Conversely, in a bidirectional ring, normal routing of the working traffic is done so that *both* directions of a connection travel along the ring through the same ring nodes, but in two *opposite* directions. A 2-fiber, bidirectional SONET ring is shown in Figure 14.5. Note that physically, it is indistinguishable from its unidirectional cousin. The difference occurs in the traffic flow.

In this case, when a user on the NE1 node is to talk to the NE2 node, Span # 1 is used, as before. But when the user on the NE2 node wants to reply to the user on the NE1 node, Span # 5 is now used for working traffic, instead of Spans # 2, # 3 and # 4 as before. Note that in a bidirectional

Figure 14.4
A unidirectional ring

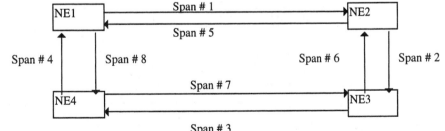

Traffic from user on NE1 to user on NE2 uses NE1 to Span # 1 to NE2. (Clockwise)

Traffic from user on NE2 to user on NE1 uses NE2 to Span # 2 to Span # 3 to Span # 4 to NE1. (Clockwise)

Figure 14.5

A bidirectional
SONET ring

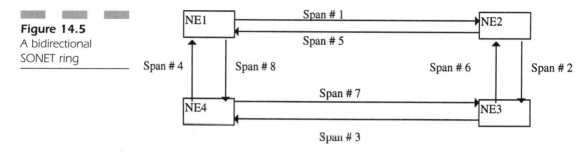

Figure 14.5

A bidirectional
SONET ring

Traffic from user on NE1 to user on
NE2 uses NE1 to Span # 1 to NE2.
(Clockwise)

Traffic from user on NE2 to user on
NE1 uses NE2 to Span # 5 NE1.
(Counter-clockwise)

ring, one transmission path is clockwise, while the other path is
counter-clockwise. That is, for a complete 2-way connection, both directions, or *bidirectional* paths, have been used between the NEs. Protection
switching now uses the spans between NE2 and NE3, NE3 and NE4, and
NE4 and NE1, in case the links between NE1 and NE2 were to fail.

Two-fiber Versus Four-fiber Rings

Both types of rings (i.e., unidirectional and bidirectional) may have two
or four fibers run between each SONET NE node. However, it is quite
uncommon to have four fibers run for unidirectional rings (two fibers
are robust enough for smaller geographical areas); therefore, the discussion here will be limited to a bidirectional ring.

Not surprisingly, a 2-fiber bidirectional ring requires two fibers for
each span of the ring. Each fiber span carries both the working-traffic
channel and the protection channel. This means that on each fiber, only
a maximum of *half the channels* are defined as working channels and the
other half the channels must be defined as protection channels. This is
shown in Figure 14.6.

Figure 14.6

Two fiber, bidirec-
tional ring protection

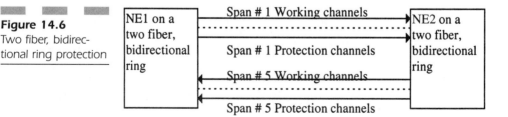

In Figure 14.6, one fiber, which may carry an STS-n signal and has only one set of SONET transport and path overhead bytes, is assigned for the working channel as well as for the protection channel. Each channel takes half of the bandwidth (capacity) of the STS-n signal.

Contrast this operation with a 4-fiber bidirectional ring. This type of ring requires four fibers for each span of the ring. Working and protection pairs are carried over different fibers. That is, two fibers, each with some form of loop diversity and transmitting in opposite directions, carry the working channels; two other fibers, also diverse and in opposite directions, carry the protection channels. A set of overhead is dedicated either to working or protection channels for the 4-fiber ring. This can be seen in Figure 14.7.

An advantage of using four fibers over two can be immediately seen. Each fiber carrying an STS-n can be used to its full capacity, instead of half as before. Of course, twice as much fiber cable is required, and it must be diversely routed or the whole scheme makes no sense; however, the real advantage of 4-fiber rings is that they may suffer multiple failures and still function. Usually, two failures will effectively segment a 2-fiber ring of any type. This makes 4-fiber rings very attractive for interexchange carriers, with wide geographical areas to cover, and competitive carriers, who typically have fewer maintenance personnel and staff relative to incumbent LECs.

Line Switching Versus Path Switching

The last major ring characteristic to be explored is the type of protection switching employed. SONET rings are either line switched or path switched. Much ink has been spilled debating the relative merits of each. SONET documents mention "ring" and "span" switching as well, but only line and path switching are useful concepts in SONET network.

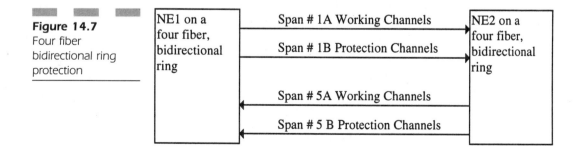

Figure 14.7
Four fiber
bidirectional ring
protection

However, both the line and path forms will work. Switching type is only one characteristic of SONET rings.

Line switching works by restoring all working channels of the entire OC-*n* capacity as a single protection operation. Note that the protection capacity is idle, and must be idle, when the ring is operating normally. Line switching is illustrated in Figure 14.8.

In Figure 14.8, live user traffic is always sent on the working channel. In the event of a failure, the live traffic is switched to the protection fiber at both ends of the span. Channels within the "line" are switched this way, which is why it is called *line switching*.

Conversely, path switching can restore all working channels at a level below the entire OC-*n* capacity. That is, the protection switching can be done on an STS-1 in an OC-3, or even at the VT level if that is what the STS payload is carrying. In order to do this, the signals are sent onto both the working and protection channels across a span. The receiving end constantly monitors both and selects the best signal to be used. Figure 14.9 shows how path switching works.

Path switching is almost universally done on 2-fiber unidirectional rings. Line switching is characteristic of 2- or 4-fiber bidirectional rings. Because line switching can operate at the line level of SONET, line switching has a reputation for being more efficient than path switching. Switch the line, and all paths on the line are protection switched as well. Path-switched rings have a reputation for having less fundamental capacity than line-switched rings, but the fact is that the aggregate bandwidth on both types of rings is pretty much the same.

It is nonetheless true that *unprotected* traffic can and often is carried on any type of SONET ring. Such unprotected traffic, a service usually offered at a deep discount, rides the spare capacity reserved for protection purposes on the SONET ring. In case of a fiber or node failure, the protected traffic is shunted onto the protection links, while the unprotected traffic is simply dropped for the duration of the outage. Today such outages rarely exceed four to eight hours, depending on location.

Line-switching systems must be able to restore service and reverse direction in 50 milliseconds or less (for comparison, the human eye

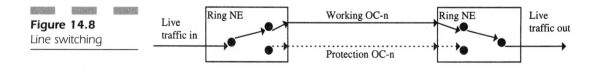

Figure 14.8
Line switching

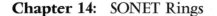

Figure 14.9
Path switching

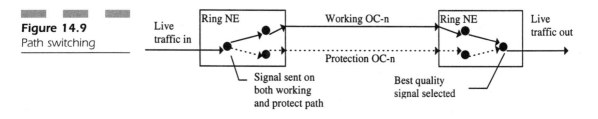

blinks in 100 milliseconds) for rings with circumferences of 1,200 kilometers (750 miles) or less.

SONET rings are also limited to 16 "nodes" (in reality, SONET ADMs) on a ring. There used to be no limit at all, and huge rings with 32 or more nodes were constructed in the early 1990s. Then the structure of the K1/K2 APS bytes was defined, and only four bits were reserved for ring node "address" or position of the node in the ring. The four-bit field limited the number of nodes on the ring to 16 or less. Now rings have grown large enough and ring nodes are common enough so that there are proposals for working around this apparent ring circumference and node count limit.

A summary of the major differences between the three major SONET ring types is shown in Table 14.3. There are also OC-192 (STM-64) rings, but these are not common and suffer from a lot of limitations due to optical nonlinearities at 10 Gbps (e.g., dispersion is 16 times worse at OC-192 than at OC-48, although the speed is only four times higher).

Although this chapter has dealt with, and will detail, SONET rings almost exclusively, ETSI has recently defined everything needed to construct similar rings for SDH.

TABLE 14.3

Major SONET ring types

Characteristic	Unidirectional Path Switched	Bidirectional Line Switched
Number of fibers between NEs	2	2 or 4
Typical speeds	OC-3 (STM-1) or OC-12 (STM-4)	OC-12 (STM-4) or OC-48 (STM-16)
Common application	Metro area, loop plant	Long distance, regional
Specifications of interest	Telcordia TR-496 ANSI T1.105.01-1998	Telcordia GR-1230 ANSI T1.105.01-1998

Real-World SONET Rings

The discussion of SONET rings to this point has been rather theoretical in nature. Perhaps a more detailed look at the types of SONET rings used in real-world SONET networks would be enlightening. In all cases, clear illustrations are used to point out the important aspects of SONET ring operation.

SONET 2-FIBER UNIDIRECTIONAL RINGS SONET is usually deployed today not in point-to-point configurations, but in a ring configuration. Many SONET rings used two fibers between SONET nodes to produce the "2-fiber, unidirectional, self-healing ring" shown in Figure 14.10. This arrangement is called a UPSR because protection switching is done on an end-to-end path basis.

In Figure 14.10, all traffic circulates on the working or "active" fiber path. If the fiber between any two SONET ADM nodes were lost, then the signal would "wrap" at the SONET nodes adjacent to the failure "path," and the protect or "standby" fiber span is used to keep the communications between all of the nodes open. This exact concept is used in the Fiber Distributed Data Interface (FDDI) standard. The nice thing about UPSR is that only two fibers are needed for the ring.

Most UPSRs will inject the transmission signal on both the working fiber and the protection fiber. So the node that handles the user traffic and passes it off to a SONET TM has a choice of processing either arriving signal. The nice thing about this aspect of the UPSR is that if a chosen signal degrades, it is a simple matter to pick up the other signal very quickly.

Naturally, the trickiest thing about any SONET ring is traffic *mapping*. The ring configuration means that not only must all the VTs and STS-*ns* that originate or terminate at a specific node be mapped to the single ring, but all of the other pass-through VTs and STS-*ns* must be

Figure 14.10
A UPSR ring

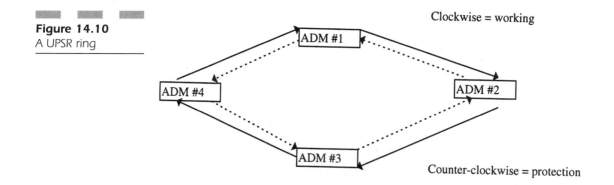

Clockwise = working

Counter-clockwise = protection

configured as well. For example, an OC-12 ring might have six nodes, and each node might have multiple OC-12 ports, but there is only one "OC-12's worth" of bandwidth on the whole ring.

This is an important aspect of SONET rings that must not be overlooked. Two SONET NEs with a point-to-point OC-12 between them can map out 12 STS-1s apiece. But put those SONET NEs on a ring with four other nodes, and an even distribution of traffic would allow only two STS-1s to be picked up or dropped at each node! In the real world, of course, things are much more complex. There is always the risk of mismatching node mappings—and much more of a risk when protection switching occurs. Much vendor and service provider training is geared not only toward equipment setup and troubleshooting, but also toward the complex task of configuring traffic on SONET rings. Interconnected rings, which are also common, just complicate the task even further. Fortunately, today there are several software packages that make the task easier, but not trivial.

Two-fiber unidirectional rings are still used today in many local access SONET ring arrangements; however, an issue arose when these unidirectional rings were used for regional and national SONET networks.

As should be obvious from Figure 14.10, when ADM A sends to ADM D, the signal arrives quickly across a single span. But when ADM D sends to ADM A, the signal must flow all the way around the ring. The traffic flow is unidirectional, by definition. Thus, 2-fiber unidirectional rings have an *asymmetrical* delay. That is, the delay outbound is not equal to the delay inbound. This never happens on point-to-point T-carrier links.

If the ring were small enough and the application were voice, this would be no big deal. With voice, the exact delay does not matter either way—that is *not* to say that voice is delay-insensitive—but data applications are different. Most data applications perform flow control through a "windowing" process. The details of this procedure are unimportant. The point is that if the ring were large enough (some span several states or countries and can have up to 16 nodes), the difference between the outbound and inbound delays could be substantial. Both ends will have a round-trip delay of (for example) 30 milliseconds, but for one it may be 5 milliseconds out and 25 milliseconds in, while the other will experience exactly the opposite.

The problem is that while one data application is trying to adjust its window size for a 15/15 millisecond "split" in delay, data keeps arriving early (5 milliseconds)! Of course, the other end is busily adjusting the same way, but data keeps arriving late (25 milliseconds). Bidirectional rings, which have symmetrical delays as in T-carrier networks, exhibited no such strange behavior.

SONET FOUR-FIBER BIDIRECTIONAL RINGS Most large-scale SONET networks today use 4-fiber, bidirectional, path-switched rings, especially in large regional or national ring configurations. SONET 4-fiber, bidirectional, path switched rings send active traffic in both directions on both pairs of fibers around the ring. The delays are always the same inbound and outbound.

The whole 4-fiber arrangement is more robust than the 2-fiber rings, and 4-fiber BLSR or BPSR configurations are the most popular SONET installations today, especially for regional or IEC SONET rings.

Of course, many service providers offer other ring variations and SONET configurations.

Two-Fiber Protection

Two-fiber BLSRs provide protection against the failure of any individual fiber pair on the ring. Two-fiber BLSRs are most often used by Local Exchange Carriers in major metropolitan areas. The signal is often (but not always) propagated in both ways around the ring to provide path protection as a two-fiber BPSR. The destination picks the "best" signal to deliver to the user. The bidirectionality gives a symmetrical delay for data users. These 2-fiber rings are quite economical to deploy because only two fibers are used to close the ring.

However, a potential drawback with 2-fiber BLSRs is that spare protection capacity must be reserved in all fibers to provide for the protection switching feature. This means that 2-fiber BLSRs are not especially efficient in using installed bandwidth and must be "loaded" and configured carefully.

For example, in an OC-12 ring, the first 6 STS-1s are used for live traffic (the working fiber) and the last 6 STS-1s are used for the protection capacity. Under normal operating conditions, the SONET equipment will send the traffic on both outbound fibers; however, in the vast majority of cases, both fibers are run in the same conduit. Therefore, both will fail at the same time.

When this happens, STS-1s 7 through 12 are used for the protection-switched traffic. The general method by which 2-fiber BLSRs provide protection switching is shown in Figure 14.11. There is a lot going on in the figure, so a few words of explanation are in order.

Figure 14.11 shows a five-node OC-12 2-fiber BLSR. Naturally, the capacity of the fiber in each direction is 12 STS-1s. Again, the important point

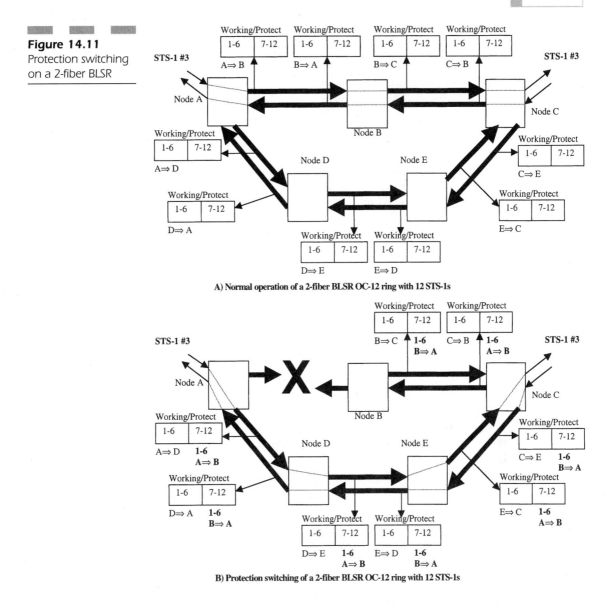

Figure 14.11

Protection switching on a 2-fiber BLSR

A) Normal operation of a 2-fiber BLSR OC-12 ring with 12 STS-1s

B) Protection switching of a 2-fiber BLSR OC-12 ring with 12 STS-1s

is that there are only 12 STS-1s for the *whole ring*. The figure makes this point by showing the third STS-1 on the ring (labeled *STS-1 #3*) as configured (mapped) between Node A and Node C. In normal operation, shown in the upper portion of the figure, this working STS-1 would be mapped bidirectionally through Node B, as is shown in the figure as well. So no user traffic could be mapped onto STS-1 #3 at Node B, since this STS-1 is obviously in use and user traffic must pass right through Node B.

The first six STS-1s in each fiber are the working channels on the ring. Figure 14.11 shows the structure of the STS-12 frames as they travel around the ring, labeling each span with the working and protect STS-1s and the nodes they connect. Only the working STS-1s are protected. Other STS-1s can be used in these protect STS-1s, but in case of a failure such unprotected traffic is preempted. So STS-1 #8 (for example) could be mapped between Node B and Node C (or Node D and Node E) and work as long as the ring was intact. In fact, STS-1 #3 is available on the working STS-1s between Node E and Node D, regardless of normal operation or not. All that is needed is careful configuration and mapping of capacity on the ring as a whole. It used to be claimed that 2-fiber BLSRs limited capacity to half of what a 2-fiber UPSR could furnish. But with careful loading and capacity management, 2-fiber BLSRs can be very efficient. Today, software packages make this *capacity planning* much easier than in the early days of SONET rings, when the whole ring-mapping scheme was kept on a piece of paper and taped to the side of a node.

The lower portion of Figure 14.11 shows what would happen on the 2-fiber BLSR if the pair of fibers between Node A and Node B were cut. Now the protection STS-1s, numbers 7 through 12, carry the protected working STS-1s between Node A to and from Node B the long way around the ring, as shown and labeled in the figure. Note that any traffic mapped to STS-1 #3 between Node D and Node E is unaffected by the protection switch. However, any traffic mapped to STS-1 #8 between Node D and Node E would now be preempted (which is why unprotected traffic services on a SONET ring are frequently offered at a deep discount compared to protected services).

An apparent oddity is also illustrated in the lower portion of Figure 14.11 when the ring is in protection mode. It might seem, since STS-1 #3 is now taking the long way around the ring, that STS-1 #3 is now available between Node B and Node C! But configuring this STS-1 would destroy the mapping scheme when service was restored to normal operating conditions. It took a while to get rings to squelch this oddity, but there are protocols in the SONET line overhead now to prevent this from happening while in protection mode.

Four-fiber Protection

Four-fiber BLSR or Path Switched Rings (BPSR) with ring switching providing protection against the failure of not only any individual fiber pair on the ring, but also both fiber pairs between any two nodes on the ring. Four-fiber BLSRs are most often used by Interexchange Carriers for regional or even national (and in one case, international) rings. The signal is propagated in both ways around the ring. The destination picks the "best" signal to deliver to the user. This gives a symmetrical delay for data users. Four-fiber rings are quite expensive to deploy because four fibers are used to close the ring; however, 4-fiber BLSRs can easily be loaded to capacity, making them very efficient once installed.

In normal operation, the signals are split and sent in both directions around the ring. If one pair were to fail, the signal would be sent *in the same direction* on the second pair of fibers (span switching). In the event that the second pair also were to fail between the same nodes, the signal now would be picked up on the protect pair of fibers on the "long way" around the ring (ring switching). Thus, 4-fiber rings can continue to operate even in the face of multiple failures.

Four-fiber line-switched rings can both *span switch* and *ring switch.* Typically, all that is needed is to swap, or span switch, the working pairs for the protection pairs between two SONET NEs. In other words, the traffic flow between ring nodes is not disrupted. But if a ring switch is necessary, then all traffic between two ring nodes is disrupted and the ring "wraps" to shunt traffic onto the protection fibers all the way around the ring. These two types of protection switching are shown in Figure 14.12.

Figure 14.12 shows an STS-*n* that is provisioned on a SONET ring (perhaps OC-12 or OC-48) between two sites attached to ring Node A and ring Node F. The fibers labeled "working" and "protect" are two fibers in each case. In normal operation, the working fibers between Nodes A and B, and Nodes B and C are used to carry this traffic. Note that the protect fiber pairs might carry unprotected ("nonprotected" in some documentation) traffic at the same time. The middle portion of the figure shows how span switching works in the case of a fiber failure between Nodes B and C. All working traffic between Node B and Node C is shunted onto the protect fiber pair. Any unprotected traffic mapped between the other nodes on the ring is unaffected by this outage. The bottom of the figure shows the loss of both pairs of fibers between Node B and Node C. Typically, the fibers would be diversely routed between these nodes. However, since even 25 feet of facilities separation can qualify as *route diversity*, road trenching and similar activities can easily wipe out both pairs of fibers. Now the ring wraps to provide ring switching and all of the protect pairs are used. All of the unprotected traffic is now preempted.

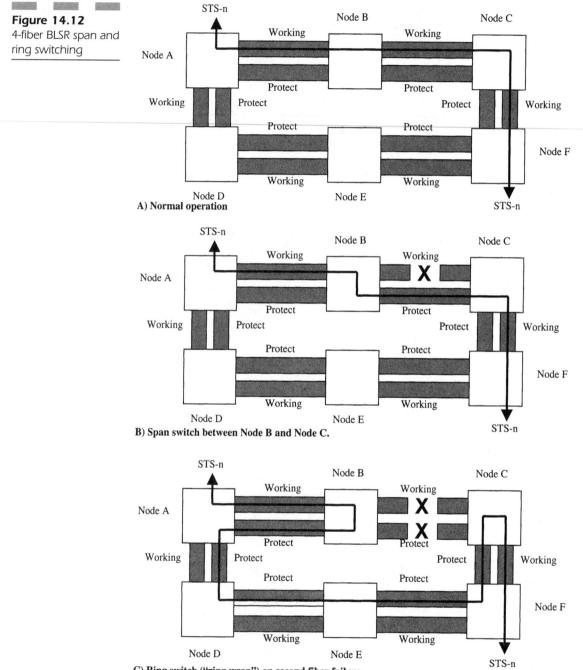

A) Normal operation

B) Span switch between Node B and Node C.

C) Ring switch ("ring wrap") on second fiber failure.

In many cases the OC-*n* traffic would be sent simultaneously on both the working and protect fiber pairs. This would make it much easier and faster to wrap the ring when ring switching is necessary. The protection switching documentation has other scenarios for 4-fiber BLSR failures, especially node failures. This section has outlined only the basics.

In all cases of ring wrap, whether on a 2-fiber BLSR or a 4-fiber BLSR, all the user sees is a very brief period of "garbage" traffic, followed by a jump in the end-to-end delay, reflecting the new flow of the traffic signal. SONET ring delay is treated more fully later in this chapter.

SONET Rings Compared

Obviously, all kinds of fiber rings are capable of the kind of protection switching that makes SONET rings attractive. But there must be some differences among them. Table 14.4 shows these differences.

Unidirectional and 2-fiber rings are usually seen in cities, while 4-fiber rings are seen in regional and larger installations. Although unidirectional rings will have asymmetrical delays, both 2- and 4-fiber rings have symmetrical delays. Although both unidirectional and 2-fiber rings will protect against single fiber failures, only 4-fiber rings can protect against multiple failures in separate geographical areas.

TABLE 14.4

SONET rings compared

	Two-Fiber Unidirectional Rings	Two-Fiber Bidirectional Rings	Four-Fiber Birectional Rings
Usually Seen in:	Cities	Cities	Regional and beyond
Symmetrical Delays?	No	Yes	Yes
More than one failure:	Problem	Problem	Usually not a problem
Bandwidth Efficiency	Medium	Medium	High
Initial Cost	Medium	Medium	High
Expansion Costs	Low	Medium	Low
Complexity	Low	High	Medium

Four-fiber rings have higher bandwidth efficiency, but their initial installation costs are higher. Because 4-fiber rings can be fully loaded, the expansion cost is less than expected. Finally, 2-fiber rings are quite complex compared to 4-fiber rings and especially unidirectional rings because of the complex spare channels used to provide protection in 2-fiber BLSRs.

Rings of Rings

As SONET deployment progresses, it will become more common not to see isolated SONET rings, or bigger and bigger SONET rings (which just increase delay, as will be pointed out later), but "rings upon rings." That is, lower speed SONET rings (access rings) will feed higher speed SONET rings (backbone rings) in a virtually unlimited array of speeds and sizes.

In fact, it is quite common today to have "low-speed" OC-3 rings feeding OC-48 rings. Both may even be 4-fiber BLSRs. In some cases, the OC-48 rings feed an even higher speed OC-192 ring spanning an entire regional service area or crossing national borders (in North America, at least). Each ring is failure resistant on its own and has all of the SONET management advantages for service providers and customers alike. The structure of such an interconnected ring architecture is shown in Figure 14.13.

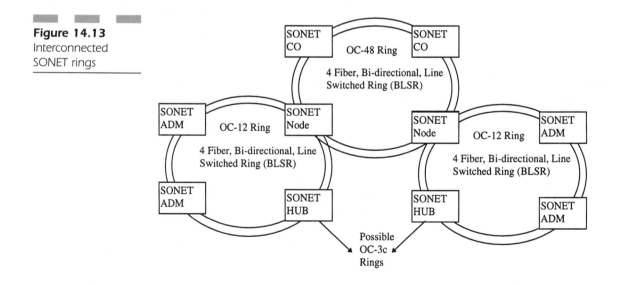

Figure 14.13
Interconnected
SONET rings

Although the example illustrated shows interconnected 4-fiber BLSRs, such interconnectivity is not limited to 4-fiber BLSRs, of course. In fact, it is possible to interconnect rings of any form, including 2-fiber UPSRs to 2- or 4-fiber BLSRs. The procedure is not trivial, but can be done and even provides the proper protection switching for traffic mapped across the mixed ring configurations.

Ring Automatic Protection Switching (APS)

There is much more to SONET ring Automatic Protection Switching (APS) than just getting the OC-*n* to detect a loss of signal (LOS) and perform some signal reconfiguration. APS takes place at the line level in SONET anyway, and LOS is an STS-level alarm, so all of the STS-*ns* inside an OC-*n* must be able to decide if and when an APS should take place. This is a very complex task, and the current ANSI document on the APS issue, ANSI T1.105.01-1998, is over 100 pages long. This section will only explore some of the basics of SONET ring APS, emphasizing the role played by the K1/K2 bytes in the SONET line overhead.

The K1/K2 bytes play a role in *linear* SONET APS as well, but only ring operation will be detailed here. When used in a ring configuration, the structure and messages of the K1/K2 bytes is as appears in Figure 14.14.

There is a lot going on in what appears at first to be a simple table. No full explanation of the entire APS message protocol can be attempted in a section of a chapter, but some of the overall goals and aims of the K1/K2 byte operation can be outlined. The full ANSI documentation contains many illustrations and should be consulted for details.

Perhaps the most important piece of the K1/K2 APS scheme is right up front. The first four bits of the K1 byte are the Switch Request bits. Sixteen messages are possible, and all are listed in Figure 14.14. In normal ring operation, no APS request is being made and the bits are 0000. There are Switch Requests for span and line switching operations, some initiated externally (e.g., MS-R), and some required to be signaled automatically (e.g., SF-R) by the nodes on the ring.

Briefly, these are the externally initiated commands carried in the K1 byte. The lockout (LP-S) disables normal span switching. A forced switch (FS-S, FS-R) is just what it sounds like: forcing working traffic to protection paths. A manual switch (MS-S, MS-R) is a type of override to force the same thing but is only supposed to be used in certain circumstances.

Figure 14.14
The SONET APS
K1/K2 bytes

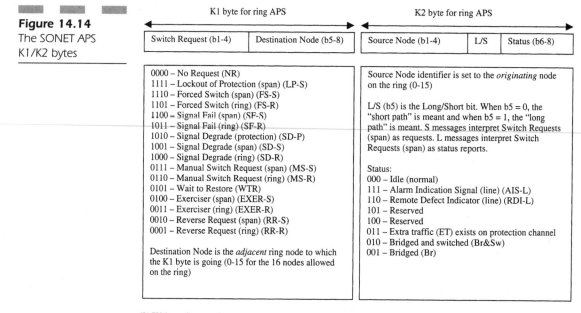

K1/K2 bytes in normal operation:

X X X X = Node IDs
No APS request, Short path, idle

The exercise commands (EXER-S, EXER-R) try out the protection-switching circuitry and operations without harming live traffic.

The automatically initiated commands can be briefly described as well. Signal failures (SF-S, SF-R) are triggered by loss of frame (LOF) or loss of signal (LOS) at the node. There is also a signal fail for protection that is the same as the lockout message. Signal degrades (SD-P, SD-S, SD-R) are caused by things like elevated BIP errors. The wait to restore (WTR) is used to make sure all nodes go back to the working channels at the same time. Reverse requests (RR-P, RR-S) are used as acknowledgments to the original Switch Request.

The last four bits of the K1 byte are the destination ring node identifier, not to be confused with the SONET NE address in the J0 Trace byte of the section overhead. By convention, the nodes are not numbered in most documentation, but given letter values from A to P, and there is no fixed relationship between node letter and ring node number. Four bits limit SONET rings to 16 nodes or less, but some efforts have been made to increase this limit. The destination node is always the *adjacent* ring node for which the K1 message is intended. But due to the failure, this message might have to be sent the long way around the ring. So on a five-node (A-B-C-D-E) ring, if a fiber failure occurs between Node B and Node C, the destination node field in the K1 bytes showing up at Node A will be for Node C, which is adjacent to Node B (but not Node A).

The first four bits of the K2 byte are the source node ring identifier. In the preceding simple example, the source node identifier for the K1 message to Node C would be Node B, the node that originated the message.

The fifth bit of the K2 byte can be confusing. This is the Long/Short path bit. When set to 0, it indicates a message sent on the short path, and when set to 1, it indicates a message sent on the long path. But the use of this bit has little to do with ring circumference, although every ring node contains a *ring map* of the location of all ring nodes. In practice, this bit is used only in 4-fiber BLSRs that can perform both span and ring switching (2-fiber rings can only ring switch). When span switching, a 4-fiber BLSR looks much like 1:1 linear protection and the L/S bit is set to a 0 (short path). Ring switching messages set this bit to a 1 (long path). Furthermore, when span switch requests are sent with the L/S bit set to 1 (long path), then all Switch Requests coded in the K1 byte are to be interpreted as status reports, not switch commands. When this bit is set to 0 (short path), and the Switch Request refers to span operation, then the Switch Request coded into the K1 byte is interpreted as a command. So this bit just helps the ring nodes figure out how to process span-related Switch Requests, which are of little interest except to the ring nodes at each end of the span.

The last three bits of the K2 byte are interesting as well. Eight Status messages are possible, and six have been defined. Normal operation is APS "idle" and coded as 000. Two other status messages are for the line-level Alarm Indication Signal (AIS-L) and the line-level Remote Defect Indicator (RDI-L), use of which was outlined in the previous chapter. There is also a message (ET) to let other nodes know that extra traffic exists on the protection channels. The ANSI specification actually has three categories for traffic on protection channels on a ring. There can be no extra traffic, preempted traffic, or non-pre-empted unprotected traffic (NUT) on the protection channels. The presence of NUT users complicates APS considerably, but this category prevents users from being thrown off the ring because of a failure (on the other hand, NUT users are never protection switched during ring failures, which is why these users are NUTs).

The final complication is the presence of the bridged (Br) and bridged and switched (Br&Sw) status messages. A full explanation would run many pages. It is enough here to note that switching always takes place with bridging, and never by itself. It is tempting to try and think of bridging as span switching and bridging and switching together as ring switching, but this is not the case. The simplest way to explain the difference between these two messages is to understand that with some ring node or fiber failures, bridging must be done before switching is added, and when the problem is clearing, the switching is removed before the bridging. This is just another way that all ring nodes are kept informed as to what they should be doing at any time during the failure.

The bottom of Figure 14.14 shows what the K1/K2 bytes for each STS would look like on a SONET ring during normal (no APS) operation. The first and last four bits of the two bytes are all 0s, indicating no APS request, short path, and APS idle. The middle 8 bits form the destination and source ring node identifiers.

Even the simplest ring fiber failure APS scenario would be far beyond the scope of this chapter to illustrate in any detail. So Figure 14.15 simply shows a simple five-node SONET ring in normal operation. By normal ring convention, Node A is shown twice for clarity. The arrows between each node are accompanied by the values of the K1/K2 bytes during normal ring operation. Instead of reproducing a bunch of 0 bits, the figure uses the No Request (NR), Short path, and IDLE status notation. Letters are used instead of bits, again by convention and to reinforce the fact that the letters and bits have no official correspondence.

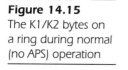

Figure 14.15
The K1/K2 bytes on a ring during normal (no APS) operation

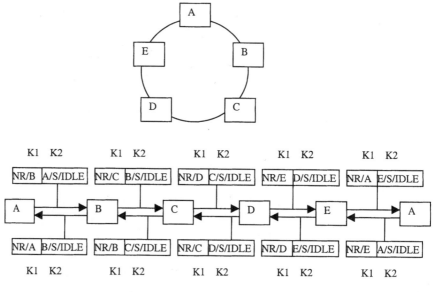

SONET Rings and Delay

An interesting aspect of SONET rings concerns the delays experienced by users whose connections are delivered by these rings. Things are not as simple as in the T-carrier world where links were simple point-to-point spans and a good idea of network delay could be worked out by the simple airline mileage between two endpoints and a few assumptions about equipment. Because all of these change with SONET rings, it is a topic worthy of some explanation.

First, it should be pointed out that there are three aspects to delay in any network: bandwidth, propagation delay, and nodal processing delay. Consider a point-to-point link operating at 1.5 Mbps. When a user is sending a 1,500 byte Ethernet frame, the receiver generally cannot process it as it arrives off the network until it is complete (the CRC error checking is at the end, memory needs to be allocated based on size, and so on). Therefore, at 1.5 Mbps, the "bandwidth delay" would be:

12,000 bits (1,500 × 8) divided by 1,500,000 bps (close enough) = 0.008 seconds = 8 millisec to get a full frame off the network

Propagation delay is due to the finite speed of light. Most media (with the notable exception of wireless) do not propagate signals at the speed of light, but at the medium's Nominal Velocity of Propagation (NVP), which is usually expressed as a fraction of the speed of light (c). Copper wire, for example, will typically have an NVP of 67 (sometimes ".67"), or 2/3 of c. This is 0.67 × 300,000 Km/sec or 200,000 Km/sec. Strangely, fiber is about the same. Fiber's NVP is given by the formula NVP = c/Index of Refraction of the fiber. Many SONET fibers have an Index of Refraction of 1.5, so the NVP is still 200,000 Km/sec (300,000 ÷ 1.5). This gives the fairly universal propagation delay of 5 microseconds/Km (200,000 ÷ 1 = 0.000005 seconds) for almost all media, including SONET fiber. (Notable exceptions are some coax cables, with NVPs of about 75.)

Finally, the nodal processing delay is the delay added by going through SONET ADMs, DSCs, and other switching office equipment. In the telecommunications world, the ITU says a signal must get in and out of a "switching node" in 450 microseconds or less (about 0.5 milliseconds, since 1,000 microseconds = 1 millisecond).

How does all of this help to calculate overall, end-to-end delay? Well, when the endpoints are 100 Km apart and there are three switching nodes between them, the end-to-end delay a user would experience is merely the sum of all three components:

$$8 \text{ millisec} + (100 \times 0.005 \text{ millisec}) + (3 \times 0.5 \text{ millisec}) = 8 + 0.5 + 1.5 = 10 \text{ millisec}$$

Note that the largest component of this delay by far is the bandwidth delay. This user application, whatever it is, is *bandwidth bound*. That is, the limiting factor for the application is the network bandwidth, not network delay (these applications are *delay bound*). Everything falls into one of these two categories. Either an application will not function properly because it does not have enough bandwidth, or it will not function properly because it does not have a low enough delay.

What about SONET rings? Because the numbers are nearly universal, they still apply. The problem SONET rings pose for bandwidth/delay calculations is to determine the following:

1. What is the actual cable mileage *around the ring?* (And how does it change on failure, when span or rings protection switching might be used?)

2. How many "switching nodes" are on the ring?

No SONET standard can answer these questions. Only the service provider or carrier can and should. Sometimes a tariff or contract will specify network delay ("not to exceed 20 msec within the LATA..."), but not always, especially with SONET rings. What can be done in this case? Either make assumptions (tricky) or push back ("tell me and I'll sign..."). Beyond that, a customer or network designer needs to apply the general guidelines above. Of course, most SONET rings will cutover to protection links in 50 milliseconds, but this is strictly a one-shot deal. Then the delay stabilizes at the new (probably higher) value.

An example of how this may work would be helpful. Consider that Carrier A has three 100 Km (cable length) SONET rings, interconnected, each with 10 switching nodes per ring (not unwarranted assumptions, especially where two OC-3 rings feed an OC-12 ring). The end-to-end user network delay is then:

$$3 \times (0.005 \text{ millisec/Km} \times 100 \text{ Km}) + 3 \times (0.5 \text{ millisec/node} \times 10 \text{ nodes}) = 1.5 \text{ millisec} + 15 \text{ millisec} = 16.5 \text{ millisec network delay}.$$

The bandwidth delay for a full Ethernet frame (assuming the 1.5 Mbps DS-1 is now delivered on the SONET ring) is still 8 millisec; thus, the end-to-end user delay is now:

$$8 \text{ millisec} + 15.5 \text{ millisec} = 23.5 \text{ millisec}$$

Note that the application is now becoming more delay bound in the sense that the network delay is higher than the bandwidth delay. There will be a 50-millisec outage "glitch" whenever any of the three rings "wrap" or "protect switch."

The point to this exercise is as follows:

1. The route on a SONET ring is not point-to-point anymore. Network-delay-sensitive applications need to know more than just the airline mileage between source and destination. But only the carrier knows the exact ring route.

2. There are generally more "switching nodes" on SONET rings than on point-to-point links, due to their very nature as a mechanism for linking offices cost effectively (the more offices, the more cost effective).

3. Some applications that are now bandwidth bound may become delay bound on SONET.

All of these points should be understood with regard to SONET rings and the services they deliver to customers.

More Ring Issues

The issue of SONET ring delay is only one of the issues that make services delivered on SONET rings fundamentally different from those delivered by T-carrier. The main point is that although more service providers are building SONET rings, customers are not necessarily buying SONET. There is little customer equipment that can handle SONET speeds anyway; therefore, most SONET links are sold as virtual tributaries delivering T-carrier services at T-1 through T-3 speeds.

In today's competitive marketplace, network survivability is no longer an option, it's an entry-level requirement. At the same time, customers expect technology advances to result in lower costs, which forces many service providers to reduce their staff levels. These two factors have forced many service providers to deploy both SONET rings and digital cross-connect systems (DCSs) to interconnect their rings. But when service providers interconnect SONET rings, they can actually create environments where protection switching can mask problems on the network. Therefore, they must be very careful to ensure that such interconnected rings do not impact user network survivability.

The problem is that some faults and alarms are not reported properly on interconnected rings. Dual node interconnect with "drop and continue" wideband DCS equipment has become the standard to guarantee that inter-ring connectivity is secure in the event of a SONET NE failure. Drop and continue can be used between path-switched, line-switched, or any other combination of ring types. In many cases, as ring deployment grows, DCSs are necessary to groom inter-ring traffic at both the VT and STS-1 levels. In other cases, M13 or VT multiplexers are used to multiplex DS-1 traffic from DS-1/VT1.5-based rings into other DS-3/STS-1s that are carried by OC-*n* interoffice rings.

The whole point of this architecture is to provide services that are at once survivable, economical, and manageable. Typically, ring interconnection is done at two points along the ring, known as the primary and secondary office. This architecture connecting an OC-3 ring with customers to an OC-48 backbone ring is shown in Figure 14.16.

However, when a SONET DCS or M13 is placed between rings in order to interconnect them, this can actually mask failures that occur in the switching office. All levels, from DS-3 to STS-1 to DS-1 signals are all vulnerable to the same problem. Here is why: A simple failure of the ADM or intra-office wiring provides the DCS with a failed DS-3, STS-1, or DS-1. The SONET Wideband Digital Cross-Connect (WDCS) grooms at the VT level and inserts VT/DS1 alarm indication signal (AIS) inside a freshly created DS3/STS-1.

Because the OC-48 ring monitors the STS-1 performance to select the better STS-1 signal, it is unaware of the failure to the DS-1 or other VT embedded in the STS-1. In the same fashion, when the M13s are at the interconnection points between the rings, these devices will still multiplex a failed DS-1 into a good DS-3 and pass it to the OC-48 ring. Once again, the OC-48 is unaware of the failed DS-1 within the higher speed signal.

An example of this situation can be illustrated using a path-switched OC-48 ring, but the same thing would happen with a line-switched ring. To select the good alternate path, a manual switch must be invoked at

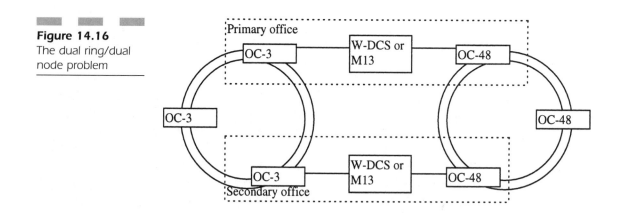

Figure 14.16

The dual ring/dual node problem

the OC-48 terminal, the failed DS-1 experiences a significant outage before this can occur.

A failure at this DS-1 level in the primary office is multiplexed into a good DS-3 or STS-1 by the W-DCS or M13 multiplexer and is sent *clockwise* around the OC-48 ring. The *counter-clockwise* direction for the signal originates from the secondary office and contains an STS-1 without any embedded DS-1 or VT failures, which is the whole point of doing rings in this fashion. The third ADM in the terminating OC-48 still sees two perfectly good STS-1s and does not protect switch, even though the still selected STS-1 contains failed DS-1 or VTs. There is just no way to know.

What can be done about this? Vendors such as Alcatel have designed a new "fault escalation" feature into many of their SONET products to provide this visibility of embedded faults at the STS-1, DS-3, or VT levels. When a piece of equipment receives a "bad" input signal at the DS-1, DS-3, or STS-1 level, it escalates the failure to the output and provides a DS-3 or STS-1 loss of signal (LOS) and companion AIS to the OC-48 ring equipment. The escalated failure causes the OC-48 ring to perform a protection switch, restoring traffic.

Because fault escalation terminates all the embedded signals, good and bad alike, fault escalation should be performed at only one of the two (in fact, the primary) W-DCS or M13 sites. If fault escalation is performed at both offices on the same DS-3 or STS-1, all of the embedded traffic would be lost. The likelihood of this "double fault" scenario occurring cannot be ignored, because each DS-3 carried on an STS-1 can receive DS-1s (as VT1.5s) from up to 28 different SONET NEs or other T-carrier facilities. The two faults must be within the same outbound DS-3/STS-1 for this to happen, but not necessarily the same inbound DS-1/VT.

In order to avoid this final double-fault situation, only the primary W-DCS is provisioned to perform the fault escalation procedure. Finally, the DS-3 or STS-1 switched rings should be provisioned to revert to the path containing the W-DCS that performs fault escalation (the primary one). Because such fault escalation is provisioned on a "per high speed circuit" basis, a single-fault escalating network element can be both primary and secondary if it supports multiple rings.

SONET Rings and Timing

One more SONET ring topic should be mentioned briefly. This is another look at network timing and synchronization. Most timing networks are built along T-carrier, linear, point-to-point architectures. The inherent looped nature of SONET poses special challenges when timing signals are injected onto SONET rings.

Recall that there are five types of timing allowed in SONET equipment:

1. *External timing:* Direct stratum 1 clock reference available.

2. *Line timing:* Derived clock from the scrambled SONET signal input.

3. *Through timing:* Also derived from input, but clocks the signal in a different outbound direction.

4. *Loop timing:* Same as Line timing, with one major difference. Line timing can be used in an add-drop mux (ADM) used as CPE. Loop timing is used for straight CPE and the bits must stop here. The ADM is configured for Terminal Mux (TM) mode only.

5. *Free running:* Internal clock only.

For SONET rings, there are two strategies. First, one of the NEs on a SONET ring may be free running, and the others are through timed. Second, external timing may be used. Either strategy avoids the creation of timing loops on the SONET rings.

The Synchronization Status Message (SSM) and SONET Rings

It is important to realize that the importance of timing on SONET rings is directly related to the need to add, drop, or cross-connect VTs. If all SONET links carried ATM cells or IP packets directly inside SPEs, timing would not be the crucial issue it is in SONET. But as long as DS-1s or DS-0s are in use on SONET links (and these DS-1s or DS-0s might themselves be carrying ATM cells or IP packets!), timing will be important in SONET systems.

One of the bytes in the line overhead of SONET is the S1 byte. The S1 byte is used for carrying the Synchronization Status Message (SSM) between SONET NEs. Although the SSM is important on all SONET links, linear or ringed, the SSM is especially relevant to SONET rings and so is discussed at the close of this chapter.

When it comes to timing networks and clock distribution, the worst thing that can happen is that *timing loops* are created. A timing loop is like someone asking someone else the time, who asks someone else, until the first person is asked! Slips causing jitter are bad enough. Jitter is annoying for voice, bad for data (resends needed), and deadly with compressed streams such as MPEG video. Timing loops are much worse as slips (jitter) build up, SONET payloads go wild with pointer adjustments, and *no alarms* are generated! This is the whole reason that the network synchronization hierarchy is arranged as a tree structure of various *strata* (layers). As long as networks consisted of linear, point-to-point links, it was easy enough to avoid the creation of timing loops (but they still did occur from time to time). However, a SONET ring is *inherently* a timing loop just waiting to happen. That is the main reason that this chapter deals with the SSM issue, although the SSM is used in linear SONET systems as well.

It is impractical to expect a large mesh of SONET rings to draw timing from a single master clock. Most systems draw from several available sources, some of which might be direct Stratum 1 sources such as GPS. The Timing Signal Generator (TSG) available to each SONET NE will each have a primary and secondary source to choose from. If a primary clock source is lost, the secondary can be used.

So far so good, but the issue becomes more complex when SONET rings are considered. If there is only one clock source injected into the ring, as discussed in the previous section, what happens when this reference is lost? And if there are multiple clock sources available to a SONET ring node, which one should be used in preference over another? And what prevents timing from passing all the way around the SONET ring and ending up referencing *itself* as a clock source?

The SSM messages in the S1 byte answer all of these questions, especially with regard to SONET rings. When SSM is used, SONET rings can have multiple clock sources, ring nodes can choose the best stratum clock available, and timing loops are still avoided. As with APS, a full discussion of the SSM protocol is beyond the scope of this chapter. But a few basics of the S1 byte and how the SSM protocol is used on a SONET ring can still be covered.

The SSM information is carried in bits 5 to 8 of the S1 byte. Previously, this byte in the SONET line overhead was required to be coded as 0000 0000, so one of the SSM codes can be 0000 for backward-compatibility. There are 15 other possible combinations of bits, and 9 others have been given values and abbreviations in SONET documentation. Each SSM is, in addition, assigned a Quality Level that helps SONET NEs figure out which timing source to use when a choice is available. These are shown in Table 14.5.

Only a few more comments about Table 14.5 are needed. The RES quality level can be assigned by users (i.e., service providers) to force timing signals arriving via a certain STS-*n* to take priority over others. Naturally, this must be done carefully. Most of the other levels follow logical preferences, but there are three noteworthy points to make. First, the

Synchronization Status Message (SSM)	S1 bits 5-6-7-8	Abbreviation	Quality Level
Stratum 1 traceable (source can be traced to Stratum 1)	0001	PRS	1
Synchronized—traceability unknown (backward compatibility)	0000	STU	2
Stratum 2 traceable	0111	ST2	3
Transit node clock traceable (a nonstandard clock level)	0100	TNC	4
Stratum 3E traceable (another nonstandard clock level)	1101	ST3E	5
Stratum 3 traceable	1010	ST3	6
SONET minimum clock traceable	1100	SMC	7
Stratum 4 traceable (no value has been assigned for this)	N/A	ST4	8
DO NOT USE for synchronization	1111	DUS	9
Reserved for network synchronization use	1110	RES	User assigned

traceability unknown (STU) value, which is basically for backward-compatibility, is given a fairly high priority, exceeded only if a Stratum 1 clock is available on the STS-*n*. Second, the transit node clock (TNC) and Stratum 3E clocks fall between Stratum 3 and Stratum 2 in quality. These two clock levels are not even mentioned in many standards documents, but since they are used in the real world, it is only proper to include them at their correct quality level. Finally, although a place-holder has been established for Stratum 4 clocks, no bit coding has been defined for this level. SONET is supposed to offer synchronous, byte-interleaved multiplexing and fast, error-free cross-connections. This would be very hard to accomplish with Stratum 4 clocks. Usually SONET rings and links are Stratum 3 or better.

The most important part of the SSM set is the DO NOT USE (DUS) message (oddly, this message in all documentation always has the "DO NOT USE" or "DON'T USE" in uppercase). This is the message that is most useful for preventing timing loops on SONET rings and is the main reason SSM is discussed at this point in the book. SSM has many applications in linear SONET systems as well, but this chapter closes with a look at SSM use on a SONET ring.

The issue with timing and clock distribution on a SONET ring is that the ring forms a natural timing loop. If nothing is done on a SONET ring, a Stratum 1 clock injected onto a SONET ring node will chase its tail right through all the ring nodes and end up back at the source. If the injected clock is lost, the source node might think "no problem, here's a clock traceable to Stratum 1 coming in right over here." The end result would be disaster as the timing jitter reinforces itself around the ring.

It would also be better, on a 16-node SONET ring, to allow nodes closer to the primary reference source (PRS) clock to choose a signal that has passed through only 4 nodes around the ring, rather than 11 nodes. In other words, the short way around is always better than the long way around the ring when a choice of timing sources is available, traceable to the same stratum level.

The trick is to configure two nodes on the SONET ring not to pass through the timing level available in the S1 byte, but to convert this to a DO NOT USE (DUS) SSM message at some point in the ring. This point should logically be halfway around the ring, but this is not a requirement. Figure 14.17 shows how the SSM protocol is used to prevent timing loops on a simple six-node SONET ring with an injected Stratum 1 clock source.

Figure 14.17
SSM on a six-node
SONET ring

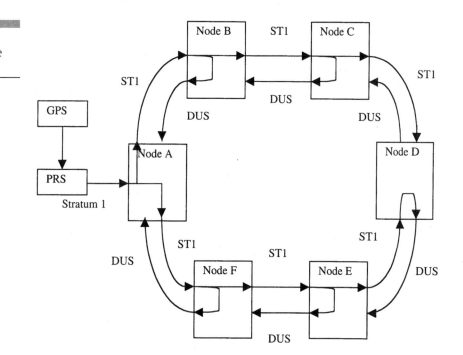

ST1 = Clock traceable to Stratum 1 (0001)
DUS = DO NOT USE for synchronization (1111)

As usual, the ring node with the injected timing (Node A) passes the clock signal in *both* directions around the ring. Yet timing loops are prevented. As shown in Figure 14.17, the whole trick is to configure one of the ring nodes (Node D in this case) *not* to pass the clock on to another node. And note that the nodes all change the returning clock SSM to DO NOT USE (DUS). These methods prevent Node F (for example) from deciding to take clock from Node E, five nodes away from the PRS, instead of from Node A, the direct PRS node. Node D, halfway around the ring in this case, has a valid choice of timing through the same number of nodes.

Figure 14.17 shows the ring under normal operating circumstances. This example is quite simple, but the SSM protocol can be shown to work properly when *two* nodes on the ring have timing sources available, perhaps Stratum 2 as a backup to Stratum 1. The interplay between the S1 bytes can then be shown to properly detect the loss of Stratum 1 clock, stabilize to Stratum 2 clock holdover, and then return to Stratum 1 when available.

PART 5

SONET
Service
Deployment

The previous parts of this book dealt with various aspects of SONET. There have been series of chapters on SONET standards, pieces of SONET equipment, SONET rings, and so forth. Little of the material, however, has given the SONET "big picture" in any amount of detail. The term "big picture" just means that no systematic attempt has been made to place SONET in its proper context as an access and trunking network to deliver high bandwidth and highly reliable services to as many customers and users as possible.

This part of the book corrects the situation. SONET is not yet everywhere. SONET deployment is an ongoing process that will still take many years to complete. This part of the book examines this deployment process in detail, showing many SONET network scenarios with the intention of putting together all of the concepts previously discussed and showing SONET networks in a real world, service-delivery environment. This is also where the international version of SONET, SDH, is detailed.

Merely installing SONET fiber everywhere will not solve any user problems. SONET, as has been pointed out in this work before, was originally part of a networking strategy where ATM provided switching and multiplexing and B-ISDN defined the services users employed for their applications. However, in the last ten years, SONET has leapfrogged ATM and especially B-ISDN in the sense that customers and users more often look to SONET for advanced network services rather than to ATM or B-ISDN.

Of course, it is always good to keep in mind that service providers with extensive SONET networks and rings in place may not be selling SONET directly to customers. But even a DS-1 or DS-3 delivered on a SONET ring offers many benefits to customers, mostly in the form of reduced pricing and increased reliability through SONET rings.

Also, the future of SONET is always of interest. For now, SONET seems to top out at the OC-192 rate, but equipment vendors will sooner or later push digital links beyond the OC-256 level that currently caps SONET. In fact, there are some techniques that promise to make the future of SONET even brighter than it currently is.

This last major section of the book is divided into four chapters. Chapter 15 deals with SONET service deployment. The chapter examines the typical SONET deployment pace in a given geographical area. It also examines where SONET is encountered in most carrier networks today.

Chapter 16 deals with SONET services provided to customers. It explores more fully just what customers do with SONET links, such as

LAN connectivity and linking private ATM network switches. Other applications, some obvious, some not, are mentioned as well. SONET offerings from various service providers also are outlined.

Chapter 17 explores the future of SONET. It details some of the new optical networking techniques for extending SONET and looks at other aspects of SONET deployment in summary fashion.

Chapter 18 is a detailed look at SDH, the international version of SONET. In all cases, an attempt is made to relate the newly introduced SDH concepts such as overhead and frame structures with their SONET equivalents.

15
SONET
Deployment

This chapter looks at SONET deployments in various network configurations. Comparisons to T-carrier networks are made at each step along the way, both for familiarity purposes and to show how SONET improves upon T-carrier scenarios.

The issue of SONET deployment is linked with the issue of fiber distribution. After all, SONET will not appear until fiber is there on which SONET can operate. That is not to say that all fiber optic cable is SONET fiber. Plenty of fiber still supports various kinds of DS-3 links and the proprietary higher levels from pre-SONET days. But the vast majority of new fiber installation miles are fully intended for SONET.

Before looking at the particular SONET deployment scenarios, this may be a good time to take a look at the constantly increasing pace of fiber installation in general and SONET in particular.

SONET/SDH, Gigabit Ethernet, DWDM, and Fiber

Not too long ago, when fiber was run by a service provider, it was pretty much a given that the fiber would be used for SONET and/or SDH. This is no longer true. As has been pointed out in an earlier chapter, DWDM and optical networking have changed the rules where SONET/SDH is concerned today. A full exploration of the future of SONET in a world dominated by newer technologies and techniques such as Gigabit Ethernet (GBE) and DWDM will be given in the last chapter of this book. For now, it is enough to point out that, although fiber deployment rates have never been higher, SONET and SDH are no longer the only game in town.

Some of the characteristics of running fiber intended for DWDM have been examined earlier and need not be repeated. Gigabit Ethernet has not been mentioned before, but is basically an extension of 10/100Base-T LAN technology to run on fiber not only down the hall, but also over fiber spans up to 20 km (13 miles) in length. In fact, commercial GBE products exist to extend the reach of GBE to 120 km (75 miles) without optical amplifiers, and up to 1,000 km (620 miles) with them.

The whole point here is that if GBE frames can be sent directly on fiber 100 km or 1,000 km, what need is there for a WAN solution like SONET/SDH when a LAN can do it all? More on the implications of this whole issue belong in the last chapter of this book. All that is needed here is an appreciation of the value of "dark fiber," as opposed to traditional service provider offerings on "lit fiber" such as SONET.

Fiber carries signals from transmitter to receiver. But who owns the transmitter and receiver? Who owns the fiber? In the past, the same entity almost universally owned both. Service providers who sold spare capacity on their SONET/SDH links and rings always controlled the transmitters and receivers (and any regenerators along the way, of course). The service provider "lit the fiber."

Today, there is often a scramble to install and control dark fiber. It is the fiber that is seen to have intrinsic value today, not so much the transmitters and receivers. End equipment ages, is upgraded, and changes capabilities quite often. And end equipment prices always seem to go down. Not so with fiber installation prices. The price of the fiber itself might tumble, but installation is still a labor-intensive activity, and although productivity might increase marginally, labor costs always escalate.

One large interexchange carrier estimates the cost per mile of new fiber installation at about $70,000 per mile, or a little over $13 *per foot*. Even when existing conduit or banks are used, costs still routinely exceed $30,000 per mile. No wonder it is the fiber *route miles* that have value today, not so much the signals the fibers carry.

The pressure today is to possess as much fiber as possible, not necessarily bandwidth. The same fiber could potentially carry much more bandwidth than in the past, especially with DWDM in the equation. It is the dark fiber that has the value, not the SONET/SDH that could be put across it. Even stranger, since fiber is a cable, just having *right-of-way* to run fiber somewhere has tremendous value based only on the potential of someday running fiber.

These points should always be kept in mind when reading this chapter. There will be plenty of SONET/SDH riding fibers for some time to come, but whenever fiber is run today, SONET/SDH is not the only possible or even probable use.

SONET's Function

Because SONET represents a lot of bandwidth, purely and simply, what is all of this bandwidth needed for? Much of the SONET bandwidth will be used for the same types of applications in use today, along with many new services. Most of these new services are defined by B-ISDN specifications but, in some cases, ATM specifications from the ATM Forum overlap somewhat with B-ISDN service definitions.

However, the use of SONET is in no way restricted to new services. Most existing services can still take good advantage of the enormous bandwidths SONET is capable of providing. Table 15.1 shows the results of a survey from the Idea Factory done in March of 1996.

Not surprisingly, most user organizations claimed that they needed more bandwidth, especially on their WAN links, for visually intensive applications, such as videoconferencing and graphics applications. Such graphics applications include, for example, CAD/CAM and scientific visualization (e.g., of molecules or seismic data).

User organizations also need more bandwidth for such mundane things as file transfer and database access. As distributed computing using client/server LAN applications becomes even more common than it is today, this need will only increase. This is often overlooked with highly publicized concerns about bandwidth for the newer services.

Even such common and relatively low-speed applications as SNA transactions and e-mail are on the list. There is just so much networked traffic that more bandwidth is a requirement, not a luxury. After all, many of the LANs to which these users are attached now operate at 100 Mbps or better. It is difficult, if not impossible, to link these together with 64 Kbps leased lines. The Web and e-commerce concerns have only helped to extend this bandwidth pressure to the residential user as well.

T-Carrier Trunking Network

Before showing how SONET is deployed in access, trunking, and regional networks, it might be a good idea to contrast SONET with T-carrier in these respects. After all, a recurring theme in this book has been to constantly compare T-carrier methods with SONET modifications and improvements in all areas of network operations. It should be further

TABLE 15.1		
Idea Factory survey of user applications needing more bandwidth	Videoconferencing	64 %
	Graphic applications (CAD/CAM, visualization)	60 %
	Batch file transfer (ftp)	58 %
	Local database queries and access	57 %
	Remote database queries and access	39 %
	Transaction processing (SNA)	36 %
	Internal e-mail	33 %
	External e-mail	31 %

noted that many of the ideas and scenarios in the following sections are based on Nortel Networks deployment plans and equipment categories. Besides acknowledging this debt, it should be kept in mind that other SONET vendors and users may differ somewhat in their major architectural SONET features at the access, trunking, and regional networks.

The traditional T-carrier trunking network has been used in North America since the 1960s. In the T-carrier network, switching nodes in switching offices are linked by T-carrier (usually T-3s on fiber) in a service area. Most are local serving offices (central offices) with users attached to a switch. Others are various, more specialized, switching nodes. For instance, a *toll office* usually has no users (i.e., telephones). Rather, these kinds of switching offices link together other switches in a given geographical area. The wire center can also link all users in the area to one or several interexchange carrier (IXC) points of presence (POP).

Each T-carrier switching node could have many DSX-3 cross-connects or DS-3 drop-and-insert devices at each site. The fiber system is symmetrical in the sense that one terminal end is the mirror image of the other. A typical T-carrier trunking network is shown in Figure 15.1.

The main point to be made is that the channels between the offices are usually limited to some number of 64 kbps channels, and all the other limitations of working with T-carrier apply. These limitations include the need for back-to-back multiplexing equipment and complex drop-and-insert arrangements and have already been detailed in earlier portions of this book.

Figure 15.1

A typical T-carrier trunking network

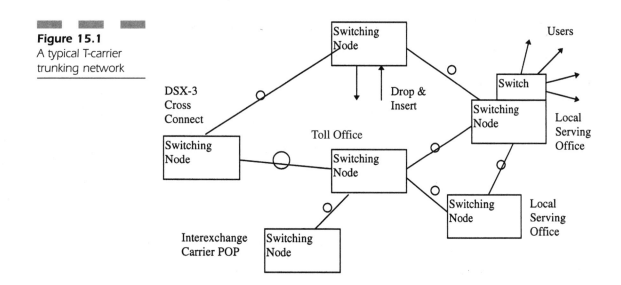

T-Carrier Access Network

In a typical T-carrier network, customer access to the T-carrier trunking network has its own distinctive features. The switching node, or local serving office, contains the endpoint of the fiber trunking network, which is normally a fiber optic termination system (FOTS) device. This device is connected to the M13 multiplexer and fiber multiplexer terminals to other devices. The M13 multiplexer is connected, in turn, to the digital switch itself and the channel banks for digitizing analog local loops leading to the users.

Access to the serving office is done with a combination of methods. Telephony end-users with analog phones are serviced by plain old telephone service (POTS) unshielded twisted-pair (UTP) copper local loops. Alternatively, a T-1 could be run to a remote digital loop carrier (DLC) remote terminal (RT) device, and twisted-pair copper wire loops extend to all locations in a carrier service area (CSA). The DLC link could also be fiber, leading to a Remote Fiber Terminal (RFT) arrangement, but the effect is the same.

UTP copper wire is also used to provide leased lines (i.e., nonswitched special services) for those users who desire these services. Finally, newer fiber optic cable may extend directly to various customers' premises for providing DS-1 leased lines or other services. This architecture is shown in Figure 15.2.

SONET Trunking Network

After taking a brief look at how T-carrier trunking and access networks are put together, this is the time to see how SONET networks are deployed to accomplish the same ends. It will now be much easier to compare and contrast T-carrier and SONET, with the intention of making SONET advantages that much more concrete.

A SONET trunking network is much simpler and yet much more robust than a similar T-carrier trunking network. SONET switching nodes, either containing add-drop multiplexers (ADM) or SONET hubs (which combine the SONET ADM capabilities with digital cross-connect system (DCS) capabilities for grooming and other purposes), are now configured as a series of interconnected SONET rings. The use of SONET rings in the trunking network assures higher availability and reliability for the trunks.

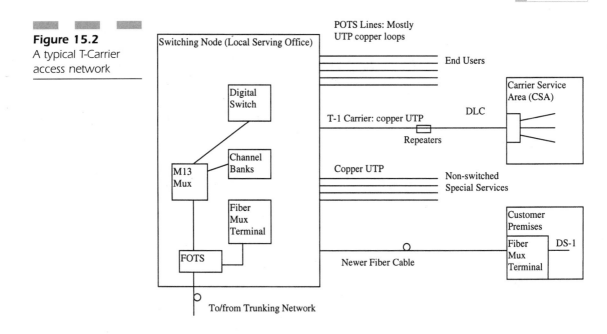

Figure 15.2
A typical T-Carrier access network

SONET OC-n links also can run directly to a customer's premises, if the customer has SONET customer premises equipment (CPE) installed. This is admittedly unusual in more current customer environments. It is much more typical for the SONET rings to deliver T-1, Fractional T-1, Fractional T-3, and full T-3 speeds rather than any direct SONET service. But the rings make the T-carrier customer offerings that much more error free and robust. A typical SONET trunking network is shown in Figure 15.3.

SONET also offers the possibility of a mid-span meet for interfacing with other service providers. The SONET mid-span meets allow for one vendor's equipment to be on one end and another vendor's equipment to be on the other.

There is a single craft technician interface for handling alarms and monitoring network performance—another plus. On the whole, SONET is simpler and more efficient than the earlier T-carrier trunking network.

SONET Regional Network

An especially nice feature of SONET is the ability to extend the trunking network from centralized service areas to regional SONET networks that include suburban and even rural areas. Such regional area networks

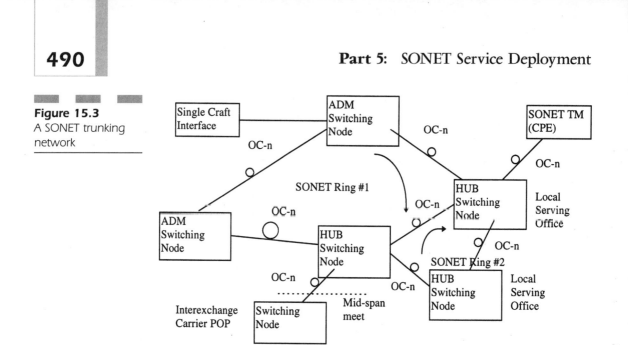

Figure 15.3
A SONET trunking
network

have no real counterpart in the T-carrier world, but are a structure and architecture that SONET allows to be deployed easily and effectively.

The SONET rings and trunking network from the previous section are still in place, of course, and the links to the interexchange carriers are still SONET as well. However, it is now possible to extend the benefits of SONET to a wider (i.e., regional) area by the addition of various OC-n links to SONET ADM and hub locations. Hubs combine SONET ADM equipment and DCS capabilities at the same location. This overall deployment for regional SONET networks is shown in Figure 15.4.

These SONET ADMs and hubs are located away from high-density traffic areas and extend the benefits of SONET to locations that were unreachable for advanced services before. In urban areas, SONET rings provide redundant and robust paths. Suburban and rural areas can take advantage of this urban structure as well. Note that SONET deployment at this level is an ongoing process, one that is based on the presence of SONET in an urban area before extending SONET services to less densely populated areas.

SONET Long-Haul Network

So far, the SONET networks mentioned have been concentrated in major metropolitan areas and, in spite of the promise of extending SONET in

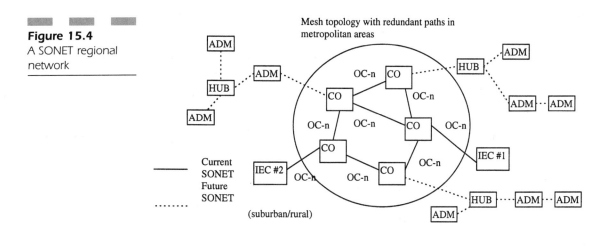

Figure 15.4
A SONET regional network

more rural areas and attaining regional coverage, these deployments mostly appeal to the LECs. But SONET's flexibility and scaleability also make it attractive to IXCs that often use SONET to create long-haul networks between major cities.

SONET is used in many long-haul networks built by IXCs or LECs entering the inter-LATA market. A long-haul SONET network deploys an OC-n (usually OC-48, or the even higher OC-192 on occasion) high-bandwidth transport network over an existing point-to-point fiber network (usually a 565 Mbps or other speed FOTS network with multiple DS-3s on a nonstandard fiber system). The main network equipment is a large (often very large) SONET ADM used in both terminal and repeater configurations. SONET broadband cross-connects are used for provisioning and managing the traffic on the long-haul network.

This overall structure is shown in Figure 15.5. In the figure, SONET equipment is shown as square symbols and the non-standard DS-3 FOTS equipment is shown as circle symbols. The SONET construction can "overlay" the existing FOTS routes or be entirely new construction of fiber intended to "close the rings." Naturally, this could be done much later, depending on the service provider's deployment schedule and capital expenditure budgets.

These other routes between locations, including many of those not previously serviced by FOTS, could be added to the SONET network to close the rings on the network. The older network may not have been extended to the other cities for a number of reasons, including cost, traffic patterns, and user densities.

Figure 15.5
A SONET long-haul
network

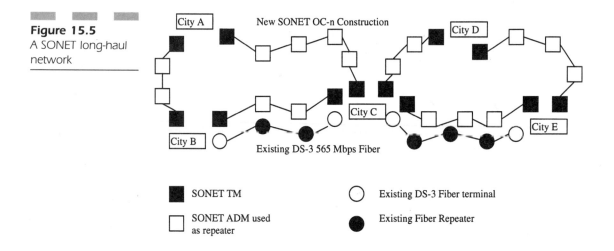

The new SONET rings offer many advantages over simple point-to-point configurations in terms of robustness.

SONET Access Network

Most SONET deployments by service providers, especially early ones, did not begin as trunking, regional, or long-haul networks, especially when SONET rings are considered. Most early SONET deployments were in the heart of downtown metropolitan areas, and rings from the very start. The intention was to take advantage of SONET's overall benefits in general, and ring benefits in particular, in the access network. The access network gathers traffic from a number of customer locations, each with distinctive services and requirements, and delivers the traffic to the service provider's switching nodes in the serving offices. Historically, the access network has been the source of most of the problems with service interruptions.

Therefore, SONET makes a lot of sense in the access network. This is especially true when new fiber is being installed and/or existing routes are becoming exhausted in terms of bandwidth. SONET will provide a means for not only supporting new network capabilities in terms of protection and management, but also in supporting new services (e.g., video). There are three main scenarios for SONET access networks. All three are illustrated in Figure 15.6.

In a simple point-to-point configuration, instead of using a pair-gain link to service a CSA containing DLC systems, SONET can be used. This

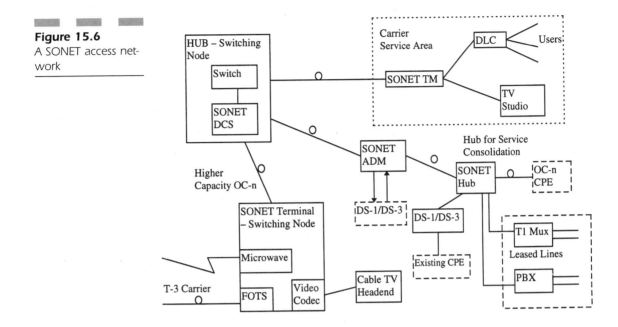

Figure 15.6
A SONET access network

situation is shown at the top of Figure 15.6. The SONET TM can still support the DLC system and yet has plenty of bandwidth to spare for supporting new services like the video feed from a TV studio.

It is worth pointing out that technically sophisticated users like TV studios can no longer be assumed to reside in totally business environments and commercial zones. Many communities allow for "high-tech" zoning which allows such businesses to operate near or even within otherwise mainly residential areas. In some areas of the country, office parks and housing developments are interspersed so closely that the dividing line between them is often no more than a backyard fence. SONET offers a very good way of reaching both types of customers, even with different servicing needs, in a cost-effective and scaleable fashion.

Another type of access arrangement is shown in lower right portion of Figure 15.6. In such a hubbing or ADM application, SONET can transport traffic from the switching node to a TM at a customer (or other) location having a SONET hub. The ADM could service existing DS-1s and DS-3s along the way, enabling them to take advantage of SONET's low bit error rates, reliability, and management capabilities. Of course, the sale of new T-carrier services could continue as before, giving customers familiar technology and services and also allowing these customers to purchase low-priced T-carrier end equipment.

The SONET hub offers a means of consolidating separate services in a cost-effective manner. The hub can support any customers with SONET OC-n CPE, although initially these may be sparse. Other DS-1s and DS-3s can be supported as well, with existing CPE, which is always attractive to customers. These customers do not even notice the arrival of SONET in an area. Finally, a customer's existing T-1 multiplexer or PBX can be supported by means of ordinary leased lines, as before, but delivered on SONET. The ADMs could ultimately be connected in a closed OC-n loop to create a SONET ring.

Last but by no means least, another type of SONET access is shown in the lower left portion of Figure 15.6. Here, a higher capacity (perhaps an OC-48) SONET link can be used to connect to another SONET switching node, but with a different function. This node could have access to an existing microwave or cellular network, and existing FOTS facilities. More intriguingly, especially in the age of local service deregulation, this node could function as the entry point for a cable TV head-end; thus, SONET would provide the distribution network for cable TV video signals. It is difficult to imagine any other technology besides SONET handling this integrated role well.

SONET Access Arrangements

The previous section presented a comprehensive array of access alternatives for SONET. This does not imply, however, that all types need to be used, or that some of them will ever even become common. So this may be a good time to detail exactly which SONET access arrangements will be seen most often.

As can be seen from the previous section, SONET has many possible access arrangements. Nevertheless, SONET will mostly be used in the access network in a relatively routine manner, at least initially. SONET will be used either to service existing DLC RT to provide service for POTS users, or to provision DS-1s and DS-3s, all within a CSA. This is illustrated in Figure 15.7.

Not all sites in a service area may be directly reachable from a given serving office. Even SONET fiber has distance limitations. In this case, SONET can be used in a straight ADM configuration, dropping and adding various DS-0, DS-1, or even DS-3 channels at hubbing sites along the way. These remote ADMs in environmental hardened enclosures may eventually loop back to the serving office to form a SONET ring.

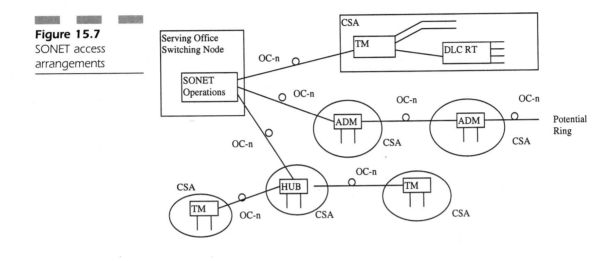

Figure 15.7
SONET access
arrangements

A single SONET hub can branch out to several SONET TM arrangements, typically in office parks or to serve large housing tracts, again dropping DS-n channels along the way.

Private SONET Rings

It is not only for public network services that service providers are busy deploying SONET rings in access networks, trunking systems, and the like. Given the right circumstances, it is even possible to find examples of totally private SONET infrastructures dedicated for one large customer.

The simple reason is that increasing numbers of users, as well as traditional applications (e.g., large file transfers) and new needs (e.g., videoconferencing), are forcing many organizations to push the bandwidth delivered by their networks to the limit. Alternatives to existing private lines like ATM and frame relay are available. But it would be expensive and time consuming for an organization to convert from a familiar private-line environment, especially when private-line solutions have been the rule rather than the exception.

ATM and frame relay service offerings sometimes do not support the access bandwidth needed (e.g., 45 Mbps) to improve application performance at an access affordable price. ATM typically is not available on SONET fiber, and neither is frame relay. Conversely, SONET is about more bandwidth, purely and simply. It offers a clear migration path to greater bandwidth. With SONET, it is relatively painless and inexpensive to go from OC-3 (155.52 Mbps) to OC-12 (622.08 Gbps) and even on to OC-48 (2.4 Gbps).

This ease of migration to greater speeds prompted the Miller Brewing Company of Milwaukee to deploy a private SONET ring infrastructure for their networking needs. Miller is the second largest brewery in the United States and the third largest in the world with an annual production of 45 million barrels (this amounts to about 19.5 billion, 12-ounce glasses). The movement to a SONET ring from copper wire was prompted by a need for more bandwidth to accommodate additional users and new applications, and the need for reliable disaster prevention. The brewery has been using its own exclusive SONET OC-3 ring since 1995 as the underlying transport backbone for its frame-relay switches.

Miller uses the SONET ring network at the Milwaukee campus for all voice and everyday data applications. The data applications include payroll, accounts payable and receivable, as well as more business-oriented applications, like internal controls, production scheduling, and shipments.

Miller's SONET ring has all the usual benefits of SONET. Although the ring is dedicated to Miller, they have access to multiple IXC offices for further disaster protection for switched services. Like many organizations, communication is vital to the company. Downtime is not necessarily fatal, but loss of data or lengthy outages can be enormously costly.

Another nice aspect to SONET is the tie between SONET and ATM. When the time comes for Miller to migrate to ATM, which the company estimates may happen in about three years, the cost of upgrading on SONET will be relatively inexpensive. SONET is not limited by any bandwidth constraints. Technology options no longer need to be evaluated in terms of bandwidth available.

SONET is ultimately cheaper than other architectures. Many users can recover any of the costs related to a SONET upgrade by way of total network savings in three to five years. Users with extremely heavy data traffic can easily compare the cost per megabit for SONET networks with copper costs. Fiber is actually cheaper that running copper in most cases.

Ameritech, who supplies the ring, rebates a portion of the monthly cost for downtime of greater than one minute for that section of the ring.

Running SONET Fiber

It should be obvious at this point of the chapter that SONET deployment is intimately tied to the presence or absence of suitable fiber for the

SONET signals. The good news is that there is plenty of fiber around from FOTS systems. The bad news is that in the first place this fiber is still used for FOTS traffic (and often filled to capacity anyway) and in the second place this older fiber may not be suitable for SONET. Fiber types and construction have evolved even more rapidly than achievable SONET speeds.

Thus, most new SONET deployments require new fiber deployments. It may be a good idea to close this chapter with a look at just where all of this new fiber is coming from. After all, the aggressive SONET deployment plans of most carriers and service providers may lead some to believe that there should be backhoes busily digging up every square inch of the United States sooner or later.

It may eventually come to that. For now, there are several ways of quickly installing the conduits and fiber cables needed for SONET services in a less disruptive fashion. This section briefly outlines two methods, one for underground fiber runs and one for undersea fiber runs, and the organizations responsible for these advanced conduit and fiber installation techniques.

"Railblazers"

The Railblazers program is the product of SP Construction Services, a business unit of Qwest Communications of Denver. They have taken the art of installing SONET fiber to new heights. SP uses a "railplow" and more than one locomotive to lay fiber in conduits alongside railroads' rights-of-way around the country, mostly in the southwest. Railroads have found that their tracks' rights-of-way for SONET fiber are a significant source of income and, in some cases, produce more profits for the company than actually running rolling stock down the line.

The "railblazer" arrangement installs up to eight miles of multiple-strand fiber per day, in conduit, using a rail-mounted trenching plow, conduit and fiber supply cars, and two locomotives. The arrangement digs the trench, lays down conduits and cable from very large reels, and buries it all in one smooth operation. Galvanized pipes are used to span bridges where trenching is not viable. The conduit system is quite secure, being buried at a depth of four to five feet. Using railroads' rights-of-way, usually with patrolled and secured access already in place for safety reasons, also protects the fiber from inadvertent contractor dig-ups.

The fiber system can be adjusted to dig parallel trenches without disturbing the first. Multiple fibers from several service providers, even competitors, can be handled in the same way.

The AT&T Fleet

AT&T has a fleet of "fiber boats" laying huge amounts of undersea, international SONET fiber, mainly to "replace" the satellite system currently in use in most cases. SONET terrestrial links have much lower delays than satellites, as well as lower bit error rates.

The SONET undersea fiber, in spite of layers of protection, is still light enough to drag on shore to termination points with an installation crew of four. A sonar arrangement alerts the captain to undersea mountains and the like, and adjusts the "slack" in the SONET fiber trailing behind the boat.

With recent fiber advances, there are now no active and powered repeaters underwater. This makes the whole process that much easier. The SONET fibers have a thirteen-year mean time to failure (MTTF) and a twenty-five-year service lifetime. Breaks are handled by hauling up the broken ends, which are located by means of an optical time-domain reflectometer (OTDR), and splicing them back together. (The difficulty is in *finding* the cable under miles of sea.)

A side benefit of this use of SONET is to make international leased lines actually affordable in many cases. Since SONET is the North American version of the official SDH international standard, it is relatively easy to have such links.

Plowing Cable

It is only fitting to close this chapter not with AT&T ships, nor SP Quest locomotives, but with a tribute to those who install fiber the old-fashioned way. They just do it because its their job and someone has to.

Consider the case of northern Nevada. In 1930, 30-foot redwood poles snaked their way across the high plains, carrying toll calls from coast to coast. During World War II, buried analog toll cable was laid between bomb-hardened repeater buildings, but the aerial copper remained. It was not unusual for line maintenance personnel to drive 150 miles just to get to a fifty-mile span of wire on the poles to service a few ranches.

The replacement of these wires with fiber has been mainly the work of one man: Art Brothers, who took over the maintenance of the copper wire because AT&T asked him to do it and no one else wanted the job. There are three exchanges with 100 customers on a 107-mile stretch of I-80. In

1992, fifteen miles of fiber were "plowed in:" in 1995, eleven miles. Another fifty-two miles got plowed in 1996, ending in the early snow.

It is not just the elements that make the plowing tough. It is also the $1,000 per mile needed for historic and cultural studies before the trenching can begin. The remains of the Chinese work camp from the days when the railroad was built could not be disturbed. Not to mention the swarm of university students troweling and screening six feet down to see whether those slivers of material turned up by the fiber plow were of prehistoric value.

Art Brothers's fiber may never be used for SONET. But the fact that it is there and may be called upon if needed is a tribute to the ingenuity of outside plant personnel everywhere. Thanks for all the bits.

16

SONET Services

This chapter explores more fully what customers do with SONET links, whether delivered on SONET rings or not. The typical and somewhat obvious applications, such as asynchronous transfer mode (ATM) and LAN interconnectivity, are considered in more detail, but from a more user-oriented approach than on a pure technology basis, as before. Less obvious applications are mentioned, such as scientific visualization, medical imaging, and other graphical applications. The most important aspect of this part of the chapter is in the area of SONET support for entirely new services that were impossible before SONET bandwidths became available.

The chapter also provides details about the SONET offerings from various carriers around the United States. Although not primarily concerned with pricing schemes and components, the major characteristics of these services are discussed. Several service offerings are investigated in detail, not to advocate these offerings, but just to give a flavor for what is available.

SONET Applications: ATM

ATM is the international standard for cell relay technology. In cell relay, information is transported through the ATM network in fixed-length cells. ATM is the multiplexing and switching mechanism for a B-ISDN network. SONET, in turn, is intended to be the trunking and access technology for ATM-based B-ISDN networks. Originally, B-ISDN standards were intended to contain the service definitions that were to be used by service providers as the basis for service contracts for private services and tariffs for public network services. However, detailed service definitions and functional descriptions never appeared to guide service providers in this area. Thus, B-ISDN has sort of languished for a number of years.

ATM itself has had modest success as an alternative desktop technology to shared-bandwidth LANs like Ethernet. ATM has been much more successful as a backbone technology but, in this respect, ATM is somewhat masked from users and this use of ATM need not be considered further here.

ATM, therefore, is used in application environments other than a public B-ISDN. Until B-ISDN is more fully defined, this must be the case. Most ATM applications use the basic definitions of services derived from the ATM Forum, a vendor consortium faced with trying to fill the service gaps left by B-ISDN on an interim basis. The majority of these are essentially LAN-based data solutions, although not exclusively.

Therefore, when it comes to linking ATM switches with SONET links and rings, the architectures used are fairly consistent. For example, some LAN equipment vendors will use ATM in a LAN hub or router. The customer LAN itself may use ATM directly to the desktop, in which case each desktop device needs its own ATM interface board, usually operating at 155 Mbps or even 622 Mbps (note the alignment with SONET speeds). More likely, the LAN itself is running some high-speed LAN technology other than ATM, such as Fast Ethernet at 100 Mbps, or even Gigabit Ethernet. ATM is used as a high-speed WAN technology because no high-speed LAN technology will link sites around the country. In either case, the problem now becomes one of effectively linking together the ATM switches attached to the LANs at different locations. After all, it makes little sense to use ATM as a high-speed LAN or WAN technology when the traffic must be funneled through very low-speed 56 or 64 Kbps links. SONET leased lines can be used, of course, but such huge amounts of bandwidth will be expensive when used exclusively for one customer, probably prohibitively so.

This use of SONET rings to interconnect ATM switches is shown in Figure 16.1.

Figure 16.1
SONET as an ATM transport

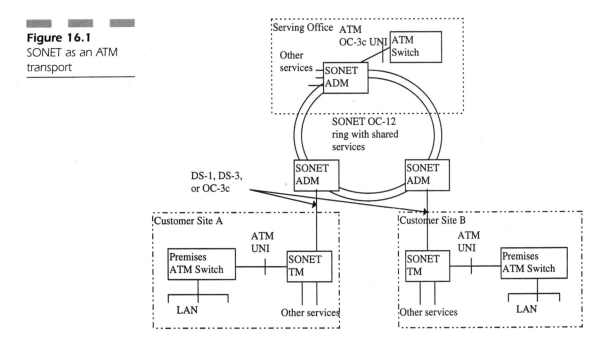

In Figure 16.1, a service provider with a SONET ring has linked the customer premises ATM switches to a SONET ring and thus provided the desired connectivity. The ring has more than enough bandwidth to satisfy many customer's bandwidth needs, especially since the ring in the figure is an OC-12 ring. If the customer wishes to comply with existing B-ISDN and ATM specifications, the access link to the ring must be OC-3c or faster.

In the figure, this access is shown as running from a SONET TM on the premises in each location. Technically, this TM would be SONET CPE and, thus, could be the choice of the customer, because SONET standards allow for vendor interoperability. In many cases, this link would be a SONET link running directly to the nearest SONET ring node. In most major cities, this would not be too far, although it may become a major failure concern to the customer without some form of diversity.

Note that other services besides ATM connectivity could be provided by the SONET TM at the customer site. These do not have to be ATM services, of course, and could be many forms of traditional voice and data applications, as well as multimedia or videoconferencing and so on.

An alternative to this method of using SONET for ATM transport would be to supply simple ATM LAN connectivity through a DS-1 fractional T-1, fractional T-3, or full DS-3 link to the service provider's SONET terminal multiplexer (TM). This would not require SONET gear to be located at the customer's premises, and most users are quite comfortable with T-carrier. The service provider's TM would be located on a SONET ring.

Note also that if the ring is connected to a serving office with an ATM switch of its own, then various ATM services could be offered to the ATM LAN users along with simple connectivity. For instance, the ATM switch could provide switched digital video (SDV) to any of the users on the ATM LANs equipped to receive it (most newer PCs would easily handle this). This, of course, opens up all kinds of new revenue sources for the service providers. However, this arrangement is not expected to become common for some time yet, as service providers wait to see exactly how common ATM equipment becomes with potential customers.

SONET Applications: LAN Interconnection

The popularity of ATM is still a very open issue. Not many suitable SONET applications are tied to ATM in any way. The vast majority of customer LANs are indifferent to ATM at best, and many position themselves as viable alternatives to ATM in any case; Fast Ethernet at 100 Mbps is a good example.

Another good application for the bandwidth that SONET provides is LAN interconnection. LANs today are no longer restricted to the older 4, 10, and 16 Mbps Ethernets and token rings. The Fast Ethernet of 100 Mbps has arrived, and 1 Gbps (1,000 Mbps) Gigabit Ethernet is already appearing. (It must be pointed out that Gigabit Ethernet frames can be put directly on fiber and SONET is not required.) Even without matching these enormous LAN speeds, the delivery of LAN interconnection on 4-fiber, bidirectional, line-switched rings (BLSR) has a lot to offer a customer.

The scenario for LAN interconnectivity is shown in Figure 16.2. Note that even in this scheme, the presence of ATM LANs and routers is not ruled out. The major difference is that the service provider is only furnishing passive transport of a customer's ATM traffic, not processing or even switching it in any way. Any ATM functionality must be provided by the ATM CPE, which may actually be a plus from the customer's perspective.

When the DS-1, or other speed link, such as fractional T-3, is supported and delivered on the BLSR, there should be very few errors and service outages. Oddly, DS-2, never a common or popular T-carrier offering, is making somewhat of a comeback as a speed delivered on these SONET rings. At about 6 Mbps, DS-2 is a good match for existing Ethernet LANs operating at 10 Mbps. For all customers needing higher speeds to link their LANs, whether with ATM LANs and routers or just faster versions of Ethernet, SONET BLSR is really the only practical means of delivery for this LAN interconnectivity in a cost-effective fashion.

Note that the customer's existing TCP/IP routers still support traditional T-carrier interfaces. This cuts down on customer expense because there is no need to migrate to SONET OC-n interfaces, and allows the

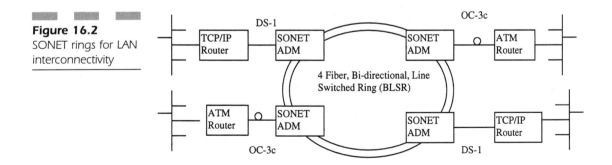

Figure 16.2
SONET rings for LAN interconnectivity

customer to retain the familiarity of working with T-carrier interfaces. No customer network management software used for internal purposes should need to be changed in spite of the added presence of the SONET ring.

SONET Applications: Video and Graphics

Enough has been said about the bandwidth demands of video and graphics applications earlier in this work. They need not be repeated here.

A brief summary of what users are doing to put such loads on LANs and WANs is worthwhile. Users run many applications that can and do heavily load LANs and WANs. Here is a summary look at the demands of two very common application types on LANs. In the late 1980s, once PC video monitor displays broke free of the severe limitations imposed by text-based DOS and UNIX interfaces, applications began to include sophisticated techniques for displaying eye-popping images and bit-mapped graphics on the monitor screen. The availability of high-resolution, noninterlaced screens at a reasonable price fueled this movement.

Today, many insurance companies and financial institutions include visual data in the form of digital photographs or scanned documents as part of their applications. These may even be stored as part of a database, as a basic field type. These images, even with compression, may be very large.

For example, a medical x-ray image may be represented by a 4,000-by-4,000-bit array. Each "gray scale" pixel is coded by 8 bits. This $4K \times 4K \times 8$ bit image yields about 128 Mbits or 16 Mbytes of data to transfer over a network. A typical 14.4 Kbps Internet link with a transfer rate of 1,500 bytes/second would need over 10,000 hours to send this image from one medical center to another. Over a high-speed network operating at 150 Mbps, like a SONET OC-3c link, the transfer would take a second or so.

The publishing industry is another large user and generator of huge graphic files. These files represent the advertisements seen in any glossy magazine today. They are not done up at 300 or 600 dots per inch, as many laser printers print today. They are done up at 2,400 dots per inch. They do not use 256 colors, but 16 million colors. Even with compression, files that are hundreds of megabytes in size are not uncommon.

When the graphic or image demands change over time—creating an animation or video sequence—the file size grows larger very rapidly. A relatively modest rendering company (this graphic evolution process is known as "rendering") of 30 or so workers may overwhelm even a 10 Mbps Ethernet quite easily with images flowing frequently between clients and servers.

However, when linking two 10 Mbps LANs connected with bridges or routers over a 64 Kbps point-to-point leased line dedicated circuit, this will automatically create a bottleneck greater than 100:1 (rounding 64 Kbps to 100 Kbps, then 10000 Kbps / 100 Kbps = 100). Clearly, many of the files and graphics cannot be adequately used in a client-server environment with a 64 Kbps speed link. The medical image sent across this network from one hospital's LAN server to another hospital's LAN client will now take more than 11 hours (7 minutes $\times$ 100 = 700 minutes > 11 hours)! The patient will hopefully wait.

Even at DS-1 speeds, a full 1.5 Mbps or so, this is a 6:1 bottleneck. This cuts down the file transfer delay to "only" 42 minutes or so, but it is still woefully inadequate for building enterprise-wide client-server networks that span the country. Even so, DS-1s have been widely embraced as a LAN interconnectivity option, to the extent that more than 3,000 corporations in America today have at least one DS-1.

Therefore, the search is on for a more effective and efficient way to both implement LAN technologies and the WANs that connect them. Lately, another class of application has demanded attention from the client-server community. This type of application not only demands high bandwidths, but low delays from the network.

The new class of applications included video (i.e., animated graphics or full-motion frame sequences) as an integral part of the application. Usually called *multimedia* to reflect the fact that these applications consist of and deliver information that is not merely read, but also looked at and listened to, these new types of multimedia applications began to spring up in the early 1990s. Multimedia put strains on the LAN's stations as well, which more or less limited their deployment on hardware platforms based on 80286 chipsets with only 1 meg or so of memory. Once 80386 chipsets became common, and memory sizes grew to the 2 to 4 meg range, multimedia applications became feasible. Today, of course, with 750- to 1000-MHz Pentium and other chips, multimedia is a given. WAN distribution is the problem.

If graphics and imaging applications produced large files than stretched PC hard drives and LANs to their limits, adding video to other applications just made matters worse. Video on a PC functions by presenting a series of still images, called frames, in rapid sequence on the PC monitor screen, usually in a Windows environment. Of course, this is exactly how TV and movies work as well. The smoothness of the motion depends on how fast the sequence changes. For PC video, this can vary from a very jerky 3 frames per second to a smooth-looking and respectable 30 frames per second. The most common video file formats can easily take up to 10 megabytes of hard disk space.

Adding video to applications stressed LANs as well. In fact, it proved so difficult to transfer video files over LANs that almost all videos come directly from a CD-ROM drive on the PC itself. The main difficulty was not even so much in terms of bandwidth: compared with large high-resolution graphics and image files, low frame-rate videos placed relatively modest demands on the LANs. Instead, the problem proved to be the variable delay in delivering the video frames from a server to a particular client. For instance, when displaying video frames at a rate of 10 per second, the LAN must deliver frames consistently at 10 frames per second. The LAN cannot deliver 8 frames in one second, and then 12 in the next, even though the net result is still 10 frames per second. This concept of *isochronicity* (i.e., things on the network must happen with a consistent delay) is an important one, especially when applied to high-bandwidth networks.

There is an important distinction between bandwidth and delay when considering high-speed networks. Confusion comes when both concepts are spoken of in terms of "speed." High bandwidth is a measure of the amount of time between the arrival of the first bit at any point on a network and the arrival of the last bit at that point. Low delay is a measure of the amount of time between the departure of the first bit from the source on a network and the arrival of this first bit at the receiver.

The point about the variability of delays on some networks, especially Ethernet-type LANs and router-based WANs, is worth emphasizing. Some applications, like voice and video, cannot tolerate more than a few milliseconds of variation in delay across a network. Otherwise, the receiver will assume that something has gone wrong and take steps to react. This has historically been a problem in the decades-long struggle to "packetize voice," that is, send digitized voice samples over the same network as is used to transmit data. The problem is that the voice samples get delayed variable amounts of time behind a large packet of data; thus, the receiver will not function correctly.

SONET, with its enormous bandwidths, stable delays (the links are essentially leased lines), and high reliability and availability, offered the best hope for productively running these types of applications across a WAN. (Other solutions using DWDM and Gigabit Ethernet exist today.)

SONET Applications: Cinema of the Future

One of the more exciting applications for SONET is to do things that were never possible before. For example, ATM switches can be combined with SONET to deliver major motion pictures to movie screens.

A nagging problem in the movie industry is that if a theater has only one screen showing a suddenly popular film, or two screens showing a suddenly "dead" film, how can this situation be handled gracefully? For instance, instead of turning away paying customers, why not just "add" another screen showing in an otherwise almost empty theater? Why screen the same show in two theatres when one will do? This is impossible today, but may not be if the Motion Picture Association of America has its way. Their plan is illustrated in Figure 16.3.

Right now, movies are usually distributed in five or six cases containing the reels to be run through the projectors. It has been done this way since the 1930s. However, if movies were digitized (most films contain a

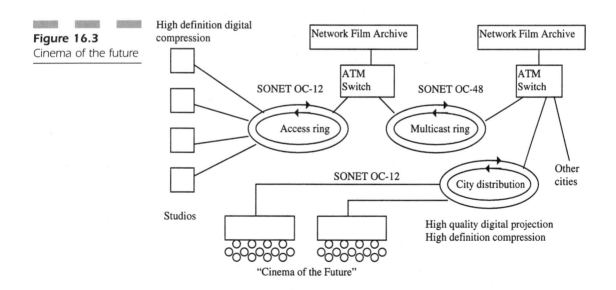

Figure 16.3
Cinema of the future

varying amount of digital "special effects" already), they could be "down-loaded" directly from studios or pulled off video archives. This process is shown in Figure 16.3.

Studios would produce not film but bits. The bits would be stored in a high-definition, digital format, and highly compressed until the feature was completed. When released, the studio would simply pipe the video to a SONET OC-12 access ring and through an ATM switch. The switch could archive the movie to allow for different time zones and so on. When shown, the movie would be multicast on a SONET OC-48 ring. Other ATM switches would distribute the signal onto various rings in cities across the country, where the theaters would be linked by multiple OC-12 access lines. The ATM multicasting switches could distribute these video streams to vast numbers of theaters simultaneously, potentially all across the country. This would involve the development of high-definition, digital-compression techniques, and high-definition, digital-projection techniques to go along with it, but the revenue potential is such that this is exactly what may happen soon. (SONET is no longer required to distribute cinema this way, but might still play a role.)

SONET Services

The first edition of this book contained a section some 12 pages long listing service providers, both local exchange carriers (LECs) and interexchange carriers (IXCs), that offered SONET services for ATM, LAN interconnection, and the like. The point is that not so long ago, SONET services were big news, and it was important that people knew that SONET was a working technology. Now, SONET offerings are greeted with a kind of shrug. Just as SONET equipment has receded from the home Web pages onto lists of SDH and DWDM vendor offerings, so SONET services have receded into the depths of the service provider's Web sites. One prominent SONET pioneer does not even list SONET as a separate service offering any longer, and searches on the Web site for "SONET" mention only its use internally (but you can still buy it from them, of course).

This is not to say that SONET service offerings are not important and that there is nothing to be said about them. But it is no longer necessary to detail each and every aspect of a SONET service offering, especially when more current information is usually available at the service provider's Web site. It is enough to list the largest service providers, no longer restricted to telephone companies, and the SONET speeds available for purchase and use from them.

This section is more of an opportunity to explore in more detail the actual structure of a major service provider's SONET backbone and access network. If nothing else, this section will provide a useful baseline for comparison with the topics in the following chapters on SDH and the future. In order to appreciate "how good it's going to be," it's always a good idea to start with how good it is.

SONET Service Providers

The first edition of this book listed eight SONET service offerings from local carriers and four from long-distance carriers. With few exceptions, all were traditional telephone companies, and a lot of space was devoted to the then almost unheard of practice of refunding money or reducing rates for service outages. Now, such service-level agreements (SLAs) are routine practice and need no further comment. Some of the service providers listed in the first edition no longer exist (e.g., NYNEX) and if nothing else this shows how quickly things can change in the telecommunications business today.

Table 16.1 shows a sampling of the SONET services offered by a current batch of service providers. All of the information was gathered from the individual companies' Web sites in early 2000. If there are inaccuracies or the information is out of date, then the Web site information might have changed. Also, in many cases the service providers offer services not listed on their Web sites. There are always "special assembly" arrangements or services offered on an "individual case basis," and no effort to pursue such avenues has been taken here. Comments about service availability were also included in the first-edition lists. This is also unnecessary today. Generally, given enough money and influence, any SONET speed can be obtained almost anywhere in a service provider's territory, with a few exceptions.

Several things about this list are worthy of note. First and foremost, no effort has been made to make sure this list is exhaustive. Consider this a SONET service sampler. The absence of a particular service provider or not in no way implies they do or do not offer SONET services. Next, note the presence of several energy companies on the list. The electric utilities were among the first companies to deploy SONET, even before the telephone companies in many cases. SONET fiber was great for running alongside 400-Kv transmission towers, where even coax-

TABLE 16.1

SONET services sampler as of 2000.

Service Provider	Service Name (if known)	Speeds Available to Users
Allegheny Energy	(Allegheny Comm. Connect, Inc.) "OC-XX"	
AT&T TCG	OmniRing^{SM}	OC3, OC-3c, OC-12, OC-12c, OC-48
Bell Atlantic	Dedicated SONET Private Network	DS-1/DS-3 on SONET, STS-1, STS-3c, OC-3, OC-12, OC-48
BellSouth	SMARTRing®, SMARTPath®	DS-1/DS-3 on SONET, STS-1, OC-3, OC-12
GPU Energy	(GPU Telcom Services)	"SONET transmission"
MCI WorldCom	Private Line Services	DS-3 on SONET, OC-3, OC-3c
Montana Power Company	(Touch America)	"SONET OC-XX Service"
Qwest	Private Line Service (PLS)	DS-0, DS-1, DS-3 on SONET, up to "OC-n"
SBC		
Ameritech (SBC Telecom)	Dedicated SONET Service	DS-1/DS-3 on SONET, OC-3c, OC-12c
PacBell	FasTrak^{SM}	DS-3 to OC-48
SNET	SONET Network Service	DS-1/DS-3 on SONET, STS-1, OC-3, OC-12
Also: Southwestern Bell		
Sprint	—	OC-3, OC-12
US West	Various: Synchronous Service Transport (SST) common	DS-1/DS-3 on SONET, OC-3, OC-12, OC-48
Williams Communications	Private Line Services	DS-3 on SONET, OC-3, OC-3c, OC-12, OC-12c, OC-48

ial cable struggled with noise and twisted-pair copper was out of the question. And power companies have rights-of-way second to none. Some energy companies, notably Enron, have huge fiber networks but employ direct IP-over-fiber, and so are not listed. Others do not market to the general public but only to other service providers, and so are also excluded. Finally, mergers and acquisitions have condensed the telephone company listings, and new, aggressive telecommunications companies like Qwest and Williams are appearing for the first time.

Another change is the listing of DS-1 and DS-3 speeds. These are not native SONET speeds, of course, but deserve to be listed when the service provider claims that these T-carrier speeds are supported on SONET. After all, few customers want or can afford to have a SONET TM on the premises when all of their existing equipment supports only DS-1 or DS-3 interfaces. There are too many T-1 multiplexers on the customer premises to ignore this market. So the DS-1 or DS-3 may extend only to the basement or equipment room in a building where the service provider maintains its own TM for this customer VT support. And if a digital cross-connect (DSC) is available as the serving office (usually a telephone company central office) then hairpinning can take care of links over SONET, even over relatively short distances.

Some companies distinguish further between channelized OC-3 (for example) and unchannelized OC-3c, but many do not. The simple listing of OC-3 does not preclude the possibility of OC-3c service availability.

It is not necessary to give each service provider a separate section about these SONET services, but a few words on each are in order. For those needing more details, the individual Web sites should be consulted. However, in many cases all of the details available are in Table 16.1. Some SONET services are no longer worthy of front-Web-page news, so tariff searches and the like might be needed to uncover further information.

The list in Table 16.1 is in alphabetical order, which is as good a way to organize the information as any. Some service providers still have unique (and often protected) names for their SONET services, but the current trend is just to group these offerings under one "private line service" umbrella, which is essentially what SONET is, whether provisioned on a ring or not. Although SONET rings are quite common, it is never a given that a SONET link is actually provisioned on a ring or not. And even if a SONET ring is used at the service provider *backbone* level, the link between customer premises and ring is often (all too often) still a single fiber run. Here are a few comments on each service offering.

Allegheny Energy is one of the new breed of telecommunications service providers encouraged by the relaxed regulatory environment resulting from a number of factors in the United States in the past few years. Allegheny's service is offered by a subsidiary called Allegheny Communications Connect, Inc. All they reveal is that SONET "OC-XX" service is available. This is most likely OC-3, but Allegheny should be contacted for details.

AT&T and TCG (sometimes "AT&T Local Services") offer TCG's OmniRing[SM] SONET service. Speeds of OC-3, OC-3c, OC-12, OC-12c, and OC-48 are available. As an aside, OC-48 SONET service remains relatively scarce in many parts of the country, relying as it does on the presence of an OC-192 (or higher) speed backbone in most cases. High-speed backbones are not a problem for AT&T, of course.

Bell Atlantic absorbed NYNEX and now features SONET as a "Dedicated SONET Private network" service. Bell Atlantic will even build private SONET rings for major customers (basically, Fortune 100), and offers DS-1 and DS-3 on SONET, STS-1 (quite unusual today), STS-3c, OC-3, OC-12, and OC-48.

BellSouth still has SMARTRing®, which is its 2-fiber UPSR, in major metropolitan areas such as Atlanta, Georgia, and Orlando, Florida. There is also SMARTPath®, which is the linear version. BellSouth offers DS-1 and DS-3 on SONET, plus STS-1, OC-3, and OC-12.

GPU Energy, servicing New York, New Jersey, and Pennsylvania in the United States, and parts of the United Kingdom, Argentina, and Australia, offers SONET in the United States under the GPU Telcom Services label. The speeds are not highly publicized, and "SONET transmission" is all that is available as a description.

MCI Worldcom, a real agglomeration of networking company from LDDS to MFS, offers SONET private line services. The speeds are relatively modest, however, at DS-3 on SONET, to OC-3 and OC-3c.

Montana Power Company has a SONET service as part of its Touch America service. The speed is "SONET OC-XX Service" and is most likely OC-3. It is not surprising to see a power company offering SONET in Montana. Montana was one of the last States in the Union to get even basic T-1 digital links, so there is plenty of incentive for someone to take an initiative.

Qwest has been another one of the new breed of service providers that specialize in having fiber miles and bandwidth, and plenty of them. All things considered, this is not a bad strategy, and fiber can support more than just SONET, of course. Qwest's Private Line Service (PLS) features DS-0 (most DS-1s carry DS-0s on SONET, but Qwest is one of the few so explicit about it), DS-1, and DS-3 on SONET. They also offer "OC-n." In Qwest's case, this is probably OC-3 and OC-12. (Qwest and US West have now merged.)

SBC is another modern telecommunications company that has grown much larger by acquisition. SBC now comprises Southwestern Bell, Ameritech, Pacific Bell (PacBell), SBC Telecom (SBC plus Ameritech), Nevada Bell, and SNET (Southern New England Telephone, based in Connecticut). Cellular One is also part of the SBC family. SONET service is offered by Ameritech (SBC Telecom), PacBell, SNET, and Southwestern Bell, although the details about the SONET service offered by Southwestern Bell at the SBC Web site(s) is minimal.

Ameritech offers a Dedicated SONET Service featuring DS-1 and DS-3 on SONET, OC-3c, and OC-12c. PacBell still offers FasTrak^{SM} SONET at speeds from DS-3 to OC-48. SNET has its SONET Network Service offering DS-1 and DS-3 on SONET, STS-1, OC-3, and OC-12.

Sprint, a pioneer in the SONET ring arena, especially with regard to 4-fiber BLSRs, is oddly quiet on its Web pages when it comes to SONET. Even its former "Clearline" name has disappeared. This is probably due to the Sprint emphasis on its ION^{SM} (Integrated On-demand Network) service, but this is just speculation. In spite of the lack of detail on the Web pages, Sprint still offers SONET private lines at OC-3 and OC-12, as well as supporting a variety of VT speeds.

US West probably leads the pack when it comes to the amount of SONET information available through its Web site. There is even a picture of a SONET frame. US West SONET services fall into five types of private line transport services. First, there is US West SONET Ring Service (SRS), for DS-1 and DS-3 on SONET, as well as OC-3. Next comes US West Self-Healing Network Service (SHNS, pronounced "shins") that enhances the basic speeds by offering LAN connectivity. The workhorse is the US West Synchronous Service Transport (SST) offering, with speeds from OC-3 to OC-48. There is also US West Self-Healing Alternate Route Protection (SHARP), which not only is provisioned on a SONET ring, but offers an alternate path *to* the ring in case of failure. SHARP is available for DS-1 and DS-3 speeds, as well as the SST speeds of OC-3 to OC-48. Finally, some of these service flavors are supported by US West Network 21, which is a series of high-speed backbone rings in major metropolitan areas.

Williams Communications, much like Qwest, sees intrinsic value in rights-of-way and fiber conduits as well as in the bits and services supported by this infrastructure. Its Private Line Services support DS-3 on SONET, OC-3, OC-3c, OC-12, OC-12c, and OC-48. (Qwest and US West have now merged.)

All in all, it is a good sign that SONET service offerings have left the cozy world of the telephone companies and arrived at the door of the energy companies and "bandwidth companies" as well. This list of SONET service providers could probably be extended in several ways, but its intention is to furnish a good idea of the service providers and speeds available on SONET today.

SONET Service Architectures

The overall impression of the previous section might be taken to mean that there is a bewildering array of SONET services offerings in terms of service provider, speeds, and packages. But once the service surface is scratched, all SONET service offerings are based on two very distinct, but very consistent, architectures. That there is not a lot of variation in SONET service architectures should come as no surprise. All SONET service architectures are based on the same array of SONET building blocks such as ADMs, TMs, and related devices such as digital cross-connects (DCSs). All vendors of SONET equipment can be only so creative when making SONET devices and still stay within the firm guidelines of specifications and standards. (One of the aspects of SONET currently under assault, and explored in the next chapter, is this apparent inflexibility of SONET).

This section examines the two basic architectures for supporting SONET services. This section details the first major SONET service architecture, which has been used mostly by the traditional telecommunications service providers, the telephone companies. The next section investigates the more ambitious SONET service architecture, which integrates more types of access arrangements and even some higher-level services such as switching and/or routing into the basic SONET ring.

The first SONET service architecture can be called the "telephone company model." This is only because it was the telephone companies, most notably the former pieces of the AT&T Bell System such as Bell Atlantic or Ameritech (the Regional Bell Operating Companies, or RBOCs), that invented and deployed this SONET architecture. Nothing prevents an ISP or even a major corporation from building and using this same SONET architecture.

Until recently, this SONET service architecture was the traditional way that customers made use of SONET services. So this is also the typical model. The overall architecture is shown in Figure 16.4. There is a lot going on in this figure, so a lengthy explanation is unavoidable. The central feature is the SONET ring itself, which typically runs at OC-12 or OC-48, depending on the number of customer premises linked and the amount of traffic on the ring. There are even OC-192 rings appearing with this architecture. The ring itself could be a 2-fiber UPSR, or a 2- or 4-fiber BLSR; this does not matter to the overall architecture.

The service provider buildings on the ring, heavily outlined, are all telephone company central offices (COs), but this is not a requirement. Any suitable site for an ADM or two will do, but typically SONET rings in this architecture link existing central offices. Minimally, the CO will have a SONET ADM, but some COs will have a DCS and more elaborate arrangements, also shown in Figure 16.4. The customer premises in the figure are dotted boxes. In this architecture, the customer premises

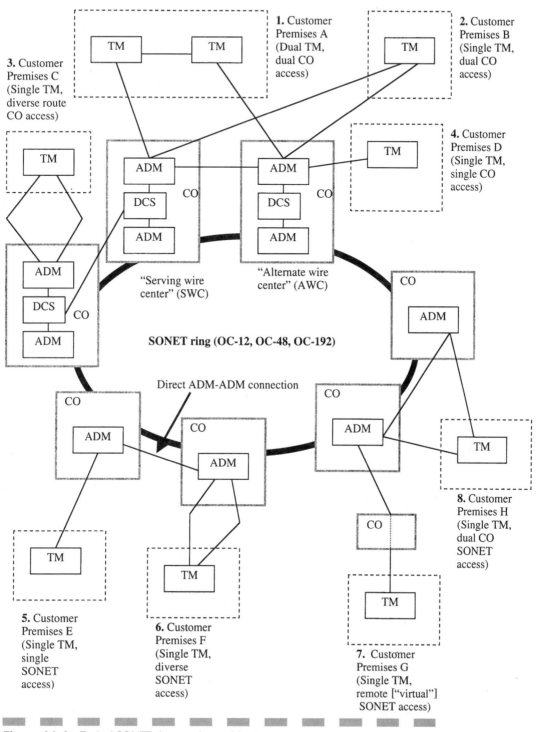

Figure 16.4 Typical SONET ring service architecture

equipment (CPE) would appear to be the SONET TM, but this need not be the case. It is more likely that the TM is owned and operated and configured and fixed by the service provider. The TM is therefore network equipment in most cases, but nothing precludes major customers from having their own TM, of course. The TM could be in a common equipment space (usually in the basement) or in a special area to service an office complex or campus. All of the links between offices and customer premises in the figure are SONET fiber links. Point-to-point links are thin lines. Note that these links can run from ADM to ADM as well as from TM (customer site) to ADM.

Figure 16.4 shows eight distinct access arrangements to the SONET ring. The upper four are used mainly when VT support is the main service provided, usually at DS-1 or DS-3. This is one good reason why the TM is usually service provider equipment: customers retain their T-1 multiplexers and see no reason why they should also have to pay for a TM. The lower four access arrangements shown in the figure are primarily for ring access at native SONET speeds. Lower-speed VTs might be present on these SONET links, but all SONET frame content is just carried through the network untouched and delivered to another customer site. The lack of a DCS function is the distinguishing feature of these access arrangements.

The eight customer access arrangements are listed in Table 16.2. The rest of this section will discuss some of the major features of each arrangement. Usually, each arrangement is covered by a separate pricing plan. Any CO might actually support all eight scenarios. The use of many COs is just for illustration purposes.

The first four scenarios consider customers whose primary use of SONET is for VT speeds—that is, mostly DS-1s and any DS-0s that ride inside the DS-1 channels, and perhaps even some DS-3s. The usage is always channelized in the vast majority of cases, so a key aspect of these scenarios is the ability to cross-connect at the VT level. This is the primary function of the DCS, of course. Now, the DCS might be a separate unit distinct from the SONET ADM, or the DCS could be a shelf or series of cards in the SONET ADM itself. No matter, it is the function that counts. The COs in each of these first four scenarios show *two* ADMs with a DCS in between. This is certainly a possible arrangement, since a small ADM might gather traffic from a number of TMs, cross-connect and groom as needed, and then pass traffic on to a larger ADM with higher-speed SONET ports attached to the ring. Or the ADM could be one unit with the DCS function built in or separate. There is a lot of flexibility in precise equipment configurations.

TABLE 16.2

The eight access arrangements

Scenario	Characteristics	Comment
1	Dual TM, dual CO access	Premises has two TMs, linked together and linked separately to two COs (SWC, AWC). Hairpinning of VTs possible.
2	Single TM, dual CO access	Premises has single TM, but two links to SWC and AWC. Hairpinning of VTs possible.
3	Single TM, diverse route CO access	Premises has single TM and access to single CO. Diverse routing reduces link failure risk. Hairpinning VTs possible.
4	Single TM, single CO access	Premises has single TM and single link to CO. Hairpinning of VTs possible.
5	Single TM, single CO SONET access	Premises has single TM and access to ring ADM is direct. No VT cross-connecting possible.
6	Single TM, diverse route SONET access	Premises has single TM; diverse routing to CO reduces link failure risk. No VT cross-connecting possible.
7	Single TM, remote "virtual" SONET access	Premises has single TM, and ring ADM can be reached only through "tail-end" CO. No VT cross-connecting possible.
8	Single TM, dual CO SONET access	Premises has single TM, but links to two SONET COs. No VT cross-connecting possible.

The main difference between the first four SONET access scenarios is in the details of how the TMs on the customer premises are linked to the CO with the SONET ADM. A few words about each are all that is necessary.

1. There are two TMs on the customer premises, not only linked together, but also linked to two physically separate COs. So the TMs can access the ring through two separate ring nodes. The ADMs might also be linked by a point-to-point fiber run. Also, the DCS might be linked directly to another DCS in case of ring failure. This is the most bulletproof SONET access arrangement. The TMs on the customer premises are redundant, as are the point-to-point links to the COs. Since the COs are physically separate, route diversity is more or less automatic, except perhaps at the entrance facility of the building itself (even this is sometimes

possible to diversify). If the ring fails, links mapped through DCS hairpinning (a cross-connection that turns right around and goes back the way it came) still works. If the main ring node fails, the other CO provides access. Usually, one of the COs is the *serving wire center* (SWC) for the customer and the other is the backup *alternate wire center* (AWC). The only real weak link is that the SWC and AWC are usually a few miles apart, so a major flood (for example) throughout the area can take out both the SWC and AWC.

2. There is only one TM on the customer premises, but the route diversity and SWC/AWC structure is preserved. As long as the TM is reliable enough, this arrangement might be almost as good as the first. Hairpinning of VTs is also possible here.

3. There is again only one TM on the customer premises, and now only one CO and ADM at the other end. However, the SONET links between TM and ADM are two, and diversely routed. This can be much more inexpensive than running links to two COs, and many customers come to realize that if the outage condition is serious enough, even being hooked up to two COs in the same general area might not help. Hairpinning of VTs is possible here also.

4. The last scenario essentially for VT support is the simplest. There is one TM with one SONET fiber link to one CO with ADM. Cross-connecting and grooming still take place, but the weakest link here is the single-access fiber run, especially right at the premises entrance facility, where a lot of construction (stairways, ramps, etc.) is constantly taking place. Hairpinning of VTs is also possible here.

The last four SONET service architecture scenarios emphasize native SONET speeds. There is no need for a DCS at the CO, because access is presumed to be at native OC-3 or OC-12 speeds (OC-48 is also possible). The OC-3 and OC-12 could be OC-3c or OC-12c, but it is the speed that is important. It is important to realize that the ADMs might also groom and perform normal add/drop functions on the traffic inside an OC-3 and OC-12 on an STS-by-STS basis.

1. There is a single SONET TM on the customer premises, and a single link to the CO ADM. This is quite similar to scenario #4, listed previously.

2. There is a single TM with two route-diverse links to the CO ADM. Similar to scenario #3 listed previously.

3. This scenario is a little different from anything discussed before. The TM is at a customer premises not ordinarily serviced by the

CO with the SONET ring node ADM. The CO for the customer premises is not on the SONET ring. How can this customer receive SONET service? Simply by routing the fiber through the intermediate CO to the SONET ring node CO. This is *backhauling* at its best and has been done for ISDN and other services for many years. This arrangement is sometimes referred to as *virtual SONET* or *remote SONET,* but it is still basically backhauling. The risk here is that if the CO in the middle is hit by an outage, even if TM, ADM, and ring are otherwise in fine condition, service might be disrupted.

4. There is a single TM on the customer premises, the access to two COs with SONET ADMs. Similar to scenario #2, listed previously.

The architecture described in this section has been challenged recently by a more integrated architecture. This approach is interesting enough to deserve a section of its own.

SONET Integrated Architectures

One reason that the tradition architecture for SONET services emphasizes VTs (DS-1 and DS-3) and native OC-*n* private lines is that telephone companies have been historically restricted due to regulation to offer only "basic transport" services. In this environment, "enhanced services" such as routing IP packets have been strictly off limits. While this climate has allowed ISPs to grow and flourish even in the shadow of the telecommunications giant, the trend today is toward a more permissive structure where service providers are allowed more latitude than ever before. So the trend today with SONET service architectures is to create a ring structure that includes more switching and routing capability than ever before.

Service providers that favor the more integrated architecture can try to sell more than just private lines with lots of bandwidth or ring protection for circuits. These service providers try to integrate at least some routing of IP packet, or switching ATM cells, right into their network infrastructure. The services they sell are more inclusive than simple connectivity. Many ISPs have been actively pursuing this approach for years. In fairness to the telephone companies, many traditional telephone service providers, especially the RBOCs, have been limited by state and federal regulations with regard to the types of data services they can offer. So BellSouth.net the ISP (as an example) is not BellSouth the telephone company and care must be taken to keep the two entities separate. Whether right or wrong, it is a reality, although things are slowly but surely changing.

The general structure of a more integrated SONET services architecture is shown in Figure 16.5. While not quite as busy as the traditional SONET architecture, there is enough going on in the figure to deserve some explanation.

The central feature is still the SONET ring, running at OC-3, OC-12, OC-48, or even OC-192 speed. Traditional model ADM and customer premises TM might still be part of the overall picture. VT support and

Figure 16.5 A more integrated SONET services architecture

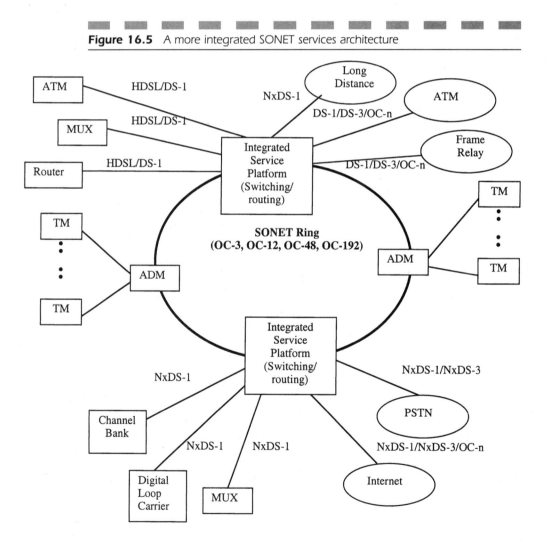

OC-*n* private lines, as well as all of the access arrangements detailed in the previous section, might still be present as well.

But the important features of the integrated SONET services architecture are the presence as SONET ring nodes of devices labeled "Integrated Service Platforms." These devices could also serve as SONET ADMs, but their real purpose is to add value to the basic bit transport service that the SONET ring provides. These devices are usually capable of both switching any ATM cells and routing any IP packets that arrive in the payloads of the SONET frames. No separate routers or switches are needed: these functions are integrated into the fabric of the devices themselves. But it is what the Integrated Service Platform attaches to that is the most important part of Figure 16.5.

Each Integrated Service Platform in the figure attaches to a variety of customer premises equipment (CPE), shown as boxes. The platforms also have access to a number of different types of networks, shown as circular clouds. That is the power of the Integrated Services Platform—this bringing together of many types of CPE and many types of network services, all in the same package.

Consider CPE types first. Customers can link routers, T-1 multiplexers, and even more exotic CPE such as ATM switches to the integrated ring node. The links need not be fiber links; those shown are DS-1s or even High-bit-rate Digital Subscriber Line (HDSL), a more cost-effective way of provisioning 1.5-Mbps T-1 rates on twisted copper pairs (HDSL2 requires only one pair of copper). Since only 5% of commercial office space in the United States has fiber access, this nonreliance on fiber is a nice feature. There are some CPE arrangements that normally require more than one DS-1 speed access link. These NxDS-1 arrangements, such as a digital voice channel bank, digital loop carrier, and larger T-1 multiplexers, are also shown. Naturally, there are variations in CPE access line speed requirements. Figure 16.5 is just a general diagram of connectivity, not a blueprint.

On the network service side, the Integrated Service Platform can link by means of the SONET ring to a voice long-distance network, an ATM network separate from the ring, or even frame relay. The link speeds are appropriate to the service type: NxDS-1 (or DS-3 in some cases) for long distance, and DS-1 or DS-3 or OC-*n* for ATM and frame relay. Local voice services can be provided by direct links to the PSTN, and data services (e.g., Web access) are provided by the Internet. PSTN links usually would run at NxDS-1 or NxDS-3, depending on traffic load. Internet access usually would use NxDS-1, NxDS-3, or even OC-*n*, depending on the traffic load and the ISP at the other end of the link.

Admittedly, SONET rings could provide such connectivity in the traditional model. The attractive feature of the integrated services model is that the SONET ring nodes take a more active role in the services arena, not just in the connectivity arena. No additional routers, switches, or the like are needed.

It should be noted that there is nothing exclusive about either the "telephone company model" or the "integrated services model" of SONET service architecture. That is, many telephone companies are busily enhancing their legacy SONET architectures with integrated models, and many new service providers are seeking to capture new customers by offering private-line services.

This chapter has explored SONET service offerings and the architectures typically deployed to deliver them. But the SONET scene has been complicated by the arrival of more advanced optical networking techniques, trends introduced in an earlier chapter. It is time to consider the impact of DWDM and optical networking on SONET. Some observers feel that the rise of the IP protocol due to the popularity of the Web and Internet, coupled with the appearance of Gigabit Ethernet (GBE) and DWDM products, has left the future of SONET (and SDH) very much up in the air. This is the topic of the next chapter.

CHAPTER 17

SONET Futures and Issues

This chapter takes a look at a few topics that are of definite interest, but did not fit into the overall outline and plan of this work particularly well. Therefore, they have been placed near the end of the book, not as an afterthought, but in order to give the reader with a some summarized thoughts and directions about SONET.

The first and foremost issue to be considered is the future of SONET, with regard to current developments in optical networking. The future of SONET considerations will be in two main areas. First, the important developments of dense wavelength division multiplexing (DWDM) and optical networking are examined. This technology is not exclusive to SONET, but is generally applicable to all fiber optic communications. However, this chapter investigates how SONET fits in with optical networks.

Second, this chapter examines the position of SONET (and SDH) with regard to the IP protocol and the use of Gigabit Ethernet (GBE). IP is the protocol of the Internet and Web, and GBE is intended for fiber networks. Neither in any way requires SONET/SDH. What is the future of SONET/SDH in an IP/GBE world?

The second set of issues to be discussed have to do with the testing of SONET links to make sure that they are within specifications. This does not appear to be much of a problem. SONET standards with regard to OAM&P, protection switching, and above all SONET rings, would appear to make this a nonissue—but it is.

This is because SONET has been marketed and sold as a more advanced "private-line solution." Because this means that an organization sees SONET as another way to build private networks, most organizations want their network management personnel to monitor the SONET links, as they would any other leased line. Yet few customers have SONET equipment on their premises. What can be done with regard to SONET performance in this case?

Finally, some speculative thoughts on the implications of a possible Unlimited Bandwidth Network (UBN) are offered.

This chapter is intended to give a sense of closure to these SONET issues and the future of SONET in general.

Optical Networking and SONET/SDH

SONET is *not* an optical network, as has been pointed out previously. SONET and SDH are still firmly electrical networks with optical links. Anytime anything happens worth mentioning in SONET/SDH, the electrical frame structures, not the lightwaves themselves, are needed.

Optical networking, by definition, has NEs and components that perform at least some of the essential networking tasks at an optical level. Optical network components currently include regenerators (really optical amplifiers, or REDFAs), optical ADMs (OADMs), and optical cross-connects (sometimes misleadingly called optical switches). REDFAs were discussed previously and will not be considered further. This section will deal with OADMs and optical cross-connects, however. True optical switching has been a dream for many years, and optical switches in DWDM systems are almost, but not quite, ready to go.

This is not to say that DWDM is not an essential part of an optical network. DWDM allows for the packing of many wavelengths onto the same physical path, the optical equivalent of FDM. But if all there was to optical networking was DWDM, then optical networking would be as exciting as watching 100 TV channels whiz by on a coaxial cable. The excitement of optical networking is what happens before, during, and after the wavelengths come together on the fiber.

One of the problems when mentioning DWDM and optical networking in the same section is that the constantly evolving capabilities of DWDM in terms of channels and aggregate speeds make written statements almost instantly obsolete. Fortunately, this section is not so much concerned with feeds and speeds as how DWDM and optical NEs come together to make a true optical network. If readers need more detail, assume in the following that the DWDM is a 100-channel (current maximum is over 1,000) system, with each channel running at 10 Gbps, the OC-192/STM-16 rate (current serial maximum is 40 Gbps), for a total throughput of 1 terabit per second. Later in this chapter a terabit router will be introduced as the perfect companion for this configuration.

After the principles of operation of the OADM and optical cross-connect are outlined, an overall architecture for optical networking will be introduced. The position of SONET in this architecture is of paramount interest for the purposes of this book. OADMs in particular make use of the in-fiber Bragg grating and optical coupler to selectively separate and merge wavelengths on a single strand of fiber. The details of these components have been examined in Chapter 3 and need not be repeated here.

The Optical ADM (OADM)

The fundamental principle behind the optical add-drop multiplexer (OADM) is simple enough. A filter is used to isolate, or drop, the desired wavelength from the multiple wavelengths arriving on a fiber. Multiple wavelengths can be dropped, of course, or allowed to pass, depending on the physical configuration of the filters. Once a wavelength is dropped, another channel employing the same wavelength can be added, or inserted, onto the fiber as it leaves the OADM. This is shown in Figure 17.1.

This simple OADM has only four input and output channels, each with four wavelengths (λs). Within the OADM, the absence of a dropped wavelength is indicated by a space in the wavelength profile on the fiber. There is no absolute connection between dropped or inserted wavelengths and the ports on the OADM. This is a configuration issue. Naturally, two channels using the same wavelength cannot share an output fiber. On their way through the OADM, the wavelengths might be amplified, equalized, or further processed. The important point is that all of this takes place in optical form, with no electrical/optical (E/O) conversion required.

All an OADM does is basically isolate wavelengths for some purpose. It is useful to be able to access a wavelength selectively, but the real need is to be able to rearrange wavelengths from fiber to fiber. Since the basic transport link is still the physical fiber itself, the OADM must have some means of rearranging the wavelengths from input fiber to output fiber. This is where the optical cross-connect, or switch, comes in.

Figure 17.1
An optical ADM
(OADM)

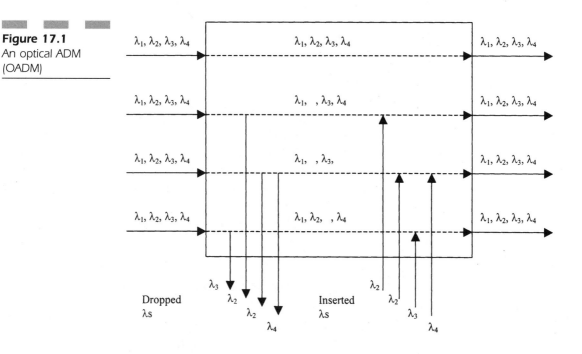

Changing Fibers in the Middle of the Stream

Electricity easily can be made to follow paths through electronic components, a fact of nature that makes electrical cross-connecting a trivial exercise. It is not a problem to make electricity jump from wire to wire or from path to path. Electricity always follows the easiest path when confronted with a choice. But the whole idea behind fiber optics in general is to make light *stay* in the fiber. How can lightwaves be induced to change fibers without changing the light into electricity? This is what the optical cross-connect does.

An optical cross-connect can take four input fibers (for example), each carrying 4 wavelengths (not a lot for DWDM, but used for ease of illustration), and rearrange the 16 wavelengths onto four output fibers. Naturally, if all four fiber channels use the same wavelengths, then there is a risk that two input channels of the same wavelength will want to share the same output fiber. In this case, a simple *transponder* inside the optical cross-connect will shuffle one of the wavelengths to an available channel.

The optical cross-connect process is sometimes called "switching," but this is misleading. Switching (and routing, for that matter) takes place by examining some header or label inside the arriving traffic stream. Needless to say, cross-connecting does no such thing. It is a simple port-by-port rearranging process, related much more to multiplexing than to switching and routing. The proof is that if an output fiber can carry only four channels, then only four channels can be cross-connected onto the output fiber. In the example, 16 input channels are rearranged, and wavelengths reassigned, but always become 16 output channels. True switches can switch all arriving traffic onto one output port, if need be, and sometimes it seems like this is always the case. The optical cross-connect principle applied to DWDM is shown in Figure 17.2.

Note that a wavelength can arrive on one fiber and leave on another, as long as that is the way the cross-connection table has been configured. The wavelength used can also change, and often does. This "wavelength shift" just requires a simple transponder to detect one wavelength and transmit another (wavelengths can even shuffle within a fiber).

The optical cross-connect performs exactly the same port-by-port reconfiguration functions as the electrical cross-connect. The traffic streams are not STS-1s (or whatever) inside an STS-12 (for example). The traffic streams cross-connected are wavelengths on the fibers. The trick is to do all this entirely in the optical domain. The key component that allows this to happen is the *optical cross-connection element*, or *fiber switching junction* (which is one reason the entire device is sometimes called a switch).

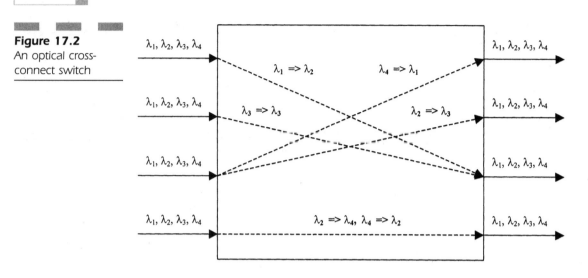

Figure 17.2
An optical cross-connect switch

Ironically, there is plenty of electricity involved in the optical cross-connection element—not so much in terms of electrical power, but just in terms of role.

The trick, if there is one, is simply to fabricate the fibers (in some cases they are actually optical channels etched onto the silicon substrate of a chip) in the optical cross-connection element so that the fiber cores actually come in contact with each other for a length. Usually, the length is no more than 3 to 4 millimeters, but there is a lot of variation here. Since light is optical, but still within the realm of electromagnetic phenomena, the proper application of a modest voltage (or absence of voltage) across this junction area can make or prevent light from changing cores from one fiber to the other. The principle behind the optical cross-connection element is shown in Figure 17.3.

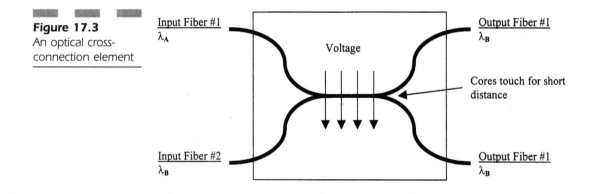

Figure 17.3
An optical cross-connection element

The wavelengths input into the optical cross-connect element are shown as λ_A and λ_B because optical cross-connects function at the same wavelengths. But the two input streams do flip-flop onto the output fibers.

The basic optical cross-connect element can be combined into a simple optical cross-connect device. This is shown in Figure 17.4. No DWDM is shown, just four wavelengths being shuffled around among input and output fibers according to how the device is configured. Only the path for λ_A is shown as a dashed line, but the others are routed in a similar fashion.

DWDM capabilities can be added to the basic optical cross-connecting, perhaps by the creative use of transponders to shift wavelengths into and out of the optical cross-connect elements. This is a giant step closer to the optical "switch" since what is being "switched" are multiple traffic streams on each input fiber. Care must be taken, of course, to prevent two wavelengths from trying to share an output port.

Whenever SONET/SDH is wanted as a wavelength in DWDM systems featuring OADMs and optical cross-connects, transponders are almost always used. This is because SONET/SDH signals all want to use the standard wavelength established for the SONET/SDH, 1310 nm. But DWDM operates in the 1550-nm range. So transponders not only shift the native SONET/SDH wavelength into the DWDM range, but also prevent collisions when two or more OC-12/STM-3 (for example) inputs are being carried.

Figure 17.4

A simple optical cross-connect in detail

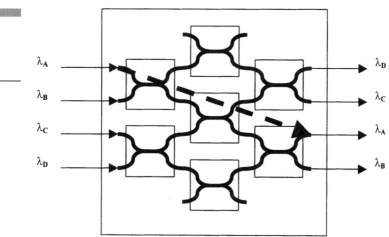

Putting It All Together

The basic components of optical networks—DWDM, OADMs, transponders, and optical cross-connects (sometimes called ODCSs, although there is nothing really digital about them)—can be combined to create an optical node on a fiber ring. This fiber ring is no longer a pure SONET or SDH ring, of course, although SONET/SDH ADMs and the like might still be present on the ring. The fiber is now carrying DWDM wavelengths and might even be new fiber installed just for DWDM (although if the spans are short enough, SONET/SDH single-mode fiber works just fine). This basic node structure is shown in Figure 17.5.

The important part of Figure 17.5 for the purposes of this book is the role that SONET/SDH plays in the overall architecture. The ring architecture can still be a four-fiber BLSR or other ring type, and often is. Protection switching is still a key element of any ring. The nice things about the continued presence of SONET/SDH on the new DWDM ring are twofold. First, transponders can be used to allow one or more of the SONET/SDH

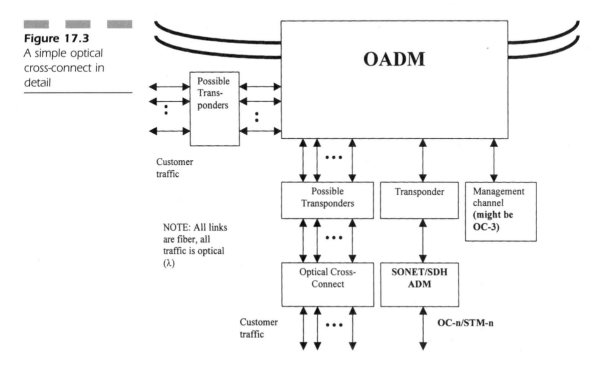

Figure 17.3
A simple optical cross-connect in detail

ADMs to use the DWDM ring. In fact, ring capacity for SONET/SDH is greatly enhanced, since there are many wavelengths available. This also allows legacy SONET/SDH equipment to share the ring, of course. Second, since there is still only a single physical fiber ring to protect, once the SONET/SDH line level overhead makes the protect switching decision to use another span or even ring-wrap, then all of the other DWDM channels go along for the ride.

What about all of the other wavelength channels? There is no need to fill these with SONET/SDH streams, although that is always possible. Perhaps other *direct-to-fiber* interfaces could be used, and this has always been a goal of DWDM and optical networking systems. Many frame structures other than SONET/SDH use fiber as a transport medium. For instance, the Fiber Distributed Data Interface (FDDI) is a 100-Mbps fiber ring architecture, but FDDI frames are not SONET frames and fit very poorly into SONET STS-3c payloads (about 55 Mbps are "wasted"). Fibre Channel is another architecture not aligned with SONET, and even 10-Mbps Ethernet has run on fiber for a number of years (10Base-F). None of these are SONET or benefit from the presence of SONET directly (SONET protection switching is really an indirect benefit, since FDDI and some others have their own forms of protection switching). The position of Gigabit Ethernet (GBE) will be considered in the next section. The point here is that to allow direct-to-fiber traffic to most efficiently use a fiber ring, the requirement to use SONET/SDH framing should be dropped. If SONET/SDH nodes are present, just plug them into an OADM, assign a wavelength, and stand back.

Optical networking offers a way for service providers just to *sell a wavelength* to a customer for almost any form of traffic. The customer could use the wavelength for anything at all, SONET/SDH or not. The OADMs, transponders, optical cross-connects, and so on, take care of making sure the traffic stream carried by the chosen wavelength finds its way through the ring nodes, across rings, between rings, and so forth, properly. The customer need not know (and need not care) what wavelengths are used for the traffic internally, although this knowledge might be provided routinely for troubleshooting and network management purposes.

This flexibility is the real promise of optical networking to customers and service providers. SONET/SDH runs at fixed speeds, but many things are not aligned with SONET/SDH speeds. SONET/SDH frames can carry many things, but in a very real sense both SONET and SDH are optimized for VT voice channels. Other things fit, but some only poorly, with much wasted bandwidth and many redundant overhead functions.

If all channels of a DWDM optical network become non-SONET/SDH, how then do the customer and service provider gain any benefit from SONET/SDH protection switching and network management functions? If there is no SONET/SDH present on the optical network, then standard protection switching and network management must be added to each and every direct-to-fiber technology, right?

Probably not, although there have been proposals just along these lines. The DWDM standards allow for a *network management channel* at 1310 nm, outside the normal DWDM traffic-bearing range. This channel coincides with the standard SONET/SDH wavelength, and use of SONET/SDH for this network management channel makes a lot of sense, but is not required. The presence of SONET/SDH on this channel protects and manages all of the other services on the single physical fiber.

This SONET/SDH management channel need not carry any live user traffic at all. This might seem quite wasteful, since only the overhead is needed for protection and management purposes. However, some DWDM vendors have made creative use of this channel by placing forward error correcting (FEC) code information derived from blocks of bits on the other channels. So there is the added benefit of cutting down on the need for resends, which is the whole point of FECs in the first place. Some vendors even carry this FEC on in-band SONET/SDH channels, simply replacing idle stretches in the SONET/SDH payloads with the FEC.

Summing Up

So is there a place for SONET/SDH in the optical networking world? If services are provisioned and sold by wavelength, who needs SONET/SDH? For now, the answers are "yes" and "anyone concerned with standard telecommunications network management" based on the role that legacy SONET/SDH equipment and services can play in the optical networking world. But this might not always be the case. There is one other consideration that needs to be examined in the next section.

There are many transport frame structures besides SONET/SDH: FDDI, Ethernet, and Fibre Channel are but a few. But given the domination of the Internet and Web in the world of networking today, most frames exist for one purpose only: to carry IP packets between client (e.g., Web browser) and server (e.g., Web page). The IP packets themselves can carry voice, video, or data.

It is time to consider the impact on SONET/SDH of IP packets coupled with Gigabit Ethernet (GBE) direct-to-fiber interfaces.

SONET/SDH, IP, and Gigabit Ethernet (GBE)

The rise of the Internet and Web has all but eliminated concerns about OSI-RM Layer 3 (Network Layer) protocols other than IP. IP packets have become the Layer 3 protocol data unit (PDU) of choice. Once there were multiprotocol routers and multiprotocol networks and even "dual stack" clients and servers that understood more than one protocol. Today, with the importance of the Web and the Internet, intranets for remote workers, and extranets for e-commerce (for consumers) and e-business (for all intercompany networking), there is little room or need for anything else but IP.

Other Layer 3 protocols are often tolerated only if they look like IP and tunneling (placing packets within packets) is a direct result of this need in many cases. Most routers just handle IP packets, period. The Internet is not a multiprotocol network; the Internet is a single-protocol network. IP only, please. Want to use the Internet and Web for sales, marketing, e-business and so on? Do you run IP? If so, welcome. If not...well, nice of you to ask...

What has all of this to do with SONET/SDH? Both SONET and SDH are merely OSI-RM Layer 1 (Physical Layer) transports, and as such are spectacularly unconcerned with the exact format of the information their frames carry. All that is needed is a standard way to package voice, video, ATM cells, or IP packets inside SONET/SDH frames, and that should be that.

But today that is not the whole story. The dominance of IP means that the very existence of the SONET/SDH frame structure, especially when DWDM and OADMs are factored in, has been called into question. Fiber links are more and more used not to connect SONET/SDH ADMs or ATM switches, but IP routers. Routers are the engines of the Internet in the same way that telephone company central offices were the engines of the analog voice network. The question today is not, "Can IP routers be connected using SONET/SDH fiber links?" but rather, "What is the *best* way in terms of cost and efficiency for fiber networks to be used to connect IP routers?" If it turns out to more cost effective and efficient to use DWDM fiber directly for IP than SONET/SDH, then so be it.

Why Not IP Direct-to-Fiber?

This section would be a lot shorter if there were a simple and easy way to put IP packets directly onto a fiber link. This direct-to-fiber interface for IP packet transmission would quickly replace every other form of WAN interface on a router. There would only be two remaining issues. First, how the routers could keep up with such high speeds (up to 40 Gbps per channel, and many channels with DWDM). And second, how could many such point-to-point links (paid for by speed and usually by the mile) ever be cost-justified (more later on these related issues).

However, the most widely used current definition of IP—IP version 4 (IPv4)—makes it difficult, but not impossible, to put IP packets directly onto a fiber link. This has not stopped people from trying some ingenious methods for direct-to-fiber IPv4. It is also true that it is easier to place IPv6 packets onto fiber directly. However, this discussion will not consider direct IP-to-fiber methods, since most observers see something else needed between IP and the fiber, whether IPv6 is used or not.

There are two problems with trying to put IP packets directly onto any physical transport, not just fiber. The same is true of wireless, coaxial cable, twisted pair copper, and anything else. The problems are:

1. The IP packet header contains no "start of packet" indicator.
2. The IP standards do not define what happens on the line *between* IP packets (true of both IPv4 and IPv6).

Both problems are of equal impact and consideration.

Packets, or any other PDU, need some way for a receiver to determine the beginning and end of the PDU (it is always assumed that the sender, which originates the PDU, knows the length!). Because packets are, by definition, of variable length between some minimum and maximum, receivers must know, if nothing else, whether they have room in a buffer (memory area for communications) for the packet. The packet would need a way to tell receivers "this is the start of the packet" and "this is the end." This process is called *delineation,* or *framing.*

It might seem that simply adding start-end indicators to carry IP packet into a fiber would solve the whole issue. But this is really only half the problem. It is just as important to know what happens on the fiber *between* the IP packets. Packets are neither TDM time slots nor ATM cells.

Both time slots and ATM cells (once called "labeled time slots") are of the same length on a line. Once delineated (receiver locates head-tail boundary), both time slots and ATM cells are easy to distinguish. Time slots use timing (naturally) to accomplish this, while ATM uses a different method of no concern in detail. The issue with time slots and ATM cells is whether the time slot of a cell has useful information, or is filled with an *idle pattern* which need not be processed.

IP packets, on the other hand, *always* contain information. Otherwise there is no reason to compose and send a packet. So packets are as bursty as the information they represent. Even if some packet streams carry 64-kbps voice, there are ways to make this into a bursty IP packet stream. So a link carrying IP packets, IPv4 or IPv6, will sometimes be totally quiescent and sending no packets at all, if only for a short while.

Thus the second issue boils down to the fact that *there is no standard idle pattern for IP*. No IP document spells out what the line is doing when not sending IP packets. Moreover, it is best to have the line doing *something* when not sending IP packets. Otherwise, long gaps between IP packet arrivals can be confused with outright line failures. This is not a good idea when network management is concerned. When idle patterns are used, line failures are instantly obvious.

And the issue runs deeper than just picking a bit pattern to fill the gaps between IP packets. Any idle pattern is obviously just a string of bits. What happens if this string of bits occurs *in the middle* of an IP packet? Would not a receiver think that a *new* IP packet had begun? So there is always the related issue of *transparency* when the idle pattern can appear within a packet.

Fortunately for networking, the related issues of packet length, idle pattern, and transparency have all been solved to everyone's satisfaction. They just have not been solved at the IP packet layer (Layer 3 of the OSI-RM). These issues and functions are all properly concerns of the OSI-RM Layer 2 (data link layer). This is what the Layer 2 *frame* is for.

All of the current IP documentation specifies placement of an IP packet into some form of OSI-RM Layer 2 frame structure (even PPP) before transmission on a line. Frames have two main purposes for the purpose of this discussion. First, frames always have a length code. Second, frames always define an *interframe fill* pattern to be used on the line when there are no frames to send (there are ways of making this interframe fill transparent as well, but this is not important here). Frames also serve useful purposes for multiprotocol support and error control, but these aspects are not the focus in this section.

So IP by itself need not worry about the length and interpacket bit patterns. Everything works just fine if the IP packet is placed in a standard frame. So for the rest of this section, it is assumed that IP packets are placed into some form of frame before transmission. The question is, now, what kind of frame? And how many frame levels are needed?

How Many Layers Between IP and the Fiber?

How many layers are needed between IP and the physical fiber network? It would appear that only three layers are needed: the IP packet layer itself, the frame that contains the packet, and, finally, the physical fiber link itself. However, in most implementations of networks, there is more than just one framing layer between IP packet and fiber.

The time has come to examine specifics. Initially, international SONET/SDH standards did not recognize IP as a legitimate, standard protocol at all. So the only way to carry IP packets inside a SONET/SDH frame was to carry IP packets inside either a VT (low speeds) or a stream of ATM cells (high speeds). Eventually, Packet over SONET (POS) allowed IP packets to use a PPP frame to perform the same task at high speeds without requiring ATM cells.

A lot of attention has focused recently on Gigabit Ethernet (GBE), which is basically a series of GBE switches (or *hubs* to some) connected by fiber links. The wavelength used by GBE is usually 850 nm, but 1300 nm can be used as well. DWDM transponders can easily shift these wavelengths onto DWDM systems with OADMs and optical cross-connects. No SONET/SDH required! Finally, there is the promise of direct-to-fiber IP interfaces (if standard "interframe fill" and transparency rules are established for IPv6).

These alternatives are shown in Figure 17.6.

The real point when discussing SONET and SDH is that only the first two protocol stacks even have room for SONET/SDH. Once it is realized that the important parts of the stack are the IP top and fiber bottom, the only important consideration is how best to get from IP to fiber.

The third method uses the GBE frame to carry IP packets directly onto the fiber. This is fine and works very well, but one potential drawback is that there can be only one GBE stream on the fiber. The trend today is toward DWDM and OADM optical networking. So the fourth protocol stack makes use of DWDM and OADM to allow many GBE streams on the same physical fiber.

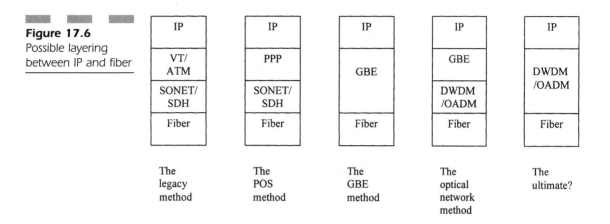

Figure 17.6
Possible layering between IP and fiber

IP	IP	IP	IP	IP
VT/ATM	PPP		GBE	DWDM/OADM
SONET/SDH	SONET/SDH	GBE	DWDM/OADM	
Fiber	Fiber	Fiber	Fiber	Fiber

The legacy method | The POS method | The GBE method | The optical network method | The ultimate?

Finally, there is nothing magical about the GBE frame. It's a package for IP packets, just like SONET/SDH. If an easy way can be found to place delimited IP packets directly onto an optical network, then the ultimate stack might be the last. However, for the time being, IP in and of itself cannot replace all frame functions. So GBE will be needed for some time to come.

In fact, there are many attractions of the GBE frame for IP packet transmission. Consider a normal 10-Mbps Ethernet frame the maximum 1,500 bytes in length. These 12,000 bits take 1/833rd of second to send, or 1.2 milliseconds to send. In 1.2 milliseconds, a GBE link can send 1,200,000 bits, or 150,000 bytes.

Now, an IP packet can be about 64,000 bytes long, but it made no sense to even try and generate an IP packet that large, since a packet so huge would have to span many 10-Mbps Ethernet frames. This packet fragmentation imposes a large performance penalty on the processor. Many IP protocol implementations tune the IP packet size to the 10-Mbps Ethernet frame size by default for this very reason. The largest IP packets routinely produced (by Sun's Network File System application) are about 10,000 bytes in length. Most IP stacks will fall on their knees if asked to produce or process an IP packet larger than 10,000 bytes or so. (Hackers routinely take advantage of this fact.)

But since many files and audio and video clips, are many megabytes in length, the larger the packet, the more efficient the transfer process in terms of overhead and effort. With a frame size in GBE on the order of 150,000 bytes, *multiple* very large IP packets can be sent in a single GBE frame over a fiber optic link, with no SONET or SDH required. Indeed, this practice is encouraged in GBE.

Once GBE is combined with DWDM and OADM, some amazing things can happen. A GBE switch (switching GBE frames) can send and receive multiple IP packet streams on separate wavelengths over the same fiber, up to 120 km in many cases. This is sometimes called *photonic networking* or *IP over* λ or the like, but the key is that this application of optical networking requires no SONET/SDH at all.

Can the Routers Keep Up?

It is all well and good to explore a world where service providers sell wavelengths and GBE frames carry larger and more IP packets than ever before. But all of these huge IP packets must come and go from an IP router. A GBE switch is not necessarily a router. How can IP routers possibly keep up with as many as 2,000 maximum-size IP packets per second arriving and departing on a *single* GBE DWDM *channel*? Even 10 channels used for this purpose would require the router to process and route 20,000 maximum-size IP packets per second per link!

Fortunately, there are terabit (1,000-Gbit) IP routers available today. These new routers are sometimes called *carrier-class* routers to distinguish between these fiber backbone nodes and the simple IP routers used as CPE for the enterprise. These routers are claimed to support link speeds as high as 20 to 60 Gbps, and extreme claims as high as 150 Gbps have been made. This is on a per-port basis, but all of these units support multiple links, of course. Most terabit router boxes are deployed in clusters and can be made to look like one unit operating at up to 19 terabits per second in aggregate. The distance between units in a cluster can vary from "right here in this room" to 15 km (about 10 miles), or even more, away, and so can be spaced much like ring nodes.

Ironically, most terabit routers still support SONET/SDH rates, from OC-3/STM-1 to OC-192/STM-16, and are not particularly aimed at GBE or DWDM. However, it is still impressive to see a router handle 15 or 16 OC-192/STM-16 ports running at 10 Gbps each. Table 17.1 shows a sampling of this new class of terabit routers.

Eventually, the terabit router, GBE switch, DWDM channels, and OADMs will come together to create a new form of optical networking. Whether this is called photonic networking or IP over λ or something else is not important. What is important is that there is little room for SONET/SDH in many of the new equations. SONET/SDH might remain useful as a network management channel for TMN-compliant OSSs, but the IP world uses other OAM&P techniques such as SNMP and Java.

TABLE 17.1 Terabit router sampler

Company	Product	Number in Cluster	Cluster Through-put	Max. OC-3 Ports	Max. OC-12 Ports	Max. OC-48 Ports	Max. OC-192 Ports	Cluster Distance (miles)
Avici	TRS	14	1.4 Tbps	640	160	40	NA	NA
Lucent	NX64000	16	2.5 Tbps	128	64	64	16	1.25
Charlotte's Web	Aranea	32	5.1 Tbps	NA	128	64	NA	~3
Pluris	20000	128	19.2 Tbps	NA	NA	60	15	(feet)
Cisco	12016	16	2.39 Tbps	240	60	60	15	<0.1
Nortel	Verslar	96	4.8 Tbps	96	24	24	NA	~1

Of course, none of this may happen if the cost of optical networking components such as GBE direct-to-fiber interfaces remains higher than paying for the equivalent SONET/SDH capacity. Oddly, this is not the case. Newer fiber line cards such as GBE cost significantly *less* than their SONET/SDH counterparts, and offer higher speeds. How can this be so?

The Cost Issue

Consider a customer or service provider given a choice between configuring a router/switch with an OC-12 port (622 Mbps) or a GBE port (1,000 Mbps). Both will work, since either one can be carried on a separate wavelength over a DWDM system. In other words, the unit purchased and offered for sale is the λ, not the speed. In that case, the choice will more than likely boil down to the cost per bit and the cost per port (usually several ports will be desired, of course).

An OC-12 interface for a SONET ADM can cost up to $12,000. A GBE interface can go for about $2,500. And once the overhead is stripped off the OC-12, the throughput available is about 600 Mbps, a little more than half available on GBE.

Why the difference? Because to comply with specifications, the OC-12 board must be able to handle all SONET OAM&P, timing distribution, VTs, and so on, just in case anyone would want to put channelized voice onto the SONET link. GBE suffers from no such requirement to support channelized voice. Want to do voice over GBE? Use voice over IP (VoIP) and forget about the need to cross-connect!

Strangely, this GBE-versus-SONET price differential closely mimics the early price differential between 10-Mbps Ethernet and 4-Mbps Token Ring LANs. Token Ring was and is arguably the better technology: there were no collisions, delays were predictable, and so on. But Ethernet won the war. Ethernet boards cost about half (or less) what Token Ring boards cost, and the speed difference more than made up for the limitations of Ethernet. Why did not Token Ring vendors just lower prices to be competitive? They really could not, because the extra circuitry needed to perform token distribution, beaconing, and so on, as required to be compliant with the Token Ring specification. Wherever chip prices fell for Token Ring, the same economies applied to the much simpler Ethernet chips, so the differential held until Ethernet dominated the LAN marketplace. This same phenomenon might repeat itself with GBE and SONET/SDH.

This might be a good place to examine the issue of the relationship between OSI-RM frames, such as GBE and PPP frames, and transmission frames, such as SONET/SDH and DS-1 frames. Essentially, SONET/SDH frames exist because the type of traffic that both SONET and SDH are optimized for—namely 64 kbps DS-0 voice—has no frame structure in and of itself. And the OSI-RM protocol concepts of the frame and packet apply to *data* networks, not voice networks. So in order to get all of the benefits of frames in the voice world, such as error control and the like, it was necessary to invent a voice frame structure.

This is hardly a SONET/SDH innovation. The T-1 and E-1 frames are just voice application frame structures to allow for voice access and cross-connections. Naturally, the issue is now whether such a frame structure for voice is really needed if VoIP becomes the rule, as appears to be the case.

The whole point is that since GBE has a perfectly adequate frame structure for VoIP and even video over IP and anything else over IP, there is no longer any need to provide SONET/SDH framing on a fiber link. The potential absence of SONET/SDH might require the use of OAM&P methods other than TMN for the network, of course, but TMN had been under attack for other reasons all along.

Sunset for SONET?

Will the rising sun of DWDM, optical networking, GBE, and the like, be mirrored by the sunset of SONET (and SDH)? All technologies give way to newer methods eventually. This section looks at the possible deployment of the missing piece of optical networking, the optical switch/IP router. But even in the photonic networking world, SONET/SDH might still have a place as a legacy, or even consumer, network method.

Optical cross-connects are sometimes called optical switches in the DWDM world, but that is somewhat misleading. The shuffling of wavelengths and fiber links that occurs within optical cross-connects is done by configuration, not packet by packet or cell by cell based on header content as true routing and switching does. But there are architectures that actually route or switch packets through the device without ever changing the input photons to electricity, at least the payloads. One such architecture for a terabit router is shown in Figure 17.7.

The trick of photonic switching and routing is to do it with mirrors—literally! Figure 17.7 shows how. Note that the routed traffic enters and emerges are a stream of light, and yet individual packets present in the stream can be routed to different output ports. This fulfills the definition of photonic routing, but that is not the end of the story. True photonic switching and routing can occur only when a true *optical computer* and optical circuitry control the entire process. This development is still a ways off. In the meantime, the best that can be done is to control the whole process with electrical circuits and table lookups.

Here is how the optical router works:

1. Light of a given wavelength enters an input port. The input fiber could have many wavelengths. If so, each is first isolated for routing purposes. When a packet header is detected in the input stream of photons, the optical signal is split through a simple process so there

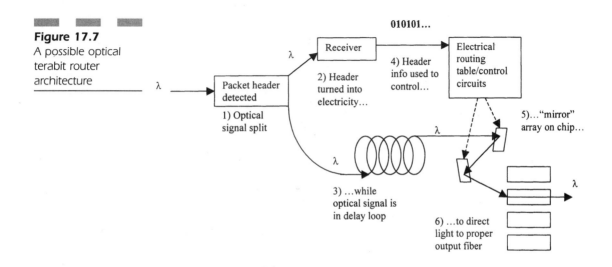

Figure 17.7
A possible optical terabit router architecture

are two copies of the header portion of the optical packet. The header can be detected by several means, including fixed-length data units (e.g., cells) or a special optical pattern used as a delimiter.

2. The light representing the packet header is converted to electricity by using a standard optical receiver. Only the header need be converted in this fashion. The process is quite efficient since the header is presumably much smaller than the packet payload. This is sometimes called the electrical wrapper for the optical packet.

3. Meanwhile, the optical signal representing the packet is fed into a simple delay loop. Routing takes time, and this time is gained by delaying the optical packet until the path through the device is set up.

4. The electrical header information, now a string of 0s and 1s, is used to perform a normal routing table lookup. Routing table maintenance can be performed by existing routing protocols, either electrically or optically, since most routing information flows between adjacent systems and is not routed itself.

5. Once the proper output port has been determined for the optical packet, an array of mirrors etched onto a silicon substrate is adjusted—electrically—so that the optical packet emerging from the delay loop is "aimed" at the proper output fiber.

6. Finally, the optical packet is directed to the proper output. Naturally, a stream of packets is conducted through the optical router very quickly, and capacities are in the terabit range.

Once the optical terabit router is added to the optical networking mix, many things change. Transport networks now carry IP by wavelength and legacy traffic such as SONET/SDH by normal time division multiplexing (TDM). The basic access node is the DWDM/OADM device. The services provisioned and sold are either IP over λ, good for almost everything from VoIP to video to data, or legacy voice, frame relay, and ATM services, good for backward-compatibility. GBE is easily added to the IP side of the ledger as a framed transport.

Once placed onto the DWDM fiber(s), the traffic is passed through an optical cross-connect node and onto a fiber ring (or mesh). This photonic core ring differs from SONET/SDH rings in the sense that it is the optical cross-connects that are the ring nodes, not the DWDM/OADMs themselves. This makes sense because the basic transport unit of the photonic core is the wavelength, not the frame that rides the wavelength. This type of fiber network of the future is shown in Figure 17.8.

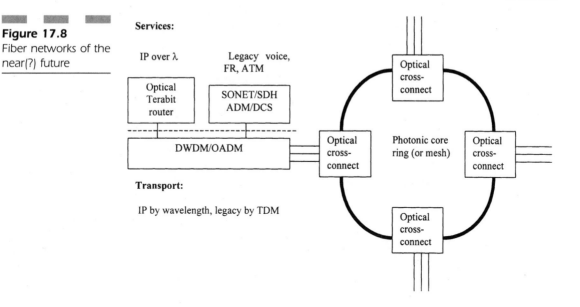

Figure 17.8
Fiber networks of the near(?) future

It is highly debatable how near or how far in the future this architecture might be. Some see it around every corner, some see such a bandwidth-rich infrastructure as total overkill except in a handful of countries around the world (the United States is one of the handful). One skeptic noted that plans to place a gigabit fiber link to the Fiji Islands was like running a gigibit link into a 30,000-person community in the United States—the economies are about the same size. Even with Internet demand what it is, it would be hard to cost-justify the link in the United States, let alone to the Fiji Islands.

So SONET/SDH will provide enough bandwidth for present needs for some time to come. As for the future, many technologies make their way from service provider backbone to access network to consumer product. Modems have certainly made the trip, and DS-1 might follow (most plans for new residential networks feature speeds of 1.5 Mbps or better, at least downstream). Why not SONET/SDH?

Stripping off the VT and timing support in SONET/SDH could reduce the price of an OC-12/STM-4 interface to about $3,300 for 622 Mbps. Perhaps the backbone SONET/SDH ADM once used as the ring node, and now replaced by the OADM, could make its way to the end of the access fiber link serving someone's house. This might be the best long-term hope for SONET/SDH, but there will be competition at every step along the way.

Miscellaneous SONET Issues

The positioning of SONET as an advanced private-line solution has posed special challenges to the organizations that employ SONET in this fashion. SONET is optical, but most test equipment is electrical in nature. A customer may still buy a DS-3, but it is delivered over SONET fiber. How can this be tested without enormously expensive fiber optic test equipment all over the place? And how are the optical test results to be squared with the end-to-end electrical observations?

When it comes to bit errors, SONET fiber all but eliminates them. How can technicians measure the bit error rates on links that have so few? The differences between SONET operating at a BER of 10^{-12} and 10^{-13} may seem trivial, but it may still be important for billing and service level purposes. It is impractical to just wait for these rare bit errors to occur. They may be literally weeks apart. Yet some method of measurement must be developed.

What about jitter and wander? SONET is supposed to be synchronous, true, but these effects still exist. How should timing defects in a system designed to eliminate them be detected? This final section briefly explores some of these areas.

SONET Electrical Testing

The simple hierarchical structure of SONET signal levels has been a real benefit for maintenance and installation personnel: A technician simply needs to access the information embedded in the SONET overhead to find all the data needed to identify things like payload types and to accurately detect error conditions. Access to SONET overhead has been one of the main challenges.

Many organizations with SONET links cannot afford to invest in equipment designed to access optical signals, particularly in the early stages of SONET use and deployment. Ironically, even though SONET rings typically are used with a traditional point-to-point structure, especially to the customer's premises, using a lot of existing test equipment would mean the loss of many of SONET's real advantages, such as the capability to cross-connect an embedded lower speed signal without demultiplexing the entire signal.

Fortunately for SONET users, almost all of the most critical overhead information needed to effectively troubleshoot and isolate problems on a SONET link circuit can be found and readily accessed at an electrical rather than optical level. This allows approaches to SONET testing to be both economical as well as practical.

In SONET, any STS-1 signal can be extracted directly from any other level of the hierarchy. The payload of each STS-1 signal may be composed of multiplexed virtual tributary (VT) signals containing lower-speed asynchronous signals, such as from T-carrier. Different VT sizes are defined to accommodate different types of signals within the STS-1 format. This is what makes SONET so much more attractive than earlier asynchronous T-carrier (DS-1/DS-3) networks, where all "tributary channels" (usually DS-0s) had to be demultiplexed before any cross-connect or switching operation. With SONET, only the channels being acted upon need to be extracted. Other channels pass through the SONET network element undisturbed.

The most important part of the SONET STS-1 frame for testing purposes is the transport overhead bytes. These contain the most critical information regarding the performance of the SONET link on the section and line levels. The section overhead provides the diagnostic information needed to support the transportation of the user information in the synchronous payload envelope between repeaters on the link. The section overhead contains information about framing, error monitoring, data communications, and a local orderwire for point-to-point voice communication within the section.

The line overhead is also part of the transport overhead. These bytes are concerned with the synchronous payload envelope's (SPE) movement with the STS frame as it moves between adjacent network elements and nodes. It also includes information for monitoring individual STS-1s in a higher level STS-n, line alarm indications, and far-end receive failure (FERF) indicators.

The SPE itself contains the path overhead. Path overhead is concerned with the end-to-end SONET circuit between termination points, such as add/drop multiplexers (ADMs) and channel service units. It includes its own remote alarm indicators as well as information on VT framing. All of these bytes allow for effective in-service testing of a SONET system because the equipment passes failure indicators between these sections, lines, and paths.

These OAM&P functions have all been dealt with before and need not be repeated here in detail. The point is that the SONET overhead bytes provide a lot of information that allows a technician to reconstruct the sequence of alarms initiated by the failure and to trace the trouble back to the source. This information is not only useful for the maintenance technician troubleshooting the SONET link, but also for the installation technician performing routine adds, drops, and changes to the network.

The key to this whole process is to allow these technician easy access to the SONET overhead bytes. This can be very difficult to do at the optical level. Here, individual elements and pieces of the signal cannot be easily extracted without disturbing the rest of the signal on the link. Special signal "splitters" can be used, but this involves expensive components and also requires great care in installation to avoid disrupting the network. It also introduces the potential for annoying SONET signal degradation.

However, when the SONET optical signal is converted to an electrical environment, testing becomes much simpler. None of the original information is lost because there has been no intermediate signal processing. Testing at the SONET electrical level also can be done with familiar methods from the T-carrier world. This can occur at commonly deployed service monitoring points. Because it does not need to be designed to function with often quite delicate fiber optics, this test equipment can be much less costly and more hardened for frequent use.

A natural point for electrical testing on SONET is through the DCS and/or ADM at central office locations. Most of these DCSs will include access to the STS-1 signal level through an STX-1 cross-connect point. This can be inexpensively added to existing DS-3 bays just using cabling, connectors, and hardware. The use of the STX-1 point allows very simple access to SONET overhead byte information for fault location. All of the technicians can get as much information in the overhead bytes as they need for the problem.

Many SONET overhead byte functions also translate directly into messages or LED indicators that are both common and familiar from the T-carrier world. Things like loss of signal or loss of pointer conditions are easily indicated with LEDs, and so are available on most transmission test sets. Where more detail indications are needed, test sets with full displays of overhead bytes at all levels can be used.

Many traditional tests, such as loopbacks, can be conducted from the STX-1 access port to verify simple circuit integrity. This can be done at the DS-1 or DS-3 interface unit to verify performance not just of the SONET link, but of the demultiplexed VTs that carry the DS-1 or DS-3 service. This is very useful in situations where an STS-1 frame is not being directly delivered to the customer, but is broken down into multiple DS-1s or a DS-3 before it is sent out to the customer site.

Electrical SONET testing puts the testing power where it is needed the most. Most SONET devices include built-in optical diagnostics that are always monitoring the status of the optical link. However, connection to the customer premises almost universally occurs at the electrical level. A technician equipped with both STS-1 and DS-1/DS-3 capabilities is better prepared to find and isolate a fault within any of the common digital speeds on the circuits actually delivered to the customer.

Accessing optical circuits can be very complex; thus, such testing is not a day-to-day activity. Everyday and routine add/drop activity does not need a testing program built around optical access. However, electrical testing is perfect for routine operations. It is similar to traditional T-carrier transmission testing, so the learning curve is shorter and technicians can be more effective more quickly.

Electrical test sets for SONET, like many others, often are designed as hand-held packages with battery operation to simplify usage for field technicians. By making electrical SONET testers available to every technician, and by providing a shared optical tester for installation purposes, a service provider or even end user can put together a sophisticated testing package. This also provides a significant cost savings compared with investing in strictly optical test equipment.

Bit Error Rates and SONET

The previous section should have convinced everyone that traditional electrical testing still has a role in SONET. This section explores this issue with respect to one important aspect of testing: bit error rates (BER) and BER testing with BER testers (BERTs).

BERTs are a basic tool for testing not only SONET systems, but any transmission system, for signal quality in terms of bit errors. The BERT is the first tool used by a designer, and the last tool used by an installer to establish that the basic, physical transmission path is acceptable. A BERT consists of two parts, the transmitter and the receiver. Sometimes these are housed in separate devices for ease of use. The task of the BERT device is very simple: It feeds bit-by-bit test patterns across the transmission system and confirms that the transfer was error free or, if not, what the BER was for that particular test.

At heart, a BERT is 0s and 1s. Generally, BERT operates entirely on this level. This type of BERT is used to develop and test a new and as yet unspecified protocol. The tester can control the frequency and output ranges, and can adjust these values in order to *stress* the system. The process of stressing a system helps to determine the marginal conditions and boundaries of acceptable operation. With this type of test, there is no structure to the data transmitted.

However, many BERTs are also designed to test link quality at the higher layers of a particular interface or electrical specification, such as T-1 or E-1. They can also use a higher level protocol, such as ATM or ISDN. These testers will operate at the correct operating frequencies and tolerances for the standard chosen. Such BERTs are able to generate and recognize frames and/or packets for that protocol and generate them. Typically, once a protocol standard is firmly established and common, the BERT device is replaced by a full protocol analyzer.

What has all this to do with SONET? Well, a BERT's most basic measurement is the BER. This is just defined as the number of bit errors divided by the total number of bits sent during the test. This is a very small, positive number, between 0 and 1. Usually, the BER is in the range of 10^{-6} to 10^{-12}. The first number is typical of copper networks, and the second is more typical of SONET networks.

BERT tests are very simple to perform. The technician just connects the BERT to both ends of the transmission system or subsystem to be tested, and then the transmitter part of the BERT sends test signals across the system. At the other end, the receiver part listens to the incoming signal and determines the correct data level, voltage level, and timing used by the transmitter. The receiver can also compensate for things like for phase delay. The testing works by having both ends generate the test pattern locally using the same algorithm. Then the receiver synchronizes the transmitted pattern with the correct receive pattern.

Other measurements that BERTs can perform include a time interval within the test. SONET standards define things like Errored Seconds (ES). The ES is the number of seconds during which there was at least one bit in error. Severely errored seconds (SES) have a BER of 10^{-3} or greater. A measurement called Consecutive SES is the measurement of SESs for which the previous two seconds were also SESs. Some BERTs will graph these for ease of interpretation.

Here is where the SONET comes in. Every time the BERT detects an error, the device logs it. However, how long should a BERT device run before a high level of confidence is achieved about the BER? When no errors are found, this does not mean that the link is perfect. There is no such thing. Consider that with a SONET link engineered for a BER of 10^{-15} (not impossible today), an OC-24 system operating at about 1 Gbps will only encounter an error every 11.5 days, on average.

A valid test certainly requires more than a single error to offer statistically relevant results. The key is that most BERT tests involve random sampling and is, by definition, random. A SONET span may encounter a burst of errors that will not occur again for another three months, but when the sample period was not large enough, the test may indicate that the link is unacceptable. Conversely, a very bad link may appear to be good.

Statistically, testing until ten errors have accumulated offers only a 68% confidence in the BER, and even 100 errors offers only 90% confidence. On very low BER SONET links, imagine how long it would take to accumulate 100 bit errors, even at 1 Gbps. Well, at 1 Gbps with a BER of 10^{-15}, 100 errors will be logged in about 3.17 years. Therefore, how can SONET links get accurate BER information?

SONET BIP checks can be used to determine BERs internally, but this is not the answer. Suppose the whole point is to verify that the BIP checks are working correctly? How could they be reliable? A better answer is to stress the system and increase the error rate. Error rates measured while a system is under stress can be extrapolated to lower error rates for the system when it is under no stress. These can be collected for the system under different levels of stress, compared to normal testing results, and used to produce a performance curve.

For example, under specific conditions of stress, a SONET link or ring may produce a BER of 10^{-5}, which can then be extrapolated to a BER of 10^{-10} for the link or ring without any stress. The goal of this stressing of the link is not to determine the absolute error rate, but rather the upper bound on the error rate. Stressing the system to increase the error rate can take significantly less time, and still offer an accurate picture of the link's quality.

There are three sources of error in any digital system: noise, jitter, and intersymbol interference. Noise is not a good option for stressing optical systems, however, because they are fairly impervious to noise is the first place. Jitter errors are caused by the active components within the system, such as the timing electronics and their associated errors. SONET is also designed to minimize these effects as well. Intersymbol interference errors are caused when particular patterns (i.e., symbols) of bits interfere with one another, leading to "smearing" of signals. Fortunately, fiber systems are not invulnerable to this type of stressing. Intersymbol interference in SONET can be generated by cycling specific patterns through the design.

One major difficulty with using BERTs in SONET is getting the receiver side of the BERT hooked up to the other side of the link. Usually a person is needed at both ends to set up a test. One possible solution to the distance problem is to loop the signal back. The disadvantage here is that if an error were to occur, it would not be obvious whether the error occurred outbound or inbound. However, by looping the signal around, it is always possible to determine whether it is necessary to put another person at the far end in the first place.

Another possible solution is to permanently integrate the BERT into the central office of the system and remotely test lines. Most BERTs can be controlled by way of a modem from a PC. By using a digital access cross-connect, it is possible to drop any embedded SONET circuit into a test port. This method avoids requiring to have a person at each end of the line. Some BERTs can even be programmed to automatically test links one by one.

SONET Timing Errors

The path from the primary reference source (PRS) clocks to the network elements in a SONET network may be many. There are the PRS clocks themselves, Building Integrated Timing Supply (BITS) clocks, and derived DS-1 clocks. Both the American National Standards Institute (ANSI) and Bellcore specify limits on the wander (the official term for what most people call "jitter") of each. The standards specify things like time deviation (TDEV) and maximum time interval error (TIE) masks, and frequency offset and drift rate limits.

This section briefly describes typical synchronization distribution systems, explains various wander measurement techniques, and discusses a little about how the specified limits apply to a SONET system. Synchronization needs proper maintenance, like any other network component. SONET synchronization makes it possible to multiplex from an STS-1 at 51.84 Mbps up to an STS-48 operating at 2488.32 Mbps and back in one device. However, the synchronization system itself can cause problems when it is not properly monitored and maintained.

Probably the most common problem with SONET synchronization is faulty provisioning of the synchronization configuration itself. Every once in a while, clocks will fail and may go unnoticed. As a SONET network of many links and rings grows, wander will accumulate. Sometimes, wander may grow until it affects the payloads themselves. Thus, several points in a synchronization system must be monitored for preventive maintenance and possible trouble.

When a SONET network element (NE) receives bits and frames that have excessive wander relative to the synchronization clock provided to that network element, the SONET pointer adjustments are intended to be used to make up for any phase displacement. The greater the wander in the timing, the more frequent the pointer adjustments. When the times comes to extract the payload from the frame, these repeated pointer adjustments cause jitter in the payload. Excessive pointer adjustments cause excessive jitter and produce SES.

Not surprisingly, applications most sensitive to delay variations, such as voice and video are the most sensitive payload traffic. If a corporate PBX were to use a DS-1 payload as a timing reference, pointer adjustments could cause the local clock to lose lock on the payload bits. Therefore, service providers routinely warn customers not to use a DS-1 carried over SONET for timing purposes. Sometimes, however, the customer has no viable alternative, and the problems persist. In the case of video, pointer-induced jitter on the DS-3 payload used to carry the video can cause the colors to shift.

SONET rings can be a concern. Usually, a SONET ring will have two PRSs. Normally, the timing information from one PRS is carried clockwise on the ring to the half of the other SONET NEs. The timing information from the other PRS is carried clockwise to the other half of the devices. This use of two PRSs allows the timing of all the SONET NE devices to be traceable to a PRS when there is one "cut" in the ring. Even when there are no cuts, one PRS can act as a backup in case the other PRS fails.

The SONET network elements on the ring are usually ADMs. In each ADM, a SONET minimum clock (SMC) can be provisioned to receive timing information from either the line signal on the ring or from an "external" DS-1 reference. In some nodes on the rings, a Building Integrated Timing Supply (BITS) clock takes care of the timing from the incoming DS-1, filtering the phase and providing some *holdover* capability. This means that the BITS clock "remembers" the timing frequency should the derived DS-1 clock fail. Because the SMCs provide some holdover capability themselves, it is not always necessary to have a BITS clock at every node. This is what SONET calls the "line timing" mode. A BITS clock also could be used to receive timing from a DS-1 signal not associated with the SONET ring.

Suppose a SONET rings passes traffic to another ring. Each has the same timing arrangement just described; thus, each is an "island" of timing. Whenever such "islands" of SONET pass payloads from one to another, payload jitter, due to pointer adjustments, is bound to accumulate. As the number of interconnected SONET "islands" grows, jitter could become large and ultimately excessive if wander and the associated pointer adjustments were not limited in some way. Bellcore SONET documents specify how SONET equipment handles wander, and ANSI standards have limits for wander that is found at interface points.

Naturally, standards are in place to ensure the integrity of SONET synchronization. Test equipment also is available to test conformance to these standards. Excessive pointer activity is an indication of excessive wander. It is very useful to acquire data about the synchronization of a SONET link or ring system at installation. Then, if timing problems or concerns were to arise later, a known baseline of information would be available. Measurement and consideration of jitter and wander will become increasingly important as links are made between SONET "islands" (especially rings) and the complexity of the SONET network grows.

The Unlimited Bandwidth Network (UBN)

As this book draws to a close, this is a good place to speculate on the future of networking in a world where bandwidth might well be considered to be unlimited. Not "infinite," as some proponents of optical technologies like DWDM, but enough so that bandwidth concerns need never be an issue again. This section will suggest a working definition of an *unlimited bandwidth network* (UBN). A UBN does not have infinite bandwidth, but enough bandwidth so that no one need ever worry about running out or limiting application capabilities in any way. Along the way to describing the UBN, several related concepts will be introduced, including *flying storage* and *pancake packets*. What follows might not be accurate, but perhaps it might be entertaining.

Bandwidth had always been the limiting characteristic in any network, voice or data, analog or digital. But advances in the areas of fiber optic communications systems such as SONET/SDH, self-healing rings, direct packet mappings, and dense wavelength division multiplexing (DWDM) have made it possible to contemplate a not-so-distant future where networking with practically unlimited bandwidth might not just be a dream, but a reality. This section will explore some of the implications of networking with unlimited bandwidth, particularly the impact on the current debate about the ability of networks to deliver adequate quality of service (QOS) for all applications from voice to video to data. The conclusion is that a type of flying storage could replace all other forms of storage and that unlimited bandwidth could make special network mechanisms to deliver bounded and stable delays, the required error performance, and even reliability and security, totally unnecessary.

Unlimited Bandwidth Defined

The emphasis on bandwidth as the premier network resource is understandable. Digital networking has become so common, most forget how new all of this really is. Bandwidth was a scarce and precious commodity until very recently. There was plenty of bandwidth on the public voice network, of course, but only available to end users in the form of 56/64 kbps channels. Multiple channels could be used, and full 1.5 Mbps T1s existed, but were cost effective only in some circumstances and only with considerable effort on the part of user and carrier alike. Digital data networking, as opposed to analog modem use, was new and uncharted territory less than 20 years ago. A pamphlet issued internally by AT&T in October 1980 describing AT&T's "computer and telecommunications" services dedicates only 2 out of 16 pages to digital services. "Dataports" for hooking up digital devices directly to AT&T channel banks were only two years old. And the pamphlet looked forward to the day when services with "speeds as high as 1.5 megabits per second... may be expanded to 95 major population centers." For the vast majority of customers in 1980, such bandwidth speeds were just a dream. Today, we do not need 1.5 Mbps in 95 cities, we need 1.5 Mbps in all 95 homes in my own housing development! The common use of the term *broadband* for such high bandwidth networks just reflects the intense preoccupation with the amount of bandwidth available to users on a network today.

Most university and professional network design courses still revolve around bandwidth determination techniques, either as a goal or as a starting point. Many techniques address the related concept of network *throughput* rather than merely bandwidth, but with the types of network nodes and switches available today, most network throughput restrictions are due to bandwidth restrictions. This point will become clearer later on.

When more bits arrive on input ports than can be disposed of through output ports at a network node, queuing theory was employed to determine to amount of buffers needed to keep all the bits without throwing any away (loss). Yet they had to make sure end-to-end delays did not soar through the roof. In the unlimited bandwidth future, such exercises and traffic management techniques could be of historical interest only. During periods of gasoline shortages, elaborate queuing rules evolved to keep the lines of cars at gas stations to a minimum. Today, if there is a line at the gas pump, just go to another station. Bandwidth calculations may soon go the way of these old "odd license plate, odd day" rules. The new networking paradigm might not be "determine the bandwidth first" but rather "assume all the bandwidth you could ever consume."

This section will use the following working definition of unlimited bandwidth. Unlimited bandwidth is public network bandwidth on all links that greatly exceeds (by perhaps a factor of 10 or more) the bandwidth needs of not only any single application, but all applications on the network and is generally available at no incremental cost beyond the small, fixed basic fee. So if 100 servers access a network at 1.5 Mbps, if the backbone has a lot more than 100×1.5 Mbps = 150 Mbps of bandwidth available at all times, this is essentially a UBN. The characteristics of networking with unlimited bandwidth are listed in Figure 17.9.

Note the emphasis in the definition on *practically* unlimited bandwidth rather than *theoretically* unlimited bandwidth, which is another matter altogether. The definition, while not the most elegant, is at least descriptive of the future world of DWDM networking.

Unlimited Bandwidth and Solitons

Many of the concepts presented in this section are speculative, some highly so. But one of the ideas that is absolutely real is terabit fiber optic transmission links. While these links have been mostly of the demonstration variety, the short time to market for previous fiber-based transmission equipment means that it will not be long before fiber systems operating at 1 Tbps or above over 100 km or more are not curiosities. These terabit links will form the backbone of every major carrier's core network all over the world.

Figure 17.9
Characteristics of
networking with
unlimited bandwidth

Unlimited Bandwidth means:
Bandwidth exists on the public network links (not private networks).
Bandwidth greatly exceeds (by factor of 10?) applications total bandwidth needs.
Exceeds not only any single application, but all network applications.
No bandwidth limit applies to any user.
No additional cost for more bandwidth: just small, fixed basic fee.

The next wave of improvements is just dawning. Soliton DWDM technology promises to extend the reach of these new systems without requiring the reshaping of the optical signal due to the effects of dispersion. Solitons are a form of standing wave and when used in optical networks are essentially laser pulses inserted onto a fiber with intentional nonlinearities. Usually such nonlinearities are to be avoided, but solitons are much like sonic booms in the fiber core that keep their pulse shape over long distances. Solitons do not disperse, but they do attenuate, a fact not usually played up.

But these terabit figures are just numbers. Is this really unlimited bandwidth as defined here? Not on a single terabit link all by itself. But as part of a national or global network, there could probably be a UBN soon. Here is why.

Each step along the way from kilo- (10^3) to mega- (10^6) to giga- (10^9) to tera- (10^{12}) to peta- (10^{15}) is just an increase of three powers of 10. It is easy to forget that each step represents a 1,000-fold increase over the previous step. It is even easier to forget that the relationship between 1 and 1,000 is about the same as the relationship between one day and three years. One of the most attractive things about new fiber optic systems is that they now can operate at bit error rates (BERs) of 10^{-12} (1 in a trillion) when less than 10 years ago the accepted BER was 10^{-9} (1 in a billion). This means that when a link is upgraded from a BER of 10^{-9} to one with a BER of 10^{-12}, all of the errors that were seen on the link yesterday will not be seen for the next three years. And when an old copper link with a BER of 10^{-6} is upgraded, yesterday's errors take 3,000 years to occur.

As with errors, so with bandwidth. A copper link operating in the 10^6 "mega" bits per second range upgraded to a fiber link operating in the 10^{12} "tera" bits per second range carries as many bits on its first day of operation as it could have carried in the past 3,000 years. Of course, this stretches back a good fraction of the history of civilization on this planet—about half—which is the whole point. In the whole of history, IBM has said that less than a petabit (10^{15}) of bits has ever been sent on all networks combined, and this included older "digital" systems such as optical telegraphs, Morse code, smoke signals, and the like. Given the fact that megabit systems like T-carrier and E-carrier have been around for less than 30 years, this petabit figure is probably high.

Now suppose a network of 1,000 terabit links was built and put into operation. This hardly seems a stretch given that any trade show worth its salt probably sports a couple of terabit links already. The aggregate bit rate would be 1 Pbps. Which means that *every second* the network would carry more bits than have ever been sent in recorded history. This is unlimited bandwidth in action. With such a network the question will no longer be, Where do we get bandwidth for our bits?, but, Where will we get bits for our bandwidth? With unlimited bandwidth, the need for a buffer or queue might be done away with altogether. (Buffering and queuing serve more purposes in a network than just storing excess bits, such as traffic shaping and the like. So these practices might still be retained for these other purposes.)

Bandwidth mongers are fond of measuring network capacity in terms of the Library of Congress. The reason for this particular choice of metric is unclear, save that most people have an image of the Library of Congress as a place with lots of books, and books have lots of information which could be a source of bits on a network. The actual size of the Library of Congress in bits is subject to debate, but based on some simple assumptions about the size of an average book and the 17 million books held, a figure of around 100 terabits (or 100,000 gigabits) should not be too far off. This excludes maps, manuscripts, and other holdings, but books are what people think about when a library comes up. So a 10-Gbps OC-192/STM-16 serial fiber link could conceivably transmit all the books in the entire Library of Congress in about 10,000 seconds, or about 2 hours and 45 minutes. A single terabit link carries the Library of Congress across it about every minute and a half. The 1,000-terabit links do the job in 0.1 seconds. And so forth.

Now the idea of unlimited bandwidth seems less outlandish. Suppose a petabit system of 1,000 terabit links was used to support a book reading application to send every person in the United States a new book to read each and every day. This is in the near future, of course, so the population of the United States might be about 300 million people by then. A book of 300 pages has about 100,000 words, and each word has about 5 letters, plus a trailing space. If each word is represented by 8 bits, and each word has 8 overhead bits for formatting, fonts, and so on, then the whole book can be represented by 5,600,000 bits, or 5.6 megabits (this figure is actually quite accurate). This book distribution application could not consume more than 1,680 petabits of capacity going to 300 million avid readers. A petabit network could distribute all of the books in 1,680 seconds, or about half an hour. Presumably, readers would be patient enough to tolerate the delay between first reader and last.

Traditional data applications would not begin to tax the resources of a national terabit/petabit backbone. The upper speed available for individual networked devices is likely to remain the 1-Gbps speed of Gigabit Ethernet for some time to come. This is mainly because of two things. First, any further increase in speed could not be handled on unshielded twisted pair (UTP) copper wire over any reasonable length. But most people would be reluctant to completely abandon UTP in favor of fiber everywhere just yet. Second, it will be some time before any individual device will be able to generate or consume more than 1 Gbps worth of information, and many would argue that this is pretty much impossible even today. In any case, it seems likely that if a single terabit link can handle 1,000 Gigabit Ethernet sources at the same time, this should be enough to be considered unlimited bandwidth.

Even video could not consume all of the bandwidth on a large enough terabit/petabit system. By most accounts, high-definition digital TV (HDTV) should not require more than 20 Mbps of bandwidth. Suppose every person—not just every home—in the United States were to have access to all HDTV stations, but only one at a time (more demand is possible, but should not necessarily be encouraged). The total bandwidth demand would be about four times more than with the reading application used above. All this means is that there should be more than four TV systems with terabit/petabit capabilities to deliver these bits. Since there are about 6,000 video headends in the United States today, this hardly seems a stretch.

More examples could be given, all extended to an extreme. But in no case would the total network capacity needed exceed the amount anticipated to be available once the new fiber optic systems become common. And the principle of "no incremental cost beyond the small, fixed basic fee" means that the cost of unlimited bandwidth would be along the toilet tissue model. No one charges more for toilet paper depending on how much is used by any individual. The price per roll is the sole cost involved, and is very low at that. Most important, any effort to conserve toilet tissue use would cost more than the savings derived. So billing for bandwidth consumption beyond a flat minimum charge on an unlimited bandwidth network would cost more than any incremental revenue derived from the process.

Implications of the Unlimited Bandwidth Network

As fascinating as surveys of fiber optic transmission capabilities and capacities may be, that is not the main purpose of this section. The intent here is to explore the implications of this unprecedented bandwidth on networking in general. What are the implications of having unlimited bandwidth in the first place? There are two main areas the rest of this section will explore: storage of information and delivery of network QOS for any application.

The first area, storage of bits, might seem odd. What does a network, which exists for the transport of bits from place to place, have to do with storage, which by definition is the accumulation of bits in one place? In fact the use of a network or "line" to store bits goes back to the early days of computing. Alan Turing, for example, used the concept of a delay line as storage in the first large-scale British computer in the early 1950s. The idea was that bit patterns could be repeatedly generated onto a line, given an engineered delay. The delay determined both the size of the storage and the interval that the bits could either be read off of the line (and replaced with new information) or recycled once again.

Interest in this type of storage waned with the introduction of magnetic storage for computers, which was the predecessor of the integrated circuits memories and electro-optical disk drives of today. The result of basing storage on semiconductors is that the price per bit of both random access memory (RAM) and secondary storage (disk, CD-R, and CD-RW) has fallen dramatically in the past few years. As recently as 1979, a Texas Instruments paper put the price of storing a *bit* at $0.001 for "static RAM," and older (but more reliable) core memory at about $0.003 per bit. This translates to $8,000 and $24,000 per *megabyte*, respectively. No wonder the first PCs in 1981 had 64 kilobytes of RAM!

In the past, an acceptable guideline in network design was that it cost at least twice as much to send a bit as it did to store a bit in a buffer or on disk. In fact, this observation was at the heart of the whole delay versus bandwidth/trunking network design issue. A network designer could always save money on a network by buffering and queuing whenever possible and applying an absolute minimum of expensive bandwidth. Only when users constantly screamed about delays was it essential to increase network bandwidth. If the designer could get away with it, they just added buffers to handle traffic increases.

But as the capacity of broadband networks increases into the terabit and petabit ranges, the cost of sending a bit compared to storing a bit might change in a fundamental way. In the unlimited bandwidth future, *adding bandwidth will become cheaper than adding storage!* The issue appears not to be "may be cheaper" but simply "when will it be cheaper." One text on data network design, written by Darren Spohn, states: "Why store the information on a local hard drive or CD ROM, when the network will store it and send it to you on demand for a fraction of the cost?" It was just such thinking that led IBM and Sun to embrace the idea of a *network computer* which had no local storage at all.

But we can go one step better. Consider networking in the world of unlimited bandwidth where it will be less expensive to send a bit than to store a bit. It is no longer a question of storing a bit locally or storing it remotely on the network. There is no longer any storage at all! Why bother with expensive hard drives and CDs? The network has enough capacity to have several Libraries of Congress flying by every second or so. If users can be patient for a few minutes (not a given), all of the bits in recorded history will be along soon. Just grab what you need.

What about original bits not yet present on the network? No problem. The concept is the same as a filing system without file cabinets. Just put the folders in an envelope and mail them to yourself. As long as the mail—all of it—comes every day, you can find what you need and just pass the rest along until tomorrow. I call this use of an unlimited band-width network to store bits *flying storage.*

There could even be "short-term networks" and "long-term networks" for things needed more frequently than others and materials needing archiving. In fact, the current trend toward fiber optic rings makes the idea of flying storage even more attractive. The rings always bring the bits back around. The concept of using an unlimited bandwidth net-work as flying storage is shown in Figure 17.10.

A further implication of unlimited bandwidth involves network QOS. Many network technologies, from IP to ATM, are currently addressing the need for a network to deliver a set of basic service characteristics to user applications. Some networking techniques, such as IP, have tradi-tionally left the QOS issue up to the end user applications themselves in favor of network architecture speed and simplicity. Such networks are considered to be unreliable by proponents of network architectures that prominently feature QOS as the centerpiece of their own architecture, such as ATM. The characterization of IP as unreliable can be confusing and has nothing at all to do with network availability. With regard to IP

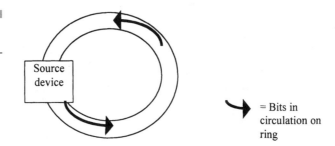

Figure 17.10
Flying storage

it means that even though IP can dynamically reroute around failures, an IP router network cannot deliver *reliable* QOS performance such as stable and minimal delays.

Even the designers of such "unreliable" architectures routinely refer to their creations as "best effort" networks. It would be better to call networks using protocols such as IP *relative QOS* networks. There are fields in the IP packet header that would allow a router to favor one stream of packets over another in terms of queuing delay and so on. But this would not guarantee *absolute QOS* in terms of bounded delay through the network. Other network architectures such as ATM feature a set of mechanisms designed to deliver a range of QOS parameters directly from network to application. QOS parameters such as absolute bounded delay and limits on information loss (errors) are included in ATM.

So there are really two philosophies regarding QOS and networks. The IP approach is to leave this up to the end applications. How could the network presume to know precisely what delay is adequate for a given application anyway? The ATM approach is to release the end applications for having to deliver their own QOS as well as run themselves. All applications fall into only a handful of categories anyway. Neither argument, the IP's smart-application-simple-network or ATM's simple-application-smart-network, is right or wrong, just different.

However, the world of unlimited bandwidth will lead to a philosophy of simple-application-simple-network. In this regard, the simple-network IP approach might make the most sense.

The Six QOS Parameters

One of the biggest problems with QOS discussions is that there is no universal agreement of just what network quality of service means. Some define QOS quite narrowly, while others define it more broadly. For the purposes of this discussion, which will seek to show that, given unlimited bandwidth, network QOS is trivial to deliver no matter what the issue, a broader definition is more desirable.

One major network consulting firm, run by Daniel Briere and Christine Heckart, has defined QOS as "the ability of a service to define delay, jitter, cell/packet loss ceilings, security and bandwidth on an application-by-application basis." While oddly placing bandwidth last on the list, this definition is very good and comprehensive. The list that will be used here will put bandwidth first, for reasons that will become obvious. Jitter is just delay variation, or how much the end-to-end network latency varies from time to time due to effects such as network queuing and link failures, which cause alternate routes to be used. The cell/packet loss ceiling is just another spin on the effect of network errors. Some applications can recover from network errors by retransmission and related strategies. Other applications, most notably voice and video, cannot realistically resend information and must deal with errors in other ways, such as the use of forward error correction codes. Either way, the application must be able to rely on the network to lose only a limited amount of information, either to minimize resends (data) or to maximize the quality of the service (voice/video).

There is only one other characteristic that might be added to broaden the definition even further. This additional QOS parameter is *reliability*. The term is used here not in the IP sense of "no absolute QOS," but in its more usual sense: just how much time per year is the network unavailable?

So there are now six different QOS parameters on the list. In the order they will be considered, these are:

1. Bandwidth

2. Delay

3. Delay variation (jitter)

4. Errors (cell/packet information loss)

5. Reliability

6. Security

Most discussions of QOS focus on the first four on the list. But reliability and security also belong with the others, for a number of reasons. Security appears in the Type of Service (TOS) field of the IP packet header. The TOS bit configuration 1111 means that the router handling the packet should "maximize security" for this packet. Security concerns play a large part in much of IPv6. And reliability can be maximized in OSPF routing tables. There are several other areas where security and reliability impact QOS parameters; the ones presented here are just a few examples. And the intent here is to make the QOS list as broad as possible for unlimited bandwidth purposes, not narrow it down.

This section will now consider how each of the six QOS parameters can be provided with no other network resource besides unlimited bandwidth delivered on fault-tolerant, state-of-the-art fiber optic rings using DWDM.

1. BANDWIDTH The first QOS parameter that has to be satisfied with unlimited bandwidth (UB) networking is the bandwidth need itself. "Unlimited bandwidth bandwidth" sounds like a tautology, and to some degree it is. However, there is more to an overall DWDM and SONET/SDH ring fiber optic network upon which a UB network (UBN) is based than just a lot of bandwidth. So this section will take a look at the overall architecture of a UBN and provide yet another working definition for UB itself.

The overall architecture for a simple UBN is shown in Figure 17.11. Note that there are no real internal network nodes such as IP routers or ATM switches. All ring-to-ring connectivity is handled by optical cross-connects with no need for electro-optical conversion at all. The optical cross-connects can shuttle bits from one channel to another and from one ring to another directly. All end-device connectivity is in the form of permanent virtual circuits (PVCs) since the bandwidth usage optimization afforded by switched virtual circuits (SVCs) is just a needless complication to a UBN. So a UBN is as channelized as T-carrier, E-carrier, or SONET/SDH. No switching of bits or examination of the bit stream is done at the cross-connection nodes. In the United States, the rings could be based on SONET, while outside of the United States, the rings could be based on SDH, but the use of neither is required.

Any switching or examination of the bit stream is done at the endpoints of the UBN. Note that the endpoint of the UBN might not necessarily be the ultimate source or destination of the bits themselves. In the example UBN, the endpoint devices of the UBN are IP routers, which are almost certainly not the source and destination of the bits themselves. What the UBN does then is just provide huge bit pipes between the routers or ATM switches or Layer 3 switches or whatever.

Figure 17.11
The basic architecture of an unlimited bandwidth network (UBN)

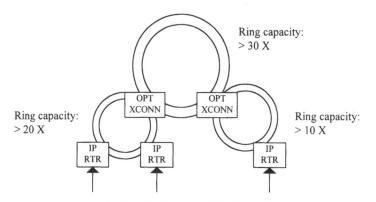

Ring capacity:
> 30 X

Ring capacity:
> 20 X

Ring capacity:
> 10 X

OPT XCONN

OPT XCONN

IP RTR

IP RTR

IP RTR

Total input load on each: X bits/second

Perhaps the endpoint device of choice in a UBN will be one of the new family of direct-to-DWDM devices now appearing on the market. There are several devices being demonstrated today that allow the direct connection of a Gigibit Ethernet hub or a Cisco switch router to a DWDM device to allow the creation of a virtual Gigabit Ethernet or virtual private network (VPN) across up to 20 km today and probably more dispersed areas soon.

The size of these pipes is an indication of the bandwidth available on a UBN. The bandwidth available between any two sites should be more than 10 times the total number of bits that could ever be sent from the endpoint device onto the UBN, in order to allow for future growth. There could be multiple endpoint devices, but this $10\times$ ratio must still hold true to permit the growth of the backbone to keep up with the growth of the number of applications.

This "10 times" characteristic can be used to provide yet another definition of a UBN. That is, a UBN is a network with enough bandwidth such that assigning more bandwidth to a given application will not and cannot improve the perceived performance of the application in terms of QOS. So why bother using more bandwidth? The concept of "enough bandwidth" is not quite the same as "the maximum amount of serial bits" that a device on a UBN can generate. Other QOS parameters might require two or more copies of the bit stream.

One other point should be made before moving on to other QOS considerations. It is the ring architecture of a UBN that makes the idea of flying storage practical in the first place. All that is needed is to define circular PVCs of varying lengths, as shown in Figure 17.12.

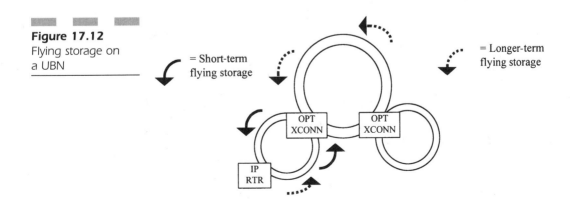

Figure 17.12
Flying storage on a UBN

2. DELAY   Although bandwidth and delay are often treated as if they were totally independent of each other on a network, they are, in fact, intimately related. One of the most common definitions of the term *throughput,* the number of packets/frames/cells per second that a network can handle, is actually based on the relationship between the twin QOS parameters of bandwidth and delay. The relationship between delay and bandwidth on a network link is shown in Figure 17.13.

So the delay is measured from first bit *in* (T_1) to first bit *out* (T_2) of a network, while bandwidth is measured from *first* bit out to *last* bit out (T_3) of the network. In fact, one definition of network *latency,* which no one uses in practice, is from *first bit in* to *last bit out,* which neatly combines the effects of both delay and bandwidth on network throughput. Some people speak of applications like voice as being *delay bound* in the sense that giving voice more bandwidth will not make it "better" if the network delay is not acceptable. Other applications such as Web page access are *bandwidth bound* in the sense that the delay due to propagation or other effects through the network could be zero ($T_1 = T_2$), but the application is abandoned because there is just not enough bandwidth ($T_3 \gg T_2$) to load the page effectively. Data applications in general, with their error checking at the rear of the frame, commonly do absolutely nothing with frames until all of the frame's bits have arrived at the destination.

Once this relationship between network bandwidth and delay is established, it is easy to see how a UBN can provide delays as close as possible to the basic propagation delay across the fiber mileage between the two endpoints on the network. And this would not be just the first bit in to first bit out delay, but the first bit in to last bit out "latency" or "throughput," or whatever term is chosen.

Figure 17.13
Network delay and network bandwidth

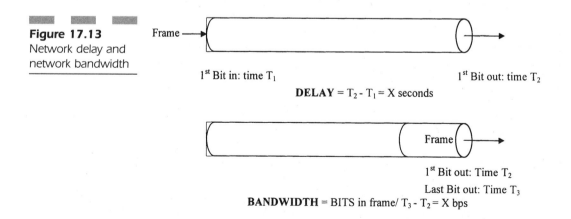

Frame

1st Bit in: time T_1 1st Bit out: time T_2

DELAY = $T_2 - T_1$ = X seconds

Frame

1st Bit out: Time T_2
Last Bit out: Time T_3

BANDWIDTH = BITS in frame/ $T_3 - T_2$ = X bps

The net result of a UBN is to make a frame (and the packet within it) look not a like a *cylinder* on the network due to the bit duration, as shown in Figure 17.13, but more like a *pancake* when measured from first bit out to last bit out. The UBN concept of *pancake packets* is how a UBN can deliver any delay QOS needed by any application. The destination need wait only a tiny interval for the whole frame and its packet content to arrive after the first bit appears.

Table 17.2 illustrates the concept of pancake packets on a UBN. In all cases, the nominal velocity of propagation (NVP, or link speed) is considered to be 0.67, or two-thirds the speed of light (c = about 300,000 km/second). This signal speed of 200,000 km/second is fairly valid for all media, including fiber optic cable. The major exception is wireless communications, which is closer to the speed of light itself. The whole table is built upon the observation that signals propagate on fiber optic cable at about 4.5 microseconds (μsec) per km. One bit per baud is assumed. Each row gives the time and size of the frame packet using the following example. The default IP packet size is about 570 bytes, or 4,560 bits. The bit duration of a 1.536-Mbps DS-1 link is about 6.5 μsec (1/1,536,000 of a second). Ignoring overhead, these 4,560 bits stretched out across 6,586 km (4,560 bits × 6.5 μsec/bit/4.5 μsec/km) on a DS-1 link. Certainly an IP packet that was formerly a 6,000-km cylinder should now be considered a pancake packet if it is less than 10 meters long on a UBN. The term could be stretched for the 1-km terabit IP packet, but compared to the five-times-to-the-moon size at 1.5 Mbps, this is still a pancake. The same logic is used for all the other speeds, with some approximations. Note that times and sizes for the UBN basic terabit links are all in nanoseconds (1 billionth of a second) and meters.

TABLE 17.2 Pancake packets on a UBN

Frame/Packet Size:	At 1.5 Mbps:	At 45 Mbps:	At 1 Gbps:	At 1 Tbps:
570B (4,560 bits) (default IP packet)	Time: 6.5 μsec/bit Size: 6,586 km	Time: 0.2 μsec/bit Size: 202.67 km	Time: 0.009 μsec/bit Size: 9.12 km	Time: 0.009 nsec/bit Size: 9.12 meters
1500B (12,000 bits) (Max. Ethernet frame)	Time: 6.5 μsec/bit Size: 17,333 km	Time: 0.2 μsec/bit Size: 533.33 km	Time: 0.009 μsec/bit Size: 24 km	Time: 0.009 nsec/bit Size: 24 meters
9180B (73,440 bits) (Max. AAL5 unit)	Time: 6.5 μsec/bit Size: 106,080 km	Time: 0.2 μsec/bit Size: 3,264 km	Time: 0.009 μsec/bit Size: 146.88 km	Time: 0.009 nsec/bit Size: 146.88 meters
64KB (512 kbits) (Max. IP packet)	Time: 6.5 μsec/bit Size: 739,556 km	Time: 0.2 μsec/bit Size: 22,756 km	Time: 0.009 μsec/bit Size: 1,024 km	Time: 0.009 nsec/bit Size: 1,024 meters

Over a UBN spanning 2,000 km, there could actually more than one maximum-size IP packet in flight at a time, and literally hundreds of Ethernet frames (although Gigabit Ethernet is another matter). This certainly makes everything into a delay-bound application. The first bit of any frame would take about 9 milliseconds (9,000 µsec) at 4.5 µsec/km to traverse the distance anyway. At 1 Tbps, the end of a 73,440-bit frame would arrive 0.66 µsec later anyway! So every application is now one in which attempts to lower delay by adding bandwidth are totally unnecessary. So a UBN totally satisfies the delay QOS parameter.

3. DELAY VARIATION (JITTER)   Jitter can be defined as the variation in end-to-end delay that an application experiences. So this type of jitter differs from the low-level, line coding jitter that results from timing variations on a network, when jitter is usually given in parts-per-million. This was a real concern in the early days of networking. Today, in the era of inexpensive atomic clocks and GPS receivers at many points throughout a network, this type of jitter is no longer the major problem it once was. In fact, without very fast and very stable network clocks, networks based on terabit links would be impossible to build at all.

The type of delay variation considered in this section applies to the highest level of a network. The effects of jitter have usually been ignored in data applications. Who cares if the end of a file goes through a network slower than the beginning? No one can realistically do anything with the file until it is all there anyway.

But jitter is quite important in streaming voice and video applications. Delay variations in voice can cause a deepening or raising of a speaker's voice, sometimes dramatically. Jitter can cause a video stream to freeze momentarily or to jerk and sputter like an old-time movie. The point is that jitter is an important part of the QOS suite as multimedia applications combining voice and video and data become more common and the same network must be used for all applications.

What is the source of delay variations that cause application-level jitter in a network? The leading cause of jitter today is variations in the nodal processing delay caused by varying traffic loads. This is almost exclusively a concern of packet-switching networks rather than circuit-switching networks. In digital carrier systems like T1 and E1, transmission frames did not back up on a switching office's trunking network. Nor do transmission frames back up in a PVC-based optical cross-connect network. But packets back up in routers, frames back up in frame relay switches, and cells back up in ATM switches all the time as the bursty loads on the network vary during operation. The simple fact is that network nodes can route or switch traffic faster than the traffic can be gotten rid of, a consequence of optimizing the use of relatively scarce bandwidth for bursty traffic patterns.

The varying packet/frame/cell loads are handled by the judicious use of buffers. The traffic queues up in a buffer area until it can be sent. The more variable the traffic load, the larger the queuing buffers needed for efficient bandwidth use and the larger the potential delay variation. This is the basis for yet another common network design trade-off today: queue depth versus delay variation.

So sometimes there might be 10 packets/frames/cells ahead of a particular traffic unit in a network node, and sometimes there might be 100. If there are no buffers available for an arriving traffic unit, the traffic unit is usually dropped, although exact mechanisms vary among network architectures.

It should be apparent by now that a UBN would make such buffering and queuing unnecessary. The optical cross-connect is a totally circuit-oriented device with minimal delay even at terabit speeds. The routers or switches that feed the UBN can always get rid of traffic as soon as it arrives, given the vast amount of bandwidth available on the UBN.

And in any case, even given common queue sizes, the delays of UBN pancake packets will always amount to a small fraction of the propagation delay across the network. This was the whole idea behind the invention of cell relay technology: the small delay variations resulting from both small traffic units and fast network nodes.

Table 17.3 shows the effects of buffering on nodal processing delays at the ingress point of a network for a common frame relay maximum

TABLE 17.3	Frame = 4096B = 32,768 bits	At 1.5 Mbps	At 45 Mbps	At 1 Gbps	At 1 Tbps
Delay variation in a UBN					
Q = 0	Time % of 9 ms	21 ms > 2x	0.7 ms 7.7%	32.7 μsec 0.36%	3.27 μsec 0.036%
Q = 10	Time % of 9ms	210 ms > 20x	7 ms 77%	327 μsec 3.6%	32.7 μsec 0.36%
Q = 100	Time % of 9 ms	2.1 seconds > 200x	70 ms > 7x	3.27 ms 36%	327 μsec 3.6%
Q = 1000	Time % of 9 ms	21 seconds > 2000x	0.7 seconds > 70x	32.7 ms >3x	3.27 ms 36%

frame of 4,096 bytes at various speeds and queue depths. The UBN figures are in the last column. The propagation delay for a 2,000-km fiber optic network is set at 9 milliseconds, as before. For each set of speeds, the table shows the maximum queue depth encountered, the time it would take to send the last frame in the buffer, and the percentage or ratio of the 9-millisecond propagation delay this represents. The table shows the effect of rounding in some calculations.

As the UBN column shows, the maximum delay variation of a long frame relay frame with a queue depth of 100 at the ingress device is ± 327 λ sec, less than 4% of the propagation delay of 9 ms. As long as queue depths are kept to about 100 or less, the delay variation even at the ingress point of the network should satisfy the jitter needs of most, if not all, applications. And if jitter remains a problem, the use of multiple network nodes should eliminate higher queue levels. Note that this queuing is not done by the UBN proper, but rather the endpoint device on the network. So a UBN totally satisfies the delay variation (jitter) QOS parameter.

4. ERRORS   In the context of QOS, the term *errors* does not refer to individual bit errors. These bit errors are almost a non-issue in modern fiber networks where bit error rates (BERs) of 10^{-13} (1 in 10 trillion) or better are common. With regard to QOS, the term *errors* means packet/frame/cell information loss, since most modern networks respond to detected bit errors or congestion by discarding whole traffic units of whatever form they might be. And in most networks today, the effects of congestion account for most of this information loss. In fact, most bit errors are probably caused more today by a network device's internal memory errors than actual bit errors on the link.

In some ways, the issues of delay variation (jitter) and errors in terms of traffic unit loss on a network are as closely related as the issues of bandwidth and delay. That is, errors and delay are traded off against one another. The more buffers, the less chance of information loss (errors) and the fewer buffers, the more chance of this loss occurring.

However, as the previous section has pointed out, such trade-off issues are of no concern to the UBN itself, and the endpoint device need only be constrained by such considerations in certain configurations. The UBN has more than enough bandwidth to handle any amount of traffic. If the endpoint device allows more input than can be output onto the UBN, then that not only violates the UBN definition, but it is entirely the concern of the endpoint administrator, rather than the UBN itself.

So if there is no concern about information loss due to errors (given the low BERs of the modern fiber networks upon which a UBN is based) or congestion-driven discards (given the UBN itself), then the UBN completely satisfies the error QOS parameter.

5. RELIABILITY In the context of QOS, *network reliability* refers to the overall availability of a path through the network between a source and a destination. In an SVC environment, paths might not be available even when the network is available in the general sense. An SVC network, just as the public telephone network, is subject to *blocking*. In a PVC environment, once a path is set up, it is expected to be available constantly. Any service outage on this path in the network is considered a reflection of the overall reliability of the network. Network reliability is most often measured in local terms, for a particular user, and not in global terms, for the network as a whole, since complete network failures are rare (but not unknown even today, as a major frame relay service provider found).

Reliability is typically measured in annual available percentage. That is, a network path that was unavailable for 24 hours out of a year (any combination of long and short outages adding up to 24 hours) is considered to be 99.726% reliable (364/365). This figure is actually quite horrible for a modern network. Today such outages are usually measured in minutes per year, as more and more of the network infrastructure is based on self-healing fiber optic rings.

The self-healing and protection-switching features of modern fiber rings are sufficiently well-known to make further discussion here unnecessary. It is enough to say that service providers with such ring backbone networks routinely offer service level agreements (SLAs) that offer "four 9s" (99.99%) or "five 9s" (99.999%) availability on network paths.

It is not always possible to determine the exact QOS reliability represented by a *fault-tolerant* network or device as opposed to a *fault-resilient* network or device. Attempts have been made to translate these typical vendor and service provider claims to availability numbers. Table 17.4 shows one such attempt. Although the table applies to computers, it is the percentages and downtime duration that are of interest.

So "five 9s" SLAs are being written today for 2.628 minutes (157.68 seconds) of downtime per year on networks that do not consist exclusively of ring architectures. In fact, the weakest point on modern fiber optic networks is the continued presence of point-to-point links with inadequate protection switching. Once a UBN with only rings can be built, perhaps this availability figure would increase by a factor of 10, to "six

TABLE 17.4

Availability and
downtime

System Availability Levels	Total Availability	Approximate Total Downtime
Continuous availability	99.999% to 99.9995%	2.628 to 5.256 minutes/year
Commercial fault tolerant	99.99% to 99.995%	26.28 to 52.56 minutes/year
Fault resilient	99.99%	52.56 minutes/year
High availability	99.9%	8.76 hours/year

9s." Surely, a UBN with only about 15 seconds of downtime per year should be reliable enough to satisfy the most sensitive application. And if this still is perceived as a problem, enough bandwidth exists on the UBN to allow a concerned source to send a *second* copy of the information on an alternate network path. The probability of having *two* 15-second failures per year coincide on the same network span is vanishingly small.

So a UBN completely satisfies the reliability QOS parameter.

6. SECURITY   Security is the last of the six QOS parameters to be investigated. Of the previous five, some have seemed fairly obvious to address with a UBN (bandwidth, delay, jitter) and others are more aspects of the fiber optic nature of the UBN than of the UBN itself (errors, reliability). One security aspect is part of this fiber optic UBN nature as well: fiber taps always add loss to a fiber link. Security is maintained of fiber optic networks by constantly monitoring signal attenuation. Sudden and abrupt changes are sure signs of physical damage due to accidents or intentional tampering.

But security seems an odd thing to address with pure bandwidth. How can an unlimited amount of bandwidth add to the basic security that a network can afford an application?

The answer is simple. A UBN can provide *fragment security* to applications. This type of security (actually an *enhancement* to many existing security methods) has always been available, and is sometimes used as the *wheat and chaff* security method. But the impact of other QOS factors such as bandwidth, errors, and so on, has made this option quite inefficient to exploit. A UBN makes fragment security not only possible, but attractive.

A full discussion of security through encryption techniques is far beyond the scope of this section. Suffice it to say that encryption can provide security features such as privacy, authentication, and so on to a greater degree than ever before, largely thanks to the availability of more and more powerful computing devices.

However, for anyone intercepting an encrypted message, whether detected or not, the same computing power is available to crack the code. It is only a matter of time before a code is broken. And detection methods cannot prevent the theft of the information in the first place, they can only report it.

A UBN offers a way out of this "too late" scenario. The simple fact is that messages can be cracked only if a substantial portion of the message is intercepted, and usually a contiguous portion. Encrypted files are normally sent whole for this very reason. Any missing information usually makes it totally impossible to decode the message, even by the intended receiver.

It only makes sense to direct the whole message along the same path when bandwidth is limited. Even with dynamic routing, routers attempt to use the same paths to minimize delay variations and produce predictable traffic patterns to conserve scarce bandwidth.

But a UBN has unlimited bandwidth and rings with multiple possible paths to any destination. Why not *intentionally* send fragments of an encrypted message on many different paths through the network? Neither bandwidth consumption, nor delay variations (beyond different fragment propagation delays due to different length paths), nor errors, nor any other QOS parameter need be a concern. As noted, this approach has been tried before, and has always been available, just not often used.

If any interception of secure traffic could not yield enough of a message to enable its decryption in a reasonable time frame, regardless of computing power applied to the effort, then the UBN could provide any level of security desired. For lesser security needs, fragment bytes. For greater security needs, fragment down to the bit level. There will be added effort to perform this, but only marginally greater use of network bandwidth! So a UBN can completely satisfy the security QOS parameter.

Conclusion

This section began by defining a new type of networking based on unlimited bandwidth. The overall characteristics were presented and then the implications of creating unlimited bandwidths networks (UBNs) were explored. Six QOS parameters were introduced and the architecture of a UBN was applied to each to show that each QOS parameter could be satisfied to any extent desired by the UBN. The conclusion was that other networking techniques for delivering a desired level of QOS to applications are totally unnecessary, given enough bandwidth. So efforts to develop network nodes that enforce QOS at the application level are superfluous. Such efforts should be redirected to research into DWDM and other bandwidth-enhancing network techniques. The QOS issues will ultimately take care of themselves.

18
Synchronous
Digital Hierarchy

It seems quite odd to devote the last chapter of this book to the Synchronous Digital Hierarchy (SDH) because SONET and SDH are really quite compatible, as this chapter will show. SONET and SDH are designed and intended for exactly the same purposes and uses, and so are fully compatible; however, SONET is the product of the North American carriers who grew up in the world of T-carrier, while SDH is the product of the ITU-T and the carriers around the world who grew up in an E-carrier (or CEPT) environment. Thus, SONET and SDH are the same, but different.

This awkward terminology need not be confusing. Many people are familiar with some other common "same but different" examples in networking. For instance, Ethernet and the IEEE 802.3 LAN standard are often cited in this "same but different" category. In fact, most people say "Ethernet" when the term "802.3" is more appropriate. The truth is that the differences between Ethernet and IEEE 802.3 are much more pronounced and significant than are the differences between SONET and SDH. Therefore, people who are comfortable in dealing with Ethernet and 802.3 should also be equally at ease with SONET and SDH.

Nevertheless, this does not imply that the differences between SONET and SDH are trivial or unimportant. Everything works exactly the same way in SDH as it does in SONET, but the names are changed. A new set of acronyms apply, and annoying details of function are interpreted slightly differently. This chapter is intended as a kind of guide for those who must deal with both SONET and SDH, typically in an international networking situation when dealing with digital fiber optic links from the United States to almost anywhere else in the world. For the rule is that when one end of a link is SONET and the other end is SDH, the SDH rules and terms apply to the whole link. Fortunately, the bits do not care what they are called on each end of the link.

It is not misleading to say that SDH is the international version of SONET, and that SONET is the North American version of SDH. In fact, there is probably a lot of sense to positioning the technology in precisely this framework. It will surely make this chapter more understandable when seen in this light.

As time goes on, SDH is growing in importance and popularity. Not too long ago, SONET was the main concern of equipment vendors and service providers, and SDH the variation. Now SDH has become the main concern, while SONET, although certainly not shrinking in importance at present, has become the variation. This is not necessarily a bad thing, but requires a few words of explanation.

The recent rise of SDH can be explained by five major factors. First and foremost, SDH is technically the international standard, not SONET. Although compatible, whenever an international boundary is crossed, the structure of the link is required by the ITU to conform to SDH, not SONET. Second, the major efforts at standardizing management systems for interoperable optical environments have been spearheaded by TMN. Third, the demand for bandwidth has all but relegated STS-1/OC-1 speeds to a footnote. Most SONET links function at OC-3 or other speeds that align quite well with SDH. Fourth, the differences between SONET and SDH are for the most part minor, so equipment vendors, seeking to address the greater global market, tend to emphasize SDH and not SONET. Finally, SDH documentation, which has lagged somewhat behind the specifications available for SONET rings, has now stabilized to the point where SDH rings can be installed and managed with confidence.

This chapter is not intended as a repeat of the full discussion of SONET overhead with an SDH spin, or ring protection, and so on. The emphasis here is on the relatively minor, but still important, differences between SONET and SDH, mostly in terminology. In many cases, the SDH definitions of some overhead messages are more general in nature, and their SONET equivalents are more tuned to the operating conditions within North America. This is exactly as it should be.

This chapter describes SDH in terms of signal types, signal rates, and signal formats. In order to do so, this section first summarizes the Plesiochronous Digital Hierarchy (PDH) signal types, signal rates, and signal formats. Although much of this has been covered before, the concepts are important enough to cover in this final chapter. Also, the transition from the international standard PDH to SDH by using the concept of containers (C-n; n = 11, 12, 2, 3 and 4) is discussed.

The SDH concepts of "path," "multiplex section," and "regenerator section" of an SDH link are detailed and compared to their SONET counterparts. Differences between SDH and SONET signal frames are described, as well as the SDH section and path overhead byte functions. In this chapter, "bytes" will now be called "octets," in keeping with ITU-T standards. Both are still in groups of eight bits. Finally, the various SDH signals, such as C-n, VC-n, TU-n, TUG-n, AU-n, and AUG (these correspond with Container, Virtual-Container, Tributary Unit, TU Group, Administration Unit, and AU Group, respectively) and their relationships are explored.

There is more, much more, to SDH; however, because the vast majority of SDH operations, functions, and use correspond exactly to their SONET counterparts, the emphasis here is on details of definition and terminology.

Basic SDH Signal Rates

Currently, there are four SDH signal rates defined. These are designated as Synchronous Transport Module (STM) rates mainly to distinguish them from their SONET Synchronous Transport Signal siblings. The signal rates of the defined STM-1, STM-4, STM-16 and STM-64 transports in the SDH hierarchy are shown in Figure 18.1.

The workhorse of the SDH is the STM-1 operating at 155.52 Mbps, the same as a SONET STS-3. In fact, they are pretty much the same, especially when the SONET link is an STS-3c. The SONET equivalent speed is also shown in Figure 18.1. An STM-1, which is the basic SDH signal, is typically used to transport a 139.264 Mbps PDH signal (usually called an E-4 or CEPT-4) or its equivalent, such as H4 ISDN signal. With the additional overhead, this 139 Mbps signal will form a 155.52 Mbps STM-1 signal. Besides a 139 Mbps signal, there are many lower-order tributaries (just as in SONET) that can also be transported by an STM-1 signal.

It should be noted that whether it is called an STM-1, STS-3, STM-4, or STS-12, the optical line bit rate is exactly the same in SONET and SDH. Now it seems that there may be a source of confusion when it comes to speaking of optical line rates. The question is, when an STS-3 runs on an OC-3 optical carrier, what does an STM-1 run on? If the optical carrier were now an OC-1 for the STM-1, this would easily be con-

Figure 18.1
The SDH hierarchy

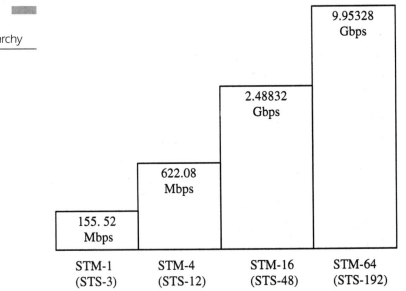

fused with an OC-1 carrying an STS-1. Conversely, if an STM-1 were to run on an OC-3, why would it be called an STM-1? Fortunately, the terminology seems to be all straightened out. An STS-3 runs on an OC-3 optical carrier link, and an STM-1 runs on an STM-1O optical carrier link. Naturally, an STM-4 runs on an STM-4O (O for optical) link, and so forth. Sometimes STME-n (E for "electrical") is used to avoid confusion.

When an E-4 or H4 signal is transported inside an STM-1, the E-4 or H4 content is now referred to as a C-4 SDH signal. The concept is the same as when a DS-1 rides inside a SONET frame as a VT-1.5. Because SDH is the same as SONET, a DS-1 can ride inside an SDH frame as easily as inside a SONET frame. Of course, it is no longer a VT-1.5: That is the SONET word. The SDH term is C-11. It is the same thing, but different. Get used to this when discussing SDH and SONET. The lower-level SDH tributaries are shown in Table 18.1, which "maps" the existing PDH to SDH containers.

To continue the SDH hierarchy, four STM-1 signals are byte-interleaved multiplexed (just as in SONET) into an STM-4 signal at 622.08 Mbps (which is 4 × 155.52 Mbps), and four STM-4 signals are multiplexed into an STM-16 signal, which has a rate of 2,488.32 Mbps (which is 16 × 155.52 Mbps). Oddly, as ITU-T standards for ISDN consistently

TABLE 18.1

SDH containers

Digital Level	North America	Japan	Europe	ISDN	SDH Signal
0	64 Kbps	64 Kbps	64 Kbps	—	—
1	1.544 Mbps	1.544 Mbps	—	H-11	C-11
	—	—	2.048 Mbps	H12	C-12
2	6.312 Mbps	6.312 Mbps	—	—	C-2
	—	—	8.448 Mbps	—	—
3	—	32.064 Mbps	—	—	—
	—	—	34.368 Mbps	H31	C-3
	44.736 Mbps	—	—	H32	C-3
4	—	95.728 Mbps	—	—	—
	—	—	139.264 Mbps	H4	C-4
	(274.176 Mbps)	—	Mbps —	—	—

mentioned rates of "thousands of Kbps" (such as 2,048 Kbps) and seemed reluctant to admit Mbps links existed, ITU-T standards for SDH consistently mention "thousands of Mbps" and seem reluctant to admit that Gbps links exist. Finally, four STM-16 signals are then multiplexed into an STM-64 signal, which is currently the highest rate SDH signal. The STM-64 has a rate of 9,953.28 Mbps (often referred to as 10 Gbps signal).

The two basic parameters of an STM-N (where N = 1, 4, 16 or 64) are signal rate and signal format (frame structure). The derivation of the signal rate and the description of the frame structure, as well as the exact relationship between an STM-1 SDH signal and an STS-1 SONET signal, is discussed later in this section. Just as in SONET, however, an STM-N signal is *modular*. This means that an STM-N has a signal rate that is exactly N times the rate of an STM-1 signal. Thus, the rate of an STM-N signal = N × rate of an STM-1 signal = N × 155.52 Mbps.

As can be seen from the previous table, there are three regional signal hierarchies used for global telecommunications. One is for North American digital networks, one is for Europe, and the third is for Japan. These three regional digital hierarchies were presented in some detail earlier in this work and need not be repeated here. These signals are not synchronous, of course, but are quasi-synchronous in most cases. Thus, they are usually called plesiochronous signals. Together, they form the PDH.

In North American digital networks, 24 digitized voice signals of 64 kbps each (or the equivalent) are multiplexed into a 1.544 Mbps DS-1 signal. There is only about 0.5% of overhead capacity that is added to the required 1.536 Mbps (which is 24 × 64 kbps per voice channel) in order to form this DS-1 signal. Four DS-1 signals may be multiplexed into a DS2 signal with a rate of 4.312 Mbps. Naturally, additional overhead must be added. Seven DS-2 signals are then multiplexed into the highest digital signal among PDH signals for North America, DS-3.

Typically in North American PDH, DS-3 signals are transmitted as a unit over the fiber optic transmission system (FOTS). For example, AT&T can transmit either nine DS-3 signals at about 417 Mbps or 36 DS3 signals at 1.7 Gbps. If WDM were employed, the capacity could easily be doubled, or multiplied even more. All this has been discussed before, but little emphasis has been placed on the European PDH hierarchy.

For European systems, 32 digitized voice signals (or equivalent signals) at 64 Kbps are multiplexed into a 2.048 Mbps digital signal, known as a CEPT-1 or E-1 signal. Four E-1 signals are multiplexed into an E-2 signal operating at 8.448 Mbps. Four 8.448 Mbps signals form a 34.368 Mbps E-3 signal. Finally, an E-4 operating at 139.264 Mbps consists of four E-3s.

For Japan, the DS-1 and DS-2 signals are the same as those in North

America. However, five DS-2 signals are multiplexed into a 32.064 Mbps signal. Three of these 32.064 Mbps signals form a 97.728 Mbps signal. It should be noted that for SDH purposes, Japanese digital signals may carry E-1 or E-4 signals for ease of implementation.

SONET is really intended only for ease of transition from the North American PDH (T-carrier) hierarchy. SDH, however, must accommodate all three regional standards, including all ISDN signals. Otherwise, international use may become a nightmare (as it is now with only PDH). For this express purpose, SDH has defined several containers of various types. These are designated as "C-n", where n = 11, 12, 2, 3, or 4, as shown in Table 18.1. These are intended for all the various PDH signals. These containers are described in detail later. It should also be noted that ISDN H11 signal is best known as the *primary rate ISDN signal*. This has a format of 23B + D (B for information *Bearer* channel and D for *Data* communication channel. Both types of channels have the same rate of 64 Kbps in North America. The counterpart of H11 in ITU-T digital networks is known as N12.

The C-11 (read as "container level 1, type 1") is designed to carry a 1.544 Mbps DS-1 signal or an H11 ISDN signal. A C-12 (container level 1, type 2) is designed to carry a 2.048 Mbps E-1 signal or an H12 ISDN signal. There were initially two types of second-level container C-2 signals (C-21 for transporting a DS-2 signal and C-22 for transporting an E-2 signal). However, the lack of common E-2 deployments has diminished the need for C-12 containers. Thus, current SDH standards define only one type of second-level container, the C-21 for transporting a DS-2 signal. Confusingly, it is now merely called a C-2, and not a C-21. This leads to an interesting problem: What is an E-2 on SDH to be called? The designation "P-22" can be used to indicate a PDH level 2 and type 1 signal and used to transport an E-2 signal, if ever the need arises. But this is not standardized. These containers described, C-11, C-12 and C-2 are known in SDH as the *lower order tributaries*.

At the third level of the PDH hierarchy, another interesting discussion arose with regard to SDH. Obviously, an STM-1 operating at 155.52 Mbps can easily handle four E-3s operating at 34.368 Mbps each; however, the STM-1 line rate is the same as the STS-3 in SONET. An STS-3 also only carried three DS-3s, the third level of the T-carrier digital hierarchy. For the sake of consistency, it was decided to carry only three E-3s in an STM-1, just as three DS-3 were carried in an STS-3. Therefore, where two types of third-level containers were defined—C-31 for an E-3 signal and C-32 for a DS-3 signal—there is now just C-3 for both. An SDH STM-1 carrying three containers (now just C-3s) can be used to transport either three E-3 signals or three DS-3 signals.

As far as the fourth level of the PDH hierarchy is concerned, the "DS-4" signal that was proposed for the North American digital hierarchy was never widely implemented. This was mainly due to the rapid deployment of higher-speed fiber optical communications systems. Thus, DS-4 never got the chance to be deployed in North American digital networks except for a very few isolated locations. For these reasons, in SDH standards, it was decided to have only one type of fourth-level container, C4. This is used to transport an E-4- or an H4-ISDN signal. Note that there is no provision for any "DS-4" container. These containers, C-3 and C-4, are known in SDH as the *higher order tributaries*.

It is important to realize that STM-1 SDH frames can carry multiple DS-1s just like an STS-1 in SONET, just more of them. These are sometimes called "DS-1As" to reflect the fact that A-law digital coding is used to digitize the voice, not μ-law digital coding as is used in T-carrier. When an STM-1 is structured in this fashion to carry virtual tributaries, called *virtual containers* (VCs) in SDH, the STM-1 pointers have a particular structure that conforms to what is called an *Administrative Unit* type 3, or AU-3. When an STM-1 frame does not carry VCs and is used for traffic like ATM cells or IP packets (an STS-3c), the pointers are structured according to the AU-4 specification.

This AU-3 or AU-4 (AU-n is the generic term) takes some getting used to in SDH, but is no more complex than the difference between an STS-3 (channelized for VTs) and an STS-3c.

SDH End-to-End Links

In SONET, an understanding of network operations and network management requires an understanding of the SONET overhead bytes. Naturally, in order to understand SDH network operations and network management, an understanding of the SDH overhead octets (bytes) is necessary. Just as in SONET, SDH has framing octets, parity-check octets, and so on. These network OAM&P (which is just OAM in SDH) functions are performed by three sets of SDH overhead octets. These are the *path, multiplex, section, and regenerator* section overhead (SOH) octets. Figure 18.2 shows a typical SDH end-to-end connection. This link connects two PDH network links but, of course, other configurations are possible. Some of the terms defined in this SDH figure have the same meaning as in SONET. When the definitions used in SDH are identical to those used in SONET, it is will be so stated, and left at that. Detailed information is available in previous chapters of this book.

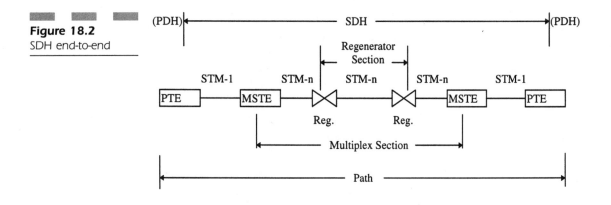

Figure 18.2
SDH end-to-end

In SDH, a path is a logical connection between the point at which the standard SDH frame format for the signal at a given rate is assembled and the point at which the standard SDH frame format for the signal is disassembled. The Path Terminating Equipment (PTE) is the SDH network element (NE) that multiplexes and demultiplexes the VC-n payloads. VC-ns will be discussed in more detail later. The PTE can originate, access, modify, or terminate the VC-n path overhead, in any combination of these actions. The definitions of path and PTE in SDH are exactly the same as for SONET.

In SDH, the following NEs are typical PTE devices, using their proper SDH names:

- Low-order (or low-speed) multiplexer
- Wideband cross-connect system
- Subscriber loop access system

As is also shown in Figure 18.2, an SDH *multiplex section* is defined as the transmission medium, together with the associated equipment, required to provide the means of transporting information between two consecutive NEs. One of the NEs originates the line signal and the other NE terminates the line signal in each direction. An SDH Multiplex Section Terminating Equipment (MSTE) is a network element that originates and/or terminates STM-N signals. The MSTE can originate access, modify, or terminate the multiplex SOH, or can perform any combination of these actions on the STM-N signal. The definition of an SDH multiplex section is the same as the "line" definition used in SONET. Examples of SDH MSTEs are called:

- Optical line terminal
- Radio terminal

- High-order multiplexer

- Broadband cross-connect system

Finally, a regenerator section in SDH is the portion of a transmission facility, including terminating points, between either a terminating network element and a regenerator, or two regenerators. The Regenerator Section Terminating Equipment (RSTE) is the network element that regenerates an STMN signal for long-haul transport. The RSTE can originate, access, modify, or terminate the regenerator SOH, or can perform any combination of these actions. The SDH definition of regenerator section is the same as the "section" definition used in SONET:

Because these terms are fundamental to SDH, and yet differ slightly from SONET, these differences are shown in Table 18.2.

Ironically, the three terms used in SDH actually have better names to represent the meanings of the terms than those used in SONET. For instance, a regenerator section in SDH simply and clearly means an SDH link connecting two adjacent network elements, and at least one of these two elements must be a regenerator. The SONET "section" term is less distinctive. Likewise, a multiplex section obviously means an SDH link connecting two adjacent SDH multiplexers (or two adjacent terminating equipment devices performing multiplexing functions). Thus, a digital cross-connect system or a radio terminal device that acts as a multiplexer may serve as an MSTE, too.

SDH rings are becoming more common as well. Most of the specifications on SDH rings have been done by the European Telecommunications Standards Institute (ETSI). Their work closely recapitulates the work done on SONET rings by ANSI. A full list of ETSI SDH ring documentation appears in the Bibliography.

As usual, most of the terms used in SDH rings are different from their SONET counterparts. Instead of the two- or four-fiber BLSR, ETSI establishes the two- or four-fiber *Multiplexer-Section Shared Protection Ring*, or MSPR. There are also UPSRs in SDH, but the emphasis is on the two- or four-fiber MSPR. Many of the figures in the ETSI documentation closely follow their ANSI cousins.

When SDH rings are used to carry VC traffic (i.e., voice channels), the rings can be configured to provide *Subnetwork Connection* (SNC) protection.

TABLE 18.2	SONET calls it...	Section	Line	Path
SONET and SDH terminology	SDH calls it...	Regenerator Section	Multiplex Section	Path

SDH Frame Structure

Any description of SONET would be lacking without an examination of the SONET frame structures for various STS-n levels. In the same way, any discussion of SDH must include the details of the SDF frame structures for various STM-N levels. This section details the SDH frame concept and briefly describes the SDH STM-1 frame. Then a description of an STM-N (where N ≥ 4) frame also is examined. The differences between SDH frames and the frames used in SONET are investigated as well.

The same two-dimensional frame format and overhead structure used SONET also are followed in SDH. Actually, the ITU-T nearly invented this format for E-1 and ISDN's primary rate (23B+D) format. In current networking standards, the two-dimensional frame structure used in SONET and SDH standards is all but universal. Just as in SONET, an SDH frame will have a given number of rows and columns and have a portion of overhead bytes that will occupy the "left-hand" side of an SDH frame. This basic structure of an STM-1 frame is shown in Figure 18.3.

As can be seen in Figure 18.3, an SDH STM-1 frame is very much the same as the SONET STS-3c frame. Both have a 9-row-by-270-column structure. The octet in the first column of the first row will be transmitted first. This is followed by the octet to its right, also in the row 1. The transmission sequence is exactly the same as in SONET: that is, from left to right, and from top to bottom. A frame is sent 8,000 times a second, or once every 125 microseconds.

All overhead in an SDH frame is just SOH or pointer octets. However, the SOH in SDH is divided into two main parts by the pointer octets. In fact, an STM-1 frame consists of three main parts:

Figure 18.3
The STM-1 frame

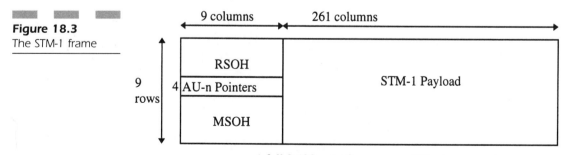

1 full 2,430 octet frame every 125 microseconds

1. *The regenerator section overhead (RSOH) octets.* This occupies an area of 3 octets (rows) by 9 octets (columns) and has a capacity of 27 octets and operates at 1.728 Mbps.

2. *The multiplex section overhead (MSOH) octets.* This can be seen as the MSOH octets plus the *AU-n pointer* octets. There are good reasons for including the AU-n pointers in the MSOH, and these will be discussed shortly. The AU-n pointers (where n = 3 or 4) serve the same purpose as in SONET: to locate the synchronous payload in within the frame. These pointers occupy the fourth row of the frame and, thus, consist of 9 octets. The MSOH proper occupies rows 5 through 9, and thus has a capacity of 45 octets.

3. *The STM-1 payload (envelope) itself.* This frame structure is 9 octets (rows) by 261 octets (columns) and has a capacity of 2,349 octets and operates at 150.336 Mbps. The payload is designed to carry an ITU-T E-4 signal operating at 139.264 Mbps or its equivalent.

STM-N (N = 4, 16, or 64) Frame Structure

Just as in SONET, the basic SDH STM-1 frame structure may be extended by a few simple rules to yield the frame structures of all remaining STM-N signals. Only STM-Ns where N = 4, 16, or 64, need be of concern. An STM-4 frame is formed by simply octet (byte) interleaved multiplexing four STM-1 frames into a frame that is still 9 rows, but now 1028 columns wide. The STM-4 signal has a line rate of 622.08 Mbps (which is 4×155.52 Mbps, naturally).

The four STM-1 frames should be frame-aligned before multiplexing, but this is not a big deal. Frame alignment implies that the first 12 bytes of the STM-4 signal are A1 framing octets. These are drawn initially from STM-1 #1, the next three from STM-1 #2, then three from STM-1 #3, and finally three from STM-1 #4. These 12 A1 framing octets are followed by 12 A2 framing octets, which are obtained from the four input STM-1s in the same fashion as the A1 bytes.

Of course, because an STM-4 signal has a line rate that is four times the rate of an STM-1 signal, there are now 36 columns (which is 4×9 octets, where the 9 octets are required for each STM-1 signal) reserved for SOH plus the AU-n pointers. There are now 1,080 (which is 4×270) columns in an STM-4 signal frame. This STM-4 frame structure is shown in Figure 18.4.

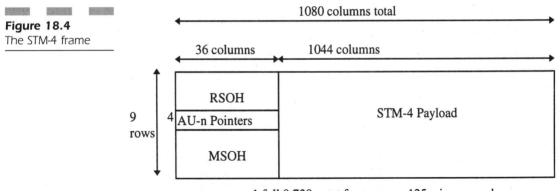

Figure 18.4
The STM-4 frame

1080 columns total

36 columns | 1044 columns

RSOH

AU-n Pointers

MSOH

STM-4 Payload

9 rows | 4

1 full 9,720 octet frame every 125 microseconds

The STM-16 frame structure is formed by the same approach. The frame would have 144 columns of SOH with pointers, and a total of 4,320 columns. In SDH, there are two ways to form an STM-16 signal: first, by multiplexing four STM-4 signals; second, by multiplexing 16 STM-1 signals directly. The first way is known as *two-stage multiplexing*, while the second method is called *single-stage multiplexing*. In either case, all low-speed signals must be frame-aligned before multiplexing. An STM-64 frame can be derived in the same fashion.

SONET and SDH Revisited

It has been casually mentioned that SONET is "tuned" to the T-carrier world and SDH is "tuned" to the E-carrier (CEPT) world. This does not mean, however, that SDH cannot be used for transporting T-carrier tributaries in an STM-N frame. In fact, a whole structure of SDH Tributary Units (TUs) has been established for lower levels of the T-carrier and E-carrier worlds alike. A full discussion of these is unnecessary at this point. Table 18.3 lists the various lower levels of the T- and E-carrier hierarchies. Note the absence of E-3 support in SONET and DS-1C support in SDH; neither is especially common and can always be carried at their constituent rates. As an aside, a designation, such as TU-21, is read as "tributary unit level 2, type 1," as were the container designations.

As shown in Table 18.3, the *sub-rate* (below STS-1 rate) signals for SONET are VT-6, VT-3, VT-2, and VT-1.5. Those for SDH are TU-3, TU-2, TU-12, and TU-11. A TU-2 has the same line rate as a VT-6, a TU-12 has the same rate as VT-2, and a TU-11 has the same rate as VT-1.5.

TABLE 18.3

SONET and SDH
tributaries

Tributary Signal	Tributary Bit Rate	SONET Name	SDH Name
DS-1	1.728 Mbps	VT-1.5	TU-11
E-1	2.304 Mbps	VT-2	TU-12
DS-1C	3.456 Mbps	VT-3	—
DS-2	6.912 Mbps	VT-6	TU-2
E-3	49.152 Mbps	—	TU-3

Of course, significant differences remain between SONET and SDH. SONET has a basic rate (the STS-1) of 51.54 Mbps. This signal in its electrical form is designated as STS-1 or rarely as an STS-1E, and in its optical form as OC-1. This STS-1 signal is primarily intended to transport a PDH DS-3 or equivalent signal, which has a line rate of 44.736 Mbps. Conversely, SDH has a basic rate (STM-1) of 155.52 Mbps, which is identical to the rate of SONET STS-3. The optical form of this signal is designated as STM-1O, where the "O" suffix stands for optical. This STM-1 signal is for transporting an E-4 or equivalent signal with a line rate of 139.264 Mbps.

In SONET, the basic STS-1 frame structure is used to build up all the higher level frame structures. In SDH, the basic STM-1 frame structure is used for the same purpose. This building-up process is shown for both SONET and SDH in Figure 18.5.

As can be seen in Figure 18.5, all frames, both SDH frames and SONET frames, have 9 rows. This applies to STM-1, STM-4, STM-16, and STM-64 for SDH, and STS-1, STS-3, STS-12, STS-48, and STS-192 for SONET. All SDH and SONET frames have the same frame interval of 125 microseconds. This 8,000-frames-per-second rate is derived

Figure 18.5
Building STM-n and
STS-n frames

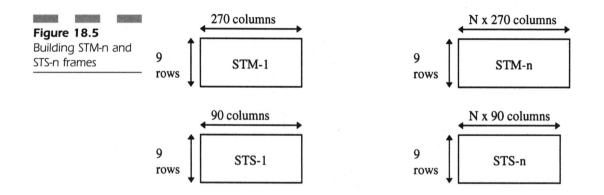

from the basic Nyquist voice sampling rate of 8,000 Hz. This is the sampling frequency for a voice frequency signal with a bandwidth of 4 kHz.

With regard to SDH, one STM-N has three times the capacity of STS-n. For example, an STM-1 has 270 columns. This is three times as many as an STS-1 signal; therefore, the relationship between the rate of an STS-n signal for SONET and the rate of an STM-N signal for SDH can be simply expressed as follows:

$$\text{Rate of an STM-N} = N \times (\text{Rate of STS-3})$$

where the available values for N are 1, 4, 16, or 64. For instance, the line rate of an STM-1 is 155.52 Mbps, which is just 3×51.84 Mbps for an STS-1. An STM-4 has a line rate of 622.08 Mbps, which is just 4×155.52 Mbps for an STS-3, and so forth.

SDH Transport and Path Overhead Bytes

A large part of understanding SONET involves understanding the purpose and use of the overhead bytes. The same is true of SDH. In order to understand the STM-1 SOH, it is best to recall the SONET STS-3c overhead because an STM-1 frame is equivalent to an STS-3c frame. Both have the first 9 columns of the frame reserved for Transport Overhead (TOH). The Path Overhead (POH) "floats" within the frame itself and consists of 9 bytes. The STS-3c frame with all overhead bytes is shown in Figure 18.6. The POH appears as a column for the sake of simplicity.

Figure 18.6

STS-3c frame with overhead

9 columns										POH	
A1	A1	A1	A2	A2	A2	J0	R	R		J1	
B1	R	R	E1	R	R	F1	R	R		B3	
D1	R	R	D2	R	R	D3	R	R		C2	
H1	H1	H1	H2	H2	H2	H3	H3	H3		G1	
B2	B2	B2	K1	R	R	K2	R	R		F2	
D4	R	R	D5	R	R	D6	R	R		H4	
D7	R	R	D8	R	R	D9	R	R		F3	
D10	R	R	D11	R	R	D12	R	R		K3	
S1	Z1	Z1	Z2	Z2	M1	E2	R	R		N1	

9 rows

For a SONET STS-3c frame, the first row contains 3 A1 bytes, 3 A2 bytes, 1 J0 byte and 2 R (i.e., reserved for future use) bytes. The second row contains 1 B1 byte, 2 R bytes, 1 E1 byte, 2 R bytes, 1 F1 byte, and 2 R bytes. Rows 3, 6, 7, or 8 have 1 DCC byte (D1, D4, ..., or D10) followed by 2 R bytes, another DCC byte followed by 2 more R bytes, and a third DCC byte followed by 2 more R bytes. Row 4 has 3 H1, 3 H2, and 3 H3 bytes. Row 5 has 3 B2 bytes (known as BIP-24 bytes), 1 K1 byte, 2 R bytes, 1 K2 byte, and finally 2 more R bytes. The last row, Row 9, has 3 Z1 bytes, 3 Z2 bytes, 1 E2 byte, and 2 R bytes. The first Z1 has been revised and assigned as the Synchronization Status Message (SSM, or S1) byte. Also, the third Z2 has been assigned (from its original growth byte function) as the "line" Far End Block Error (FEBE, or M1) byte for SONET or as the "multiplex section FEBE octet" for SDH.

Besides the 9 columns of TOH bytes in a SONET STS-3c frame, there are 9 bytes or 1 column within the STS-3c SPE that are assigned as the STS-3c POH bytes. They are J1, B3, C2, G1, F2, H4, F3, K3 and N1, and all the details of their functions have been discussed previously. Recall that F3 is now assigned as Path User Channel (PUC), but was previously a growth byte (Z3). N1 is now assigned as Tandem Connection Maintenance (TCM) byte in SONET or the Network Operator Byte (NOB) in SDH. (This was a growth byte (Z5) also.)

The N1 path overhead byte is used to determine error source when a SONET/SDH link passes from one service provider to another. There might be only two service providers involved, one for the local SONET/SDH access and a second for the SONET/SDH backbone connection, but there could be even more. Usually, SONET/SDH path hand-offs occur at a service provider facility known as a *tandem*. The tandem is just a demarcation point between service provider portions of the overall link. Naturally, there is only one path between endpoints, and the end users see the overall errors on the path, not whether the errors have crept in from one service provider or another.

One of the problems is that the service provider in the middle has no access to the ends of the path and so cannot "see" the path overhead bytes directly. This makes troubleshooting, problem isolation, and even simple performance monitoring that much more difficult for the service provider.

So both SONET and SDH allow for *tandem connection monitoring* (TCM) using the N1 byte. There is also an N2 byte for monitoring tandem connections at the virtual tributary or virtual container level. All SONET/SDH NEs can reset and examine the B3 BIP errors that occur on the path segments to help determine just which service provider's equipment is contributing the errors on the overall path that the end-user equipment sees.

Compare the SONET STS-3c frame in the previous figure with the STM-1 frame shown in Figure 18.7. It can be seen that an STS-3c frame is the equivalent of, and virtually identical to, an STM-1 frame. The apparent differences are easily explained be the more international scope and intention of SDH compared to SONET.

Just as in SONET, an SDH STM-1 frame has two parts of STM-1 TOH. However, these are now the RSOH, which is located at rows 1 to 3 and columns 1 to 9, and the MSOH, which is located at rows 5 to 9 and columns 1 to 9, as shown in Figure 18.7. This seems to leave out row 4. Row 4, in the STM-1 frame, consists of the AU-n pointers from column 1 to 9. More precisely, these are either AU-4 or AU-3 pointers. Theoretically, these pointer octets can be considered to be part of the MSOH and, in fact, good reasons exist for doing so. Most importantly, the SONET pointer bytes are part of the Line Overhead (rows 4 through 9).

Just as in SONET, these pointers and the transport SOH have fixed positions within a 125-microsecond frame. In other words, the pointer octets have a fixed phase with respect to the STM-1 frame. Conversely, the VC-3 or VC-4 POH may occupy any column from column 10 to 270 of the STM-1 frame. The POH is "floating" with respect to the whole STM-1 frame; therefore, the VC-3 or VC-4 POH has a variable phase with respect to the STM-1 frame (actually, the A1 and A2 framing octets). The whole concept of "floating" POH is the same as in SONET. Instead of the Synchronous Payload Envelope (SPE) of SONET, SDH had either a

Figure 18.7

STM-1 frame with overhead

										VC-3/VC-4 POH	
	|A1	|A1	|A1	|A2	|A2	|A2	|J0	|*	|*	|J1	
	|B1	|**	|**	|E1	|**	|R	|F1	|*	|*	|B3	
9	|D1	|**	|**	|D2	|**	|R	|D3	|R	|R	|C2	
	|G1										
	AU-n Pointer									|G1	
r	|B2	|B2	|B2	|K1	|R	|R	|K2	|R	|R	|F2	
o	|D4	|R	|R	|D5	|R	|R	|D6	|R	|R	|H4	
w	|D7	|R	|R	|D8	|R	|R	|D9	|R	|R	|F3	
s	|D10	|R	|R	|D11	|R	|R	|D12	|R	|R	|K3	
	|S1	|Z1	|Z1	|Z2	|Z2	|M1	|E2	|*	|*	|N1	

9 columns — Regenerator SOH / Multiplex SOH

* = Reserved for National Use

** = Media Dependent Octet

VC-3 or VC-4 payload. Naturally, when the format of the payload is a VC-3, the pointers are AU-3 pointers, and when the format of the payload is a VC-4, the pointers are AU-4 pointers. A more detailed explanation of the differences is not needed here.

Because they have already been detailed for SONET, the names and functions of each TOH octet and each VC-3/VC-4 POH octets will be examined here by comparing them with those of an STS 3c SONET frame. By cross-referencing the two previous figures, it is possible to observe the following similarities and differences between the SDH and SONET overhead standards.

SDH and SONET Overheads Compared

The first row of the RSOH of an STS-3c and an STM-1 are virtually identical (A1, A1, A1, A2, A2, A2, J0, ·), with the exception of the last 2 bytes or octets (indicated by ""). For SONET, these 2 bytes are not yet assigned and designated as "R" bytes, while for SDH they are assigned for national use. This means that each country gets to assign their meaning. In the United States, they are "R" bytes in SONET, which is perfectly valid usage.

In SDH, the second row has three octets (second, third, and fifth columns) assigned for media dependent usage (indicated by ""). This means that the usage varies depending on whether the STM-1 is carried on fiber optic links (normal) or on electrical media, such as coaxial cable (for short runs between equipment). The last two octets of this row are also assigned for national use. Conversely, in SONET standards, these bytes are unassigned or are reserved for future use.

In the third row, the only difference is the two media-dependent octets in columns 2 and 3 for SDH. These 2 bytes are reserved in SONET. All of the other octets have the exact same functions for both SDH and SONET.

The pointer row in both SONET and SDH is exactly identical, but this is difficult to see. The only difference is the application of the pointers. For SONET, an STS-3 signal can be used to carry three DS-3 signals. For this SONET application, these 9 pointer bytes would be split up into 3 groups. Each group would have a set of H1, H2, and H3 pointers and each set is used to locate the payload of one DS3 signal. This arrangement is also used in SDH. That is, an STM-1 can also carry 3 DS-3 signals. This helps in international situations. In SONET, an STS-3c can also be used to transport a B-ISDN signal, such as a stream of ATM cells, or even an FDDI 100 Mbps signal. In such cases, only one set of H1, H2 pointers is required. In the same fashion with SDH, an STM-1 can be used for the same purpose and so has the same arrangement of pointers. Because the original purpose of STM-1 is to carry an E-4 signal at 139.264 Mbps, only one set of H1, H2, H3 octets is required.

Rows 5 through 8 are identical for both SDH and SONET. However, in SDH the last row has two octets assigned for national use. In SONET STS-3c, these 2 bytes are reserved.

As far as POH is concerned, there is only 1 byte or octet that is different for SDH and SONET. N1 is assigned as Network Operator Byte (NOB) for SDH. This is to be used for network operators to communicate with SDH equipment. In SONET, this is the Tandem Connection Maintenance (TCM) byte.

STM-4, STM-16, and STM-64 Overhead

Enough has been said to this point about the locations and names of the TOH and VC-3/VC-4 POH bytes of an STM-1 frame, but what about the STM-4, STM-16, and STM-64? These are briefly discussed in this section.

First, consider an STM-4 frame. Everything is basically four times bigger. There are 12 (four times as many as in an STM-1) A1 octets, followed by 12 A2 octets, 4 J0 octets, and 8 octets for national use. This makes up a 36-octet row, which is just four times as many as in the 9-octet rows in an STM-1. Most of the byte assignments are similar to those of an STM-1 signal. A "condensed" picture of the STM-4 overhead is shown in Figure 18.8.

Figure 18.8
STM-4 Overhead Octets

Column:

1	2		12	13	14		24	25	26	27	28	29		36
A1	A1	-----	A1	A2	A2	-----	A2	J0	J0	J0	J0	*	-----	*
B1			E1					F1	*	*	*	*	-----	*
D1			D2					D3						
AU-n Pointers														
B2	B2	-----	B2	K1		-----		K2						
D4			D5					D6						
D7			D8					D9						
D10			D11					D12						
S1	S1	-----	Z1	Z2	Z2	-----	M1	E2	*	*	*	*	-----	*

* = Reserved for National Use

----- = Columns of same octet types

It should be noted that the "blank" octets are all reserved for future use. National usage octets are indicated by "". Finally, the "Al-----A1" column notation indicates that there is a string of Al octets (in this case) between the two locations indicated.

Although not shown in a separate figure, the STM-16 overhead is easy to describe. There are now 144 columns in an overhead row, which is 16 times the size of the STM-1 overhead. In the first row, for example, there are 48 Al octets followed by 48 A2 octets, 16 J0 octets, and finally 8 octets for national use bytes. Most of the other octet assignments are similar to those of the STM-1 signal and need not be detailed here.

The overhead for the STM-64 frame may be derived in exactly the same fashion. Overhead assignments are also essentially the same.

VC-3/VC-4 Path Overhead

There are some things of interest in the use of POH octets in SDH that should be briefly discussed. Only the differences between SDH and SONET usage are emphasized here.

As in SONET, there are 9 POH octets in SDH. These are path trace JI, path error monitoring B3, path signal label C2, path status G1, path user channel F2, multi-frame indication H4, growth byte Z3 (which has now been assigned as Path User Channel: PUC; the new name is F3), growth byte Z4 (the new name is K3, standard on this byte is under study by the committee), and growth byte Z5 (which has been assigned as Network Operator Byte: NOB; the new name is N1). These octets can be classified into four functional groups:

1. End-to-end communication bytes: II, B3, C2 and G1 (independent of payload).
2. Payload-specific bytes: H4, F2, and F3.
3. Future standard use byte: K3.
4. Path NOB: N1.

First, consider the path trace octet, J1. The purpose of this octet is to transmit repetitively a high-order path access point identifier. This is so an SDH-path-receiving terminal can verify its continued connection to the intended transmitter. This path access identifier may be in one of two formats. First, it can be a 64-octet, freely formatted string. In this respect, it is similar to the SONET Common Language Location Identifier (CLLI) code, which is allowable. Second, it may be a 16-octet E.164 formatted identifier within the national network. This is shown in Figure 18.9.

There is a good reason to be aware of this difference with SONET J1 usage. At an international boundary, only the E.164 format can be used. When the 16-octet format is to be transferred in the SONET-type, 64-byte field, it is to be repeated four times. Note the presence of the frame start marker and the CRC-7 calculation for error control.

The B3 octet can be dealt with quickly. This octet is allocated in each VC-3, VC-4, or VC-4-Xc—a concatenated signal, not discussed further—for a path-error performance-monitoring function. It uses even parity, as B1 and B2 do. B3 is calculated before scrambling over all bits of the previous VC-3, VC-4, or VC4-Xc payload. The B3 is placed, before scrambling, in the B3 octet of the current VC-3, VC-4, or VC-4-Xc.

The third octet of the SDH VC-3/VC-4 POH is the signal label, C2. This octet in SDH standards has the same function as in SONET standards. The codings allow a receiver to determine the format of the payload content (e.g., ATM cells, E-3).

Next, the path status octet, G1 is used to convey back to the originating equipment the path status and the path performance (in terms of errors). This allows the status and performance of the complete duplex path to be monitored at either end or at any point along the path. Use of the bits is essentially the same as in SONET.

The path user byte, F2, and the growth byte are designed for user communication purposes between path elements and are payload-dependent. The multi-frame indication octet, H4, is used to provide a generalized position indicator for payloads and can be payload-specific. The old (it no longer exists) second growth byte of the POH, Z4, is now known by the new name of K3, and is assigned for future use. For now,

Figure 18.9
Coding of the SDH J1 path trace octet

Bits:		Octet:
1CCC	CCCC	Frame Start Indicator
0XXX	XXXX	Octet 1 of Sequence
:	:	:
:	:	:
0XXX	XXXX	Octet 16 of Sequence

0XXX XXXX = IA5 Character of E.164 string

CCC CCCC = Result of CRC-7 over previous frame

the receiver is required to ignore the value contained in this octet. The old third growth byte, Z5, is now known by the new name of N1 and has been partially used for specific SDH network management purposes. Details are not necessary for the purposes of this survey.

STM-1 Frame and Signal Composition

The last major topic that needs to be discussed with regard to SDH is the issue of what is inside the STM-N frame. Figure 18.10 shows a comprehensive, but very complex, graphical representation of STM-N signal

Figure 18.10
Makeup of the
STM-N signal

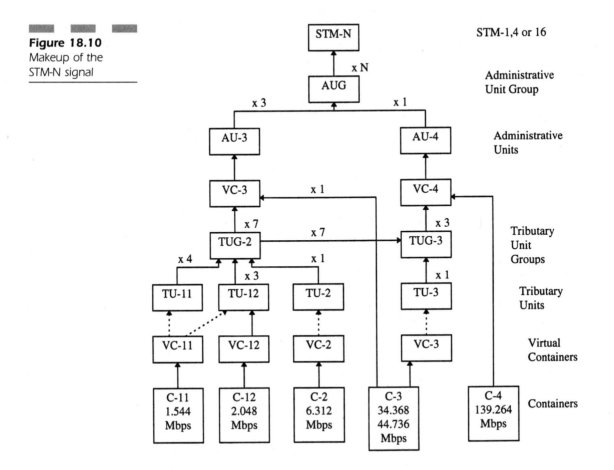

composition. The details of all of the mappings and multiplexing from low-speed containers to SDH signals at the STM-N are not discussed in detail. However, it is a good idea to briefly describe the figure.

It is easiest to start at the bottom. Then the various "paths" up the figure are explored in a little more detail in the rest of this tour through SDH.

What's in the Containers?

The previous figure started with the containers themselves. Currently, in the SDH standards, there are six sizes of containers. These are C-11, C-12, C-2, two C-3s, and C4, as shown earlier. The six containers are classified into two groups. The higher-order containers are C-4 and the two C-3s; while the lower-order containers are C-2, C-12 and C-11. In some documentation, C-11 and C-12 are simply referred to together as C-1.

Either one DS-1 signal of 1.544 Mbps or one ISDN H11 signal can be placed in one C-11. The 2.048 Mbps E-1 (CEPT-1) signal or an ISDN H12 signal can be placed in one C-12. One DS-2 signal of 6.312 Mbps is mapped into one C-2, while a DS-3 signal of 44.736 Mbps or one ISDN H32 is carried inside one C-3. Note that the container for a 34.368 Mbps E-3 (CEPT-3) signal or an ISDN H31 signal also is inside a C-3. However, the mappings of these two C-3s are different. This is why the C-3 container in Figure 18.10 has two arrows. An E-3 (CEPT-3) takes the right-hand path (i.e., VC-3, TU-3, TUG-3, VC-4, AU-4, AUG, and then STM-1). Conversely, a 44.736 Mbps DS-3 signal takes the left-hand path (i.e., VC-3, AU-3, AUG, and then STM-1). The highest rate ITU-T PDH signal of 139.264 Mbps, an E-4 (or CEPT-4), or an ISDN H4 signal is placed in one C4.

It should be mentioned that the DS-1C, which is carried by the T-carrier digital system, is not considered an SDH container at present. This is also true for the 8.448 Mbps E-2 (CEPT-2) signal. Neither will the 32.064 Mbps signal nor the 95.728 Mbps signal used in Japan be mapped into any of the currently assigned containers.

"Paths" Through SDH

It is easier to start at the bottom right of Figure 18.10. Here, a C-4 (read as Container-level 4) for transporting a 139.264 Mbps signal, such as a E-4 (CEPT-4) signal is directly mapped, with all needed VC-4 POH bytes, into a VC-4 (read as Virtual Container-level 4). The simple arrow indicates there is no multiplexing involved, just the proper overhead added to go from a C-4 to VC-4. Thus, a VC-4 is basically a C-4 with the relevant overhead. Next, the AU-4 pointers—these are AU-4 pointers for a VC-4, naturally—are added to the VC-4 to form an AU-4 (read as Administrative Unit-level 4). This AU-4 is also the same as an AUG (AU Group), but this is not always true. Finally, the MSOH octets and the regenerator SOH octets are added to this AUG to form an STM-1 frame. A number N of STM-1 frames may now be multiplexed into an STM-N frame. All of the STM-1s must be frame-aligned before multiplexing.

The C-3 is a little more complex. A C-3 (Container-level 3) can be used for 34.368 Mbps signal mapping and is directly mapped, with the proper VC-3 POH bytes, into a VC-3. This is indicated by the rightmost arrow from the C-3 box in Figure 18.10. Three octets for TU-3 pointers are now added to the VC-3 to form a TU-3 (Tributary Unit-level 3). This TU-3 is the same as the TUG-3 (TU Group-level 3) in this case, because no multiplexing of TU-3s into a TUG-3 is involved with C-3s. This distinction is discussed further a little later. Next, 3 TUG-3s are multiplexed and mapped (indicated by the "x3" next to the arrow in the figure) into a VC-4 which, just as in the C-4 case above, is mapped into 1 AU-4, 1 AUG, and 1 STM-1.

However, a C-3 is more complicated than the C-4. A C-3 can also be used for 44.736 Mbps signal mapping (a DS-3). The DS-3 is directly mapped into a VC-3, as before, but this time following the leftmost arrow out of the C-3 box in the figure. This time, it is AU-3 pointers that are added to this VC-3, not AU-4 as before. Generally, AU-3s are used for T-carrier signals and AU-4s are used for ITU-T E-carrier or CEPT signals. This now forms an AU-3 signal frame. In this case, 3 AU-3s are multiplexed and mapped into an AUG, which is finally mapped into an STM-1. Note that in this case the use of TUs and TUGs is only for C-3s that follow the ITU-T digital (E-carrier) hierarchy.

Continuing the excursion through the figure, the next box is the C-2. A C-2 (Container-level 2) is intended for 6.312 Mbps signal mapping (a DS-2). The DS-2 is directly mapped, along with the VC-2 POH octets, into a VC-2. The TU-2 pointers are added to this VC-2 to form a TU-2 (Tributary Unit-level 2). Because there is no further multiplexing, a TU-2 is the same as the TUG-2 (TU Group-level 2); however, 7 TUG-2s are multiplexed and mapped into a VC-3. Then the AU-3 pointers are added to this VC-3 to form an AU-3 signal. Finally, 3 AU-3 signals are multiplexed and mapped into an AUG which, exactly as before, is mapped into an STM-1.

Next in the figure, a C-12 (Container-level 1, type 2) is intended for 2.048 Mbps signal mapping (E-1). The E-1 is directly mapped, with VC-12

POH bytes, into a VC-12. The TU-12 pointers are added to this VC-12 signal to form a TU-12. In this case, however, 3 TU-12s are multiplexed and mapped into a TUG-2, but a TUG-2 with E-1s inside cannot form a VC-3 or AU-3. These are basically for T-carrier. Thus, in the case of C-12, the arrow must next point to the TUG-3 box. Seven TUG-2s are now essentially the TUG-3. From TUG-3 to an STM-1 for this C-12, the process is exactly the same as in the C-2 case.

The C-11 (Container-level 1, type 1) is intended for 1.544 Mbps signal mapping (a DS-1). The DS-1 is directly mapped, with VC-11 POH octets, into a VC-11. The TU-11 pointers are added to this VC-11 to form a TU-11. Next, 4 TU-11s are multiplexed and mapped into a TUG-2. Alternatively, the VC-11 may be mapped into a TU-12 and 3 TU-12s are mapped into the TUG-2. In either case, the path from TUG-2 to an STM-1 for the C-12 is exactly the same as in the C-2 case.

The figure is quite complex, but the important relationships among all the formats and signals that have been discussed can be defined in a more compact form. These relationships are more fully explored in subsequent sections.

The path from a generalized C-a to VC-a can be summarized as:

$$\text{C-a} + \text{VC-a POH} = \text{VC-a; where } a = 11, 12, 2, 3 \text{ or } 4$$

The path from VC-b to TU-b can be summarized as:

$$\text{VC-b} + \text{TU-b pointers} = \text{TU-b; where } b = 11, 12, 2, \text{ or } 3$$

And the path from VC-c to AU-c can be summarized as:

$$\text{VC-c} + \text{AU-c pointers} = \text{AU-c; where } c = 3 \text{ or } 4$$

Finally, the general path from an AUG to STM-1 is:

$$\text{AUG} + \text{RSOH} + \text{MSOH} = \text{STM-1}$$

It is important to note that in the four relationships given above, it does not follow that the left-hand signal frame is identical to the right-hand signal frame. For instance, all that a relationship like the third one implies is that in order to form an AU-c signal, the AU-c pointer octets need to be added to the VC-c signal.

C-4, VC-4, AU-4, AUG, and STM-1 "Path"

This brief tour of SDH concludes with a little more detail about the "paths" in SDH from container to STM-1 frame. A C-4 occupies 9 rows by 260 columns in a 125-microsecond frame interval. Thus, a C-4 has a rate of

149.76 Mbps (which is 9×260 octets $\times$ 64 kbps). By adding one column of VC-4 POH to a C-4, the result is a VC-4. This a 9-by-261 frame structure.

A 9 octet AU-4 pointer is then added to a VC-4 to form an AU-4. Next, 27 octets of RSOH and 54 octets of MSOH are then added to the AUG (which is the same as the AU-4) signal to form the SDH STM-1 signal. This whole process is shown in Figure 18.11.

In this case, the AUG is the same as the AU-4. Then what is the purpose of even inventing an AUG? This is because an AUG (Administration Unit Group) can be used to carry either 1 AU-4 or 3 AU-3s. In typically T-carrier-based, AU-3 applications, a group of (actually 3) AU-3 signals will form the AUG. The case of 3 AU-3 applications is discussed later in this section. For now, it is enough to point out that when an AU-4 is involved, the AU-4 and the AUG are one and the same.

C-3, VC-3, TU-3, TUG-3, VC-4, AU-4, AUG, and STM-1 "Path"

It has already been mentioned that either a 34.368 Mbps E-3 (CEPT-3) signal or a 44.736 Mbps DS-3 signal can be mapped into a C-3 container. First,

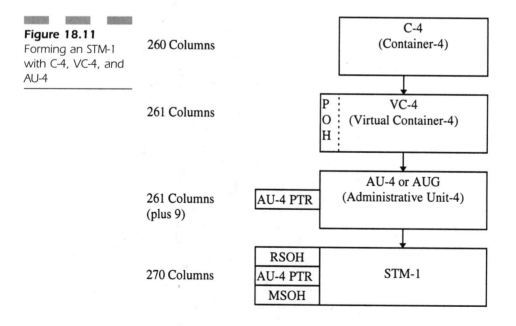

Figure 18.11
Forming an STM-1 with C-4, VC-4, and AU-4

the relationship between a C-3 container, which carries a 34.368 Mbps E-3 (CEPT-3) signal or an ISDN H31 signal, and an STM-1 is discussed.

A C-3 occupies 9 rows by 84 columns in a 125-microsecond frame interval. Thus, a C-3 has a rate of 48.384 Mbps (which is $9 \times 84 \times 64$ Kbps). Nine octets of VC-3 POH is added to C-3 to form a VC-3; therefore, a VC-3 has an organizational structure of 9 rows by 85 columns and a 48.96 Mbps rate. In order to form the next level of signal, the TU-3, 3 octets of TU-3 pointer must be inserted into the eighty-sixth column. This column, 9 octets long, has 6 octets of fixed stuffing octets occupying the remaining locations. These are shown in the bottom left (and other locations) in Figure 18.12, which illustrates the process.

This completes the 9-by-86 TUG-3. Note that a TU-3 signal has 85 columns plus only the 3-octet TU-3 pointers, while the full TUG-3 has 86 full columns. So a TU-3 with a rate of 49.152 Mbps is not quite identical to a TUG-3, which has a rate of 49.536 Mbps. Why is a separate name needed? Well, because one TUG-3 can be formed from one TU-3 (as in this case) or from a whole *group* of TUG-2 signals. More precisely, a TUG-3 contains 7 TUG-2 signals. Therefore, the TUG refers to the *group* signal.

Three TUG-3 signals are then multiplexed and mapped into an VC-4 which, besides these 3 TUG-3s, contains 2 fixed-stuffing columns, and one column of VC-4 POH. Next, 9 AU-4 pointer octets are added to the 1 VC-4 signal to form an AUG signal. Then 27 octets of RSOH and 54

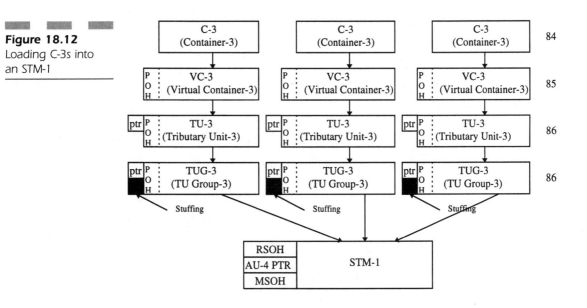

Figure 18.12
Loading C-3s into an STM-1

octets of MSOH will be added to an AUG signal to form the SDH STM-1 signal This is also shown in Figure 18.12.

C-3, VC-3, AU-3, AUG, and STM-1 "Path"

There is another path from C-3 to STM-1 as well. A C-3 may also carry a 44.736 Mbps DS-3 signal. The C-3 and VC-3 for the DS-3 has the same structure as the previous case. After forming virtual container VC-3, which has 85 columns, 2 columns of fixed stuffing and 3 octets of pointer are added to construct an AU-3. The major difference between a VC-3 and an AU-3 is the 3 added pointer octets.

Next, after having constructed 3 AU-3s, 3 octets of AU-3 pointer from each of these 3 AU-3s are multiplexed into the 9 AU-4 pointer locations. The payload of these 3 AU-3s also are multiplexed into the AUG payload location. Finally, just as all the previous cases, RSOH and MSOH are then added to the AUG to form an STM-1.

C-2, VC-2, TU-2, TUG-2, VC-3, AU-3, AUG, and STM-1 "Path"

A C-2, carrying a 6.312 Mbps DS-2 or equivalent signal, has a payload capacity of 106 octets in a 125-microsecond interval. One octet of VC-2 POH (either V5, J2, N2, or K4) is added to form virtual container VC-2. Another octet, acting as the TU-2 pointer (either V1, V2, V3, or V4), is then added to form the 108 octet capacity of the TU-2. It is worth noting that there are 4 octets in total, 1 per each 125-microsecond interval, serving as the TU-2 pointer. For this application, the TU-2 is the same as TU Group-2 (TUG-2). In fact, a TUG-2 can be equivalent to 1 TU-2, 3 TU-12s, or 4 TU-11. More details are given at the end of this SDH investigation. Seven TUG-2s will then be multiplexed into the payload location of the VC-3.

The VC-3 has a structure of 9 rows by 85 columns. One column is VC-3 POH, and 84 columns are evenly distributed among the 7 TUG-2s. That is, because 84/7 = 12, the capacity for each TU-2 is 12 columns. In other words, a TU-2 requires 108 octets of capacity (12 columns × 9 octets/column = 108 octet). All of these are discussed later. For now, it is enough to note the similarity between a TUG-2 and the SONET VT Groups (7 in an STS-1). From VC-3 to AU-3, then to AUG and STM-I, the process is the same as in the previous case.

C-12, VC-12, TU-12, TUG-2, TUG-3, VC-4, AU-4, AUG, and STM-1 "Path"

The C-12 to STM-1 path is quite complex. A C-12, carrying a 2.048 Mbps E-1 (CEPT-1) or equivalent signal, such as an ISDN H12 signal, has a payload capacity of 34 octets in a 125-microsecond interval. One octet of VC-12 POH (either V5, J2, N2, or K4) is added to C-12 to form a VC-12. Another octet (either V1, V2, V3, or V4), acting as the TU-12 pointer (sometimes known as TU-1 pointer for simplicity), is then added to the TU-12 signal to form the TU-12 itself with a 36-octet capacity. As in the case of C-2, it should be mentioned that there are 4 octets in total, 1 per each 125-microsecond interval, serving as the TU-12 pointer.

It is worth noting that this 36-octet unit of the TU-12 signal is equivalent to 4 columns in an STM-1 frame. Three TU-12s are multiplexed into a TUG-2, which now has a capacity of 12 columns. Seven TUG-2s will then be multiplexed into the payload location of a TUG-3. This seems odd at first, but note that the TUG-3 has a structure of 9 rows by 86 columns. There are 84 columns of the 7 evenly distributed TUG-2s (12 columns each); 1 column of something called NPI (Null Pointer Indication; not discussed further) plus fixed-stuffing octets, 1 column of fixed stuffing (which may be used for payload transport if needed in future standards), for a total of 86 columns. Because 84/7 = 12, the capacity for each TUG-2 is 12 columns. Each TUG-2 carries 3 TU-12s, each of which occupy 4 (which is 12/3) columns. Thus, a TU-12 requires 36 octets of capacity (4 columns × 9 octets/column = 36 octets). All of these TUs are discussed later.

From TUG-3, to VC-4, to AU-4, then to AUG and STM-1, the process is the same as the previous case.

C-11, VC-11, TU-11, TUG-2, VC-3, AU-3, AUG, and STM-1 "Path"

The last case that needs to be discussed is from C-11 to STM-1. A C-11, carrying a 1.544 Mbps DS-1 or equivalent signal, has a payload capacity of 25 bytes in a 125-microsecond interval. One octet of VC-11 POH (either V5, J2, N2, or K4) is added to form VC-11. Another octet (either V1, V2, V3, or V4), acting as the TU-1 pointer, is then added to VC-11 to form a TU-11 signal with a 27-octet capacity. Note that there are 4 octets in total, 1 per each 125-microsecond interval, serving as the TU-11 pointer. This 27-octet capacity is equivalent to 3 columns in an STM-1 frame.

Next, 4 TU-11s are multiplexed into a TUG-2, which has a capacity of 12 columns. Seven TUG-2s will then be multiplexed into the payload location of VC-3. Note that the VC-3 has a structure of 9 rows by 85 columns. One column is VC-3 POH, and 84 columns are evenly distributed among the 7 TUG-2s. Because 84/7 = 12, the capacity for each TUG-2 is 12 columns. Each TUG-2 carries 4 TU-11s, each occupying 3 (which is 12/4) columns. Therefore, a TU-11 requires 27 bytes of capacity (3 columns × 9 octets/column = 27 octets).

The path from VC-3 to AU-3, then to AUG and STM-1 is the same as the previous cases.

Tributary Unit

The previous section dealt with the container to STM-1 possibilities. But more details on the several types of intermediate signals formed that are important for the applications should be examined. For example, to map a C-11 into an STM-1, A TU-11 and TU-2 must first be formed. Therefore, a thorough understanding of these TUs would be in order. The counterpart of SDH TU-n in SONET is the VT.m. That is, a VT-1.5, VT-2, and VT-6 are identical to the SDH TU-11, TU-12, and TU-2, respectively. The purpose of the TU is to transport and switch a container, which can be defined as a sub-STM-1 signal of various forms.

With the TUG-2, VC-3 and AU-3, either with 84 TU-11s or 21 TU-2s, can be mapped and multiplexed into one AUG. These applications are designed for North American digital systems using T-carrier.

With the TUG-2, TUG-3, VC-4, and AU-4, 63 TU-12s can be mapped and multiplexed into one AUG. Also, 3 TU-3s can be multiplexed and mapped into an AUG signal through a TUG-3, VC-4, and AU-4 sequence. This set of applications is designed for ITU-T digital systems.

The last SDH topics to be discussed are the purpose and capacity of each TU-n: the relationship between a TU-n signal and a TUG-2 or a TUG-3 signal; and the relationship between a TU-n signal and an STM-1 signal.

Purpose and Capacity of TU

The purpose of each individual TU-n, except TU-3, can be seen in Table 18.4.

The TU-1 is designed to transport a DS-1 with a signal rate of 1.544 Mbps. Each column of an STM-N (N = 1, 4 or 16) or AUG has a rate of

TABLE 18.4

TU types and their uses

TU Type	Columns	Octets/frame	Octets/4 frames	Rate	Use
TU-2	12	108	432	6.912 Mbps	DS-2 (6.312 Mbps)
TU-12	4	36	144	2.304 Mbps	E-1 (2.048 Mbps)
TU-11	3	27	108	1.728 Mbps	DS-1 (1.544 Mbps)

0.576 Mbps (64 kbps/octet x 9 octets/column). Therefore, it requires 3 columns of AUG or 27 octets per frame (or 125 microseconds) for a TU-11 to accommodate a DS-1.

The TU-12 is designed to transport an E-1 (CEPT-1) that has a rate of 2.048 Mbps. Therefore, it takes 4 columns (or 36 octets per frame) of AUG for a TU-12 to accommodate an E-1.

The TU-2 is designed to transport a DS-2 signal with a rate of 6.312 Mbps. It takes 12 columns (108 octets per frame) of AUG for a TU-2 to accommodate a DS-2.

With all TUs, 4 125-microsecond frame intervals are defined as a 500-microsecond *multiframe* or *superframe*. That means that in 1 super-frame there are 4 125-microsecond frames. Each of these 4 frames is started with 1 overhead octet (either V1, V2, V3, or V4). A superframe starts with the V1 byte. The multiframe indicator octet, H4, of the VC-n (where n = 3 or 4) POH is used to indicate the *phase* of this 500-microsecond multiframe. Thus, if H4 = (XXXXXX00), where "X" is a "don't care" bit, this 125-microsecond frame is the first frame of a 500-microsecond multiframe. The first octet in this 500 microsecond frame, therefore, is V1, the first TU-n pointer octet. In the same way, if H4 = (XXXXXX01), this 125-microsecond frame is the second frame of a 500-microsecond frame. The first octet in this 125-microsecond frame is V2, the second TU-n pointer octet. The same is true for H4 = (XXXXXX10) and H4 = (XXXXXX11) and works the same as in SONET.

The multiframe capacity of each TU type also is of interest. For example, the TU-2 has a capacity of 12 columns or 108 octets for every 125-microsecond frame. Because there are 4 frames per superframe, the TU-2 can be said to have a capacity of 432 (which is 4 × 108) octets per 500-microsecond multiframe. Similarly, the TU-12 has a capacity of 144 (which is 4×36) octets per multiframe and the TU-11 has a capacity of 108 (which is 4×27) octets per superframe.

Relationship Between TU-n (n = 11, 12, or 2) and TUG-2

In addition to the TU-n (where n = 11, 12, or 2) signals, there is another intermediate signal that is required for the mapping and multiplexing of a C-11, C-12, or C-2 into an STM-1. This is known as the TUG-2 (Tributary Unit Group-level 2).

It has already been shown that a TU-11 requires 3 columns of an AUG frame to accommodate a DS-1 operating at 1.544 Mbps. Likewise, a TU-12 needs 4 columns to accommodate an E-1 signal operating at 2.048 Mbps and a TU-2 needs 12 columns to accommodate a DS-2 signal operating at 6.312 Mbps. This means that all three types (i.e., TU-11, TU-12, or TU-2) can all be carried as groups of 12 columns because the least common multiplier of 3, 4, and 12 is 12.

That means that if an AUG frame (or an STM-1 frame) has 12, 24, 36, or any n $\times$ 12 (where n = any positive integer) columns, this frame can be used to transport any TU-n (where n = 11, 12 or 2). The TUG-2, which is a special 12-column group of an AUG frame, was introduced for this very purpose. Therefore, a TUG-2 signal is equivalent to a VT group signal in SONET.

A TUG-2 has 12 columns and can transport either (1) one TU-2 signal (one DS-2), (2) three TU-12 signals (three E-1s), or (3) four TU-11 signals (four DS-1).

TU and STM-1

Just to emphasize the identity of SONET and SDH, this section closes with a look at an SDH STM-1 carrying DS-1s. The same principle can be extended to TU-12s and TU-2s, but using SDH for DS-1s is still startling to some.

Each STM-1 can carry 84 TU-11s. This is the same as an STS-3 carrying three DS-3s in SONET, of course. Both have 84 DS-1s (3×28 in the case of the STS-3). In the STM-1, the first column after the POH column is allocated for the first TU-11; the second column for the second TU-11, and so on. This is done because SONET/SDH standards must apply byte-interleave multiplexing; therefore, frame alignment is critical. The AU-4 pointer will indicate the starting point of the STM-1 payload in the frame that is started out by the VC-4 path overhead. Once this POH is found, the 3 columns for the first TU-11 are uniquely located.

Just as in SONET, the VC-4 POH has a floating phase with respect to the STM-1 frame, but the phase between the 3 columns of the first TU-11 signal has a fixed-phase relationship with respect to the VC-4 POH.

The overhead octets (or bytes because these are DS-1s) associated with each TU-11 within every 125-microsecond payload frame is either a V1, V2, V3, or V4. Each of these overhead bytes begins the 27-byte frame of the TU-11. The V1 and V2 bytes form a pointer that will point to the starting location of a VC-11 within the TU-11. This starting byte is called V5. The 500-microsecond interval of a multiframe headed by the V5 byte is called a VC-11 for this C-11 container. Thus, 84 DS-1 signals are carried by 84 C-11s, 84 VC-11s, 84 TU-11s, 21 TUG-2s, and 3 VG-3s (AU-3s), all inside 1 STM-1 signal.

This just points out the identity of SDH and SONET, but with all the names changed.

SONET and SDH Compatibility

It is only fitting that this section, and chapter, and book should come to a close with a consideration of SONET and SDH compatibility. As national networks grow and merge into a fiber-based, global Internet and Web for e-commerce, e-business, information, and even entertainment, this issue grows in importance. No equipment vendor wants to end up with a product line limited in marketability, and no service provider wants to end up with a backbone that cannot attach seamlessly to anyone else's. This is not only inconvenient today, it is all but suicidal.

A lot of discussion on the differences between SONET and SDH revolves around the structure of the AU-3 and AU-4 pointers. There are some "extra" bits defined in the SDH AU-3 H1 pointer that need to be ignored by SONET when interfacing SONET and SDH. These are bits 5 and 6, called the "ss" ("s" is for "size") bits, that SDH has used to define the structure of some payloads. Typically in SDH, these bits are set to 10, and all SONET equipment need do is ignore them. Other reports about pointer incompatibilities have been greatly exaggerated.

Of course, overall frame structures are one thing, but payload compatibility is something else. It makes no sense to allow SONET and SDH to connect if the arriving payloads are incomprehensible to the equipment at the far end. The first step is to present the SDH values for payload mappings. These occupy the C2 Path Overhead byte, as usual, but differ from their SONET counterparts. For comparison purposes, the SONET meaning is all given. As can be seen in Table 18.5, only the expected minor differences due to different PDH structures appear, with the exception of the final value.

All of the specific overhead structures and messages (K1/K2 APS, S1 SSM, etc.) that differ between SONET and SDH appear in Appendix B (for SONET) and Appendix C (for SDH).

TABLE 18.5

SDH C2 payload type coding

Code in Hex	SDH C2 Payload Mapping	SONET Equivalent
00	Unequipped or supervisory—unequipped	Unequipped (not used for live information)
01	Equipped—nonspecific	Equipped—nonspecific payload payload
02	TUG-structure (VCs inside)	Virtual Tributaries (VTs) inside ("default")
03	TUs in locked mode	VTs in locked mode (no longer supported)
04	Asynchronous E-3 or DS-3	Asynchronous DS-3 mapping inside C-3
12	Asynchronous E-4 is inside	Asynchronous DS-4NA mapping C-4
13	ATM cell mapping	ATM cell mapping
14	DQDB cell mapping	DQDB cell mapping
15	Asynchronous FDDI mapping	Asynchronous FDDI mapping
16	HDLC-over-SDH mapping (used for IP)	HDLC-over-SONET mapping (used for IP)
CF	"Experimental" value for IP inside PPP	"Experimental" value for IP inside PPP
FE	Q181 test signal	Test signal mapping (see ITU Rec. G.707)
FF	Virtual container alarm	—

TUG-2 = 4 TU-11   VC-3 = STS-1

VC- = TU-

TABLE 18.6

SONET/SDH-compatible payload mappings

Payload	Payload Rate	In STS-1	In STS-Nc	In AU-3	In AU-4
DS-1	1.544 Mbps	**VT1.5**	Not defined	**VC-11** (VC-12*)	**VC-11** (VC-12*)
E-1	2.048 Mbps	**VT2**	Not defined	**VC-12**	VC-12
DS-1C	3.152 Mbps	VT3	Not defined	Not defined	Not defined
DS-2	6.312 Mbps	**VT6**	Not defined	**VC-2**	VC-2
E-3	34.368 Mbps	Not defined	Not defined	VC-3	VC-3
DS-3	44.736 Mbps	**STS-1**	Not defined	**VC-3**	VC-3
FDDI	125 Mbps	Not defined	**STS-3c**	Not defined	**C-4**
DS-4NA	139.264 Mbps	Not defined	**STS-3c**	Not defined	**VC-4**
DQDB	149.760 Mbps	Not defined	**STS-3c**	Not defined	**C-4**
ATM cells	149.760 Mbps	**STS-1**	**STS-3c**	**VC-3**	**VC-4**

* It is possible to carry DS-1 voice channels in a VC-12, as long as the conversion to A-law digital voice (E-1) is done.

It might also be helpful to list the possible payloads of SONET and SDH, with their respective terminologies, and establish which payload mappings are compatible and which are not. In this case, "compatible" means "a payload mapping done in this fashion may pass between SONET and SDH since the structure is identical." These mappings are shown in Table 18.6.

In Table 18.6, the last four columns indicate whether the payload mapping is defined for an STS-1 or STS-Nc (for example, an STS-3c), or their SDH equivalents, AU-3 and AU-4, respectively. The **bold** entries indicate SONET/SDH compatibility. The easiest way to use the table is to interpret a line, such as line 2, as saying "an E-1 can be carried between SONET/SDH as a VT2 to SONET and a VC-12 to SDH." Line 4 would be interpreted as "A DS-2 can be carried between SONET/SDH as a VT6 to SONET and a VC-2 to SDH, but only in an AU-3 based STM-1."

APPENDIX A

ACRONYM LIST

AAL5	ATM Adaptation Layer type 5
ACSE	Association Control Service Element
A/D	Analog to Digital (conversion)
ADM	Add-Drop Multiplexer
AIS	Alarm Indication Signal
AIS-L	Line-level Alarm Indication Signal
AIS-V	Virtual-tributary-level Alarm Indication Signal
AIU	Access Interface Unit
AM	Administration Module
ANSI	American National Standards Institute
APD	Avalanche PhotoMode (or PhotoDetector)
APS	Automatic Protection Switch
ATM	Asynchronous Transfer Mode
AU	Administrative Unit
AU4-Nc	N AU4 signals concatenated
AU-n	Administrative Unitlevel n (n = 3 or 4)
B-DCS	Broadband Digital Cross-connect System
BER	Bit Error Rate (Ratio)
BIM	Byte Interleaved Multiplexer
BIP	Bit Interleaved Parity (check)
B-ISDN	Broadband ISDN (also, B-ISDN)
BITS	Building Integrated Timing Supply
BLSR	Bidirectional Line Switch Ring
BRI	Basic Rate Interface
C-n	Container-level n (n= 11, 12, 2, 3, or 4)
CBR	Constant Bit Rate
CC	Composited Clock
CCC	Clear Channel Capability
CCITT	International Telegraph and Telephone Consultative Committee
CEPT-n	Conference of European Posts and Telecommunications level-n
CEPT-1	see E1

CEPT-2	see E2
CEPT-3	see E3
CEPT-4	see E4
CLLI	Common Language Location Identifier
CLMP	Connectionless Network Layer Protocol
CMISE	Common Management Information Service Element
CO	Central Office
CORBA	Common Object Request Broker Architecture
COT	Central Office Terminal
CPE	Customer Premises Equipment
CRC	Cyclic Redundancy Check
CU	Channel Unit
CV	Coding Violations
CV-P	Path-level Coding Violaton
CV-S	Section level Coding Violaton
CV-V	Virtual tributary level Coding Violaton
DACS	Digital Access Cross-connect System
DBR	Distributed Bragg Reflector
DCB	Digital Channel Bank
DCC	Data Communication Channel
DCOM	Distributed Common Object Model
DCS	Digital Cross-connect System
DDS	Digital Data Services
DFB	Distributed FeedBack laser
DIU	Digital Interface Unit
DQDB	Distributed Queue Dual Bus
DS-SMF	Dispersion Shifted SingleMode Fiber
DS-0	Digital Signal-level 0 (64 kbps)
DS-1	Digital Signal-level 1 (1.544 Mbps)
DS-1C	Digital Signal-level 1C (3.152 Mbps)
DS-3	Digital Signal-level 3 (44.736 Mbps)
DS-3C	Digital Signal-level 3C (91.053 Mbps)
DSX-m	Digital Signal Cross-connect point for DSm signals
DS-n	Digital Signal-level n
DWDM	Dense Wavelength Division Multiplexing
E1 (E-1)	ITU-T digital signal-level 1 (2.048 Mbps)
E2 (E-2)	ITU-T digital signal-level 2 (8.448 Mbps)
E3 (E-3)	ITU-T digital signal-level 3 (34.368 Mbps)

E4 (E-4)	ITU-T digital signal-level 4 (139.264 Mbps)
ECC	Embedded Communication Channel
EDFA	Erbium-Doped Fiber Amplifier
ELED	Edge Light Emitting Diode
E/O	Electrical (signal) to optical (signal) conversion
EOC	Embedded Operations Channel
ES	Errored Seconds
ESF	Extended Super Frame
ETSI	European Telecommunications Standards Institute
FBG	Fiber Bragg Grating
FCS	Frame Check Sequence
FDDI	Fiber Distributed Data Interface
FDM	Frequency Division Multiplexing
FEBE	Far End Block Error
FERF	Far End Receive Failure (now RDI)
FOTS	Fiber Optic Transmission System
FWHM	Full Width Half Maximum
FWM	Four Wave Mixing
GBE	Gigabit Ethernet
GPS	Global Positioning System
GRIN	GRaded INdex (fiber)
HCDS	High-Capacity Digital Services
HDLC	High-level Data Link Control
HDTV	High-Definition TV
HVC	High-order Virtual Container (VC3/VC4)
ID	Identification
IDLC	Integrated Digital Loop Carrier
IETF	Internet Engineering Task Force
IM	Inverse Multiplexer
IP	Internet Protocol
ISDN	Integrated Service Digital Network
ISM	Integrated (Intelligent) Synchronous Multiplexer
ISO	International Standards Organization
ISP	Internet Service Provider
ITU	International Telecommunication Union

ITU-T	ITU-Telecommunications Standardization Sector (Formerly CCITT)
IXC	Interexchange Carrier
LAN	Local Area Network
LAPD	Link Access Procedure—D channel
LCM	Least Common Multiplier
LCV	Line Coding Violations
LEAF	Large Effective Area Fiber
LED	Light Emitting Mode (as a light source)
LES	Line Errored Seconds
LLC	Logical Link Control
LOF	Loss Of Frame
LOH	Line Overhead
LOM	Loss Of Multiframe
LOP	Loss Of Pointer
LOP-P	Path-level Loss Of Pointer
LOP-V	Virtual-tributary-level Loss Of Pointer
LOS	Loss Of Signal
LSB	Least Significant Bit
LSES	Line Severe Errored Seconds
LTE	Line Termination Equipment (also MSTE)
LVC	Low-order Virtual Container (VC-I 1/VC-12/VC-2)
MAF	Management Application Function
MAN	Metropolitan Area Network
MCF	Message Communications Function
MD	Mediation Device
MF	Mediation Function
MO	Managed Object
MS-AIS	Multiplex Section AIS
MSB	Most Significant Bit
MS-FERF	Multiplex Section FERF
MSOH	Multiplex Section Overhead
MSPR	Multiplex Section Shared Protection Ring
MSTE	Multiplex Section Terminating Equipment (also LTE)
NDF	New Data Flag
NDF-P	Path-level New Data Flag

NDF-V	Virtual-tributary-level New Data Flag
NE	Network Element (or Equipment)
NEF	Network Element Function
NLSS	Nonlocally Switched Special
NNI	Network Node Interface
NOB	Network Operator Byte
NPC	Network Parameter Control
NPI	Null Pointer Indication
NRM	Network Resource Management
NUT	Nonpreemptive Unprotected Traffic
NZ-DSF	Nonzero Dispersion-Shifted Fiber
OA&M	Operations, Administration, and Maintenance
OAM&P	Operations, Administration, Maintenance, and Provisioning
OC-N	Optical Carrier for Nth level
O/E	Optical (signal) to electrical (signal) conversion
OFS	Optical Fiber System
OFS	Out-of-Frame Second
OH	Overhead
OLC	Optical Loop Carrier
OOF	Out-Of-Frame
ORL	Optical Return Loss
OS	Operation-System
OSF	Operations System Function
OSI	Open System Interconnection
OSI-RM	Open System Interconnection Reference Model
OTDR	Optical Time Domain Reflectometer
PC	Personal Computer
PC	Protection Channel
PCM	Pulse Code Modulation
PDA	Personal Digital Assistant
PDH	Plesiochronous Digital Hierarchy
PIN	Positive-Intrinsic-Negative photodiode
PJ	Pointer Justification
PJC	Pointer Justification Count
PM	Performance Monitoring
PMO	Present Method of Operations
POH	Path Overhead

PON	Passive Optical Network
POS	Packets on SONET(SDH)
POTS	Plain Old Telephone Service
PLM-P	Path-level Path Label Mismatch
PLM-V	Virtual-tributary Path Label Mismatch
ppm	parts per million
PPP	Point-to-Point Protocol
PRC	Primary Reference Clock
PRS	Primary Reference Source (for clocks)
PS	Protection Switching
PSC	Protection Switching Count
PSD	Protection Switching Duration
PTE	Path Terminating Equipment
PTR	Pointer
PUC	Path User Channel
PVC	Permanent Virtual Circuit

| QA | Q-Adapter |
| QOS | Quality Of Service |

RDI	Remote Defect (Degrade) Indication (formerly FERF)
RDI-L	Line-level Remote Defect Indication
RDI-P	Path-level Remote Defect Indication
RDI-V	Virtual-tributary-level Remote Defect Indication
REDFA	Rare Earth Doped Fiber Amplifier
Reg	Regenerator
REI	Remote Error Indication
REI-L	Line-level Remote Error Indication
REI-P	Path-level Remote Error Indication
REI-V	Virtual-tributary-level Remote Error Indication
RFC	Request For Comment
RFI	Remote Failure Indication
RFI-V	Virtual-tributary-level Remote Failure Indication
RI	Refractive Index
RM	Resource Management
ROSE	Remote Operations Service Element
RSM	Remote Switch Module
RSOH	Regenerator Section Overhead
RSTE	Regenerator Section Terminating Equipment (also STE)
RSVP	Resource Reservation Protocol
RT	Remote Terminal

SBS	Stimulated Brillouin Scattering
SCV	Section Coding Violations
SD	Signal Degrade
SDH	Synchronous Digital Hierarchy
SECB	Severely Errored Cell Block
SEFS	Severely Errored Frame Seconds
SES	Section Errored Seconds
SF	Signal Fail
SF	Super Frame
SLC	Synchronous Line Carrier
SLED	Surface Light Emitting Diode
SLM	Single Longitudinal Mode
SLM	Synchronous Line Multiplexer
SM	Switching Module
SMDS	Switched Multimegabit Data Service
SMF	Single-Mode Fiber
SMN	SDH Management Network
SMN	SONET Management Network
SMS	SDH Management Subnetwork
SNAP	Subnetwork Access Protocol
SNC	Subnetwork Connection
SNMP	Simple Network Management Protocol
SOH	Section Overhead
SONET	Synchronous Optical Network
SPE	Synchronous Payload Envelope
SRS	Stimulated Raman Scattering
SSM	Synchronization Status Message
STE	Section-Terminating Equipment (also RSTE)
STM	Synchronous Transfer Mode (see ATM)
STM	Synchronous Transport Module
STM-N	Synchronous Transport Module-level N (Nx 155.52 Mbps: N = 1, 4, 16, or 64)
STS	Synchronous Transport Signal
STS-N	Synchronous Transport Signal-level N (Nx 51.84 Mbps: N = 1, 3, 12, 48, or 192)
SVC	Switched Virtual Circuit
TCM	Tandem Connection Monitoring
TCP	Transmission Control Protocol
TDM	Time Division Multiplexing
TIM	Trace Identifier Mismatch

TIM-L	Line-level Trace Identifier Mismatch
TIM-P	Path-level Trace Identifier Mismatch
TIM-V	Virtual-tributary-level Trace Identifier Mismatch
TMN	Telecommunications Management Network
TOH	Transport Overhead (SOH + LOH)
TS	Time Slot
TSG	Timing Signal Generator
TU-n	Tributary Unit-level n (n= 11, 12, 2, or 3)
TUG-n	Tributary Unit Group n (n = 2 or 3)
T-1	A digital carrier system for DS-1 signal

UAS	Unavailable Second
UAT	Unavailable Time
UBN	Unlimited Bandwidth Network
UM	User-to-Network Interface
UNEQ	Unequipped
UNEQ-P	Unequipped at the Path level
UNEQ-V	Unequipped at the Virtual tributary level
UPC	Usage Parameter Control
UPSR	Unidirectional Path Switch Ring

VBR	Variable Bit Rate
VC	Virtual Channel
VC-n	Virtual Container-level n (n= 11, 12, 2, 3 or 4)
VCC	Virtual Channel Connection
VF	Voice Frequency signal
VT	Virtual Tributary

WAN	Wide Area Network
WDCS	Wideband Digital Cross-connect System
WDM	Wavelength Division Multiplexing

| X.25 | ITU-T Recommendation for packet-switch user-network interface |

APPENDIX B

SONET QUICK REFERENCE

This appendix presents very little new information on SONET. It is mostly a gathering of many figures and tables that appear scattered throughout the text. It might be helpful to have all this information on SONET overhead, frame structures, and so on in one place, and this is the intent of this appendix. There are some details not presented in the text (G1 Path Status byte coding, V5 VT Signal Label coding, and Z7coding), but these are the exceptions.

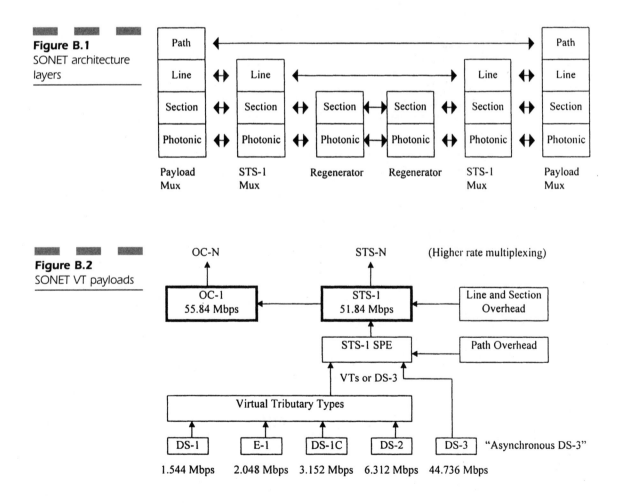

Figure B.1
SONET architecture layers

Figure B.2
SONET VT payloads

Figure B.3
SONET section, line, and path overhead

A1 Framing	A2 Framing	J0/Z0 (STS-ID) Trace/Growth
B1/undefined BIP-8	E1/undefined Orderwire	F1/undefined User
D1/undefined Data Com	D2/undefined Data Com	D3/undefined Data Com
H1 Pointer	H2 Pointer	H3 Pointer Action
B2/undefined BIP-8	K1/undefined APS	K2/undefined APS
D4/undefined Data Com	D5/undefined Data Com	D6/undefined Data Com
D7/undefined Data Com	D8/undefined Data Com	D9/undefined Data Com
D10/undefined Data Com	D11/undefined Data Com	D12/undefined Data Com
S1/Z1 Sync Status/ Growth	M0 or M1/Z2 REI-L/ Growth	E2/undefined Orderwire

Section and Line Overhead

J1 Trace
B3 BIP-8
C2 Signal Label
G1 Path Status
F2 User Channel
H4 Indicator
Z3 Growth
Z4 Growth
Z5 Tandem Connection

Path Overhead

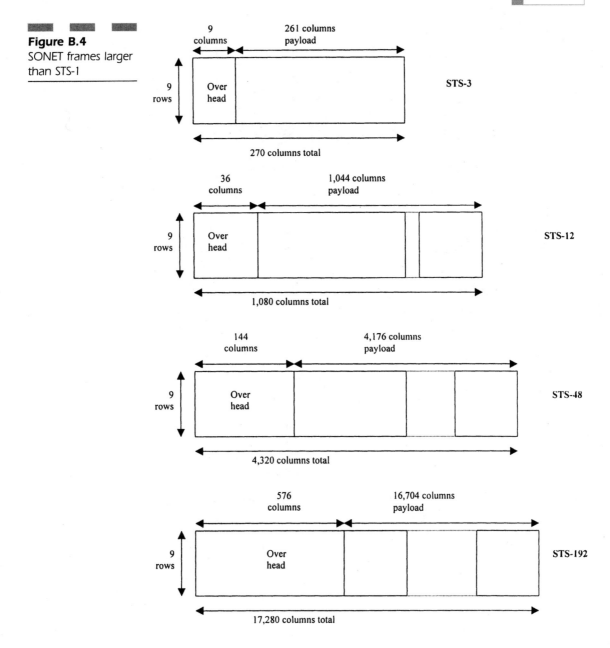

Figure B.4
SONET frames larger than STS-1

Figure B.5

STS-3c frame with overhead

9 columns   POH

A1	A1	A1	A2	A2	A2	J0	R	R		J1	
B1	R	R	E1	R	R	F1	R	R		B3	
D1	R	R	D2	R	R	D3	R	R		C2	
H1	H1	H1	H2	H2	H2	H3	H3	H3		G1	
B2	B2	B2	K1	R	R	K2	R	R		F2	
D4	R	R	D5	R	R	D6	R	R		H4	
D7	R	R	D8	R	R	D9	R	R		F3	
D10	R	R	D11	R	R	D12	R	R		K3	
S1	Z1	Z1	Z2	Z2	M1	E2	R	R		N1	

9 rows

TABLE B.1

C2 payload type coding

Code in Hex	STS SPE Content
00	Unequipped (not used for live information)
01	Equipped—nonspecific payload
02	Virtual Tributaries (VTs) inside ("default")
03	VTs in locked mode (no longer supported)
04	Asynchronous DS-3 mapping
12	Asynchronous DS-4NA mapping
13	ATM cell mapping
14	DQDB cell mapping
15	Asynchronous FDDI mapping
16	HDLC-over-SONET mapping (used for IP)
CF	"Experimental" value for IP inside PPP
FE	Test signal mapping (see ITU Rec. G.707)

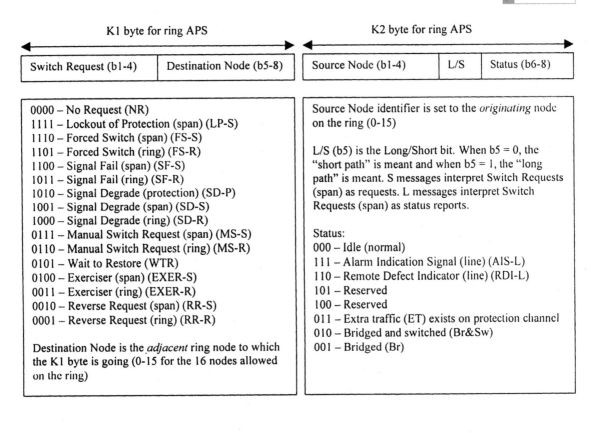

K1 byte for ring APS · K2 byte for ring APS

Switch Request (b1-4)	Destination Node (b5-8)

Source Node (b1-4)	L/S	Status (b6-8)

0000 – No Request (NR)
1111 – Lockout of Protection (span) (LP-S)
1110 – Forced Switch (span) (FS-S)
1101 – Forced Switch (ring) (FS-R)
1100 – Signal Fail (span) (SF-S)
1011 – Signal Fail (ring) (SF-R)
1010 – Signal Degrade (protection) (SD-P)
1001 – Signal Degrade (span) (SD-S)
1000 – Signal Degrade (ring) (SD-R)
0111 – Manual Switch Request (span) (MS-S)
0110 – Manual Switch Request (ring) (MS-R)
0101 – Wait to Restore (WTR)
0100 – Exerciser (span) (EXER-S)
0011 – Exerciser (ring) (EXER-R)
0010 – Reverse Request (span) (RR-S)
0001 – Reverse Request (ring) (RR-R)

Destination Node is the *adjacent* ring node to which the K1 byte is going (0-15 for the 16 nodes allowed on the ring)

Source Node identifier is set to the *originating* node on the ring (0-15)

L/S (b5) is the Long/Short bit. When b5 = 0, the "short path" is meant and when b5 = 1, the "long path" is meant. S messages interpret Switch Requests (span) as requests. L messages interpret Switch Requests (span) as status reports.

Status:
000 – Idle (normal)
111 – Alarm Indication Signal (line) (AIS-L)
110 – Remote Defect Indicator (line) (RDI-L)
101 – Reserved
100 – Reserved
011 – Extra traffic (ET) exists on protection channel
010 – Bridged and switched (Br&Sw)
001 – Bridged (Br)

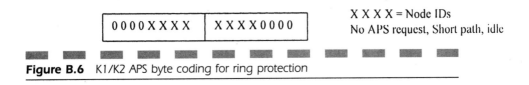

0 0 0 0 X X X X	X X X X 0 0 0 0

X X X X = Node IDs
No APS request, Short path, idle

Figure B.6 K1/K2 APS byte coding for ring protection

TABLE B.2

The SONET
synchronization
status messages
(SSMs)

Synchronization Status Message (SSM)	S1 bits 5-6-7-8	Abbreviation	Quality Level
Stratum 1 traceable (source can be traced to Stratum 1)	0001	PRS	1
Synchronized—traceability unknown (backward-compatibility)	0000	STU	2
Stratum 2 traceable	0111	ST2	3
Transit node clock traceable (a non-standard clock level)	0100	TNC	4
Stratum 3E traceable (another non-standard clock level)	1101	ST3E	5
Stratum 3 traceable	1010	ST3	6
SONET minimum clock traceable	1100	SMC	7
Stratum 4 traceable (no value has been assigned for this)	N/A	ST4	8
Do not use for synchronization	1111	DUS	9
Reserved for network synchronization use	1110	RES	User-assigned

TABLE B.3 SONET events and meanings

Overhead Byte	Event	Meaning	Comment
N/A	LOS	Loss of signal	All 0s or below receive threshold
A1/A2	LOF	Loss of framing	Framing pattern lost
J0	TIM-S	Regenerator section trace identifier mismatch	The J0 network address of the regenerator is not what was expected
B1	CV-S	Coding violation—section	BIP errors detected in B1 byte
K2	AIS-L	Alarm indication signal—line	Bits 6-7-8 of K2 byte set to 111
B2	CV-L	Coding violation—line	BIP errors detected in B2 byte
M1	REI-L	Remote error indication—line	Count of B2 BIP errors
K2	RDI-L	Remote defect indication—line	Bits 6-7-8 of K2 byte set to 110
H1/H2	AIS-P	Alarm indication signal—path	Excessive New Data Flags (NDF)
H1/H2/H3	LOP-P	Loss of pointer—path	Invalid payload pointer values
C2	UNEQ-P	Unequipped STS at path level	All 0 in C2 byte: no user information
J1	TIM-P	Trace identifier mismatch at path level	The J1 network address at the path level is not what was expected
B3	CV-P	Coding violation—path	BIP errors detected in B3 byte
G1	REI-P	Remote error indication—path	Bits 1—4 of G1 report B3 error count
G1	RDI-P	Remote defect indication—path	Bits 5-6-7 of G1 code path events*
V1/V2	AIS-V	Alarm indication signal—VT	Excessive VT pointer adjustments
V1/V2/V3	LOP-V	Loss of pointer—VT	Invalid VT pointer values
H4	LOM	Loss of multiframe—VT	0 thru 3 payload sequence in H4 is lost
C2	PLM-P	Path label mismatch—path	Payload content not what expected
V5	UNEQ-V	Unequipped VT at VT level	Bits 5-6-7 of V5 set to 000
J2	TIM-V	Trace identifier mismatch at VT level	The J2 network address at the VT level is not what was expected
V5	CV-V	Coding violation—VT	BIP errors detected in bits 1-2 of V5 byte
V5	REI-V	Remote error indication—VT	Bit 3 of V5 reports any BIP error
V5	RDI-V	Remote defect indication—VT	Bit 4 of V5 reports VT event
V5	AIS	Alarm indication signal—VT	Alarm reported to affected tributary

*See Appendix B for details.

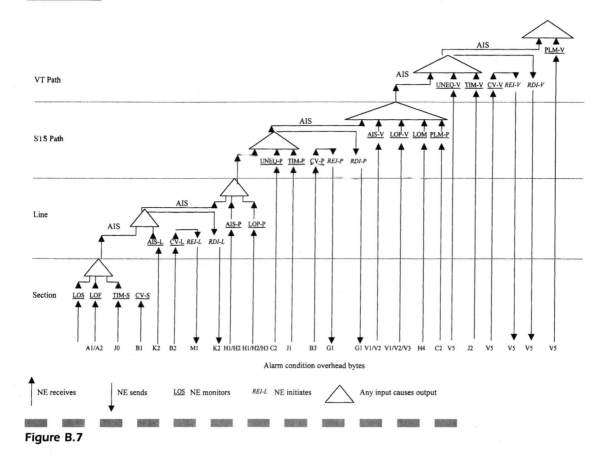

Alarm condition overhead bytes

Figure B.7

The **G1 byte** provides path performance information.

REI-P				RDI-P			Undef.
b1	b2	b3	b4	b5	b6	b7	b8

REI-P: bits 1—4 of G1 byte convey count of the interleaved bit block errors, based on B3 count, to upstream path termination equipment.

RDI-P: bits 5—7 of G1 byte convey defect details according to Table B.4.

TABLE B.4

G1 path status
coding

G1 b5-b7 (RDI-P)	Meaning	Event triggers
0 0 0	No remote defect	No defect
0 0 1	No remote defect	No defect
0 1 0	Remote payload defect	LCD-P, PLM-P
0 1 1	No remote defect	No defect
1 0 0	Remote defect	AIS-P, LOP-P
1 0 1	Remote server defect	AIS-P, LOP-P
1 1 0	Remote connectivity defect	TIM-P, UNEQ-P
1 1 1	Remote defect	AIS-P, LOP-P

The **V5 byte** provides BIP-2 error checking, signal labels for VTs, and VT path status.

BIP-2 b1 b2	REI-V b3	RFI-V b4	Signal Label b5 b6 b7	RDI-V b8

TABLE B.5

V5 (Virtual Tributary)
signal label coding

Signal Label (b5-b7)	Meaning
0 0 0	Unequipped
0 0 1	Equipped, nonspecific payload
0 1 0	Asynchronous mapping in use
0 1 1	Bit synchronous mapping (no longer used for DS-1)
1 0 0	Byte synchronous mapping
1 0 1	Unassigned
1 1 0	Unassigned
1 1 1	Unassigned

The **Z7 byte** provides additional information about RDI-V (bit 8 of V5) conditions in bits 5-6-7.

TABLE B.6

Z7 Byte Coding

Z7 bit 5	Z7 bit 6	Z7 bit 7	Meaning	Event triggers
0	0	0	No remote defect	No defects
0	0	1	No remote defect	No defects
0	1	0	Remote payload defect	PLM-V
0	1	1	No remote defect	No defects
1	0	0	Remote defect	AIS-V, LOP-V
1	0	1	Remote server defect	AIS-V, LOP-V
1	1	0	Remote connectivity defect	UNEQ-V
1	1	1	Remote defect	AIS-V, LOP-V

APPENDIX C

SDH QUICK REFERENCE

This appendix presents very little new information of SDH. It is mostly a gathering of many figures and tables that appear in Chapter 18. It might be helpful to have all this information on SDH overhead, frame structures, and so on in one place, and this is the intent of this appendix. There are some details not presented in the text (G1 Path Status byte coding, V5 VT Signal Label coding, and Z7 coding), but these are the exceptions.

Figure C.1
SDH architecture layers

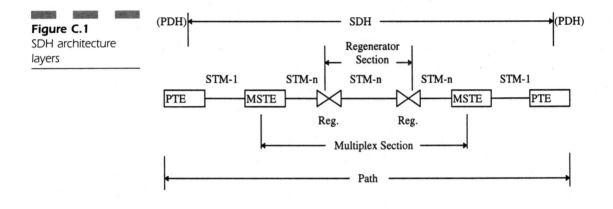

Figure C.2
SDH VT payloads

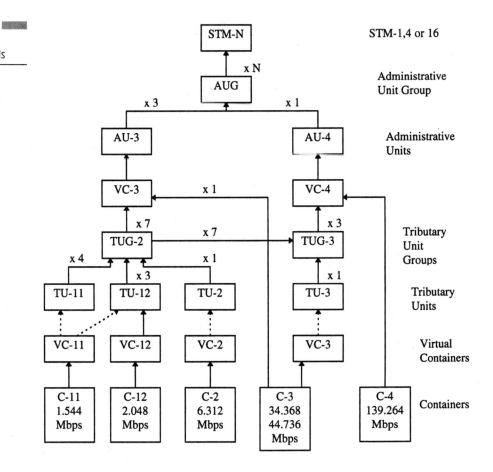

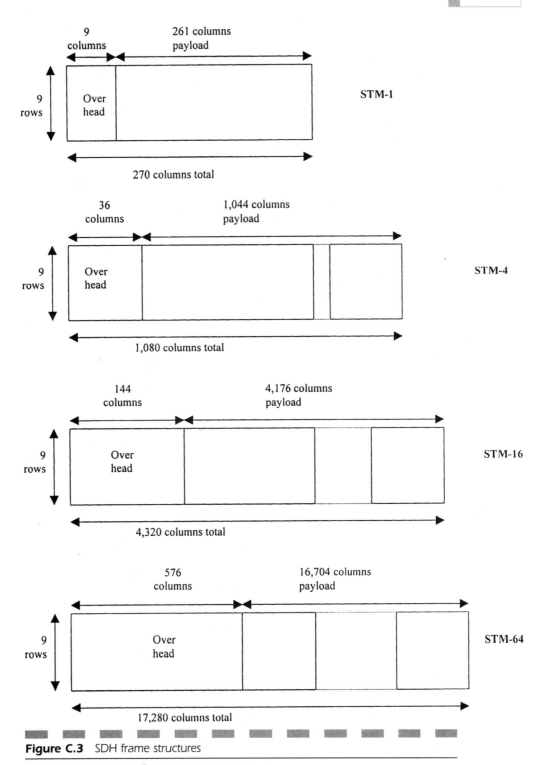

Figure C.3 SDH frame structures

Figure C.4
STM-1 frame with
overhead

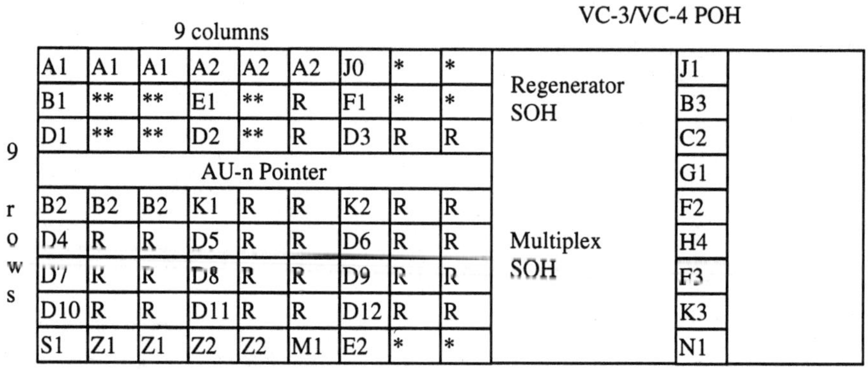

9 columns

VC-3/VC-4 POH

	A1	A1	A1	A2	A2	A2	J0	*	*	Regenerator SOH	J1	
	B1	**	**	E1	**	R	F1	*	*		B3	
9	D1	**	**	D2	**	R	D3	R	R		C2	
	AU-n Pointer										G1	
r	B2	B2	B2	K1	R	R	K2	R	R		F2	
o	D4	R	R	D5	R	R	D6	R	R	Multiplex SOH	H4	
w	D7	R	R	D8	R	R	D9	R	R		F3	
s	D10	R	R	D11	R	R	D12	R	R		K3	
	S1	Z1	Z1	Z2	Z2	M1	E2	*	*		N1	

9 rows

* = Reserved for National Use

** = Media Dependent Octet

TABLE C.1

C2 payload type
coding

Code in Hex	STM Content
00	Unequipped or supervisory unequipped
01	Equipped—nonspecific
02	TUG structure inside
03	Locked mode TU
04	Asynchronous DS-3 or E-3 mapping (C-3)
12	Asynchronous E-3 mapping (C-4)
13	ATM cell mapping
14	DQDB cell mapping
15	Asynchronous FDDI mapping
16	HDLC-over-SONET mapping (used for IP)
CF	"Experimental" value for IP inside PPP
FE	Test signal mapping (see ITU Rec. G.707)
FF	Virtual Container Alarm Indication Signal

K1 byte for ring APS

K2 byte for ring APS

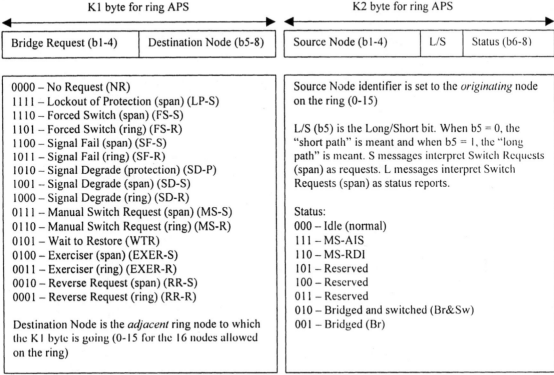

Bridge Request (b1-4)	Destination Node (b5-8)

0000 – No Request (NR)
1111 – Lockout of Protection (span) (LP-S)
1110 – Forced Switch (span) (FS-S)
1101 – Forced Switch (ring) (FS-R)
1100 – Signal Fail (span) (SF-S)
1011 – Signal Fail (ring) (SF-R)
1010 – Signal Degrade (protection) (SD-P)
1001 – Signal Degrade (span) (SD-S)
1000 – Signal Degrade (ring) (SD-R)
0111 – Manual Switch Request (span) (MS-S)
0110 – Manual Switch Request (ring) (MS-R)
0101 – Wait to Restore (WTR)
0100 – Exerciser (span) (EXER-S)
0011 – Exerciser (ring) (EXER-R)
0010 – Reverse Request (span) (RR-S)
0001 – Reverse Request (ring) (RR-R)

Destination Node is the *adjacent* ring node to which the K1 byte is going (0-15 for the 16 nodes allowed on the ring)

Source Node (b1-4)	L/S	Status (b6-8)

Source Node identifier is set to the *originating* node on the ring (0-15)

L/S (b5) is the Long/Short bit. When b5 = 0, the "short path" is meant and when b5 = 1, the "long path" is meant. S messages interpret Switch Requests (span) as requests. L messages interpret Switch Requests (span) as status reports.

Status:
000 – Idle (normal)
111 – MS-AIS
110 – MS-RDI
101 – Reserved
100 – Reserved
011 – Reserved
010 – Bridged and switched (Br&Sw)
001 – Bridged (Br)

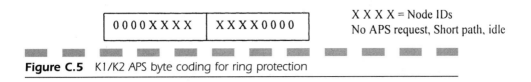

0 0 0 0 X X X X	X X X X 0 0 0 0

X X X X = Node IDs
No APS request, Short path, idle

Figure C.5 K1/K2 APS byte coding for ring protection

TABLE C.2

The SDH
synchronization
status messages
(SSMs)

Synchronization Status Message (SSM)	S1 bits 5-6-7-8
Quality unknown (backward compatibility)	0000
Reserved	0001
G.811	0010
Reserved	0011
G.812 transit	0100
Reserved	0101
Reserved	0110
Reserved	0111
G.812 local	1000
Reserved	1001
Reserved	1010
Synchronous Equipment Timing Source (SETS)	1011
Reserved	1100
Reserved	1101

TABLE C.3 SONET and SDH events and meanings

Overhead Byte	Name	SONET Event	SDH Name
N/A	LOS	Loss of signal	LOS (same)
A1/A2	LOF	Loss of framing	LOF (same)
J0	TIM-S	Regenerator section trace identifier mismatch	RS-TIM (Regenerator section TIM)
B1	CV-S	Coding violation—section	RS-BIP (Regenerator section CV)
K2	AIS-L	Alarm indication signal—line	MS-AIS (Multiplex section AIS)
B2	CV-L	Coding violation—line	MS-BIP (Multiplex section CV)
M1	REI-L	Remote error indication—line	MS-REI (Multiplex section REI)
K2	RDI-L	Remote defect indication—line	MS-RDI (Multiplex section RDI)
H1/H2	AIS-P	Alarm indication signal—path	AU-AIS (Administrative Unit AIS)
H1/H2/H3	LOP-P	Loss of pointer—path	AU-LOP (Administrative Unit LOP)
C2	UNEQ-P	Unequipped STS at path level	HP-UNEQ (Higher Path unequipped)
J1	TIM-P	Trace identifier mismatch at path level	HP-TIM (Higher Path TIM)
B3	CV-P	Coding violation—path	HP-BIP (Higher Path CV)
G1	REI-P	Remote error indication—path	HP-REI (Higher Path REI)
G1	RDI-P	Remote defect indication—path	HP-RDI (Higher Path RDI)
V1/V2	AIS-V	Alarm indication signal—VT	TU-AIS (Tributary Unit AIS)
V1/V2/V3	LOP-V	Loss of pointer—VT	TU-LOP (Tributary Unit LOP)
H4	LOM	Loss of multiframe—VT	TU-LOM (Tributary Unit LOM)
C2	PLM-P	Path label mismatch—path	HP-PLM (Higher Path PLM)
V5	UNEQ-V	Unequipped VT at VT level	LP-UNEQ (Lower Path unequipped)
J2	TIM-V	Trace identifier mismatch at VT level	LP-TIM (Lower Path TIM)
V5	CV-V	Coding violation—VT	LP-BIP (Lower Path CV)
V5	REI-V	Remote error indication—VT	LP-REI (Lower Path REI)
V5	RDI-V	Remote defect indication—VT	LP-RDI (Lower Path RDI)
V5	AIS	Alarm indication signal—VT	TU-AIS (Tributary Unit AIS)

NOTE: Appendix B contains a table similar to Table C.3 and a graphic representation of SONET alarms and events and overhead bytes. There is no need to repeat this picture for SDH: the flows are the same, but the names are different. Table C.3 above shows the differences.

In addition, SDH calls the four layers in the Appendix B picture by different names:

SONET name	SDH name
Section Level	Regenerator Section
Line Level	Multiplex Section
STS Path Level	Higher Order Path
Virtual Tributary Path Level	Lower Order Path

REI				RDI			Undef.
b1	b2	b3	b4	b5	b6	b7	b8

The **G1 byte** provides path performance information.

REI-P: Bits 1—4 of G1 byte convey count of the interleaved bit block errors, based on B3 count, to upstream path termination equipment.

RDI-P: Bits 5—7 of G1 byte convey defect details according to Table C.4.

TABLE C.4

SDH G1 path status coding

G1 b5-b7 (RDI-P)	Meaning	Event triggers
0 0 0	No remote defect	No defect
0 0 1	No remote defect	No defect
0 1 0	Remote payload defect	LCD
0 1 1	No remote defect	No defect
1 0 0	Remote defect	AIS, LOP, TIM, UNEQ (or PLM, LCD)
1 0 1	Remote server defect	AIS, LOP
1 1 0	Remote connectivity defect	TIM, UNEQ
1 1 1	Remote defect	AIS, LOP, TIM, UNEQ (or PLM, LCD)

The **V5 byte** provides BIP-2 error checking, signal labels for VTs, and VT path status.

BIP-2		REI	RFI	Signal Label			RDI
b1	b2	b3	b4	b5	b6	b7	b8

TABLE C.5

SDH V5 (Virtual Tributary) signal label coding

Signal Label (b5-b7)	Meaning
0 0 0	Unequipped
0 0 1	Equipped, non-specific payload
0 1 0	Asynchronous mapping in use
0 1 1	Bit synchronous mapping
1 0 0	Byte synchronous mapping
1 0 1	Reserved for future use
1 1 0	Q181 test signal (TSS4)
1 1 1	Virtual container alarm indication signal (VC-AIS)

NOTE: SDH documentation contains details on the N1 ("high order path overhead," or HO-POH) Tandem Connection Montoring (TCM) byte, the N2 ("lower order path overhead," or LO-POH) TCM byte, and K4 Lower order path automatic protection switching byte.

These bytes were not detailed in the text, so no information is given in this appendix. Interested readers are referred to source documentation.

BIBLIOGRAPHY

Telcordia (Bellcore) Publications

Family of Requirements

FR-476, OTGR Section 6: Network Maintenance: Access and Testing (Bellcore, 1998 edition; a subset of OTGR, FR-439).

FR-480, OTGR Section 10: User System Interface (Bellcore, 1998 edition; a subset of OTGR, FR-439).

Generic Requirements

GR-20-CORE, Generic Requirements for Optical Fiber and Optical Fiber Cable, Issue 2 (Bellcore, July 1998).

GR-63-CORE, Network EquipmentBuilding System (NEBS) Requirements: Physical Protection (a module of LSSGR, FR-64, TSGR, FR-440, and NEBSFR, FR-2063), Issue 1 (Bellcore, October 1995).

GR-78-CORE, Generic Requirements for the Physical Design and Manufacture of Telecommunications Products and Equipment (a module of RQGR, FR-796 and NEBSFR, FR-2063), Issue 1 (Bellcore, September 1997).

GR-129-CORE, Generic Requirements for Fiber Optic Branching Components, Issue 2 (Bellcore, February 1998).

GR-149-CORE, Generic Requirements on Security for OSI-Based Telecommunications Management Network Interfaces, Issue 1 (Bellcore, September 1994).

GR-199-CORE, OTGR Section 12.2: Operations Application Messages Memory—Administration Messages (a module of OTGR, FR-439), Issue 2 (Bellcore, November 1996).

GR-303-CORE, Integrated Digital Loop Carrier System Generic Requirements, Objectives, and Interface (a module of TSGR, FR-440), Issue 2 (Bellcore, December 1998).

GR-326-CORE, Generic Requirements for Single-Mode Optical Connectors and Jumper Assemblies, Issue 2 (Bellcore, December 1996).

GR-409-CORE, Generic Requirements for Premises Fiber Optic Cable, Issue 1 (Bellcore, May 1994).

GR-418-CORE, Generic Reliability Assurance Requirements for Fiber Optic Transport Systems (a module of RQGR, FR-796), Issue 1 (Bellcore, December 1997).

GR-436-CORE, Digital Network Synchronization Plan, Issue 1 (Bellcore, June 1994) plus Revision 1, June 1996.

GR-449-CORE, Generic Requirements and Design Consideration for Fiber Distributing Frames, Issue 1 (Bellcore, March 1995).

GR-454-CORE, Generic Requirements for Supplier Provided Documentation (a module of OTGR, FR-439, LSSGR, FR-64, and TSGR, FR-440), Issue 1 (Bellcore, December 1997).

GR-468-CORE, Generic Reliability Assurance Requirements for Optoelectronic Devices Used in Telecommunications Equipment (a module of RQGR, FR-796), Issue 1 (Bellcore, December 1998).

GR-472-CORE, OTGR Section 2.1: Network Element Configuration Management (a module of OTGR, FR-439), Issue 2 (Bellcore, November 1996) plus Revision 1, July 1997.

GR-474-CORE, OTGR Section 4: Network Maintenance: Alarm and Control for Network Elements (a module of OTGR, FR-439), Issue 1 (Bellcore, December 1997).

GR-487-CORE, Generic Requirements for Electronic Equipment Cabinets, Issue 1 (Bellcore, June 1996).

GR-496-CORE, SONET Add-Drop Multiplexer (SONET ADM) Generic Criteria (a module of TSGR, FR-440), Issue 1 (Bellcore, December 1998).

GR-499-CORE, Transport System Generic Requirements (TSGR): Common Requirements (a module of TSGR, FR-440), Issue 2 (Bellcore, December 1998).

GR-765-CORE, Generic Requirements for Single Fiber Single-Mode Optical Splices and Splicing Systems, Issue 1 (Bellcore, September 1995).

GR-815-CORE, Generic Requirements for Network Element/Network System (NEI NS) Security (a module of LSSGR, FR-64, and OTGR, FR-439), Issue 1 (Bellcore, November 1997).

GR-820-CORE, OTGR Section 5.1: Generic Digital Transmission Surveillance (a module of OTGR, FR-439), Issue 2 (Bellcore, December 1997).

GR-826-CORE, OTGR Section 10.2: User Interface Generic Requirements for Supporting Network Element Operations (a module of OTGR, FR-439), Issue 1 (Bellcore, June 1994).

GR-828-CORE, OTGR Section 11.2: Generic Operations Interface—OSI Communications Architecture (a module of OTGR, FR-439), Issue 1 (Bellcore, September 1994) plus Revision 2, October 1996.

GR-831-CORE, OTGR Section 12.1: Operations Application Messages— Language for Operations Application Messages (a module of OTGR, FR-439), Issue 1 (Bellcore, November 1996).

GR-833-CORE, OTGR Section 12.3: Network Maintenance: Network Element and Transport Surveillance Messages (a module of OTGR, FR-439), Issue 2 (Bellcore, November 1996).

GR-834-CORE, OTGR Section 12.4: Network Maintenance: Access and Testing Messages (a module of OTGR, FR-439), Issue 2 (Bellcore, November 1996).

GR-836-CORE, OTGR Section 15.2: Generic Operations Interfaces Using OSI Tools—Information Model Overview: Transport Configuration and Surveillance for Network Elements (a module of OTGR, FR-439), Issue 2 (Bellcore, September 1996) plus Revisions.

GR-836-IMD, OTGR Section 15.2: Generic Operations Interfaces Using OSI Tools—Information Model Details: Transport Configuration and Surveillance for Network Elements (a module of OTGR, FR-439), Issue 2 (Bellcore, September 1996).

GR-839-CORE, Generic Requirements for Supplier-Provided Training (a module of LSSGR, FR-64, TSGR, FR-440, and OTGR, FR-439), Issue 1 (Bellcore, July 1996).

GR-910-CORE, Generic Requirements for Fiber Optic Attenuators, Issue 2 (Bellcore, December 1998).

GR-1031-CORE, OTGR Section 15.6: Operations Interfaces Using OSI Tools: Test Access Management, Issue 2 (Bellcore, October 1997) plus Revision 1, December 1998.

GR-1042-CORE, Generic Requirements for Operations Interfaces Using OSI Tools Information Model Overview: Synchronous Optical Network (SONET) Transport Information Model, Issue 3 (Bellcore, December 1998).

GR-1042-IMD, Generic Requirements for Operations Interfaces Using OSI Tools Information Model Details: Synchronous Optical Network (SONET) Transport Information Model, Issue 3 (Bellcore, December 1998).

GR-1093-CORE, Generic State Requirements for Network Elements, Issue 1 (Bellcore, October 1994) plus Revision 1, December 1995.

GR-1230-CORE, SONET Bidirectional Line Switched Ring Equipment Generic Criteria (a module of TSGR, FR-440), Issue 4 (Bellcore, December 1998).

GR-1244-CORE, Clocks for the Synchronized Network: Common Generic Criteria, Issue 1 (Bellcore, June 1995).

GR-1250-CORE, Generic Requirements for Synchronous Optical Network (SONET) File Transfer, Issue 1 (Bellcore, June 1995).

GR-1253-CORE, Generic Requirements for Operations Interfaces Using OSI Tools: Telecommunications Management Network Security Administration, Issue 1 (Bellcore, June 1995).

GR-1309-CORE, TSC/RTU and OTAU Generic Requirements for Remote Optical Fiber Testing (a module of OTGR, FR-439), Issue 1 (Bellcore, June 1995).

GR-1332-CORE, Generic Requirements for Data Communications Network Security, Issue 2 (Bellcore, April 1996).

GR-1377-CORE, SONET OC192 Transport System Generic Criteria (a module of TSGR, FR-440), Issue 5 (Bellcore, December 1998).

GR- 1400-CORE, SONET DualFed Unidirectional Path Switched Ring (UPSR) Equipment Generic Criteria (a module of TSGR, FR-440), Issue 2 (Bellcore, January 1999).

Technical References

TR-NWT-000057, Functional Criteria for Digital Loop Carrier Systems (a module of TSGR, FR-440), Issue 2 (Bellcore, January 1993).

TR-NWT-000078 (see GR78-CORE).

TR-NWT-000170, Digital Cross-Connect System (DSC 1/0) Generic Criteria, Issue 2 (Bellcore, January 1993).

TR-NWT-000357, Generic Requirements for Assuring the Reliability of Components Used in Telecommunication Systems (a module of RQGR, FR-796), Issue 2 (Bellcore, October 1993).

TR-NWT-000418 (see GR-418-CORE).

TR-NWT-000468 (see GR-468-CORE).

TR-NWT-000835, OTGR Section 12.5: Network Element and Network System Security Administration Messages (a module of OTGR, FR-439), Issue 3 (Bellcore, January 1993).

TR-NWT-000917, SONET Regenerator (SONET RGTR) Equipment Generic Criteria (a module of TSGR, FR-440), Issue 1 (Bellcore, December 1990).

TR-NWT-000930, Generic Requirements for Hybrid Microcircuits Used in Telecommunications Equipment (a module of RQGR, FR-796), Issue 2 (Bellcore, September 1993).

TR-NWT-001112, Broadband-ISDN User to Network Interface and Network Node Interface Physical Layer Generic Criteria, Issue 1 (Bellcore, June 1993).

TR-OPT-000839 (see GR-839-CORE).

TR-TSY-000454 (see GR-454-CORE).

TR-TSY-000458, Digital Signal Zero, "A" (DS-OA 64 kb/s) Systems Interconnection, Issue 1 (Bellcore, December 1989).

TR-TSY-000782, SONET Digital Switch Trunk Interface Criteria (a module of' LSSGR, FR-64, and TSGR, FR-440), Issue 2 (Bellcore, September 1989).

TR-TSY-000824, OTGR Section 10.1: User System Interface—User System Access (a module of OTGR, FR-439), Issue 2 (Bellcore, February 1988).

TR-TSY-000825, OTGR Section I0.A: User System Interface—User System Language (a module of OTGR, FR-439), Issue 2 (Bellcore, February 1988).

TR-TSY-000827, OTGR Section 11.1: Generic Operations Interfaces NonOSI Communications Architecture (a module of OTGR, FR-439), Issue 1 (Bellcore, November 1988).

Technical Advisories/Framework Technical Advisories

TA-NPL-000286 (not available).

TA-NPL-000464, Generic Requirements and Design Considerations for Optical Digital Signal CrossConnect Systems, Issue 1 (Bellcore, September 1987).

TA-NWT-000487 (see GR-487-CORE).

TA-NWT-000983, Reliability Assurance Practices for Optoelectronic Devices in Loop Applications, Issue 2 (Bellcore, December 1993; replaced by GR-468-CORE).

FA-NWT-001345, Framework Generic Requirements for Element Manager (EM) Applications for SONET Subnetworks, Issue 1 (Bellcore, September 1992).

TA-NWT-001385 (not available).

TA-TSV-001 294, Generic Requirements for Element Management Layer (EML) Functionality and Architecture, Issue 1 (Bellcore, December 1992).

Special Reports

SR-104, *Bellcore Digest of Technical Information,* Volume 14, Issue 12 (Bellcore, December 1997).

SR-NWT-002224 (not available).

SR-TSV-002671, *EML Applications for Fault Management: Subnetwork Root Cause Alarm Analysis,* Issue 1 (Bellcore, June 1993).

SR-TSV-002672, *EML Applications for Fault Management: Intelligent Alarm Filtering for SONET,* Issue 1 (Bellcore, March 1994).

SR-TSV-002675, *EML Applications for Configuration Management: Resource Provisioning Selection and Assignment—Functional Description,* Issue 1 (Bellcore, December 1993).

SR-TSV-002678, *EML Applications for Configuration Management: Inventory Notification and Query—Functional Description,* Issue 1 (Bellcore, April 1994).

SR-NWT-002723, *Applicable TL1 Messages for SONET Network Elements,* Issue 1 (Bellcore, June 1993).

NOTE: All Telcordia (Bellcore) documents are subject to change, and this list reflects the most current information available. Readers are advised to check current status and availability of all documents. Documents can be ordered from Telcordia's online catalog or by contacting Telcordia.

To Contact Telcordia:

Telcordia Customer Service

8 Corporate Place, Room 3A-184

Piscataway, NJ 088544156

1-800-521-CORE (2673) (USA and Canada)

1 (732) 699-5800 (all others)

1 (732) 336-2559 (FAX)

To order documents online:

1. Enter the URL line: **telecominfo.telcordia.com.**

2. Click on the **Search** button located on top.

3. In the **Keywords** field, enter the document number (or keywords), then click on **Submit Search.**

or . . .

1. Enter the above URL line.
2. Click on the **Browse** button located on top, then click on the subject of interest.

EIA/TIA Documents

EIA/TIA-455-170, Cutoff Wavelength of SingleMode Fiber by Transmitted Power.

EIA/TIA-492, Generic Specification for Optical Waveguide Fiber.

EIA/TIA-559, SingleMode Fiber Optic System Transmission Design.

FOTP 127, Spectral Characterization of Multimode Laser Diodes.

OFSTP-2, Effective Transmitter Output Power Coupled into SingleMode Fiber Optic Cable.

OFSTP-3, Fiber Optic Terminal Receiver Sensitivity and Maximum Receiver Input Power.

OFSTP-10, Measurement of Dispersion Power Penalty in SingleMode Systems.

OFSTP-11, Measurement of Single Reflection Power Penaltly for Fiber Optic Terminal Equipment.

These publications are available from:

EIA/TIA Standards Sales Office

2001 Pennsylvania, NW

Washington, DC 20006

(202) 4574963

URL: www.eia.org

American National Standards Institute (ANSI) Documents

ANSI T1.101-1994, Synchronization Interface Standards for Digital Networks.

ANSI T1.102-1993, Digital Hierarchy—Electrical Interfaces.

ANSI T1.105-1995, Synchronous Optical Network (SONET)—Basic Description including Multiplex Structure, Rates and Formats.

ANSI T1.105.01-1998, Synchronous Optical Network (SONET)—Automatic Protection Switching.

ANSI T1.105.02-1995, Telecommunications—Synchronous Optical Network (SONET)—Payload Mappings.

ANSI T1.105.03-1994, Synchronous Optical Network (SONET)—Jitter at Network Interfaces.

ANSI T1.105.03a-1995, Synchronous Optical Network (SONET)—Jitter at Network Interfaces—DS1 Supplement.

ANSI T1.105.03b-1997, Synchronous Optical Network (SONET)—Jitter at Network Interfaces—DS3 Supplement.

ANSI T1.105.04-1995, Synchronous Optical Network (SONET): Data Communication Channel Protocols and Architectures.

ANSI T1.105.05-1994, Synchronous Optical Network (SONET)—Tandem Connection Maintenance.

ANSI T1.105.06-1996, SONET Physical Layer Specifications.

ANSI T1.105.07-1996, Telecommunications—Synchronous Optical Network (SONET)—Sub STS-1 Interface Rates and Formats Specification.

ANSI T1.105.07a-1997, Telecommunications—Synchronous Optical Network (SONET)—Sub STS-1 Interface Rates and Formats Specification (Inclusion of NxVT Group Interfaces).

ANSI T1.105.09-1996, Telecommunications—Synchronous Optical Network (SONET)—Network Element Timing and Synchronization.

ANSI T1.107-1995, Digital Hierarchy—Formats Specifications.

ANSI T1.119-1994, Telecommunications—Synchronous Optical Network (SONET)—Operations, Administration, Maintenance, and Provisioning (OAM&P) Communications.

ANSI T1.119.01-1995, Telecommunications—Synchronous Optical Network (SONET)—Operations, Administration, Maintenance, and

Provisioning (OAM&P) Communications—Protection Switching Fragment.

ANSI T1.119.02-1998, Telecommunications—Synchronous Optical Network (SONET)—Operations, Administration, Maintenance, and Provisioning (OAM&P) Communications—Performance Management Fragment.

ANSI TI.204-1994, Operations, Administration, Maintenance, and Provisioning—Lower Layer Protocols for Telecommunication Management Network (TMN) Interfaces Between Operations Systems and Network Elements.

ANSI TI.210-1993, Operations, Administration, Maintenance, and Provisioning (OAM&P)—Principles of Functions, Architectures and Protocols for Telecommunications Management Network (TMN) Interfaces.

ANSI TI.231-1997, Digital Hierarchy—Layer I In-Service Digital Transmission Performance Monitoring.

ANSI TI.245-1997, Directory Service for Telecommunications Management Network (TMN) and Synchronous Optical Network (SONET).

ANSI T1.416-1999, Telecommunications—Network-to-Customer Installation Interfaces—Synchronous Optical Network (SONET) Physical Media Dependent Specification: Electrical.

ANSI T1.416.01-1999, Telecommunications—Network-to-Customer Installation Interfaces—Synchronous Optical Network (SONET) Physical Media Dependent Specification: Electrical.

ANSI T1.416.02-1999, Telecommunications—Network-to-Customer Installation Interfaces—Synchronous Optical Network (SONET) Physical Media Dependent Specification: Single Mode Fiber.

ANSI T1.416.03-1999, Telecommunications—Network-to-Customer Installation Interfaces—Synchronous Optical Network (SONET) Physical Media Dependent Specification: Electrical.

ANSI T1.506A-1992, Telecommunications—Network Performance—Specifications for Switched Exchange Network (Absolute Round-Trip Delay).

ANSI T1.508-1992, Telecommunications—Network Performance—Loss Plan for Evolving Digital Networks.

ANSI TI.508A-1993, Telecommunications—Network Performance—Loss Plan for Evolving Digital Networks.

ANSI T1.514-1995, Telecommunications—Network Performance Parameters and Objectives for Dedicated Digital Services—SONET Bit Rates.

ANSI T1.646-1995, Broadband ISDN—Physical Layer Specification for User-Network Interfaces Including DS1/ATM.

ANSI Technical Report #6, *A Technical Report on Slave Stratum Clock Performance Measurement Guidelines.*

TI Technical Report #33, *A Technical Report on Synchronization Network Management Using Synchronization Status Messages.*

ANSI X3.216-1992, Data Communications—Structure and Semantics of the Domain Specific Part (DSP) of the OSI Network Service Access Point (NSAP) Address.

These publications are available from:

American National Standards Institute, Inc.

11 West 42nd Street

New York, NY 10036

URL: www.ansi.org

U.S. Government Publications

U.S. Department of Health, Education, and Welfare; Bureau of Radiological Health

21 CFR 1040.10, Performance Standard for Laser Products.

This publication is available from:

Director, Division of Compliance

Bureau of Radiological Health

5600 Fishers Lane

Rockville, MD 20857

URL: www.access.gpo.gov/nara/cfr

ITUT and CCITT Recommendations

SDH Documents

G.703, Physical/Electrical Characteristics of Hierarchical Digital Interfaces.

G.707, Network Node Interfaces for the Synchronous Digital Hierarchy (SDH), March 1996.

G.709, Synchronous Multiplexing Structure.

G.780, Vocabulary of Terms for SDH Networks and Equipment.

G.810, Definitions and Terminology for Synchronization Networks.

G.811, Timing Requirements at the Output of Primary Reference Clocks Suitable for Plesiochronous Operation of International Digital Links.

G.813, Timing Characteristics of SDH Equipment Slave Clocks (SEC).

G.825, The Control of Jitter and Wander in Digital Networks Based on the SDH.

G.826, Error Performance Parameters and Objectives for International, Constant Bit Rate Digital Paths at or Above the Primary Rate.

G.832, Transport of SDH Elements on PDH Networks.

G.841, Types and Characteristics of SDH Network Protection Architectures.

G.842, Interworking of SDH Network Protection Architectures.

G.957, Optical Interfaces for Equipments and Systems Relating to the Synchronous Digital Hierarchy.

G.958, Digital Line Systems Based on the Synchronous Digital Hierarchy for Use on Optical Fiber Cables.

M.2101, Performance Limit for Bringing into Service and Maintenance of International SDH Paths and Multiplex Sections.

M.2110, Bringing into Service International Paths Sections and Transmission Systems.

M.2120, Digital Path, Section and Transmission System Default Detection and Localization.

M.3010, Principles for a Telecommunications Management Network.

M.3100, Generic Network—Information Model.

O.17s, Jitter and Wander Measuring Equipment for Digital Systems Which are Based on the SDH.

O.150, General Requirements for Instrumentation for Performance Measurements on Digital Transmission Equipment.

O.181, Equipment to Assess Error Performance on STM-N SDH Interfaces.

Q.811, Lower Layer Protocol Profiles for the Q3 and X Interfaces.

Q.921, ISDN Usernetwork Interface—Data Link Layer Specification 0.

X.121, International Numbering Plan for Public Data Networks.

X.226, Presentation Protocol Specification for Open Systems Interconnection for CCITT Applications.

TMN Documents.

G.772, Digital Protected Monitoring Points.

G.773, Protocol Suites for Q Interfaces for Management for Transmission Systems.

G.774, Synchronous Digital Hierarchy (SDH) Management Information Model for the Network Element View.

G.774.01, SDH Performance Monitoring for the Network Element View.

G.774.02, SDH Configuration of the Payload Structure for the Network Element View.

G.774.03, SDH Management of Multiplex Section Protection for the Network Element View.

G.774.04, SDH Management of Sub Network Connection Protection from the Network Element View (Q3 interface network element view).

G.774.05, SDH Management of Connection Supervision Functionality (HCS/LCS) (Q3 interface network element view).

G.782, Types and General Characteristics of SDH Equipment Functional Blocks.

G.783, Characteristics of SDH Equipment Functional Blocks (replaces G.781, G.782, and G.783 version of 01/94).

G.784, Synchronous Digital Hierarchy (SDH) Management.

G.SHR-1, SDH Self Healing Rings.

G.SHR-2, SDH Ring Interworking.

G.77f, Protocol Stack for F Interfaces for Transmission Equipment.

G.77qiA, Q Interface Adaptor for Transmission Equipment.

G.803, Architecture of Transport Networks Based on SDH.

G.831, Performance and Management Capabilities of Transport Network Based on the SDH.

G.ATA, Architecture of Transport Networks in the Access Application .

G.ATMA, Architecture of Transport Networks Based on ATM.

G.ATME-1, Types and General Characteristics of ATM Equipment.

G.ATME-2, Functional Characteristics of ATM Equipment.

G.TNA, Architecture of Transport Networks Based on SDH.

G.atmm, ATM Management (Q3 interface network element view).

G.mtn1, Management of Transmission Networks.

These publications are available from:

International Telecommunication Union General Secretariat—Sales Section.

Place des Nations, CH1211.

Geneva 20 (Switzerland).

+41 22 730 5285.

URL: www.itu.int

ETSI Documents.

ETS 300 147 1 (1992-03), Reference: DE/TM-03001, Source: TM3, Title: Transmission and Multiplexing (TM): Synchronous Digital Hierarchy (SDH) Multiplexing structure.

ETS 300 147 1 (1995-01), Reference: RE/TM-03013, Source: TM3, Title: Transmission and Multiplexing (TM): Synchronous Digital Hierarchy (SDH) Multiplexing structure.

ETS 300 147 ed.3 (1997-04), Reference: RE/TM-03045, Source: TM3, Title: Transmission and Multiplexing (TM): Synchronous Digital Hierarchy (SDH): Multiplexing structure.

ETS 300 232/A1 ed.1 (1996-03), Reference: RE/TM-01025, Source: TM1, Title: Transmission and Multiplexing (TM): Optical Interfaces for Equipments and Systems Relating to the Synchronous Digital Hierarchy (SDH) (ITU-T Recommendation G.957 [1995] modified).

ETS 300 272 ed.1 (1994-03), Reference: DE/NA-053032, Source: NA5, Title: Network Aspects (NA): Metropolitan Area Network (MAN), Physical Layer Convergence Procedure (PLCP) for 155.520 Mbit/s CCITT

Recommendations G 707, G.708 and G.709 SDH Based Systems Protocol Implementation Conformance Statement (PICS).

ETS 300 276 ed.1 (1994-03), Reference: DE/NA-053033, Source: NA, Title: Network Aspects (NA): Metropolitan Area Network (MAN) Physical Layer Convergence Procedure (PLCP) for 622.080 Mbit/s CCITT Recommendations G.707, G.708 and G.709 SDH Based Systems.

ETS 300 277 ed.1 (1994-03), Reference: DE/NA-053034, Source: NA5, Title: Network Aspects (NA): Metropolitan Area Network (MAN) Physical Layer Convergence Procedure (PLCP) for 622.080 Mbit/s CCITT Recommendations G.707, G 708 and G.709 SDH Based Systems Protocol Implementation Conformance Statement (PICS).

ETSI ETS 300 300 ed.1 (1995-02), Reference: DE/NA-052512, Source: NA5, Title: Broadband Integrated Services Digital Network (BISDN): Synchronous Digital Hierarchy (SDH) Based User Network Access: Physical Layer Interfaces for BISDN Applications.

ETSI ETS 300 300 ed.2 (1997-04), Reference: RE/TM-03029, Source: TM3, Title: Broadband Integrated Services Digital Network (BISDN): Synchronous Digital Hierarchy (SDH) Based User Network Access. Physical Layer User Network Interfaces (UNI) for 155 520 kbit/s and 622 080 kbit/s Asynchronous Transfer Mode (ATM) B ISDN Applications.

ETS 300 304 ed. 1 (1994-11), Reference: DE/TM-022011, Source: TM2, Title: Transmission and Multiplexing (TM): Synchronous Digital Hierarchy (SDH), Information Model for the Network Element (NE) View.

ETS 300 304 ed.2 (1997-02), Reference: RE/TM-02213, Source: TM2, Title: Transmission and Multiplexing (TM): Synchronous Digital Hierarchy (SDH), SDH Information Model for the Network Element (NE) View.

ETS 300 337 ed.1 (1995-02), Reference: DE/TM-03007, Source: TM3, Title: Transmission and Multiplexing (TM): Generic Frame Structures for the Transport of Various Signals (including Asynchronous Transfer Mode (ATM) Cells and Synchronous Digital Hierarchy (SDH) Elements) at the ITUT Recommendation G.702 Hierarchical Rates of 2 048 kbit/s, 34 368 kbit/s and 139 264 kbit/s.

ETS 300 337 ed.2 (1997-06), Reference: RE/TM-03056, Source: TM3, Title: Transmission and Multiplexing (TM): Generic Frame Structures for the Transport of Various Signals (Including Asynchronous Transfer Mode (ATM) Cells and Synchronous Digital Hierarchy (SDH) Elements) at the ITUT Recommendation G.702 Hierarchical Rates of 2 048 kbit/s, 34 368 kbit/s and 139 264 kbit/s.

ETSI ETS 300 417-2-1 ed.1 (1997-04), Reference: DE/TM-01015-2-1, Source: TM1, Title: Transmission and Multiplexing (TM): Generic Requirements of Transport Functionality of Equipment Part 2-1 Synchronous Digital Hierarchy (SDH) and Plesiochronous Digital Hierarchy (PDH) Physical Section Layer Functions.

ETSI EN 300 417-2-1 V1.1.2 (1998-11), Reference: REN/TM-01015-2-1, Source: TM1, Title: Transmission and Multiplexing (TM): Generic Requirements of Transport Functionality of Equipment; Part 2-1. Synchronous Digital Hierarchy (SDH) and Plesiochronous Digital Hierarchy (PDH) Physical Section Layer Functions.

ETSI EN 300 417-2-1 V1.1.3 (1999-05), Reference: REN/TM-01015-2-1a, Source: TM1, Title: Transmission and Multiplexing (TM): Generic Requirements of Transport Functionality of Equipment; Part 2-1. Synchronous Digital Hierarchy (SDH) and Plesiochronous Digital Hierarchy (PDH) Physical Section Layer Functions.

ETSI ETS 300 417-2-2 ed.1 (1997-11), Reference: DE/TM-01015-2-2, Source: TM1, Title: Transmission and Multiplexing (TM): Generic Requirements of Transport Functionality of Equipment; Part 2-2 Synchronous Digital Hierarchy (SDH) and Plesiochronous Digital Hierarchy (PDH) Physical Section Layer Functions: Implementation Conformance Statement (ICS) Pro forma Specification.

ETSI EN 300 417-2-2 V1.1.2 (1998-11), Reference: REN/TM-01015-2-2, Source: TM1, Title: Transmission and Multiplexing (TM): Generic Requirements of Transport Functionality of Equipment; Part 2-2. Synchronous Digital Hierarchy (SDH) and Plesiochronous Digital Hierarchy (PDH) Physical Section Layer Functions: Implementation Conformance Statement (ICS) Pro forma Specification.

ETSI EN 300 417-2-2 VI.1.3 (1999-02), Reference: REN/TM-01015-2-2a, Source: TM1, Title: Transmission and Multiplexing (TM): Generic Requirements of Transport Functionality of Equipment: Part 2-2 Synchronous Digital Hierarchy (SDH) and Plesiochronous Digital Hierarchy (PDH) Physical Section Layer Functions: Implementation Conformance Statement (ICS) Pro forma Specification.

ETSI EN 300 417-2-2 V1.1.4 (1999-05), Reference: REN/TM-01015-2-2b, Source: TM1, Title: Transmission and Multiplexing (TM): Generic Requirements of Transport Functionality of Equipment: Part 2-2. Synchronous Digital Hierarchy (SDH) and Plesiochronous Digital Hierarchy (PDH) Physical Section Layer Functions Implementation Conformance Statement (ICS) Pro forma Specification.

ETSI ETS 300 417-4-1 ed.1 (1997-06), Reference: DE/TM-01015-4-1, Source: TM1, Title: Transmission and Multiplexing (TM): Generic Requirements of Transport Functionality of Equipment; Part 4-1: Synchronous Digital Hierarchy (SDH) Path Layer Functions.

ETSI EN 300 417-4-1 V1.1.2 (1998-11), Reference: REN/TM-01015-4-1 Source: TM1, Title: Transmission and Multiplexing (TM): Generic Requirements of Transport Functionality of Equipment Part 4-1 Synchronous Digital Hierarchy (SDH) Path Layer Functions.

ETSI EN 300 417-4-1 V1.1.3 (1999-05), Reference: REN/TM-01015-4-1a, Source: TM1, Title: Transmission and Multiplexing (TM): Generic Requirements of Transport Functionality of equipment: Part 4-1 Synchronous Digital Hierarchy (SDH) Path Layer Functions.

ETSI EN 300 417-4-2 V1.1.1 (1999-06), Reference: DEN/TM-01015-4-2, Source: TM1, Title: Transmission and Multiplexing (TM): Generic Requirements of Transport Functionality of Equipment; Part 4-2 Synchronous Digital Hierarchy (SDH) Path Layer Functions Implementation Conformance Statement (ICS) Pro forma Specification.

EN 300 462-4-1 V1.1.1 (1998-05), Reference: DEN/TM-03017-4-1, Source: TM3, Title: Transmission and Multiplexing (TM): Generic Requirements for Synchronization Networks Part 4-1: Timing Characteristics of Slave Clocks Suitable for Synchronization Supply to Synchronous Digital Hierarchy (SDH) and Plesiochronous Digital Hierarchy (PDH) Equipment.

ETSI EN 300 462-4-2 V1.1.1 (1999-12), Reference: DEN/TM-01057-4-2, Source: TM1, Title: Transmission and Multiplexing (TM): Generic Requirements for Synchronization Networks; Part 4-2 Timing Characteristics of Slave Clocks Suitable for Synchronization Supply to Synchronous Digital Hierarchy (SDH) and Plesiochronous Digital Hierarchy (PDH) Equipment Implementation Conformance Statement (ICS) pro forma Specification.

ETS 300 462-5 ed.1 (1996-09; withdrawn), Reference: DE/TM-03017-5, Source: TM3, Title: Transmission and Multiplexing (TM): Generic Requirements for Synchronization Networks; Part 5. Timing Characteristics of Slave Clocks Suitable for operation in Synchronous Digital Hierarchy (SDH) Equipment.

EN 300 462-5-1 V1.1.2 (199805), Reference: REN/TM-0301-7-5-1, Source: TM3, Title: Transmission and Multiplexing (TM): Generic Requirements for Synchronization Networks, Part 5-1. Timing Characteristics of Slave Clocks Suitable for operation in Synchronous Digital Hierarchy (SDH) Equipment.

ETS 300 484 ed.1 (1996-09), Reference: DE/TM02220, Source: TM2, Title: Transmission and Multiplexing (TM): Synchronous Digital Hierarchy (SDH) Network Information Model Connection Supervision Function (Higher Order Connection Supervision / Lower Order Connection Supervision [HCS/LCS]) for the Network Element (NE) View.

ETS 300 493 ed.1 (1996-06), Reference: DE/TM02216, Source: TM2, Title: Transmission and Multiplexing (TM): Synchronous Digital Hierarchy (SDH) Information Model of the Sub Network Connection Protection (SNCP) for the Network Element (NE) View.

These publications are available from:

ETSI Web site: www.etsi.org

for download (ETSI On-Line, EOL)

or purchase on paper or CD-ROM

Address: ETSI Infocentre

        06921 Sophia Antipolis

        Cedex FRANCE

        Telephone: +33(0)4 92 42 22

ISO Documents

ISO/IEC 74981:1994, Information Technology—Open System Interconnection Reference Model: The Basic Model.

ISO/IEC 8073:1988, Information Processing Systems—Open Systems Intelconnection—Connection Oriented Transport Protocol Specifications—Addendum 2: Class Four Operation over Connectionless Network Service.

ISO/IEC 8073:1988/Addendum 2:1989, Information Processing Systems—Open Systems Interconnection—Connection Oriented Transport Protocol Specifications.

ISO/IEC 8073:1992, Information Technology—Telecommunications and Information Exchange Between Systems Open Systems Interconnection—Protocol for Providing the Connection-Mode Transport Service.

ISO/IEC 8208/CCITT X.25, Information Processing Systems—X.25 Packet Level Protocol for Data Terminal Equipment.

ISO/IEC 8327-1:1988, Information Processing Systems—Open Systems Interconnection—Connection Oriented Session Protocol Specification—Part 1: Protocol Specification.

ISO/IEC DIS 8327-2, Information technology—Open Systems Intelconnection—Basic Connection Oriented Session Protocol Specification.

ISO 8348:1993, Information Processing Systems—Data Communications—Network Service Definition.

ISO/IEC 8473-1988, Information Processing Systems—Data Communications Protocol for Providing the Connectionless-Mode Network Layer Service.

ISO/IEC DIS 8473 1:1993, Information Technology—Protocol for Providing the Connectionless-Mode Network Service.

ISO 8648, Internal Organization of the Network Layer.

ISO/IEC 8650:1988, Information Technology—Open Systems Interconnection Protocol Specification for the Association Control Service Element.

ISO/IEC DIS 8650-2: Information Technology—Open Systems Interconnection Protocol Specification for the Association Control Service Element.

ISO 8802-2/ANSI/IEEE Std. 802.2:1989, Information Processing Systems—Open Systems Interconnection—Local area Networks—Part 2: Logical Link control.

ISO/IEC 8802-2, 1989/Amd 3, Information Processing Systems—Local Area Networks—Part 2: Logical Link Control Amendment 3: Conformance Requirements.

ISO 8802-3/ANSI/IEEE Std. 802.3-1990, Information Processing Systems—Open Systems Interconnection Local Area Networks—Part 3: Carrier Sense Multiple Access with Collision Detection (CSMA/CD) Access Method and Physical Layer Specifications.

ISO/IEC 8823-1:1988, Information Processing Systems—Open Systems Interconnection—Connection Oriented Presentation Protocol Specification—Part 1: Protocol Specification.

ISO/IEC DIS 8823-2, Information Technology—Open Systems Intel Connection Connection Oriented Presentation Protocol Specification.

ISO/IEC 9542:1988, Information Processing Systems— Telecommunications and Information Exchange between Systems— End System to Intel Mediate System Routing Exchange Protocol for

Use in Conjunction with Protocol for Providing the Connectionless-Mode Network Service (ISO 8473).

ISO/IEC TR 9577, Information Processing Systems—Data Communications—Protocol Identification in the Network Layer.

ISO/IEC 10040:1992, Information Technology—Open Systems Intelconnection Systems Management Overview.

ISO/IEC 10589:1992, Information Technology—Telecommunications and Information Exchange Between Systems—Intermediate System to Intermediate System Intra domain Routing Information Exchange Protocol for Use in Conjunction with the Protocol for Providing the Connectionless-Mode Network Service (ISO 8473).

ISO/IEC ISP 10607-1, Information Technology—International Standardized Profiles AFTnn—File Transfer, Access and Management.

ISO/IEC ISP 10608-1:1992, Information Technology—International Standardized Profile TAnnnn—Connection-Mode Transport Service over Connectionless-Mode Network Service.

ISO/IEC ISP 10608-2:1992, Information Technology—International Standardized Profile TAnnnn—Connection-Mode Transport Service over Connectionless-Mode Network Service.

ISO DIS 10747:1992, Information Technology—Telecommunications and Information Exchange Between Systems Protocol for Exchange of InterDomain Routing Information among Intermediate Systems to Support Forwarding of ISO/IEC 8473 PDUs.

ISO/IEC ISP 111831, Information Technology—International Standardized Profiles AOM1n OSI Management-Management Communications.

ISO/IEC PDISP 111881, Information Technology—International Standardized Profile—Common Upper Layer Requirements—Part 1: Basic Connection Oriented Requirements.

ISO TR 10172:199 1, Information Technology—Telecommunications and Information Exchange Between Systems—Network/Transport Protocol Interworking Specification.

These publications are available from:
American National Standards Institute, Inc
1430 Broadway
New York, NY 10018
URL: www.ansi.org

IEEE Documents

IEEE 802.2, Standard for Carrier Sense Multiple Access with Collision Detection (CSMA/CD) Access Method.

IEEE 802.3, Standard for Carrier Sense Multiple Access with Collision Detection (CSMA/CD) Access Method.

IEEE P802.6_/D 14, Distributed Queue Dual Bus (DQDB) Subnetwork of a Metropolitan Area Network (MAN), September 1990.

These publications can be obtained by calling:
IEEE Standards Publications
(800) 678IEEE or (908) 9811393
URL: www.ieee.org

SONET Interoperability Forum (SIF) Documents

SIF-002-1996, Remote Login Implementation Requirements Specification, Issue 1.

SIF-009-1997, NE-NE Remote Login Implementation Requirements Specification.

SIF-027-1999, BLSR Interoperability Requirements.

SIF-AR-9806-085, BLSR Interoperability Requirements—Cross Connect, Squelch Table, and NUT (Non-preemptible, Unprotected Traffic).

SIF-AR-9807-111, Requirements for BLSR Map Generation Protocol.

The SIF is an Alliance for Telecommunications Industry Solutions (ATIS) sponsored committee. SIF-approved documents may be obtained directly from the ATIS WWW server at: http://www.atis.org/atis/sif/sifdoc.htm.

For additional Information on SIF documents, contact:
Alliance for Telecommunications Industry Solutions
Attn.: Lisa Colaianne
1200 G Street, NW, Suite 500
Washington, DC 20005
Tel: 2024348823

INDEX